AF411346

Maintenance Management of Wind Turbines

Maintenance Management of Wind Turbines

Editor

Fausto Pedro García Márquez

MDPI • Basel • Beijing • Wuhan • Barcelona • Belgrade • Manchester • Tokyo • Cluj • Tianjin

Editor
Fausto Pedro García Márquez
Castilla-La Mancha University
Spain

Editorial Office
MDPI
St. Alban-Anlage 66
4052 Basel, Switzerland

This is a reprint of articles from the Special Issue published online in the open access journal *Energies* (ISSN 1996-1073) (available at: https://www.mdpi.com/journal/energies/special_issues/maintenance_wind_turbines).

For citation purposes, cite each article independently as indicated on the article page online and as indicated below:

LastName, A.A.; LastName, B.B.; LastName, C.C. Article Title. *Journal Name* **Year**, *Article Number*, Page Range.

ISBN 978-3-03936-629-3 (Hbk)
ISBN 978-3-03936-630-9 (PDF)

Contents

About the Editor

Fausto Pedro García Márquez works at UCLM as a full professor (accredited as a full professor from 2013), Spain. He also works as an honorary senior research fellow at Birmingham University, UK, and is a lecturer at the Postgraduate European Institute. From 2013–2014, he was a senior manager at Accenture. He completed his European PhD with maximum distinction. He has been distinguished with the prizes: Advancement Prize for Management Science and Engineering Management Nominated Prize (2018); First International Business Ideas Competition 2017 Award (2017); Runner (2015), Advancement (2013) and Silver (2012) by the International Society of Management Science and Engineering Management (ICMSEM); Best Paper Award in the international journal of Renewable Energy (Impact Factor 3.5) (2015). He has published more than 150 papers (65% ISI, 30% JCR and 92% international journals), some of which are recognized as: "Renewable Energy" (as "Best Paper 2014"); "ICMSEM" (as "excellent"), "Int. J. of Automation and Computing" and "IMechE Part F: J. of Rail and Rapid Transit" (most downloaded), etc. He is the author and editor of 25 books (Elsevier, Springer, Pearson, Mc-GrawHill, Intech, IGI, Marcombo, AlfaOmega, etc.), and five patents. He is an editor for five international journals, and a committee member of more than 40 international conferences. He has been Principal Investigator in 4 European projects, 5 national projects, and more than 150 projects for universities, companies, etc. His main interests are artificial intelligence, maintenance, management, renewable energy, transport, advanced analytics and data science. He is an expert on the European Union for AI4People (EISMD), and ESF. He is Director of www.ingeniumgroup.eu.

Preface to "Maintenance Management of Wind Turbines"

"Maintenance Management of Wind Turbines" considers the main concepts and the state-of-the-art, as well as advances and case studies on this topic. Maintenance is a critical variable in industry in order to reach competitiveness. It is the most important variable, together with operations, in the wind energy industry. Therefore, the correct management of corrective, predictive and preventive politics in any wind turbine is required. The content also considers original research works that focus on content that is complementary to other sub-disciplines, such as economics, finance, marketing, decision and risk analysis, engineering, etc., in the maintenance management of wind turbines.

This book focuses on real case studies. These case studies concern topics such as failure detection and diagnosis, fault trees and subdisciplines (e.g., FMECA, FMEA, etc.) Most of them link these topics with financial, schedule, resources, downtimes, etc., in order to increase productivity, profitability, maintainability, reliability, safety, availability, and reduce costs and downtime, etc., in a wind turbine.

Advances in mathematics, models, computational techniques, dynamic analysis, etc., are employed in analytics in maintenance management in this book.

Finally, the book considers computational techniques, dynamic analysis, probabilistic methods, and mathematical optimization techniques that are expertly blended to support the analysis of multi-criteria decision-making problems with defined constraints and requirements.

Fausto Pedro García Márquez
Editor

Article

Design of a Multi-Robot System for Wind Turbine Maintenance

Josef Franko [1],*, Shengzhi Du [2], Stephan Kallweit [1], Enno Duelberg [1] and Heiko Engemann [1]

[1] MASKOR Institute, Faculty of Mechanical Engineering and Mechatronics, University of Applied Sciences Aachen, 52074 Aachen, Germany; kallweit@fh-aachen.de (S.K.); duelberg@fh-aachen.de (E.D.); engemann@fh-aachen.de (H.E.)

[2] Faculty of Engineering and the Built Environment, Tshwane University of Technology, Pretoria 0001, South Africa; dus@tut.ac.za

* Correspondence: franko@fh-aachen.de; Tel.:+49-0241-6009-52878

Received: 14 April 2020; Accepted: 12 May 2020; Published: 18 May 2020

Abstract: The maintenance of wind turbines is of growing importance considering the transition to renewable energy. This paper presents a multi-robot-approach for automated wind turbine maintenance including a novel climbing robot. Currently, wind turbine maintenance remains a manual task, which is monotonous, dangerous, and also physically demanding due to the large scale of wind turbines. Technical climbers are required to work at significant heights, even in bad weather conditions. Furthermore, a skilled labor force with sufficient knowledge in repairing fiber composite material is rare. Autonomous mobile systems enable the digitization of the maintenance process. They can be designed for weather-independent operations. This work contributes to the development and experimental validation of a maintenance system consisting of multiple robotic platforms for a variety of tasks, such as wind turbine tower and rotor blade service. In this work, multicopters with vision and LiDAR sensors for global inspection are used to guide slower climbing robots. Light-weight magnetic climbers with surface contact were used to analyze structure parts with non-destructive inspection methods and to locally repair smaller defects. Localization was enabled by adapting odometry for conical-shaped surfaces considering additional navigation sensors. Magnets were suitable for steel towers to clamp onto the surface. A friction-based climbing ring robot (SMART— Scanning, Monitoring, Analyzing, Repair and Transportation) completed the set-up for higher payload. The maintenance period could be extended by using weather-proofed maintenance robots. The multi-robot-system was running the Robot Operating System (ROS). Additionally, first steps towards machine learning would enable maintenance staff to use pattern classification for fault diagnosis in order to operate safely from the ground in the future.

Keywords: wind turbine maintenance; climbing robot; low cost; weather independent operations; condition monitoring; odometry on wind turbines

---

1. Introduction

More than 400,000 turbines with a total output power of almost 600 GW are installed globally [1]. Wind turbines are gaining worldwide importance for sustainable power supply. Wind turbine (WT) technology is still relatively young compared to fossil and nuclear power plants. Annual maintenance of wind turbines is required, increasing the total cost of ownership and influencing the competitiveness of their operation with the established power generation. The mechanical components of wind turbines, such as the main shaft, bearings, generator, and gearboxes have evolved. On the other hand, structure parts, such as towers and rotor blades, still need regular and intense maintenance [2]. Access for maintenance is problematic due to challenging weather conditions and the large dimension of wind turbines [3]. Driven by high maintenance demands, various technical solutions with different limitations, costs, and capabilities have emerged. Maintenance requires inspections every second year, including repair based on the inspection results [2]. Inspection and monitoring can be accomplished using high-resolution vision sensors, even from a distance, e.g., ground-based or airborne. However, repairs require direct access to the area of interest. State-of-the-art solutions are industrial climbers and rope-based service frames for any individual type of wind turbine. Common drawbacks are limited payload, inadequate modern measurement technologies to detect failures, and insufficient repair tools, thus confronting service companies with the most expensive alternative, e.g., installing a completely new rotor-blade.

Standard service is restricted to daytime hours, temperatures above 10 °C, a relatively low air humidity, and wind speeds not exceeding the range of 8–12 m/s [3]. All the environmental conditions have to be considered to provide stable conditions and ensure human safety. These conditions restrict the maintenance period to seven months per year and four to six hours per day. Even under stable environmental conditions, repairs remain a delicate task that require expert knowledge, as well as a degree of comfort in working at height. The set-up of such a system takes up to three hours. It has to be reinstalled every day due to general restrictions. Time frames for maintenance are tight, hence methods for efficient documentation, e.g., by monitoring the current position of the repair unit in tower coordinates for quality management, present another challenge.

Robots can be used to keep humans safe and improve maintenance quality. Digital twins can be derived from sensor data to enable predictive maintenance and long-term condition monitoring. The first step towards this goal is to develop several prototypes to validate the basics [4–7]. Flying and surface climbing systems are collaborating with the goal of large-scale inspection in wind farms and on-demand local repairs (Figure 1, Video S1). A friction-based climbing ring robot (SMART: Scanning, Monitoring, Analyzing, Repair and Transportation) can carry high payloads and is equipped with an onboard robotic arm. The SMART robot clasps the tower surface and climbs up and down with friction-based crawlers (Figure 2). Multicopters detect surface damage with vision sensors from a safe distance, even during wind turbine operation. All sensor data is referenced to a global coordinate frame; thus, localization, navigation, and mapping are of major importance. Each mobile system relies on an RTK-GPS for high positioning accuracy. Today, RTK correction signals are not available in every region, so it is still recommended to provide base stations in the wind farm to correct the GPS uncertainty.

Figure 1. Wind farm maintenance vision.

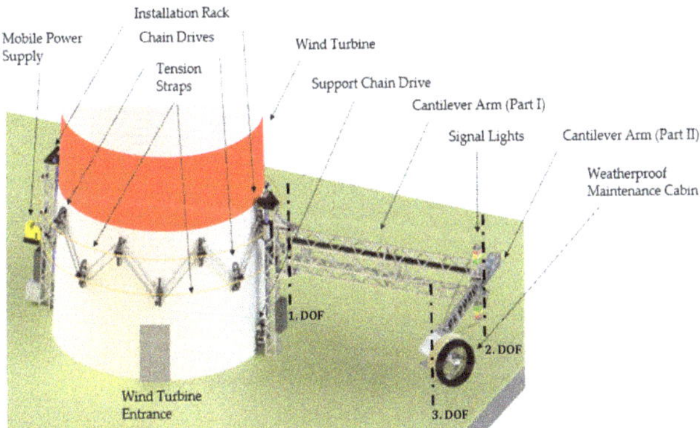

Figure 2. Climbing ring robot, SMART (Scanning, Monitoring, Analyzing, Repair and Transportation).

Smaller climbing robots carry out non-destructive testing tools, such as ultrasonic and radar with surface contact. Unlike the climbing ring robot, these systems rely on magnetic forces. These robots are faster and easier to install. In general, non-autonomous magnetic climbing robots have reached a high technology readiness level [8–11] and have been utilized in teleoperation mode on wind turbines for over a decade. This work supports the existing designs with a model-based odometry for conical-shaped wind turbine towers to increase the localization quality and thus the level of automation.

2. Climbing Ring Robot

The central robot of the multi-robot approach is the climbing ring robot (CRR), also known as SMART (Scanning, Monitoring, Analyzing, Repair and Transportation) [6]. The semi-autonomous mobile robot platform serves as a carrier for a weather-independent maintenance cabin with an industrial robotic arm inside. Figure 2 shows the final design. The CRR climbs on the tower surface while clasping the tower to increase stabilization during challenging weather conditions.

The CRR concept consists of three subsystems: locomotion, adhesion, and manipulation. The locomotion unit has two degrees of freedom (DOFs) and is able to rotate around the tower as well as move up and down along the tower axis. Therefore, multiple tracked-wheel crawlers with steering actuators are placed around the tower. A linear actuator for steering is implemented, because skid-steering on wind turbine surfaces leads to severe problems. Due to the convex shape, each crawler is fixed in an upright position and the mandatory slip for skid-steering is prevented by contracting forces [12].

This novel approach for climbing on towers relies on a four-joint gearing mechanism that connects all crawlers to combine their lifting forces. Tension straps contract the climbing ring and apply high normal forces in the radial direction from each crawler to the tower. The friction-based principle provides the highest flexibility to climb either on steel or concrete structures. The mechanism of the adhesion unit needs to adjust to different diameters to cover wind turbine towers with a conical shape. Another purpose of the connection gear is to enable an automated, efficient installation process. The purpose of this rotor blade maintenance platform is to act as a macro-manipulator with three DOFs in the vertical direction for an industrial robot arm inside a weatherproof cabin.

2.1. Material and Methods

The development process is based on several initial models on a smaller scale of 1:20. These design studies include basic functionality and provide sensor feedback for experimental investigations of the kinematics. In addition, simulation results are validated at an early stage of research. All 1:20 models operate on cylindrical shaped towers without conicity due to simplification. Thus, no mechanism for changing the diameter was implemented and the challenge of buffering ropes, belts, or any kinematic was neglected.

The initial model for the climbing principle with belts (Figure 3a) showed the negative influences of belt stretching. This made the climbing robot move downwards after each contraction and release cycle. Consequently, the development of intermittent lifting strokes was discontinued [7].

Subsequently, Lego bricks supported the rapid transition towards crawlers (Figure 3b), thus introducing continuous climbing instead of intermittent climbing. An initial test indicated that the lifting force from one crawler to another can be transferred with a rigid connection. First, diagonal ropes were employed for the connection, while rubber bands contracted the system. For the third model (Figure 3c), the ropes were replaced with "Nuernberger" scissors. At this point, a fully functional demonstrator at 1:3 scale was derived for further research and investigation. Another iteration (Figure 3d) led to the final concept of this work, which will be presented in the following subsections. The final concept was finally validated with a 1:3 demonstrator before the large-scale prototype was designed and tested on a conical-shaped tower for the first time.

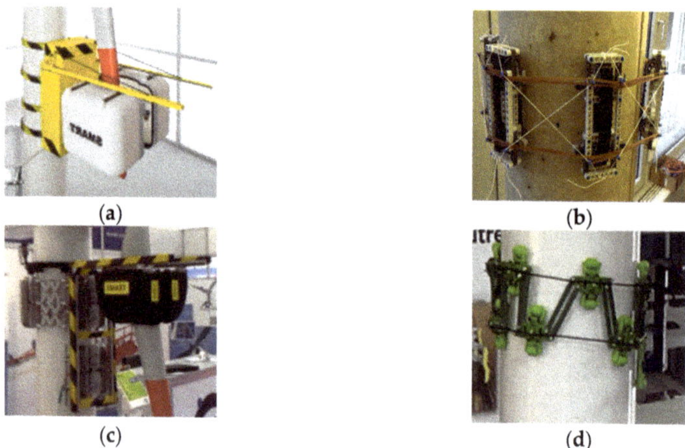

Figure 3. Design studies scaled by 1:20: (**a**) belt climbing ring robot; (**b**) Lego based design; (**c**) crawlers with scissors; (**d**) final concept.

Within the project, two systems at 1:3 scale were developed and intensively tested [5]. The crawler concept and materials were quite similar. The kinematic design has evolved and enabled a total system weight reduction of 50% from 800 to 400 kg, including the control cabinet, which was put on the ground in the previous concept (Figure 4a). The surface contact area and the combined lifting force of all 18 crawlers remained the same. Due to the novel arrangement of the crawlers, their contribution to lifting the main body increased. As part of the adhesion system, the horizontal cross connection fulfills the function of transmitting lifting forces from all crawlers to the mainframe.

The final CRR concept consists of 18 crawlers(Figure 4b), Figure 5 (1) which are coupled by a four-joint kinematic (2) and two installation racks (3) as main frames. The system is divided into two parts: seven to nine crawlers are mounted on each installation frame (Figure 11). Positioning linear actuators (2) are located in the center of the four-joint kinematic to keep the crawler in a horizontal position during the installation process. Retractable air bellow spacers (4) at the upper end of the frame support the robot on the tower during the installation process. The crawler tracks are in frictional contact with the tower. Harmonic drives (at scale 1:3 CHA-20A-160, 90 Nm; at scale 1:1 CHA-40A-160, 600 Nm) provide high torque for climbing, and have a compact design. Wire ropes are attached to the crawlers and tensioned by eight winches (7), generating the required normal force. The crawlers are pressed radially against the tower. The control system, consisting of switch cabinet (5), motor controller (8), and power supply (6) are located on the installation frames, so that no further external devices are required. A cable connects the robot's power supply units to the main supply.

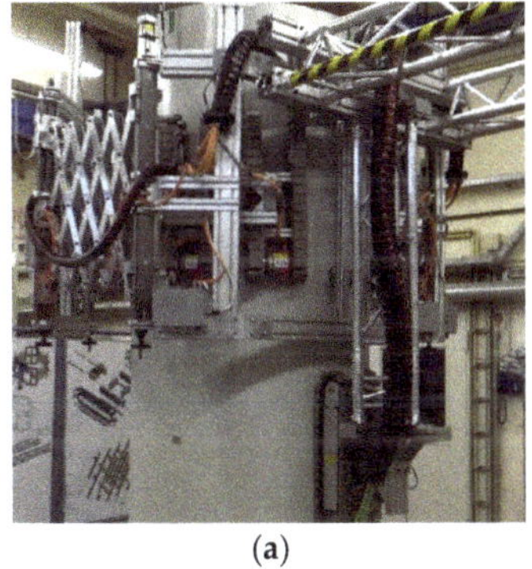
(a)

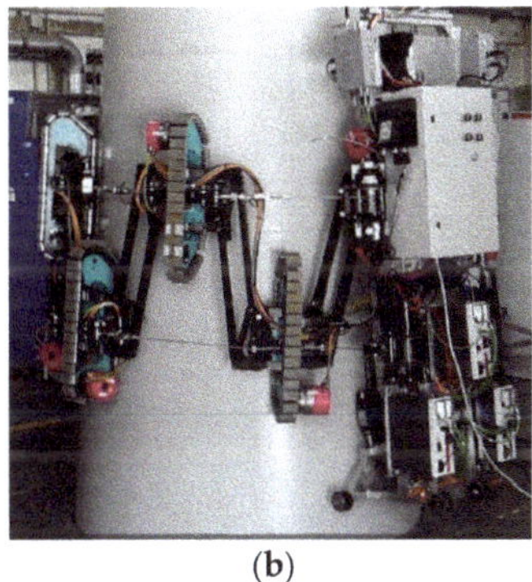
(b)

Figure 4. Climbing ring robot (CRR) demonstrator 2016 (**a**); CRR demonstrator 2018 (**b**). Both are scaled by a factor of 1:3 regarding a 2.5 MW wind turbine.

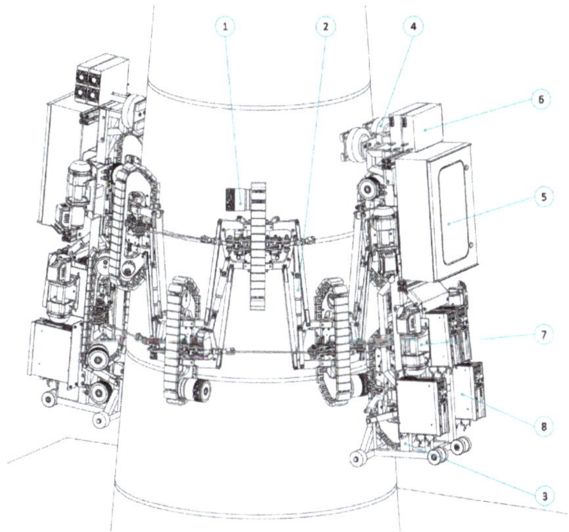

Figure 5. Climbing ring robot (scaled 1:3).

2.2. Mechanical Design

The first approach (Figure 4a) employed the concept of multiple "Nuernberger" scissors. However, finite element method (FEM) simulation indicated that a lightweight four-joint kinematic is more efficient, which was substituted for the former concept in the second approach (Figure 4b). The CRR can utilize the four-joint connections not only for transmitting the lifting forces of each individual crawler, but the actuated kinematics can also be used to support the installation process (Section 2.5).

Based on prototype parameters and size, Table 1 compares the two cross-connection concepts in favor of the four-joint kinematic. The downside of the new approach is that the pairs of crawlers that are required for steering left and right have to be split up, and thus the connection between the climbing system and adhesion system demands additional degrees of freedom (DOFs).

Table 1. Finite element method for the two cross-connection concepts.

	"Nuernberger" Scissors	Four-Joint Kinematic	Ratio	
Weight	270 kg	44 kg	6.1	
Max. Stress	240 MPa	77 MPa	3.1	
Deformation	38 mm	5.6 mm	6.8	

One DOF for steering is implemented in the center of the crawler with an IGUS disc bearing. Two synchronized linear actuators push and pull simultaneously and rotate the red part of the crawler frame vs. The black coupling towards the horizontal connection. The steering angle α is measured with an absolute rotational encoder in the center of the motion. It is mandatory to drive forward during the steering motion due to the high friction (Figure 6).

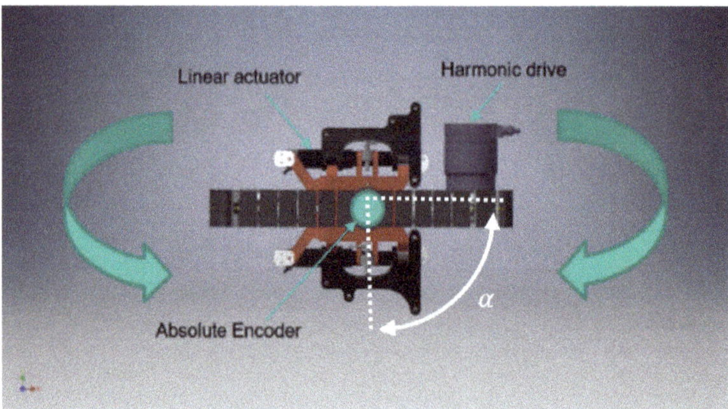

Figure 6. Top-view of the CRR crawler; arrows indicate steering axis.

A bolt bearing below the steering kinematic enables the tilt motion β for conical-shaped wind turbine towers. The horizontal connection is orientated vertically to the ground in any situation. Each crawler has to align itself to the surface. This DOF is not actuated. The total lifting force F_L and the normal force F_N are measured with a bi-directional load bolt in the center of the bearing (Figure 7). These forces are monitored to ensure balanced ascending and descending.

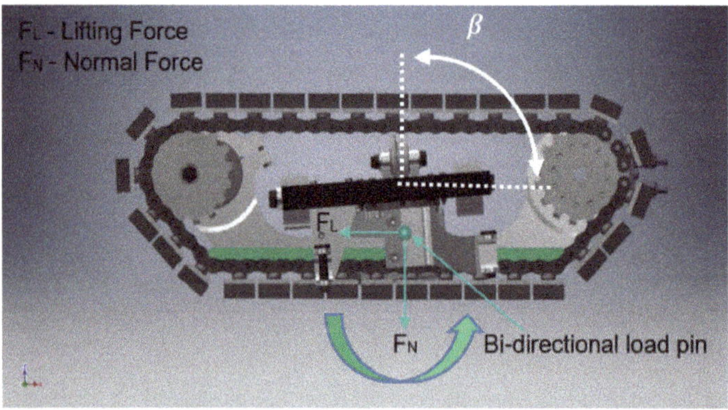

Figure 7. Side-view of the CRR crawler; arrows indicate the alignment axis for a conical-shaped wind turbine tower.

The third DOF disables the uneven contraction and expansion of the tensioning system. Thus, the black connectors to the horizontal connection can move around a vertical axis but a gear mechanism keeps both angles γ the same. Otherwise, the alignment of all grasping angles would be different. The steering motion is controlled individually, so the distance between each individual crawler may vary during the climbing process, due to slip-stick effects. Even small variations can cause a mismatch of the grasping angles. To prevent internal mechanical stress this DOF is restricted (Figure 8). Figure 8 also illustrates the measurement of the rope force to control the winches.

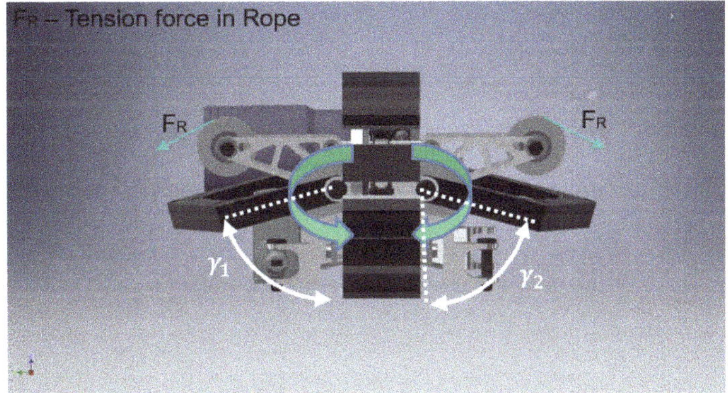

Figure 8. Front-view of the CRR crawler; arrows indicate circular tower alignment.

The scaling of the CRR design was supported by the generation of a kinematic model in ADAMS multibody simulation software (MBS, Figure 9). The kinematics of the demonstrator are simulated and experimentally validated with the measurements outlined above:

- Drive torque M_{Drive},
- normal force F_N,
- lifting force F_L,
- rope tension F_{Rope}.

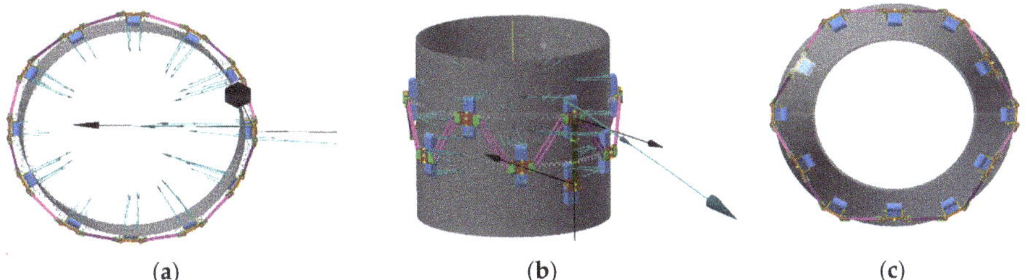

(a)   (b)   (c)

Figure 9. ADAMS Multi-Body-Dynamic Simulation: radial forces on the tower surface in top-view (**a**); radial forces side-view (**b**); set-up on a conical-shaped tower (**c**).

The MBS software was used to test different designs and concepts. Furthermore, the simulation served for estimating the forces and torques of the prototype to support the design phase. Figure 9a presents a generated model. Figure 9b outlines the distribution of the simulated normal forces in the static state of the CRR with the adhesion system switched on. Finally, Figure 9c transitions to the next set-up of simulations on a conical-shaped tower.

The simulation of the crawler tracks was carried out with the ADAMS Tracked Vehicle Module (ATV). The analysis in ATV indicates that belt-based crawler tracks load the curved tower surface unevenly [4]. Comparative investigations have shown that with disc tracks a uniform distribution of the local surface pressure is achievable. Elastomer profiles are mounted on the disc track and may be changed with regard to friction and damping properties. The experimental validation quantified the forces without (Figure 10a) and with an attached cantilever arm (Figure 10b). Thus, the torque of all crawlers was raised to a maximum. In particular, crawlers 2, 15, 10, and 16 utilized higher torque due to the higher normal force resulting from the tilting moment of the cantilever arm (Figure 11). Further data analyses might overload this comprehensive report. Therefore, this example was selected to illustrate the general method of development. Based on the data validation the simulated MBS models were improved.

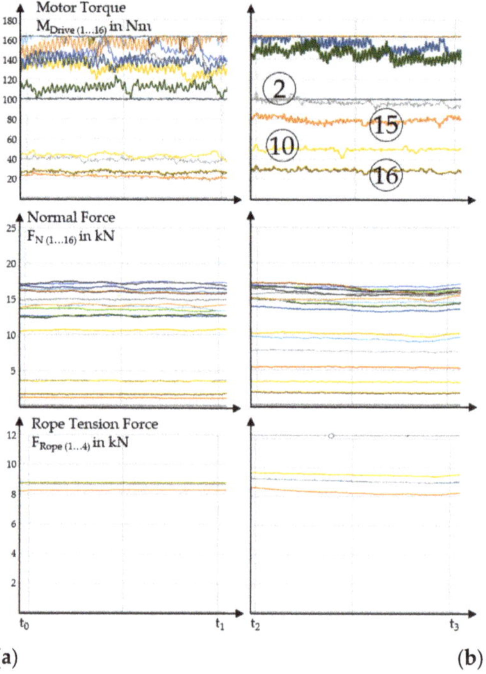

Figure 10. Experimental data from CCR: without cantilever arm (**a**); with cantilever arm (**b**).

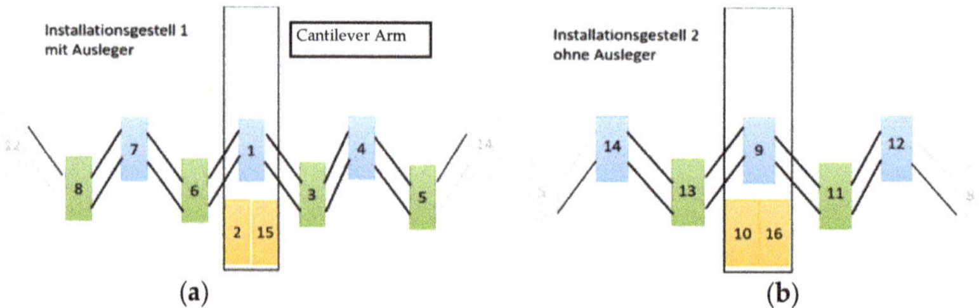

Figure 11. System overview: Installation Rack 1 (**a**); Installation Rack 2 (**b**).

2.3. Friction Testing

The system was designed considering the fragility of wind turbine towers—consisting of thin steel structures with large diameters—towards perpendicular forces. To reduce these perpendicular forces a high friction coefficient is mandatory. The coefficient describes the relationship between normal and lifting forces and has a major impact on the feasibility of the CRR in terms of total payload. A friction test bench was developed for the prototype to determine the:

- coefficient of friction between crawler and tower surface,
- the steering force of the linear actuators,
- the ratio between motor torque and normal force (rolling friction) under real conditions.

A prototype 1:1 scaled crawler (Figure 12(7)) was used for the final tests. The friction test rig consists of a welded steel frame (1) with a bolted steel plate (2). The steel plate is coated according to DIN EN ISO 12944—corrosion protection class C5—similar to the tower surface of a wind turbine. The normal force F_N is generated by the rope force F_{Rope}. The rope is tensioned by two suspension eyes (6) over deflection rollers (behind cover) of the trolley (8). The rope force is measured with a load cell (5). An electric motor generates the pulling force F_Z employing a trapezoidal thread spindle 14×7 (3). The pulling force is measured with a load cell (4). The spindle pulls the crawler over the steel plate while the rope is tensioned and breaks are activated. The quotient between the measured forces F_Z and F_N results in the coefficient of friction μ [13,14].

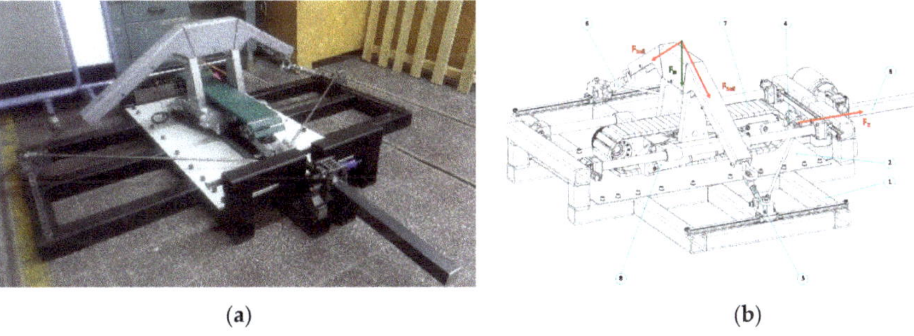

(**a**) (**b**)

Figure 12. Friction test bench: experimental assembly (**a**); system overview (**b**).

To determine the coefficient of friction, different surface conditions are examined: dry, icy, oily, and wet. These simulate possible weather and climbing scenarios. The following Table 2 provides the minimal values after five repetitions for different surface parameters for reference.

Table 2. The friction coefficients (μ_{min}) test results.

Material	Dry	Wet	Icy	Oily	Sandy
EPDM	1.702	0.679	0.307	0.299	0.382
NR	0.833	0.668	0.364	0.238	0.387
NBR	1.156	1.041	0.257	0.308	0.418
CR	0.753	0.753	0.248	0.193	0.405
SBR	0.7	0.763	0.165	0.255	0.344

The table indicates that NBR is a suitable rubber material with the best friction properties. The coefficients of friction of all types are similar, especially on sandy surfaces. The highest deviation occurs on dry surfaces, which is the standard environment for wind turbines. Therefore, EPDM was selected for the initial proof-of-concept. It should be mentioned that EPDM showed a distinctive slip-stick effect that causes the testing material to jump small distances instead of a linear sliding motion. Nevertheless, this effect only occurs in a non-static friction state, which is prevented by the contracting system. All crawler tracks are equipped with adhesive disks made of EPDM rubber (Shore hardness 40). The coefficient of friction between rubber and tower surface was estimated by $\mu_{min} = 0.8$ for the testing conditions. The coefficient is never a constant in real world applications. Nevertheless, a final product must be designed for a friction coefficient of 0.3 (icy and oily) and appropriate safety factors. The prototype design considers a significantly higher coefficient due to the fact that it is designed for testing, which results in a higher total system weight.

2.4. Proof-of-Concept

In the accomplishment of the full development process, a real size prototype served as the final proof of concept in 2019 (Figure 13a). Due to safety regulations, a simplified testing scenario was created on a wind turbine mock-up. The predefined tower size was relatively small with a 3.5 m diagonal. This size limited the 7.2 ton CRR to climb a height of 1.3 m in total before running into contraction limits. Nevertheless, the main functionality of the system was investigated and confirmed, including semi-automated installation, climbing, steering, load distribution, contact surface, navigation, and payload. All systems were designed in accordance to IP 64 standard for outdoor usage. The mobile climbing ring robot included the cantilever arm (macro manipulation) and the industrial robot arm (micro manipulation) inside the cabin, which were utilized on a rotor blade part nearby. The industrial robot arm could be teleoperated based on the Robot Operating System (ROS) [15]. The prototype was equipped with sensors, such as RGB cameras and LiDAR, to observe the working environment and to measure the surface geometry of the blade. A digital twin with a 3D surface scan and texture was matched by point-cloud library fusion algorithms. ROS was used to calculate the reverse kinematics of the cantilever for a simplified robot control by height and angle. Thus, using ROS enabled reaching the full potential of this mobile manipulator. Existing approaches for perception, localization, and collision-free path planning, together with hardware independence, created a suitable environment for future work.

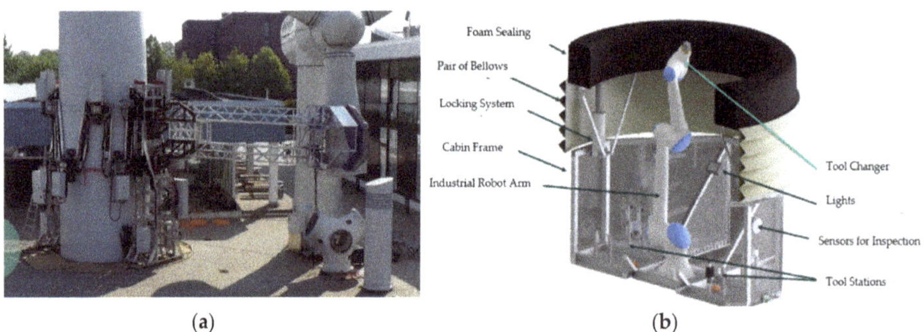

(a)   (b)

Figure 13. CRR prototype. (**a**) Test site in Cologne, Germany; (**b**) Maintenance cabin.

2.5. Installation Process of the CRR

The developed process takes the door to enter the wind turbine tower into account and includes a lifting system to contract the climbing ring above the average door height of 3.5 m. The two installation racks including the crawlers were installed with an offset of approx. 180° and 90° towards the door on the tower surface. The CRR was placed with a small truck crane. The weight of the segments was approx. 2.5 tons. After the placement of the installation racks, they were fixed with belts to the tower. One tensioning belt was placed around the installation racks at the top and bottom and around the tower, then tightened. The installation racks were pressed against the tower surface. At this point, the crawlers facing the tower surface did not yet have any contact with the tower (Figure 14a).

In the second step, the crawlers of both installation racks were guided around the tower and connected together. For this purpose, diagonal actuators in the four-joint kinematic were actuated hydraulically. These cylinders were all subjected to tensile stress, so that buckling was avoided. The crawlers still had no contact with the tower and could be checked one last time on the ground(Figure 14b).

The CRR was then lifted over the entrance door using a lifting system between the installation racks and the crawlers. Finally, the tensioning system was tightened synchronously with cable winches. The crawlers were pressed against the tower and all tensioning belts released (Figure 14c). The CRR was ready for operation.

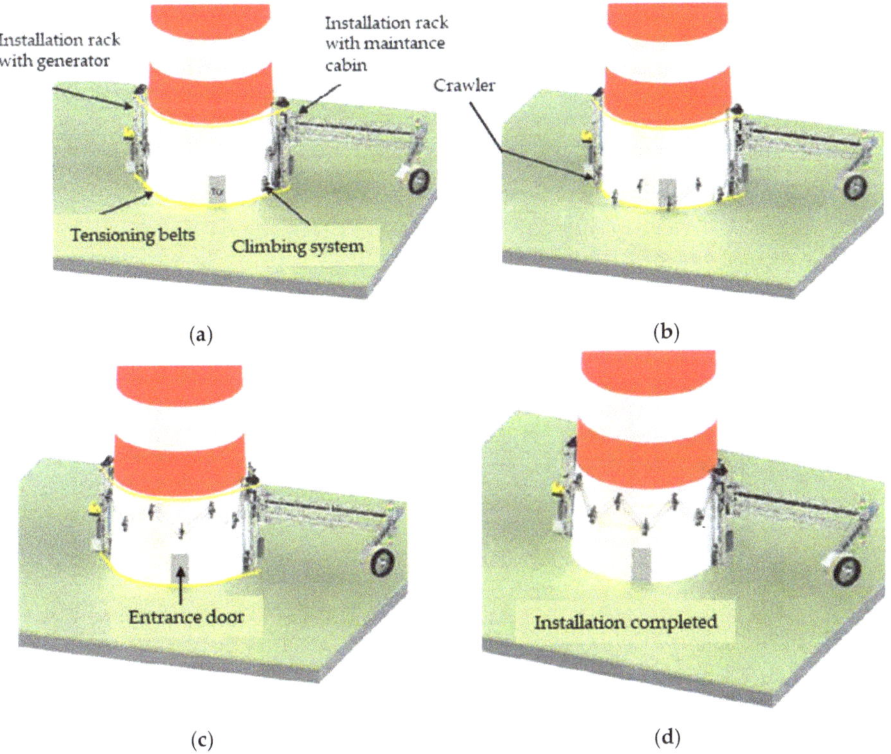

Figure 14. Steps of the CRR installation (**a**); step 2 (**b**); step 3 (**c**); installation completed (**d**).

3. Multicopter

For the regular inspection of a WT, a multicopter (Figure 15) is a cost-effective and flexible technical solution compared to the CRR. The unmannend arial vehicle (UAV) shown in Figure 15 was developed for this application and prepared for the test phase of the other prototypes. DJI's multicopter S1000 served as a platform for the sensor box and the computer. It enabled a climb and descent of the wind turbine with a flight time of 10 min. The sensor box was installed on the lower part of the frame of the S1000.

Deep convolutional neural networks (DCNN) were implemented to detect the corrosion of welding lines on the tower surface [16]. The digital twin of the tower automatically lines the image up with the tower height and angle towards the wind turbine entrance. In addition, all failures are summarized in a standard sheet and transmitted to the WT operator to calculate and conduct the required repairs. Figure 16 presents the implementation with a highlighted region of interest around the horizontal welding lines.

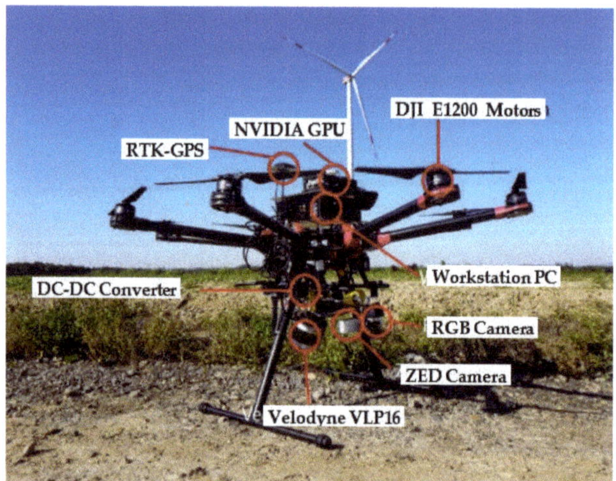

Figure 15. Unmanned aerial vehicle for WT inspection.

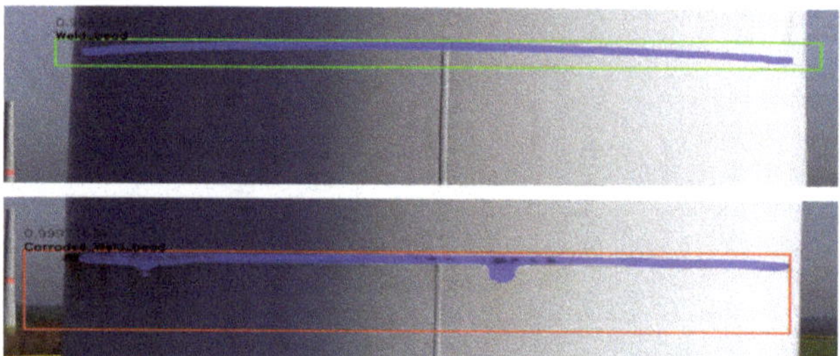

Figure 16. DCNN for the detection of corrosion.

4. Magnetic Climbing Robot

The basic function of the magnetic inspection platform is the positioning of sensors on the vertical tower surface (Figure 17). An omnidirectional approach was finally selected to be able to move on the complete curved plane in all three DOFs [17–21]. The designed magnetic climbing robot (MCR) operates without rotating the mainframe. Only the wheels may change their orientation for steering purposes. A magnet in the center of the mainframe attaches the robot to any steel surface, while a small air gap enables the motion capability.

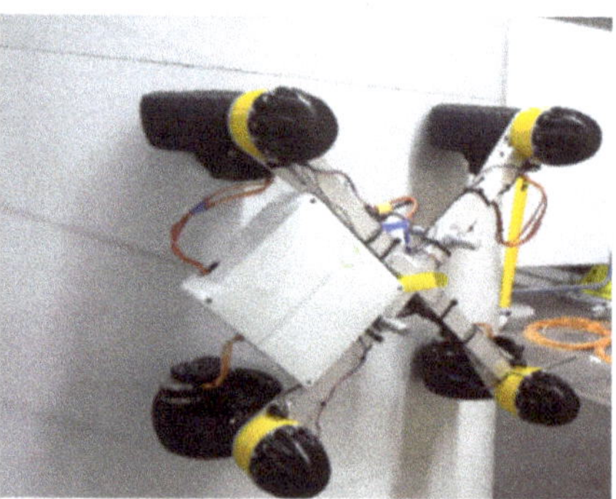

Figure 17. Magnetic platform.

The movement on the tower surface can be controlled by the following parameters (Figure 18):

- z represents the translation along the tower axis (height, H),
- $\Omega \times \vec{r}$ represents the translational movement tangentially around the tower combined with a rotation around the tower axis depending on the tower diameter,
- ω is the angular velocity around the normal vector of the tower surface.

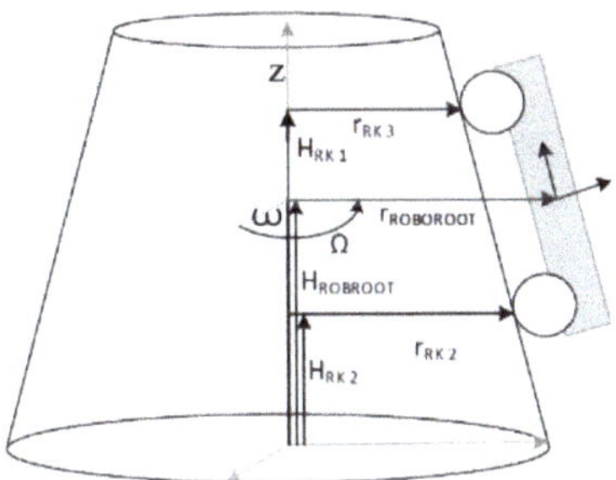

Figure 18. Tower coordinate frame.

5. Inverse Kinematics

The inverse kinematics of the robot are used to control the speed $\vec{\vartheta}$ and angular velocity Ω of the coordinate system ROBROOT and to derive the individual converted wheel coordinate system RK1 to RK4 (Figure 19). This is necessary to control the steering actuators and the traction motors [22]. Parameters of the inverse kinematic are target speed $\dot{x}$, $\dot{y}$ in the ROBROOT frame; the target angular velocity ω (Equation (6)); the position of the wheels HRK1 to HRK4; the tower geometry, consisting of HRK1 (tower height), r_b (radius bottom), and r_t (radius top); the position vectors of the robot wheels $\vec{r_1}$ to $\vec{r_4}$; and the wheel radius r_{wheel}. Based on these parameters, the inverse kinematics follow as Equations (1)–(3):

$$r_{RK} = r_0 + \frac{H_{RK}}{H_{Tower}} \cdot (r_b - r_t) \tag{1}$$

$$\dot{\Omega} = \frac{\dot{x}}{r_{ROBBASE}} \tag{2}$$

$$\vec{\vartheta} = \begin{pmatrix} \dot{\Omega} \cdot r_{RK} + \omega \cdot r_x \\ \dot{y}_{ROBROOTS} + \omega \cdot r_y \end{pmatrix} \tag{3}$$

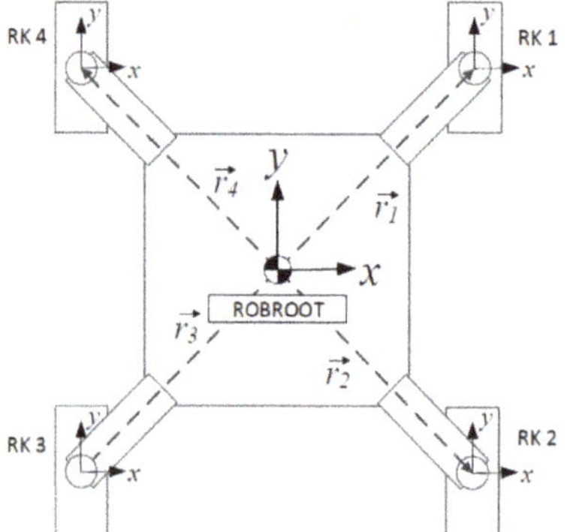

Figure 19. Robot coordinate frames.

6. Model Based Odometry

Odometry refers to the relative estimation of the pose by observing the drive sensors [23]. Odometry forms the basis for the localization of the platform. The localization is based on an integration of the velocity vectors of the forward kinematics over time (dead reckoning). In the implementation, the forward kinematics are calculated for a point in time, the calculated velocity vectors are multiplied by a time period, and the resulting change in position is summed over the total time. The input variables (Figure 20) are the steering angle of each wheel α_1 to α_4 (Equation (5)) and the speed vectors of each wheel $\vec{\vartheta}_1$ to $\vec{\vartheta}_4$ (Equation (4)).

$$|\vec{\vartheta}| = \sqrt{\vec{\vartheta}_x^2 + \vec{\vartheta}_y^2} \tag{4}$$

$$\alpha = \tan^{-1}\left(\frac{\vartheta_y}{\vartheta_x}\right) \tag{5}$$

$$\omega = \frac{|\vec{\vartheta}|}{r_{Wheel} \cdot 2 \cdot \pi} \tag{6}$$

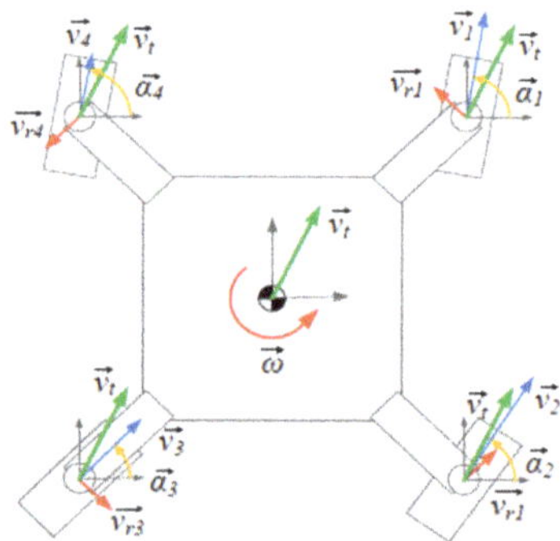

Figure 20. Motion vectors.

The accuracy of the odometry affects the uncertainty of the robot's localization [17–21]. Therefore, as with inverse kinematics, the tower curvature has to be considered. In the following, a model-based odometry for driving conical towers will be presented. The used coordinate system is the cylindrical tower coordinate system. In comparison to the odometry for plane surfaces, first it calculates the rotation of the robot. Then, the speed of the robot is calculated tangentially to the tower surface and along the symmetry axis of the tower. In the conversion of the distance covered by the robot along the y axis of the ROBROOT coordinate system to the symmetry axis of the tower, the angle of the conicity γ is used:

$$\gamma = \tan^{-1} \frac{r_b - r_t}{H_{Tower}} \tag{7}$$

The model-based odometry relies on the current rotation θ of the robot, as well as the current height H. Both angular velocity (Equation (8)) and vertical velocity (Equation (9)) are converted into absolute positioning data (Equation (10)).

$$\dot{\Omega} = \frac{1}{4} \cdot \sum_{i=1}^{4} \frac{\cos(\alpha_i + \theta) \cdot \left|\dot{\vartheta}_i\right| - r_{ix} \cdot \omega}{r_{RKi}} \tag{8}$$

$$\dot{H} = \frac{1}{4} \cdot \sum_{i=1}^{4} \frac{(\sin(\alpha_i + \theta) \cdot \left|\dot{\vartheta}_i\right|) \cdot \cos(\gamma)}{r_{RKi}} \tag{9}$$

$$\Delta\theta = \omega \cdot t_1; \qquad \Delta\Omega = \dot{\Omega} \cdot t_1; \qquad \Delta H = \dot{H} \cdot t_1 \tag{10}$$

To validate the approach of model-based odometry for conical-shaped wind turbines, two concepts were developed. Concept (Figure 21a) utilizes four wheels with individual steering actuators, while concept (Figure 21b) utilizes only two wheels in means of cost reduction. Only concept (Figure 21a) was validated with a prototype. The advantage is that a robotic manipulator can be mounted to carry out different inspection and repair tasks. System (Figure 21b) has less payload and may only be used for fast inspection.

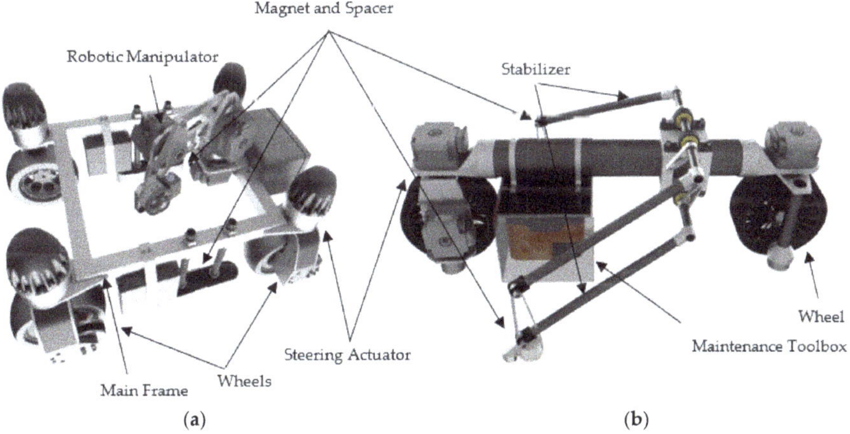

Figure 21. Concept of a four-wheel omnidirectional platform (**a**); concept of a two-wheel magnetic platform with stabilizers (**b**).

7. Conclusions and Outlook

A single system does not provide the capability to be an all-in-one replacement for conventional manned access technology. This work presents a multi-robot system that is suitable for many tasks. The multi-robot system enables maintenance staff to control inspection tasks safely from the ground. ROS is implemented for process control, especially for collision-free path planning in unstructured environments, as well as adaptive vision-based control [15]. The Bezier library within ROS provides scanning and path planning functionality for curved surfaces. Based on a 3D surface scan, an optimized tool path was generated that keeps any tool orthogonally on the freeform surface. Industrial robots are suitable manipulators for this task. This environment simplifies the process development of different thermography, radar [24], and ultrasonic applications for surface and structure inspection, as shown in Figure 22a–c, respectively. Figure 22 presents measurements carried out during this research as experimental validation. Figure 22a–c represent the same section of a 12 m rotor blade and show air gaps in a glue pattern, that occurred during the manufacturing process of this blade.

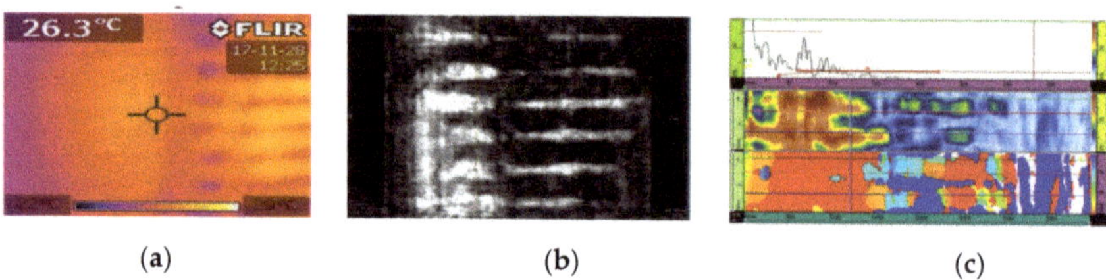

(a)   (b)   (c)

Figure 22. Non-destructive sensor feedback from a rotor blade section: thermography (**a**); radar (**b**), ultrasonic (**c**).

Future steps will include the implementation and test of a wide range of maintenance process tools based on a tool changing system. Inspection results and repair process parameters may be sampled to set up an expert system based on machine learning. A first introduction of this technique was presented with the flying system and visual inspection of welding lines via DCNN.

A future goal is the enrichment of the digital twin for wind turbines and even wind farms. This enables long-term condition monitoring, which helps to carry out predictive maintenance in order to reduce costs and increase operation time.

Furthermore, it is conceivable that several systems equipped with different tools solve the entire process chain in a swarm approach. The simplification of control by angle and height definition based on the inverse kinematics and the derived odometry is the first step towards autonomy. Any maintenance locations, due to the relatively deterministic tower structure, can be defined by height, initial diameter, and conicity. Thus, the foundations have been created to enable a single operator to efficiently control a large number of service robots in a process-oriented manner. This results in a cost advantage for the operator.

Supplementary Materials: The following are available online at https://fh-aachen.sciebo.de/s/zEROadazzhOUeBv, Video S1: Wind Turbine Maintenance with Robotics by MASKOR Institute Aachen.

Author Contributions: Methodology and concepts J.F. and E.D.; supervision, S.K. and S.D.; sensors and software, H.E.; validation and testing, J.F., H.E. and E.D.; writing—original draft preparation, J.F.; project administration, S.K. and P.D.; funding acquisition, J.F. and M.B. All authors have read and agreed to the published version of the manuscript.

Funding: This research was funded by German ministry for economy and energy (BMWi), grant number 5.3 million euro.

Acknowledgments: The research is supported by two industrial project partners: ematec AG and Gbr. Kaeufer GmbH from Germany.

Conflicts of Interest: The funders had no role in the design of the study; in the collection, analyses, or interpretation of data; in the writing of the manuscript, or in the decision to publish the results.

References

1. Durstewitz, M.; Guillaume, B.; Berkhout, V.; Buchmann, E.; Cernusko, R.; Faulstich, S.; Hahn, B.; Lutz, M.; Pfaffel, S.; Rehwald, F.; et al. *Windenergy Report Germany 2017*; Fraunhofer IEE: Kassel, Germany, 2018; pp. 19–21.

2. Rehfeld, K.; Wallasch, A.; Luers, S. *Cost Situation of Onshore Wind Energy in Germany*; Report by German Windguard; 2015. Available online: https://www.windguard.de/veroeffentlichungen.html (accessed on 13 May 2020).

3. Paulsen, T.; Thuering, H.; Fries, D. *Windenergy Service—Maintenance and Repair*; Bundesverband Windenergie, BWE: Berlin, Germany, 2012.
4. Schleupen, J.; Engemann, H.; Bagheri, M.; Kallweit, S.; Dahmann, P. Developing a climbing maintenance robot for tower and rotor blade service of wind turbines. In Proceedings of the International Conference on Robotics in Alpe-Adria-Danube Region—RAAD 2016, Belgrad, Serbia, 30 June–2 July 2016.
5. Schleupen, J.; Engemann, H.; Bagheri, M.; Kallweit, S. The potential of SMART climbing robot combined with a weatherproof cabin for rotor blade maintenance. In Proceedings of the 17th European Conference on Composite Materials—ECCM, Munich, Germany, 26–30 June 2016.
6. Kallweit, S.; Dahmann, P.; Bagheri, M.; Schleupen, J.; Engemann, H. Development of a climbing-robot for diagnose and maintenance of wind turbines. In Proceedings of the 13th Applied Automation Technology in Education and Development—AALE, Luebeck, Germany, 1 March 2016.
7. Bagheri, M.; Schleupen, J.; Dahmann, P.; Kallweit, S. A multi-functional device applying for the safe maintenance at high-altitude on wind turbines. In Proceedings of the 20th International Conference on Composite Materials—ICCM20, Copenhagen, Denmark, 19–24 July 2015.
8. Rafael, A.; Saltarén, R.; Reinoso, O. A climbing parallel robot: A robot to climb along tubular and metallic structures. *IEEE Robot. Autom. Mag.* **2006**, *13*, 16–22.
9. Rafael, A.; Saltarén, R.; Reinoso, O. Parallel robots for autonomous climbing along tubular structures. *Robot. Auton. Syst.* **2003**, *42*, 125–134.
10. Hayashi, S.; Takei, T.; Hamamura, K.; Ito, S.; Kanawa, D.; Imanishi, E.; Yamauchi, Y. Moving mechanism for a wind turbine blade inspection and repair robot. In Proceedings of the 2017 IEEE/SICE International Symposium on System Integration (SII), Taipei, Taiwan, 11–14 December 2017.
11. Waldron, K.J.; Sattar, T.P.; Rodriguez, H.L.; Bridge, B. Climbing ring robot for inspection of offshore wind turbines. *Ind. Robot Int. J.* **2009**, *36*, 326–330.
12. Morales, J.; Martinez, J.L.; Mandow, A.; Garcia-Cerezo, A.J.; Pedraza, S. Power consumption modeling of skid-steer tracked mobile robots on rigid terrain. *IEEE Trans. Robot.* **2009**, *25*, 1098–1108. [CrossRef]
13. Czichos, H.; Habig, K. *Tribologie-Handbuch*, 3rd ed.; Vieweg + Teubner: Wiesbaden, Germany, 2010; p. 85.
14. Lahayne, O.; Pichler, B.; Reihsner, R.; Eberhardsteiner, J.; Suh, J.; Kim, D.; Nam, S.; Paek, H.; Lorenz, B.; Persson, B.N. *Rubber Friction on Ice: Experiments and Modeling*; Springer: Berlin, Germany, 2016.
15. Rosen, E.; Whitney, D.; Phillips, E.; Ullman, D.; Tellex, S. Testing robot teleoperation using a virtual reality interface with ROS reality. In Proceedings of the 1st International Workshop on Virtual, Augmented, and Mixed Reality for HRI (VAM-HRI), Chicago, IL, USA, 5 March 2018.
16. He, K.; Gkioxari, G.; Dollar, P.; Girshick, R. Mask R-CNN. *arXiv* **2017**. Available online: https://arxiv.org/abs/1703.06870 (accessed on 13 May 2020).
17. Tavakoli, M.; Viegas, C.; Marques, L.; Pires, J.N.; de Almeida, A.T. OmniClimber-II: An omnidirectional climbing robot with high maneuverability and flexibility to adapt to non-flat surfaces. In Proceedings of the IEEE International Conference on Robotics and Automation, Karlsruhe, Germany, 6–10 May 2013; pp. 1349–1354.
18. Zhang, Y.; Dodd, T.; Atallah, K.; Lyne, I. Design and optimization of magnetic wheel for wall and ceiling climbing robot. In Proceedings of the IEEE International Conference on Mechatronics and Automation, Xi'an, China, 4–7 August 2010.
19. Tâche, F.; Fischer, W.; Caprari, G.; Siegwart, R.; Moser, R.; Mondada, F. Magnebike: A magnetic wheeled robot with high mobility for inspecting complex-shaped structures. *J. Field Robot.* **2009**, *26*, 453–476. [CrossRef]
20. Eich, M.; Vögele, T. Design and control of a lightweight magnetic climbing robot for vessel inspection. In Proceedings of the 19th Mediterranean Conference on Control Automation (MED), Corfu, Greece, 20–23 June 2011; pp. 1200–1205.
21. Li, J.; Wang, X.S. Novel omnidirectional climbing robot with adjustable magnetic adsorption mechanism. In Proceedings of the 23rd International Conference on Mechatronics and Machine Vision in Practice (M2VIP), Nanjing, China, 28–30 November 2016; pp. 1–5.
22. Almonacid, M.; Saltaren, R.J.; Aracil, R.; Reinoso, O. Motion planning of a climbing parallel robot. *IEEE Trans. Robot. Autom.* **2003**, *19*, 485–489. [CrossRef]

23. Information Resources Management Association. *Robotics: Concepts, Methodologies, Tools, and Applications*, 1st ed.; Hershey, P.A., Ed.; IGI Global: Hershey, PA, USA, 2013.
24. Nowok, S.; Kueppers, S.; Cetinkaya, H.; Schroeder, M.; Herschel, R. Millimeter wave radar for high resolution 3D near field imaging for robotics and security scans. In Proceedings of the 18th International Radar Symposium (IRS), Prague, Czech Republic, 28–30 June 2017; IEEE: Piscataway, NJ, USA, 2017.

Article

A New Approach for Fault Detection, Location and Diagnosis by Ultrasonic Testing

Fausto Pedro García Marquez [1] and Carlos Quiterio Gómez Muñoz [2],*

[1]   Ingenium Research Group, Castilla-La Mancha University, 13001 Ciudad Real, Spain;
     FaustoPedro.Garcia@uclm.es
[2]   School of Architecture, Engineering and Design, Universidad Europea de Madrid,
     28670 Villaviciosa de Odón, Spain
*   Correspondence: Carlosquiterio.gomez@universidadeuropea.es

Received: 27 January 2020; Accepted: 27 February 2020; Published: 5 March 2020

Abstract: Wind turbine blades are constantly submitted to different types of particles such as dirt, ice, etc., as well as all the different environmental parameters that affect the behaviour and efficiency of the energy generation system. These parameters can cause faults to the wind turbine blades, modifying their behaviour due, for example, to the turbulence. A new method is presented in this paper based on cross-correlations to determine the presence of delamination in the blades. The experiments were conducted in two real wind turbine blades to analyse the fault and non-fault blades using ultrasonic guided waves. Finally, the energy analysis of the signal based on wavelet transforms allowed to determine energies abrupt changes in the correlation of the signals and to locate the faults.

Keywords: fault detection and diagnosis; wavelet transforms; non-destructive tests; guided waves; wind turbine blade

---

1. Introduction

Wind energy is a growing renewable energy due to greater heights and powers of current wind turbines, with new and moderns installations [1]. It is expected new installations, with more than 55 GW every year until 2023 for onshore and offshore (Figure 1) [2]. Offshore installations are increasing its capacity with the associated technical issues in installation and maintenance. It is predicted to reach 2182 TW by 2030.

A wind turbine is composed of several subsystems, transforming wind energy into electric energy [3]. The wind induces the movement into the wind turbine blades (WTBs), and the main shaft transmits the mechanical energy into the gearbox. The main shaft is supported by bearings and it is connected to the generator. Different subsystems are designed for supporting the normal behaviour of the wind turbine, e.g., the meteorological unit, for controlling the pitch and brake systems. Wind farms are located in remote areas under the severe weather conditions and, consequently, each wind turbine presents problems related to ice and snow deposition on the WTBs, breakage of WTBs by impact of objects, etc. [4,5]. The wind turbine rotor, electrical devices, plant control system, hydraulic and sensors have more than 50% of total failures [6,7].

The maintenance operations have high costs, risks for workers, and energy production losses due to downtimes [8]. The wind turbine operation and maintenance (O&M) costs are between 10–25% of the total costs [9]. The efficiency and security of O&M activities are reduced due to the working conditions, and novel failure prediction techniques are required to avoid downtimes and increasing the reliability of the installations [10].

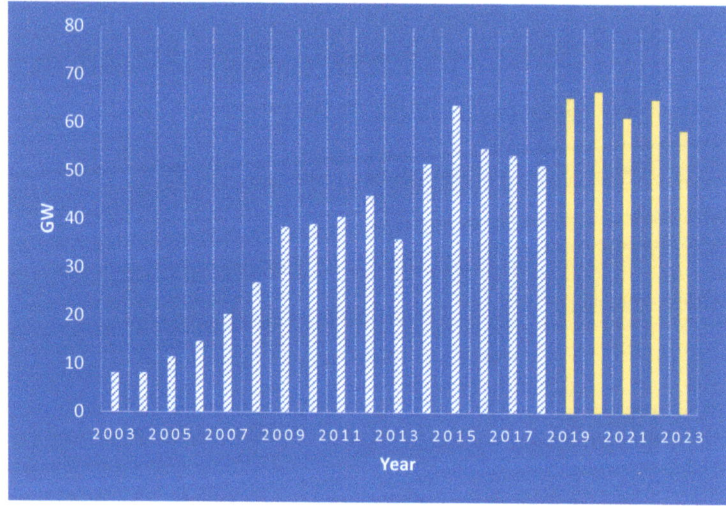

Figure 1. Wind energy capacity and projection. Data from Council, G.W.E. Global wind report [2].

In recent years, there is an important evolution in designs, materials, mechanical electronic, electrical and control of wind turbines [11]. The short-term objective of renewable energies is to increase their participation in total energy production. It is necessary to improve productivity and profitability [1,12]. The new advances in renewable energy technologies lead to this industry to be more competitive in the energy market by reducing the operation and maintenance costs. The industry requires also to improve the reliability, availability, maintainability and safety of the systems [13]. Novel technological solutions are needed to increase the competitiveness based on the maintenance cost reduction for ensuring the efficient positioning of this technology in the energy market [14]. The improvement in the maintenance management operations will help to reach the competitiveness in wind power in terms of reliability, lifetime and availability [15]. Other key factor for the evolution of an industry is the cost reduction and the system efficiency by new strategies based on advance analytics [16,17], for example, the optimization of maintenance resources or the correct use of them [18,19]. Maintenance management is considered as transcendental to improve the benefit margin [20].

The wind farm maintenance management is also complex due to the machine locations and meteorological conditions [21]. These preventive and corrective works are done over the time [22,23], but they are expensive and generate risk for the operators, working in high altitude. The generation of false alarm or deceptive signals of the monitorization system is being a fundamental issue in the management [24–26].

WTB are generally made with sandwich materials that are introduced inside its structure. These materials are based on composite skins, with the core being composed of lightweight and isotropic materials. The WTBs need to be designed according to their complexity with low weight and good mechanical properties. The WTBs must have a high resistance to fatigue, wear and tear, as well as low thermal expansion and conductivity. The WTBs are considered one of the essential components, which are subject to heavy loads and weather conditions. The WTB failures are expensive, and they can damage other WTBs or other parts of the wind turbine. It has been demonstrated that efficient non-destructive tests (NDT) processes should increase the WTB life cycle and reduce the probability of failure occurrence. The composite material is formed by long fibres that are set within a matrix, which is responsible of binding the fabrics. The composite materials depend on the order in which the fibres are stacked, the orientation, as well as their own physical properties. The sandwich structures are made up by two outer coatings covering a lightweight material called the core. They are designed to provide high rigidity whilst being lightweight. The core function is to prevent undesired movement of the outer coatings and has higher thickness and lower density compared to them.

The main interest in maintenance management is being able to employ a condition monitoring system (CMS) capable to predict failures [27,28]. The wind turbine components are exposed to physical efforts such as stress or compression, and chemical or environmental conditions such as erosion or surface degeneration [29]. They can facilitate the appearance of mechanical, electrical and structural failures [30]. It needs a correct maintenance plan and CMS to prevent almost the main failures [31].

CMS by ultrasonic guided waves is essential for the maintenance management of WTB. This system needs computing systems for continuous supervision of the main parameters to determine the condition. To achieve a correct CMS, it is required: Data acquisition: reading the physical behaviour and to convert it in a digital format to be computerized; Data processing: conversion of digital data into useful information about the condition of the WTB; Detection: determining whether the condition indicators are "normal" or "abnormal" by pattern recognition or signal processing; Diagnosis: setting the location and the severity of the fault in the WTB; Prognosis: analysing the life cycle of the fault before a failure and maintain the WTB functionality before replacement; Maintenance Management: to set what are the maintenance and correction actions to be taken and the way how they should be achieved.

Ultrasonic guided waves are employed in this paper as NDT for the WTB. They are considered to guarantee the operability of WTBs. There are different methods and algorithms that have been developed to monitor the condition of WTBs that use guided waves. Other NDT methods used in WTBs are, for example, radiography, optics, thermography and acoustics.

The tests have been designed in this paper in order to detect delamination in a real WTB. Delamination in WTBs is a structural issue that increase downtimes and costs. It consists in the separation of layers of the WTB, knowing that the WTB is made from composite materials. This separation produces points within the WTB of stress concentration, i.e., these areas are working with more traction and compression forces under normal working conditions. It would lead to cracks, and, therefore, partial or total failure in the WTB. The early detection of this phenomena is needed to prevent failures/faults in the WTB. This paper proposed ultrasonic guided waves to measure the WTB condition.

A novel approach that uses correlation analysis between a real damaged and undamaged WTB is employed for pattern recognition. The diagnosis is performed by wavelet transforms. The approach leads to detect and diagnosis faults, such as delamination, with a high accuracy.

2. Ultrasonic Testing Applied for CMS in WTBs

2.1. Non-Destructive Testing

NDT consists of non-invasive inspection techniques that are used to analyse the condition of the component studied. The techniques are also utilized to detect faults, e.g., corrosion or cracks. NDT is a safe, reliable and cost-effective inspection method of the components without damaging the part to be examined. The NDT may be carried out during or after the manufacturing process. In the case of manufacturing, inspections can determine if the parts tested are suitable for a desired function. NDT inspections can also be used to assess the current condition of equipment with faults, or to monitor any faulty parts. It allows decisions based on the information to optimize maintenance and to evaluate the remaining useful life of the equipment.

2.2. Ultrasonic Testing

Ultrasonic Testing is a technique that uses high frequency sound energy to do experiments and make measurements. Common uses for the ultrasonic inspection are in fault detection, dimensional measurements, the characterization of materials, etc. Ultrasonic inspection system consists of devices with a pulser and a receiver, which is able to produce high electrical voltage pulses to the transducers.

The transducer generates high frequency ultrasonic energy that is induced and propagated in the form of waves through the material. In the case of detecting a fault within the material, e.g., a crack in the wave path, part of the signal energy will be reflected from the fault within the material. The reflected wave signal is transformed into an electrical signal by the transducer that is sent to a computer to be analysed. This technique has the origin with the sound waves through water, where it was observed the reflected echoes to characterize submerged objects.

2.3. Long Range Ultrasonic Testing

Long Range Ultrasonic Testing (LRUT) is a cutting-edge NDT technique that is used for studying large volumes of material from a single test point. It leads to reduce the time needed to carry out multiple tests. LRUT does not require the existence of couplant gels or liquids between the transducers and the tested surface. Therefore, this technique is widely used in pipeline and plates inspections for corrosion and other fault types. The LRUT working procedure consists of fixing the transducer and generating a set of low frequency guided waves. The waves are then reflected back to the transducer whenever they reach a variation in the thickness of the wall of the plate or pipe, which would indicate the existence of corrosion, metal loss or other faults mechanisms. The LRUT has been employed because: it is sensitive to surface and subsurface discontinuities; it has high depth penetration for fault detection; it presents good accuracy for finding the reflector position, size and shape in the tested material; it requires a minimal preparation needed for testing; it gives results in real time; and the equipment is portable.

3. Case Study and Results

Two identical real WTBs have been employed in the experiments. One WTB has induced a set of delamination in the manufacturing process. The WTB has been inspected by placing the transmitter transducer on the tip and placing the transducer receiver every 100 mm along the length of the WTB. Figure 2 illustrates the WTBs on which the non-destructive tests have been conducted.

Figure 2. Wind turbine blades (WTBs) with delamination and without delamination in which the guided ultrasound wave tests were applied.

This is performed on both WTBs in order to analyse them completely. The WTBs have a length of 4000 mm and they have a honeycomb core within them.

The method employed to collect the ultrasonic signals is pitch and catch. The transducers used were Macro Fibre Composites (MFC) [14,32], specifically, the model M2814-P1 from Smart Material, and they were attached on the surface. A transducer serving as transmitter is located on the tip of the WTB. The position of the transmitter does not change, while the receiver is placed at different distances along the WTB (Figure 3). The first position of the receiver is 100 mm from the transmitter, and the experiments were done increasing the distance 100 mm until 3800 mm (38 different locations). d in Figure 3 is the distance between the transmitter and the sensor.

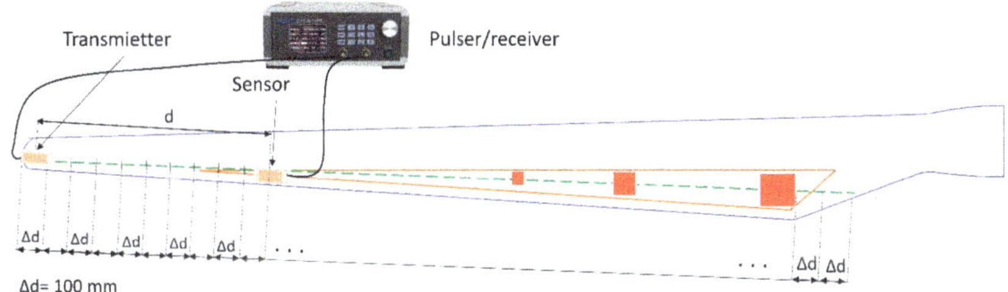

Figure 3. Location of the transducers along the WTB.

The signal generated by the transducer was a five cycles sinusoidal shaped signal, modulated by a Hanning window. At each position, a frequency sweep was deployed from 10 kHz to 100 kHz, with steps of 5 kHz, but only 50 kHz frequencies were selected because they had the best signal to noise ratio for this material. The aim of this work is to find evidence in the signal that may determine that there is a defect in the WTB, analysing the guided waves that travel through the faults.

The 50 kHz signals of the 38 distances were pooled to analyse the correlation between them. Figure 4 shows the location of the sandwich and faults locations that have been induced in the damaged WTB.

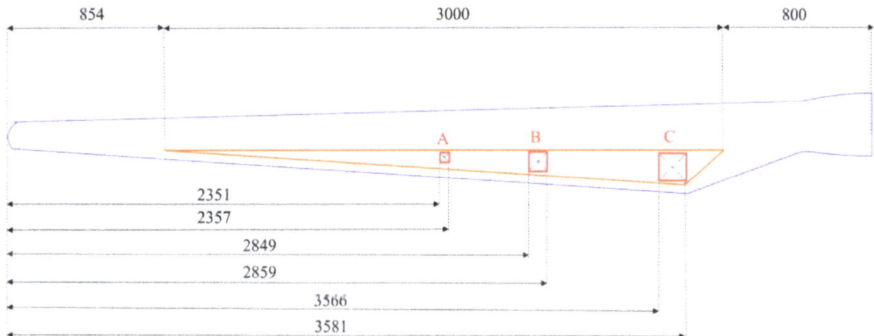

Figure 4. Defect locations in the faulty WTB. A, B and C areas are the disbonds between the honeycomb and the skin.

The approach is based on the following steps:

(1) Firstly, an autocorrelation is performed between each pair of adjacent signals, for example, the first pair are those that are located to 100 and 200 mm from the tip, the next pair are 200 and 300 mm from the tip, etc. If there are no faults on the WTBs, the shape of the contiguous signals will be almost the same and with a high correlation. However, if between two adjacent signals there is a delamination that modifies the shape of the ultrasonic wave, then there will be a lower correlation between the signals. It will indicate the possibility of the existence of a defect in the WTB. This procedure has been performed for both the healthy and the faulty WTB.

(2) Secondly, a new cross correlation was made between the signals obtained in the previous section to study the results between both WTBs. Every WTB is analysed together with WTB free of faults and, if the correlation between both areas of the WTB is high, it indicates that the WTB is healthy. On the contrary, if there is a low correlation between an area and the homologous zone of the healthy WTB, where it indicates that there is a discontinuity in that area that could be a delamination defect.

(3) Finally, the diagnosis of the delamination is done employing wavelet transforms. The energies of both healthy and unhealthy WTBs are studied together.

The flowchart of the approach is shown in Figure 5.

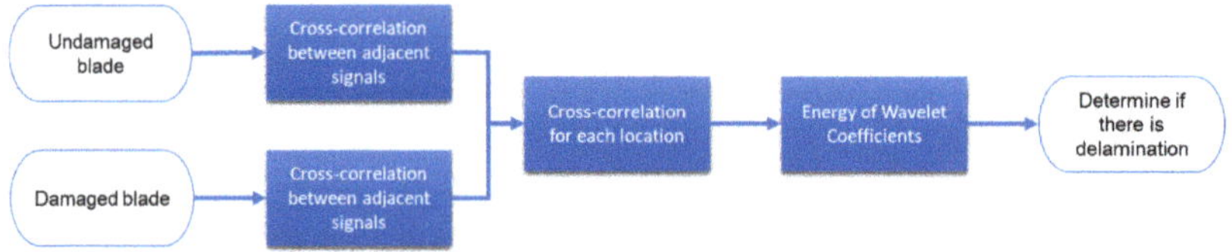

Figure 5. Approach for delamination detection in WTBs.

3.1. Undamaged WTB

The distance set between the sensors was 100 mm, and 38 acquisitions were made for each WTB. Four significant signals are shown in Figure 6.

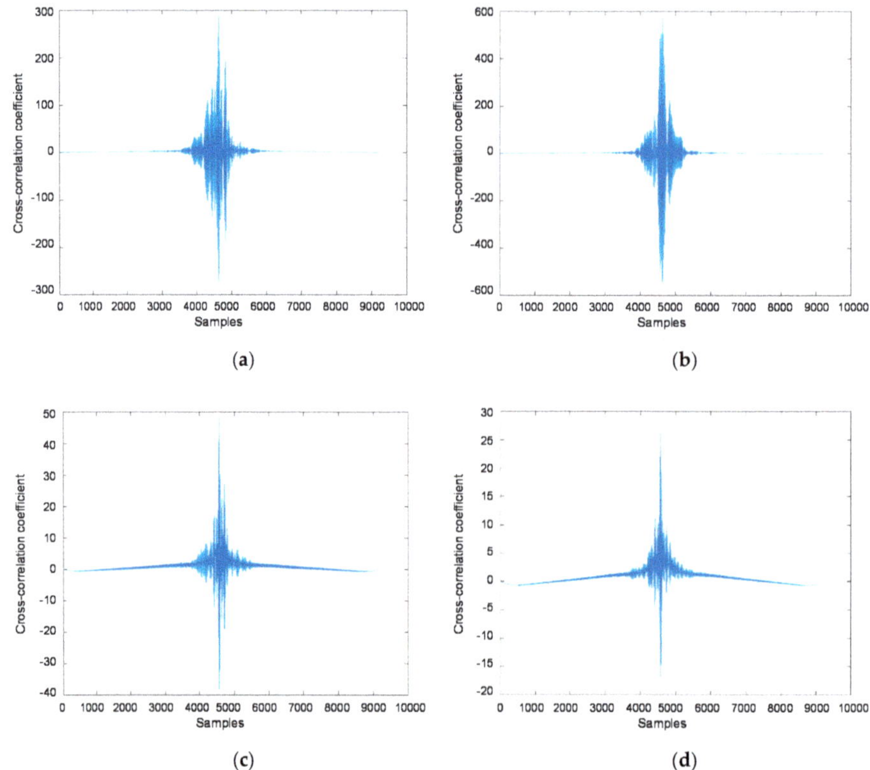

Figure 6. Cross-correlation between adjacent signals of undamaged WTB: (**a**) 200–300 mm; (**b**) 700–800 mm; (**c**) 2,300–2,400 mm; and (**d**) 2,900 and 3,000 mm.

Figure 6a shows the signal obtained between 200 and 300 mm. The signals present a high similarity due to the absence of any fault. Figure 6b also shows great similarity because of the autocorrelation signals have been performed on the signals that were acquired at 700 and 800 mm from the tip. They are located just before the beginning of the sandwich area. This similarity is found by high cross-correlation coefficients. The cross-correlation coefficients between the signals at 700 and 800 mm are even greater than those obtained at the signals collected at 200 and 300 mm from the tip (Figure 6a). This is due to the signals closest to the transmitter transducer (Figure 6a) are affected by phenomena inherent in the generation of guided waves, e.g., ringing, that slightly decrease their cross-correlation coefficients. Figure 6c shows that the similarity level presents some differences and loss of energy due to the inherent imperfections of the WTB. There are some energy losses due to the attenuation of the elastic wave. Figure 6d shows that the similarity level is different and presents energy loss because of the distance travelled by the signals. Nevertheless, the changes in the degree of similarity only occurs as a consequence of the imperfections within the WTB and not by a fault.

3.2. Damaged WTB

This section presents the main results found analysing damaged WTB. The experiments were conducted to validate the location and severity of the WTB delamination. Figure 7a shows that the signals have similar patterns between the healthy and damage WTB because there is not any fault. Figure 7b is similar to Figure 6b because this location is just before the start of the sandwich, and the ultrasonic waves have not crossed any delamination zone in any WTB. Figure 7c shows the first defect within the WTB, being the closest to the WTB tip. It has been detected due to the correlation is less compared to the WTB without fault. The correlation shown in Figure 7d is lower due to the faults and the waves are attenuated by the distance travelled.

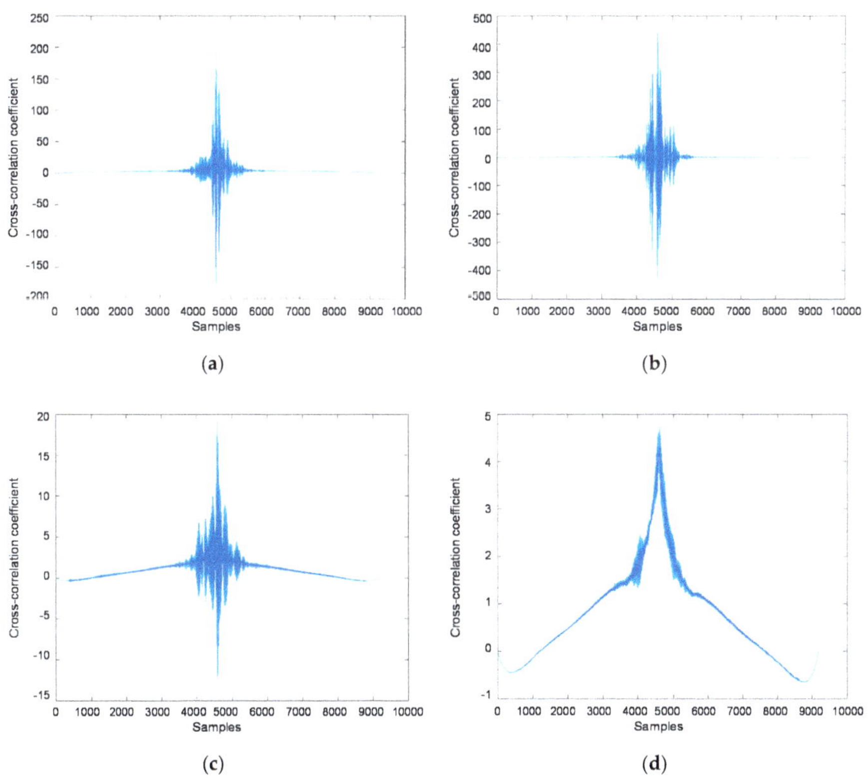

Figure 7. Cross-correlation between adjacent signals of damaged WTB: (**a**) 200–300 mm; (**b**) 700–800 mm; (**c**) 2,300–2,400 mm; and (**d**) 2,900 and 3,000 mm.

Figure 8 presents the relative error between the maximum of each correlation according to the position of the sensor in the WTB. The maximum relative error is found where the first fault is located.

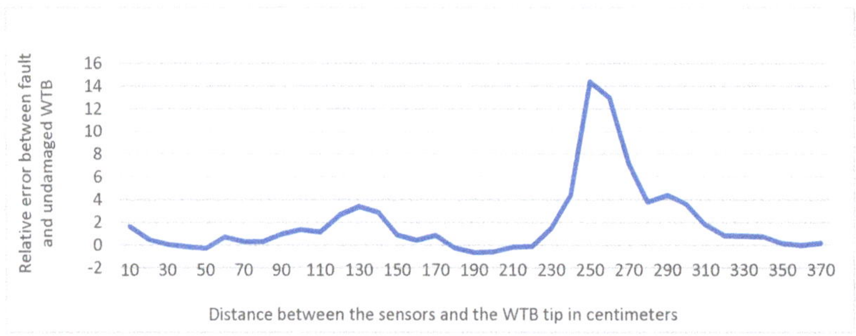

Figure 8. Relative error between damaged and undamaged WTB for each location of the sensor.

3.3. Cross-Correlation between Healthy and Damaged WTBs

The autocorrelation of each pair of signals is analysed together with the damage and undamaged WTB. The attenuation effect is partially eliminated, since each pair of signals is compared between both WTBs. Figure 9 presents the main results.

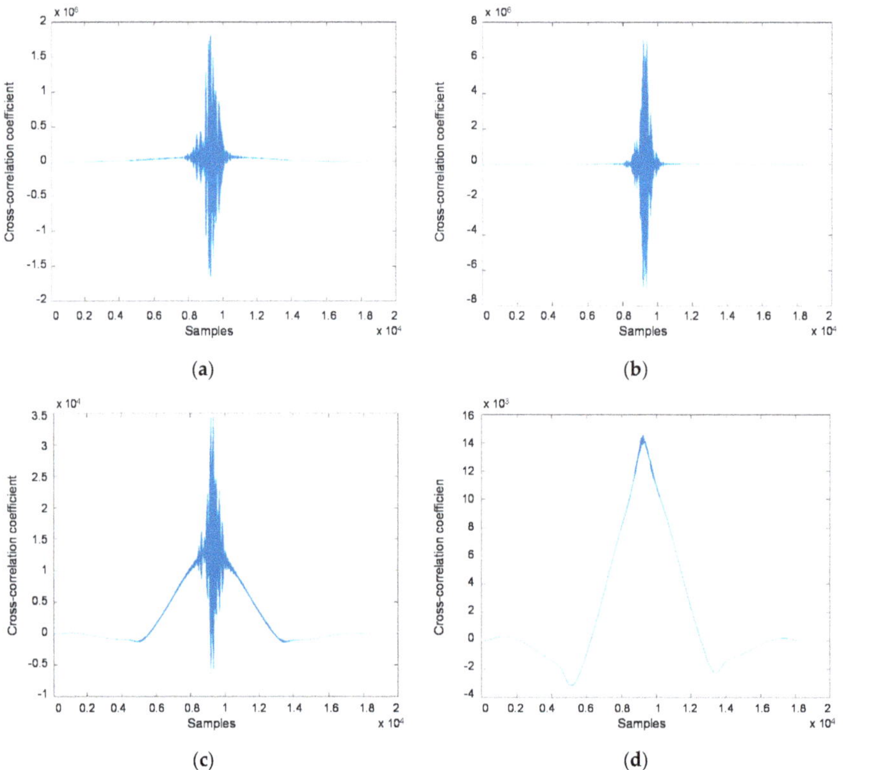

Figure 9. Cross-correlation between undamaged and damaged WTB: (**a**) 200–300 mm; (**b**) at 700–800 mm; (**c**) at 2,300–2,400 mm; and (**d**) 2,900–3,000 mm.

Figure 9a shows a high similarity because the WTB areas studied are free of faults. It is found in the experiments where there is not faults. The signals presented in Figure 9b are related to the beginning of the sandwich in both WTBs, where they are similar because of the sandwich properties are the same in both WTBs. Figure 9c shows the signals where the first defect is located, being the correlation lower due to the delamination. Figure 9d is related to the second fault, being lower than the signals showed in Figure 9c.

3.4. Energy Analysis by Wavelet Transforms

The graphic representation is not useful for a deep study of an acoustic characterization [33]. For this reason, it is necessary to support this work with a mathematical treatment. The acoustic is a non-periodic deterministic signal, characterized with a sinusoidal pulse and defined by wavelength, amplitude, frequency. The acoustic signal can be analysed mathematically in time-frequency [34], therefore, the signal analysis can be studied by mathematical tools such as Fourier Transform (FT) [35,36]. The advantages of the wavelet transform prove that this technique is efficient for acoustic signals [37]. It is defined by Equation (1).

$$S(\tau, a) = \int_{-\infty}^{+\infty} s(t) \frac{1}{\sqrt{a}} \Psi^* \left(\frac{t - \tau}{a} \right) dt \tag{1}$$

The conjugate of the mother wavelet Ψ^* is obtained, moved and scaled point to point to detect the levels of contrast of the signal s(t), being f(t) the digitized signal in the time domain, $a = f/f_0$ ($a \neq 0$) the magnitude factor or delay of the wavelet, with f_0 as central frequency and τ the translation in time [38]. Another common expression in acoustic representation is a frequency domain spectrum to study the frequency distribution range of the signal [39].

The wavelet transform is employed to obtain the signal energy decomposition divided in levels with the pyramidal algorithm and the decomposition tree. This method is employed for acoustics characterization and filtering [14,40]. There are various wavelet transforms modes, e.g., discrete or continuous [32,41]. The choice of the wavelet family is imposed by the characteristics of the signal and the nature of the application, being Dauchebies family one of the most used in the acoustic signal processing [42,43].

Signals are divided, therefore, into low frequency approximations (**A**) and high frequency details (**D**), where the sum of **A** and **D** is always equal to the original signal. The division is done using low pass and high pass filters. In order to reduce the computational and mathematical costs due to the data duplication, a sub-sampling is usually implemented, containing the half of the collected information from **A** and **D** without losing information.

In the case of the multi-level filters, they repeat the filtering process with the output signals from the previous level. This leads to the wavelet decomposition trees (Figure 10). Additional information is obtained by filtering at each level. However, more decompositions levels do not always mean more accurate results. References [44,45] show more details about the decomposition level using wavelet transform.

The objective of the signal pre-processing is to extract the most important information of the original signal before carrying out the signal de-noising. It generates new signals adjusted for the application of filters, providing more robust results and greater similarity between signals obtained under different conditions.

The energy of the signals has been calculated by wavelet transforms. The energies have been employed to study the severity of the faults. Figure 11 shows that there is a high correlation between the signals obtained from the WTBs in the first two areas analysed. However, when the ultrasonic waves reach the location of the first fault in the damaged WTB, the correlation between the healthy WTB and the damaged WTB decreases and, therefore, decreases its energy.

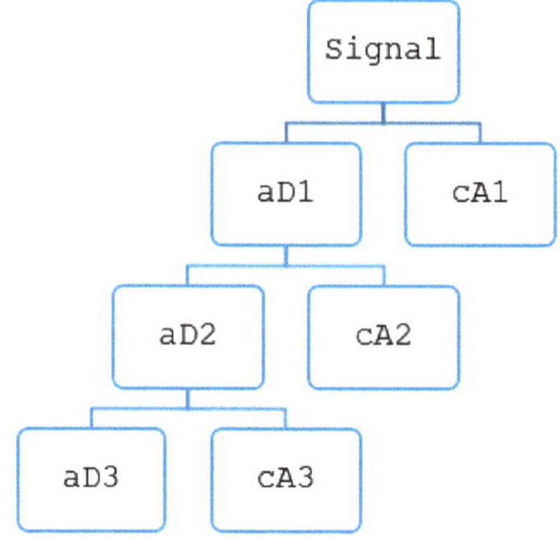

Figure 10. Wavelet decompositions tree.

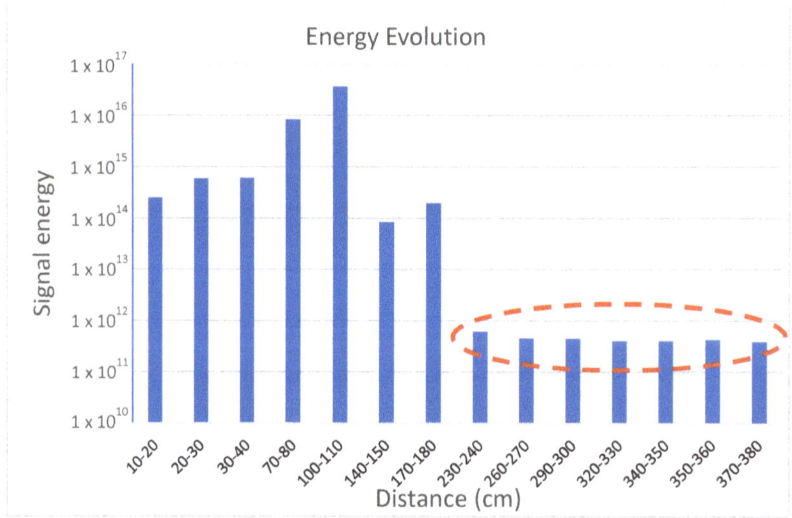

Figure 11. Evolution of energy throughout the WTBs.

The accuracy of the results has been calculated analysing them with the real scenarios according to ISO 5725-1. It has been estimated as 92%, because in certain cases the fault detection was not clear enough.

It cannot be concluded that it can be applied to WTBs in operations because it has not been tested. Nevertheless, the results have been validated in cases where they are pre-installed.

4. Conclusions

Wind turbine blades are submitted to severe mechanical and environment conditions and, therefore, they present a high failure rate and downtimes by faults/failures. Novel and robust maintenance procedures are required. This paper presents a condition monitoring system based on ultrasonic guided waves to study the structure heath monitoring. A guided wave inspection has been conducted from the tip of the blades, acquiring the signals every 100 mm. The approach is based on correlations analysis between a damaged and undamaged real wind turbine blades. Finally, the diagnosis is done by wavelet transforms. The main conclusions are:

- The approach leads to detect and diagnose faults such as delamination with a high accuracy.
- The autocorrelation of each pair of signals is analysed together with the damage and undamaged wind turbine blade. The attenuation effect is partially eliminated because of each pair of signals is compared between both blades.
- The pattern recognition presents a high correlation when a fault is not found. On the other side, the correlation found is lower when a delamination is found.
- The wavelet transform is employed to obtain the signal energy decomposition divided in levels with the pyramidal algorithm and the decomposition tree.
- The choice of the wavelet family was Dauchebies family due to they provide accuracy results in the acoustic signal processing.
- The energies have been employed to study the severity of the faults.
- The accuracy of the results has been calculated analysing them with the real scenarios. It has been estimated as 92%, due to certain cases not having a clear enough fault detection.

Author Contributions: All authors formulated the problem and methodology. F.P.G.M. contributed in the design of the tests, the analysis and writing. C.Q.G.M. performed the tests and wrote the draft of the paper. All authors have read and agreed to the published version of the manuscript.

Funding: The work reported herewith has been financially by the Dirección General de Universidades, Investigación e Innovación of Castilla-La Mancha, under Research Grant ProSeaWind project (Ref.: SBPLY/19/180501/000102).

Conflicts of Interest: The authors declare no conflict of interest.

References

1. Muñoz, C.Q.G.; Márquez, F.P.G. Future maintenance management in renewable energies. In *Renewable Energies*; Springer: Berlin/Heidelberg, Germany, 2018; pp. 149–159.
2. Council, G.W.E. *Global Wind Report*; Global Wind Energy Council: Brussels, Belgium, 2019.
3. Polinder, H.; Ferreira, J.A.; Jensen, B.B.; Abrahamsen, A.B.; Atallah, K.; McMahon, R.A. Trends in wind turbine generator systems. *IEEE J. Emerg. Sel. Top. Power Electron.* **2013**, *1*, 174–185. [CrossRef]
4. Muñoz, C.Q.G.; Márquez, F.P.G. Wind Energy Power Prospective. In *Renewable Energies*; Springer: Berlin/Heidelberg, Germany, 2018; pp. 83–95.
5. Jiménez, A.A.; Muñoz, C.Q.G.; Márquez, F.P.G. Dirt and mud detection and diagnosis on a wind turbine blade employing guided waves and supervised learning classifiers. *Reliab. Eng. Syst. Saf.* **2019**, *184*, 2–12. [CrossRef]
6. Tchakoua, P.; Wamkeue, R.; Ouhrouche, M.; Slaoui-Hasnaoui, F.; Tameghe, T.A.; Ekemb, G. Wind turbine condition monitoring: State-of-the-art review, new trends, and future challenges. *Energies* **2014**, *7*, 2595–2630. [CrossRef]
7. García Márquez, F.; Marugán, A.P.; Pérez, J.M.P.; Hillmanse, S.; Papaelias, M. Optimal dynamic analysis of electrical/electronic components in wind turbines. *Energies* **2017**, *10*, 1111. [CrossRef]
8. Walford, C.A. *Wind Turbine Reliability: Understanding and Minimizing Wind Turbine Operation and Maintenance Costs*; 2006. Available online: https://pdfs.semanticscholar.org/03dd/65b6a56a87fb9bd8e323b092b160d70c666f.pdf (accessed on 25 February 2020).
9. Irena, I. *Renewable Energy Technologies: Cost Analysis Series*; IRENA Innovation Technology Center: Dubai, UAE, 2012; Volume 13, p. 2016.
10. Pliego Marugán, A.; Márquez, F.P.G.; Lev, B. Optimal decision-making via binary decision diagrams for investments under a risky environment. *Int. J. Prod. Res.* **2017**, *55*, 5271–5286. [CrossRef]
11. Njiri, J.G.; Soeffker, D. State-of-the-art in wind turbine control: Trends and challenges. *Renew. Sustain. Energy Rev.* **2016**, *60*, 377–393. [CrossRef]

12. Spiess, H.; Lobsiger-Kägi, E.; Carabias-Hütter, V.; Marcolla, A. Future acceptance of wind energy production: Exploring future local acceptance of wind energy production in a Swiss alpine region. *Technol. Forecast. Soc. Chang.* **2015**, *101*, 263–274. [CrossRef]

13. Stapelberg, R.F. *Handbook of Reliability, Availability, Maintainability and Safety in Engineering Design*; Springer-Verlag London: London, UK, 2009.

14. De la Hermosa González, R.R.; Márquez, F.P.G.; Dimlaye, V. Maintenance management of wind turbines structures via MFCs and wavelet transforms. *Renew. Sustain. Energy Rev.* **2015**, *48*, 472–482. [CrossRef]

15. Pérez, J.M.P.; Márquez, F.P.G.; Hernández, D.R. Economic viability analysis for icing blades detection in wind turbines. *J. Clean. Prod.* **2016**, *135*, 1150–1160. [CrossRef]

16. Petković, D.; Pavlović, N.T.; Ćojbašić, Ž. Wind farm efficiency by adaptive neuro-fuzzy strategy. *Int. J. Electr. Power Energy Syst.* **2016**, *81*, 215–221. [CrossRef]

17. Márquez, F.P.G.; Lev, B. *Advanced Business Analytics*; Springer: Berlin/Heidelberg, Germany, 2015.

18. Sarker, B.R.; Faiz, T.I. Minimizing maintenance cost for offshore wind turbines following multi-level opportunistic preventive strategy. *Renew. Energy* **2016**, *85*, 104–113. [CrossRef]

19. Pliego Marugán, A.; Márquez, F.P.G. Advanced analytics for detection and diagnosis of false alarms and faults: A real case study. *Wind Energy* **2019**, *22*, 1622–1635. [CrossRef]

20. Shafiee, M.; Sørensen, J.D. Maintenance optimization and inspection planning of wind energy assets: Models, methods and strategies. *Reliab. Eng. Syst. Saf.* **2019**, *192*, 105993. [CrossRef]

21. Tavner, P.; Gindele, R.; Faulstich, S.; Hahn, B.; Whittle, M.W.G.; Greenwood, D.M. Study of effects of weather & location on wind turbine failure rates. In Proceedings of the European wind energy conference EWEC, Warsaw, Poland, 20–23 April 2010.

22. Pérez, J.M.P.; Asensio, E.S.; Márquez, F.P.G. Economic viability analytics for wind energy maintenance management. In *Advanced Business Analytics*; Springer: Berlin/Heidelberg, Germany, 2015; pp. 39–54.

23. Marquez, F.G.; Singh, V.; Papaelias, M. A Review of Wind Turbine Maintenance Management Procedures. In Proceedings of the Eighth International Conference on Condition Monitoring and Machinery Failure Prevention Technologies, Cardiff, Wales, 20–22 June 2011.

24. Marugán, A.P.; Chacón, A.M.P.; Márquez, F.P.G. Reliability analysis of detecting false alarms that employ neural networks: A real case study on wind turbines. *Reliab. Eng. Syst. Saf.* **2019**, *191*, 106574. [CrossRef]

25. Márquez, F.P.G. A new method for maintenance management employing principal component analysis. *Struct. Durab. Health Monit.* **2010**, *6*, 89–99.

26. Pedregal, D.J.; García, F.P.; Roberts, C. An algorithmic approach for maintenance management based on advanced state space systems and harmonic regressions. *Ann. Oper. Res.* **2009**, *166*, 109–124. [CrossRef]

27. Marquez, F.G. An Approach to Remote Condition Monitoring Systems Management. In Proceedings of the IET International Conference on Railway Condition Monitoring, Birmingham, UK, 29–30 November 2006.

28. Márquez, F.P.G.; Pedregal, D.J. Applied RCM 2 algorithms based on statistical methods. *Int. J. Autom. Comput.* **2007**, *4*, 109–116. [CrossRef]

29. Qiao, W.; Lu, D. A Survey on Wind Turbine Condition Monitoring and Fault Diagnosis—Part I: Components and Subsystems. *IEEE Trans. Ind. Electron.* **2015**, *62*, 6536–6545. [CrossRef]

30. Pérez, J.M.P.; Márquez, F.P.G.; Tobias, A.; Papaelias, M. Wind turbine reliability analysis. *Renew. Sustain. Energy Rev.* **2013**, *23*, 463–472. [CrossRef]

31. Benmessaoud, T.; Mohammedi, K.; Smaili, Y. Influence of maintenance on the performance of a wind farm. *Przegląd Elektrotech.* **2013**, *89*, 174–178.

32. De la Hermosa Gonzalez, R.R.; Márquez, F.P.G.; Dimlaye, V.; Ruiz-Hernández, D. Pattern recognition by wavelet transforms using macro fibre composites transducers. *Mech. Syst. Signal Process.* **2014**, *48*, 339–350. [CrossRef]

33. Gómez Muñoz, C.Q.; Jimenez, A.A.; Marquez, F.P.G.; Kogia, M.; Cheng, L.; Mohimi, A.; Papaelias, M. Cracks and welds detection approach in solar receiver tubes employing electromagnetic acoustic transducers. *Struct. Health Monit.* **2018**, *17*, 1046–1055. [CrossRef]

34. Hammond, J.; White, P. *Fundamentals of Signal Processing*; Springer: Berlin/Heidelberg, Germany, 2016; p. 145.

35. Portnoff, M. Time-frequency representation of digital signals and systems based on short-time Fourier analysis. *IEEE Trans. Acoust. Speech Signal Process.* **1980**, *28*, 55–69. [CrossRef]

36. Weller, H. *Fourier Analysis*; 2015. Available online: http://www.met.reading.ac.uk/~{}sws02hs/teaching/Fourier/Fourier_2_student.pdf (accessed on 25 February 2020).

37. Zhang, D. *Wavelet Transform, in Fundamentals of Image Data Mining*; Springer: Berlin/Heidelberg, Germany, 2019; pp. 35–44.
38. Zhang, D. *Wavelet Transform, in Fundamentals of Image Data Mining: Analysis, Features, Classification and Retrieval*; Springer International Publishing: Cham, Switzerland, 2019; pp. 35–44.
39. Pandarakone, S.E.; Mizuno, Y.; Nakamura, H. Distinct Fault Analysis of Induction Motor Bearing Using Frequency Spectrum Determination and Support Vector Machine. *IEEE Trans. Ind. Appl.* **2017**, *53*, 3049–3056. [CrossRef]
40. Gokgoz, E.; Subasi, A. Comparison of decision tree algorithms for EMG signal classification using DWT. *Biomed. Signal Process. Control* **2015**, *18*, 138–144. [CrossRef]
41. Goudarzi, M.; Vahidi, B.; Naghizadeh, R.A.; Hosseinian, S.H. Improved fault location algorithm for radial distribution systems with discrete and continuous wavelet analysis. *Int. J. Electr. Power Energy Syst.* **2015**, *67*, 423–430. [CrossRef]
42. Sheikh, J.A.; Parah, S.A.; Akhtar, S.; Bhat, G.M. Compression and denoising of speech transmission using Daubechies wavelet family. *Int. J. Wirel. Mob. Comput.* **2017**, *12*, 313–334. [CrossRef]
43. Muñoz, C.Q.G.; Jiménez, A.A.; Márquez, F.P.G. Wavelet transforms and pattern recognition on ultrasonic guides waves for frozen surface state diagnosis. *Renew. Energy* **2018**, *116*, 42–54. [CrossRef]
44. Rahim, A.A.A.; Abdullah, S.; Singh, S.S.K.; Nuawi, M.Z. Selection of the optimum decomposition level using the discrete wavelet transform for automobile suspension system. *J. Mech. Sci. Technol.* **2020**, *34*, 137–142. [CrossRef]
45. Srivastava, A.; Bhateja, V.; Shankar, A.; Taquee, A. On Analysis of Suitable Wavelet Family for Processing of Cough Signals. In *Frontiers in Intelligent Computing: Theory and Applications*; Springer: Singapore, 2020; pp. 194–200.

Article

Reducing Operational Costs of Offshore HVDC Energy Export Systems Through Optimized Maintenance

Jan Frederick Unnewehr [1,*], Hans-Peter Waldl [2], Thomas Pahlke [2], Iván Herráez [3] and Anke Weidlich [1]

[1] Department of Sustainable Systems Engineering, University of Freiburg, Emmy-Noether-Strasse 2, 79110 Freiburg, Germany; anke.weidlich@inatech.uni-freiburg.de
[2] Overspeed GmbH & CO. KG, Im Technologiepark 4, 26129 Oldenburg, Germany; h.p.waldl@overspeed.de (H.-P.W.); t.pahlke@overspeed.de (T.P.)
[3] University of Applied Science Emden/Leer, Constantiaplatz 4, 26723 Emden, Germany; ivan.herraez@hs-emden-leer.de
* Correspondence: jan.frederick.unnewehr@inatech.uni-freiburg.de; Tel.: +49-761-203-54261

Received: 12 December 2019; Accepted: 27 February 2020; Published: 3 March 2020

Abstract: For the grid connection of offshore wind farms today, in many cases a high-voltage direct current (HVDC) connection to the shore is implemented. The scheduled maintenance of the offshore and onshore HVDC stations makes up a significant part of the operational costs of the connected wind farms. The main factor for the maintenance cost is the lost income from the missing energy yield (indirect maintenance costs). In this study, we show an in-depth analysis of the used components, maintenance cycles, maintenance work for the on- and offshore station, and the risks assigned in prolonging the maintenance cycle of the modular multilevel converter (MMC). In addition, we investigate the potential to shift the start date of the maintenance work, based on a forecast of the energy generation. Our findings indicate that an optimized maintenance design with respect to the maintenance behavior of an HVDC energy export system can decrease the maintenance-related energy losses (indirect maintenance costs) for an offshore wind farm to almost one half. It was also shown that direct maintenance costs for the MMC (staff costs) have small effect on the total maintenance costs. This can be explained by the fact that the additional costs for maintenance staff are two orders of magnitude lower than the revenue losses during maintenance.

Keywords: offshore wind energy; transmission system; HVDC; voltage source converter (VSC); maintenance; missing energy export

---

1. Introduction

The generation of electricity by offshore wind power has been expanded in recent years, because of their strong reduction in electricity generation costs [1]. At the end of 2018, 18.50 GW of offshore wind power generation capacity was installed in Europe [2]. Compared to the previous year, this represents an increase of 17.2% [3]. In the next few years, a further increase of offshore wind power capacity in Europe is expected [3]. The European Union (EU) has the ambition to reach 65 to 85 GW offshore wind power generation capacity by 2030 [4]. This would represent an increase of 350% to 450% within next decade.

Due to the long distance of offshore wind farms to the coast, high-voltage direct current (HVDC) technology is in many cases used for the connection to the onshore grid. This is because it is a low-loss transmission technology compared to the classical alternating current (AC) technology [5].

Today, a commonly used HVDC technology is the voltage source converter (VSC) technology [6]. In the offshore field, the VSC is mostly applied as a modular multilevel converter (MMC), because of the reduced space required [7]. To guarantee the reliability of the HVDC system, it is taken out of service and maintained at regular intervals. Usually, the HVDC system is shut down for up to one week every year due to maintenance [8]. No wind energy can be exported to the onshore grid during these maintenance operations. For a 1 GW wind farm, this can lead to a missing energy yield of around 50 GWh per year on average. In addition, the maintenance date is often planned a couple of months ahead, while wind power generation cannot be precisely forecasted over such long periods. The main driver for the maintenance costs is the loss of income due to the missing energy export. For the mentioned 1 GW wind farm grid connection, this income loss ranges from 3 to 8 Mio. EUR per year. However, the magnitude highly depends upon the wind occurrence during the maintenance period [9–11]. In literature, the loss of income is often referred to as the indirect costs of maintenance, in contrast to the direct costs for staff, transportation and spare parts [12]. While indirect costs are usually lower than direct costs for wind turbines [13], the relation is typically inversed for energy export systems, which have higher indirect than direct costs [14].

To the best knowledge of the authors, the duration of the required maintenance outage of an HVDC energy export system has not yet been analyzed in the scientific literature. In existing approaches, maintenance of the energy export system is only accounted for by a simple non-availability value [11,12]. Given the lack of accurate wind forecasts for the outage period, it is difficult to determine the expected losses due to maintenance work in the planning process of HVDC energy export systems for offshore wind farms.

In order to improve the planning process, this study gives a detailed technical explanation of an HVDC energy export system for an offshore wind farm. In this, the different technical design aspects of the HVDC offshore and onshore stations are discussed. We quantify the MMC reliability under different maintenance strategies, and present the relationship between the maintenance period and MMC redundancy.

Based on a literature review, we show the various maintenance tasks for the HVAC and the HVDC components for the HVDC energy export system. Until today, there is no maintenance description for MMCs published. Therefore, the possible maintenance tasks for MMCs are discussed.

To analyze the influence of different design parameters and maintenance strategies on the maintenance-related losses of an HVDC energy export system, a maintenance model was built. The model is based on the previously described maintenance tasks and technical design aspects. In order to perform simulations, an energy time series is needed. Since no long-term energy data from offshore wind farms is available, a fictitious offshore wind farm and the Modern-Era Retrospective analysis for Research and Application (MERRA) dataset were both used to generate the required data [15].

In order to demonstrate the model application, a case study was performed, in which different parameters, such as maintenance period, the number of maintenance staff, and the possibility to shift the maintenance starting time, were varied. Results of different simulation runs are shown and discussed.

The paper is organized as follows: in Section 2, we described a reference HVDC energy export system. The different maintenance tasks for the HVDC energy export system are described in Section 3. The developed model is presented in Section 4. Following in Section 5, the offshore wind farm used in the case study is described. In Section 6, the case study is discussed, and finally, in Section 7, a conclusion of the findings and an outlook for further research are given.

2. Reference HVDC Energy Export System

The HVDC energy export system in this study is designed for a transmission capacity of 1 GW, which is supposed to be a typical size for an offshore wind farm in the near future [2]. Figure 1 shows the typical schematic configuration of an offshore wind farm connected to the grid by means of an HVDC energy export system [5].

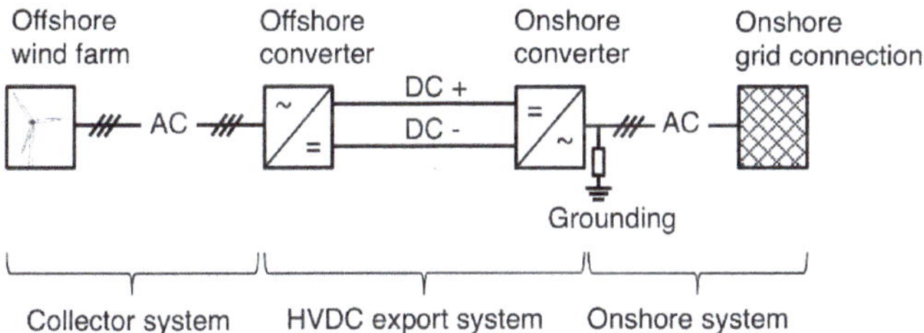

Figure 1. Schematic configuration of the offshore wind farm and the high-voltage direct current (HVDC) energy export system.

The collector system includes the offshore wind farm and the inter-array grid for the wind farm. The HVDC energy export system includes the offshore and the onshore converter, as well as the DC transmission cables. The onshore system only consists of the grounding for the HVDC energy export system and the onshore grid connection point. The offshore and the onshore converters are the main parts of the HVDC energy export system. Figure 2 shows the detailed single line diagram (SLD) of the HVDC energy export system [16].

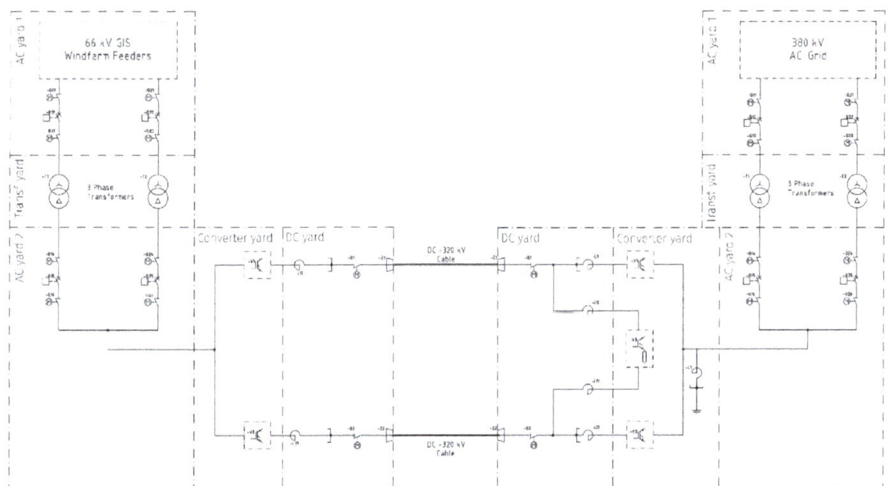

Figure 2. Single line diagram (SLD) of the HVDC energy export system and maintenance areas for the onshore and offshore station.

In Figure 2, the part to the left of the direct current (DC) cable represents the offshore station, and the part to the right represents the onshore station. On the offshore station, the AC yard 1 includes a 66 kV gas-isolated switchgear (GIS) (16 import feeders and two export feeders), that connects the wind farm to the two transformers. The two transformers (T1, T2) are located in the transformer yard, they transform the inter-array wind farm voltage from 66 kV to the transmission voltage of 320 kV. In the AC yard 2, a 320 kV GIS (Q14–Q26) connects the two transformers with the converter. In the converter yard, the MMC (V1, V2) converts the voltage from AC to DC. The DC yard includes the converter-reactors (L11, L21) and the direct current compact switchgear (DCCS) (Q1, Z1, Q2, Z2), which connects the converter to the DC cables.

Two DC subsea cables connect the offshore station and the onshore station. Today's DC subsea cables require nearly no scheduled maintenance in their planned operating time [17–19]. Therefore, the maintenance of the DC subsea cables is not further discussed in the following.

The structure of the onshore station is almost the same as the offshore station. The major difference between the two stations is that in the onshore converter yard, a chopper (V3) is installed. Using a chopper when connecting offshore wind farms with an HVDC energy export system is the most common method to avoid transferring onshore AC faults to the offshore wind farm grid [20–22]. Due to the limited available space on the offshore station, the chopper is always installed in the onshore station [16]. The advantage of using a chopper is that the wind farm can continue operating without any interruption in case a fault occurred in the onshore grid [22,23]. Today's AC grids usually have a high quality of supply. For example, Tennet's Annual Report states an AC grid availability of 99.9988%, and 17 number of interruptions for the year 2018 [24]. By using the number of interruptions as an indicator to get a sense of how often the chopper is used, we can expect that the onshore grid is in normal operation most of the time; i.e., there is no failure, such as AC three-phase short circuits, in the grid. We therefore conclude that the DC chopper is only used for a small number of times a year, compared to the rest of the HVDC energy export system. On this basis, we assume that the wear and the maintenance work on the chopper are low compared to the MMC maintenance. Therefore, the chopper maintenance is not included in this study, and the following described maintenance tasks for the onshore and offshore stations are the same.

2.1. Converter (VSC Modular Multilevel Converter)

The MMC design consists of three phase-units between the positive and the negative DC terminal. Each phase-unit consists of a positive and a negative converter-leg. Each converter-leg consist of series-connected half bridge submodules (SMs) and a single series connected converter reactor. Each phase-unit is connected to one phase of the AC terminal. Figure 3 shows the MMC structure and the detailed design of the half bridge SM [7,25].

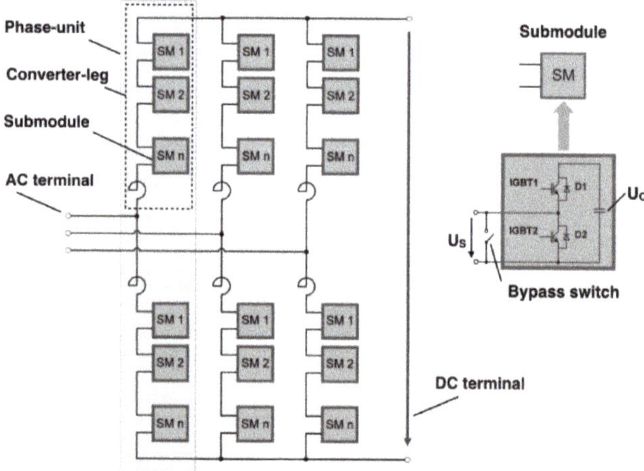

Figure 3. Schematic configuration of the three phase modular multilevel converter (MMC) and structure of the half bridge submodule (SM).

The modular design of the MMC allows the individual adaptation of the converter to the technical requirements, e.g., transmission capacity, DC voltage and maintenance design [7]. The maintenance design is mainly influenced by the number of redundant SMs. To ensure the functionality of the converter over a long period of time (maintenance period), redundant SMs are placed into each converter-leg. Two different operation schemes for SM redundancy are common: passive and active SM redundancy. In the passive operation scheme, the redundant SMs are on standby mode until a failure occurs in a normal SM. Then the faulty SM is bypassed through a bypass switch, and one of the redundant SMs becomes active.

In the active operation scheme, all SMs (redundant SMs and normal SMs) are actively operating in parallel. If a failure occurs, the failed SM will be bypassed automatically [26].

In this study, the active SM redundancy is the scheme used today. It is assumed that the active SM redundancy is the today used scheme, because the control mechanism is simpler, and therefore, it is the more reliable method compared to the passive SM redundancy scheme [27].

In order to calculate the number of redundant SMs, the minimum number of SMs n_{min} is required. The minimum number of SMs in one converter-leg results from the SM operation voltage U_{SM}, the DC voltage U_d and the AC peak voltage $\hat{U}_{AC}$ as described in [28]:

$$n_{min} = \frac{\frac{U_d}{2} + \hat{U}_{AC}}{U_{SM}} \tag{1}$$

The total number of SMs in the converter n_{Con} results from the sum of the minimum number and the redundant number of SMs n_{red} in six converter-legs:

$$n_{Con} = 6(n_{min} + n_{red}) \tag{2}$$

The redundant number of SMs depends on the maintenance period and on the reliability target for the converter. The relationship will be described in detail in Section 2.2. Table 1 shows the technical parameters of the energy export system, as seen in literature (cf. [29–31]).

Table 1. Energy export system technical parameters.

Parameter	Value
Power rating ($P_{HVDC\ System}$)	1000 MW
DC voltage (U_d)	± 320 kV
AC peak voltage ($\hat{U}_{AC}$)	320 kV
SM operation voltage (U_{SM})	2.7 kV
Submodule FIT (FIT_{SM})	1700
Minimum number of SM per converter-leg (n_{min})	238
Number of transformers	2
Transformer power rating ($P_{Transf.}$)	600 MW

2.2. Converter Reliability

The converter reliability calculation in this study follows the general equation used in the literature (cf. [32]). The equation describes the relative proportion of functioning components to be expected after a given time. Derived from this, the reliability function for one submodule, R_{SM}, is given by (3).

$$R_{SM}(t) = e^{-\lambda_{SM} t} \tag{3}$$

where t represents the time in hours and λ_{SM} represents the SM hazard rate.

The hazard rate is considered constant over the whole lifetime, thus assuming constant random failures over the lifetime of the converter (cf. [19]). The hazard rate of electrical components is often specified by the failures in time (FIT) value. FIT is the number of failures in one billion hours of operation:

$$\lambda_{SM} = \frac{FIT_{SM}}{10^9} \tag{4}$$

The reliability of the converter can be described with the binomial distribution and k out of n majority redundancy. The k out of n majority redundancy describes the reliability of a system that consists of n components, and that is working only if at least k components in the system are able to operate. The general reliability function of the system is given by (5).

$$F(k,t) = \sum_{i=0}^{k} \binom{n}{i} p(t)^i (1 - p(t))^{n-i} \tag{5}$$

where n represents the number of all components in the system, k represents the number of components that is at least needed for operating the system, $p(t)$ represents the reliability function of one component, and t represents the time.

The resulting reliability function for one converter leg is given by (6).

$$R_{CL}(t) = \sum_{i=k}^{n} \binom{n}{i} (R_{SM}(t))^i (1 - R_{SM}(t))^{n-i} \tag{6}$$

Here, n represents the total number of all SMs in one leg $(n_{min} + n_{red})$, and k represents the redundant SMs per leg (n_{red}).

The MMC can only operate if the positive and the negative legs are able to operate. That is represented by a serial configuration, and results in the reliability function for one converter phase, as given in (7).

$$R_{CP}(t) = R_{pos\_CL}(t) \cdot R_{neg\_CL}(t) \tag{7}$$

Due to the fact that the converter consists of three phases and operates only when all phases are working, the converter reliability, R_C, corresponds to the third power of the reliability of one converter phase, as given by (8).

$$R_C(t) = (R_{CP}(t))^3 \tag{8}$$

With the parameters from Table 1, the reliability of the converter can be calculated as a function of the number of redundant SMs and the maintenance period. Figures 4 and 5 show this relationship for maintenance periods of one to four years with the resulting number of redundant SM per converter-leg at a reliability target of 99%.

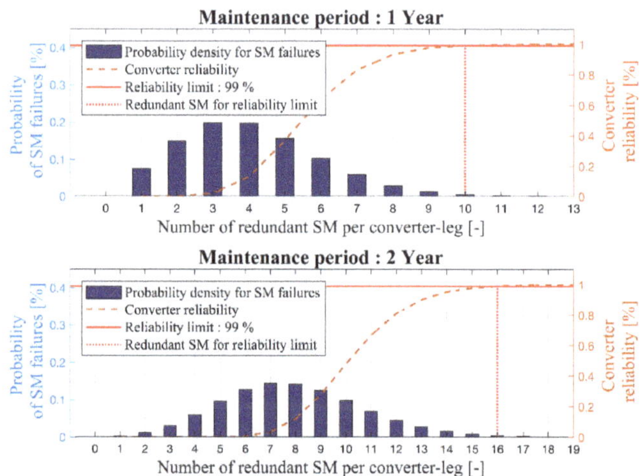

Figure 4. Converter reliability and probability of SM failure over the number of redundant SM after one and two year maintenance period.

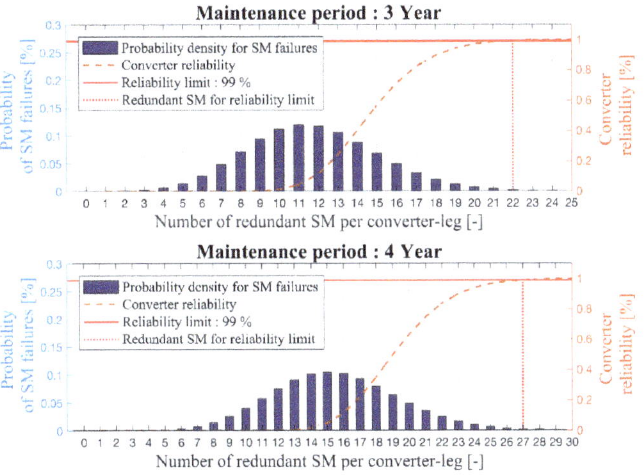

Figure 5. Converter reliability and probability of SM failure over the number of redundant SMs after the three and four year maintenance period.

The right ordinate shows the reliability function of the converter depending on the number of redundant SMs. The left ordinate shows the probability of SM failures per converter leg. In addition, a horizontal line represents the reliability barrier/target at 99%.

The dashed vertical line illustrates the resulting number of redundant SMs per converter leg that is needed to ensure the reliability target.

Table 2 summarizes the findings from Figures 4 and 5. It displays the minimum required numbers of SMs for the operation, the required redundant SMs, and the sum of two, for the different maintenance periods. It can be concluded that the number of SMs for longer maintenance periods does not rise linearly. For a maintenance period of one year, 60 redundant SMs are needed. For two years, 96 SMs are needed. That results in an increase of 2.42% in the overall number of SMs. For an extension of the maintenance period to three years, 132 redundant SMs are needed (an increase of 2.36% in comparison to two years' maintenance). For the four years' maintenance period, 162 SMs are needed (an increase of 1.92% in comparison to three years' maintenance).

Table 2. Number of redundant SMs for different maintenance periods.

Parameter	Maintenance Periods [a]			
	1	2	3	4
Minimum SM	1428	1428	1428	1428
Redundant SM	60	96	132	162
Increase of redundant SM	-	60%	37.5%	22.73%
All SM	1488	1524	1560	1590
Increase of all SM	-	2.42%	2.36%	1.92%

It turns out that four-year maintenance periods do not require four times more redundancy than do one-year maintenance periods. This means that there must be an optimum number of redundant SMs, when comparing the total costs (capital costs) associated with additional SMs with the total maintenance costs (operational costs).

3. Maintenance

In order to ensure the reliable operation of electrical components, they must be maintained in specified periods [32–34]. In this work, the transmission system is divided into different maintenance areas. Figure 2 indicates the different electrical components, and the different maintenance areas of the stations marked with a dash-dotted lines. If one of the components has to be maintained, the component must be taken out of service. This implies that during this time, no energy can be transferred.

In relation to the HVDC energy transmission system, this means that if one component needs to be maintained, and there is no redundant component (cp. transformer yard), the whole transmission system has to be taken out of service for the maintenance time. Table 3 gives an overview of the available transmission capacity in case of a maintenance in the different maintenance areas.

Table 3. Transmission capacity during maintenance for different maintenance areas.

Maintenance Area	Transmission Capacity
AC yard 1	0 MW
Transformer yard	600 MW
AC yard 2	0 MW
Converter yard	0 MW
DC yard	0 MW

In order to analyze the influence of the individual maintenance tasks on the overall system, all relevant maintenance tasks were evaluated. Table 4 shows an overview of all maintenance tasks, intervals, duration, and the assumed number of maintenance staff for the different maintenance works in the different maintenance areas.

Table 4. Maintenance work and duration for different maintenance areas.

Parameter	Maintenance Area			
	AC Yard 1	Transf. Yard	AC Yard 2	DC Yard
Interval [a]	25	1	25	5
Maintenance work	check insulating parts, cleaning electrical contacts	oil test, cooling system check, isolation test	check insulating parts, cleaning electrical contacts	check insulating parts, cleaning electrical contacts
Man-hour maintenance duration [h]	144	48	16	7
Number of maintenance staff	2	2	2	1
Sources	[35–37]	[38–40]	[35–37]	[41–43]

3.1. MMC Maintenance

For maintenance work on MMCs, there are no reliable sources available. For this reason, assumptions must be made in this study for maintenance work in the converter yard.

The converter yard contains six converter legs, as shown in Figure 3. During operation, the converter hall cannot be entered. Therefore, a certain preparation and follow-up time must be scheduled for the maintenance of the converter. The SMs are installed in towers, as described in [16]. The primary maintenance task at the converter is assumed to be the replacement of defective SMs. The exchange of defective SMs is carried out by maintenance teams consisting of two persons. For the replacement of the SM, a working time of three hours is assumed for a maintenance team. Table 5 shows the used maintenance parameters for the converter yard.

To calculate the maintenance duration for the converter yard, it is necessary to know the number of defective SMs at the maintenance time. This can be determined via the general probability function, used in the literature for electrical components (cf. [32]), that a component (SM) has become faulty up to a certain time:

$$E_{SM}(t) = p(t)\, n_{Con} = \left(1 - e^{-\lambda_{SM} t}\right) n_{Con} \tag{9}$$

where the expected value of exchanged SMs E_{SM} (i.e., defective) is calculated with t the maintenance period, n_{Con} the total number of SMs in the converter, and λ_{SM} represents the SM hazard rate as FIT value given in Table 1.

Table 6 shows the resulting numbers of defective SMs (rounded up) by performing the probability function (9), using values as defined in Table 1 for the various maintenance periods.

Table 5. Maintenance parameters for the converter yard.

Parameter	Value
Preparation time	4 h
Follow-up time	4 h
Replacement time for the SM	3 h
Persons per team	2

Table 6. Expected number of defective SMs using the probability function for different maintenance periods.

Parameter	Maintenance Period [a]			
	1	2	3	4
Expected defective SM	22	45	69	92

It can be noticed that the relation between the expected number of defective SMs and the maintenance period is almost linear, although equation (9) contains a nonlinear part. After a two-year maintenance period, there are nearly two times more defective SMs than with one-year maintenance periods.

For calculating the maintenance time, the working time per day and the number of maintenance staff must be taken into account. Table 4 shows the chosen maintenance staff for different maintenance areas.

For the calculation, a 12 h working time per day is assumed, and no shift work is scheduled. This means that within one day (24 h), up to 12 h of maintenance work can be performed. Figure 6 indicates the maintenance time for different maintenance areas over different maintenance periods, and different numbers of maintenance staff in the converter yard. In Figure 6a, the x-axis represents the different maintenance periods for the HVDC energy export system, and the y-axis shows the maintenance time in a logarithmic scale. In Figure 6b, the maintenance time is plotted over different numbers of maintenance staff in the converter yard at a maintenance period of two years. The x-axis represents the number of maintenance staff in the converter yard, and the y-axis shows the maintenance time in logarithmic scale.

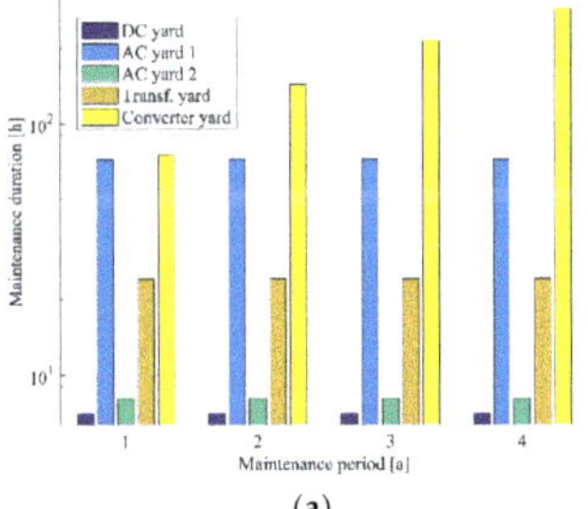

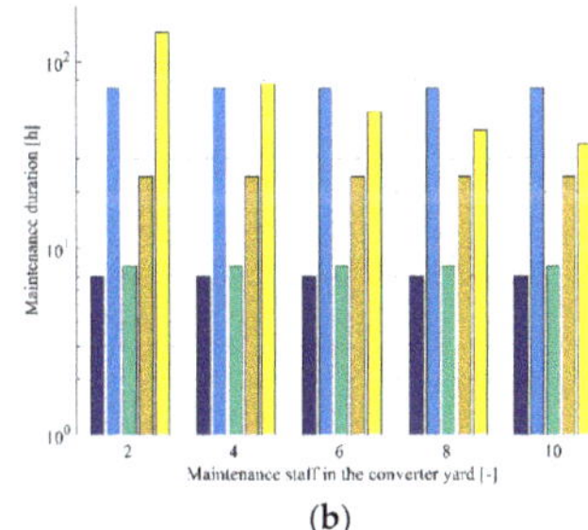

Figure 6. Maintenance time (**a**) for the different maintenance areas of the HVDC system over different maintenance periods and (**b**) for a two-year maintenance period over different number of maintenance staff in the converter yard.

In Figure 6a, it can be seen that the maintenance time for the converter yard grows by 93% if the maintenance period is increased from one year to two years. The maintenance time for the converter yard is reduced by 53% if the number of maintenance staff is increased from two to four for a two-years maintenance period, as shown in Figure 6b. It should be noted that the maintenance areas have different maintenance periods. For example, the AC yard only has to be maintained once every 25 years. This means that this area has less impact on maintenance-related losses in comparison to the transformer yard, which has a maintenance period of one year.

In conclusion, there should be an optimum for the number of maintenance staff in the different maintenance areas and the maintenance period, compared to the direct maintenance costs and the indirect maintenance costs.

3.2. Maintenance Date and Time Point

As shown in Figure 6 and explained in Section 3.1. (MMC Maintenance), the maintenance of an HVDC energy export system is associated with high personnel expenditure. Therefore, the maintenance date must be planned long-term in advance. For the planning of the maintenance date, it is essential to choose a time at which a low energy yield of the wind farm is expected. This date results from the distribution of wind speeds over the year.

At the wind measurement station FINO 1 in the north sea, the mean wind speed in June and July is 1.5 m/s below the mean annual wind speed of 10.0 m/s [44]. From this it can be deduced that a lower feed-in power can be expected in summer than in winter. For this reason, maintenance of the HVDC energy export system always takes place in the summer months of June and July [8].

3.3. Maintenance Assumtion and Model Simplifications

Some simplifications and assumptions are considered in this study, including station accessibility, individual maintenance costs and maintenance tasks on MMCs.

For offshore energy export systems, the accessibility of the two stations (onshore and offshore) is different, since the accessibility of the offshore station depends on weather conditions, such as wave height, visibility range, weather window and the availability of different ships or helicopters [12]. In our study we focused on the dependency of MMC design aspects on the maintenance-related losses. Therefore, we excluded the exact analysis of the accessibility aspect.

It is important to precisely quantify the different cost positions (staff, material or transportation costs) when analyzing maintenance for offshore wind farms. Usually, it is very challenging to quantify the costs of individual positions; hence, it is difficult to investigate their influence on each other. For example, the exact value for staff salaries or transportation costs are not available in the literature [12]. Therefore, we decided to focus on the missing energy yield. This represents only one specific maintenance cost factor, but it allows us to have a sense of how the missing energy export could behave regarding different design variations. To better understand the relation between the direct and indirect costs, a maintenance cost example was introduced in the end of Section 6.

As mentioned in Section 3.1. (MMC Maintenance), no reliable sources for maintenance tasks on MMC are available. For this reason, assumptions must be made for the maintenance tasks and durations.

Due to these assumptions and simplifications, the results of the study are considered optimistic in terms of the maintenance-related losses.

4. Model Description

The developed model, which can be seen in Figure 7, is separated into three main sections: First the data input, second the model itself, and third the data output.

The data input is represented by the specific technical parameters for the HVDC energy export system (see Tables 1–3) and a time series of energy yield data $E_{yield}(i)$ for an offshore wind farm. The time series has a one-day resolution, where i represents a specific day in the total operating time.

Instead of one representative year, we use yield data over the total expected operating time. For the case study described in Section 6 for example, we apply past weather data from 1980 until 2010 to represent a 30 year long operating period of the energy export system.

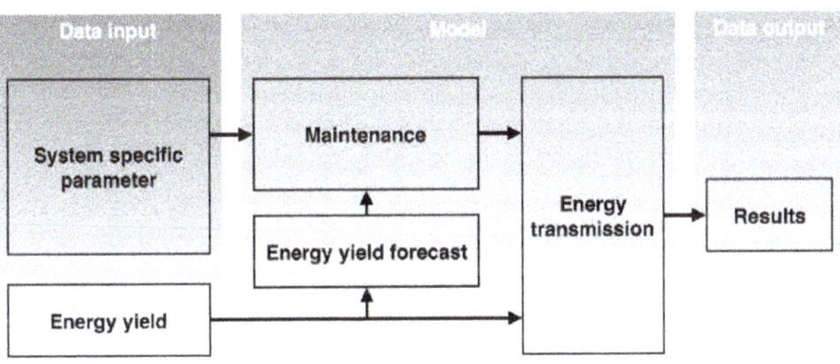

Figure 7. The overall model.

The implemented model consists out of three sub models; the energy yield forecast model, the maintenance model, and the energy transmission model. The energy forecast model provides energy yield forecasts for different time horizons (up to ten days). It is assumed that a perfect energy yield forecast exists without any forecast error, which entails the highest savings potential for an optimized maintenance time.

The duration and periods of maintenance for the different maintenance areas are calculated within the maintenance model. All maintenance intervals of the different maintenance areas (see Table 4) are matched to the chosen maintenance period of the HVDC energy export system. Therefore, the maintenance model uses the system specific input parameters and the maintenance properties. In combination with the energy yield forecast, it optimizes the starting time of the maintenance in order to minimize the losses per maintenance. Figure 8 shows one example of how the optimized maintenance is performed based on the energy yield forecast.

Possible maintenance shift [d]	Operation day	366	367	368	369	370	371
	Energy yield forecast [GWh]	1	40	0	20	2	10
0		41					
1			40				
2				20			
3				20			
4						12	

XX — Maintenance duration [d] and XX maitenance-related losses [GWh]

Figure 8. Possible maintenance shift from 0 to 4 days.

The maintenance model then creates a Boolean matrix $\beta(i)$, indicating whether any maintenance is required, or not, for each single day in the lifetime of the HVDC energy export system. This matrix is then imported by the energy transmission model.

The energy transmission model calculates, based on the original energy yield time series $E_{yield}(i)$, for every day, the losses that are caused by maintenance. The resulting maintenance-related losses for the life time of the HVDC energy export system $E_{total\ losses}$ is given by (10).

$$E_{total\ losses} = \sum_{i=1}^{10950} E_{yield}(i)\beta(i) \tag{10}$$

The model developed here can map maintenance work up to a resolution of a quarter of a day. For analyzing and optimizing, the following three input parameters can be varied in order to reduce the maintenance-related losses: Maintenance period of the HVDC energy export system, number of maintenance staff and maintenance start day in combination with the energy yield forecast.

The model was implemented by using the software environment MATLAB 2016b from MathWorks [45]. The code and data for the model is freely available online [46].

5. Offshore Wind Farm

In the case study, a fictious offshore wind farm is assumed. The offshore wind farm, with a total capacity of 1 GW, consists of 125 wind turbines, and is located in the Eastern part of the North Sea. The center of the wind farm is at N 54° 56′ 22.4736 E 6° 56′ 44.3724 in the UTM (Universal Transverse Mercator) system.

The basis for the assumed wind conditions at the wind farm site are the Modern Era Retrospective-analysis for Research and Application (MERRA) data of the National Aeronautics and Space Administration (NASA). Various publications, such as [47] and [48], show that the MERRA data is a good basis for generating energy yield forecasts for offshore and onshore wind farm projects. In order to be able to capture the annual yield variations of the offshore wind farm, wind speed data from the period 1980 to 2016 is used as an operation period in this study. In the assumed operation period, the mean wind speed at the site is 8.5 m/s from April to September and from October to March is 11.3 m/s at a height of 103 m. The main wind direction at the wind farm site is West to Southwest.

The assumed wind turbine "W8000" has a rated power of 8 MW. The technical data of the system is listed in Table 7. The wind turbines are set up in a cluster with a distance of 1000 m to the North and 1000 m to the East. By application of Pythagoras' Theorem, this results in a distance of 1414.21 m between two wind turbines in the Southwest direction (main wind direction). That corresponds to nine times the rotor diameter, and is a common distance between offshore wind turbines in a park layout [49].

Table 7. Technical data wind turbine W8000.

Parameter	Value
Rated power	8 MW
Rotor diameter	154 m
Hub height	103 m
Cut in wind speed	4 m/s
Cut-out wind speed	25 m/s

The energy yield time series for the wind farm is created with the Wind Atlas Analysis and Application Program (WAsP) [50]. Daily energy yield values were generated in the reference period from 1980 to 2016. The internal wake effects were considered with the FLaP wake model, and were taken into account when calculating energy yield [51]. The wake effects of any of the surrounding wind farms were not considered.

The availability of a wind farm as described in [52] depends on many factors, such as the maintenance concept or the location. It describes the period during which a wind turbine or an entire wind farm is ready for operation. Nowadays, the usual availability for offshore wind farms ranges between 75% and 95% [52].

Based on the greater experience of operators, we expect the availability of offshore wind farms to further increase in the upcoming years. Therefore the assumed availability of the offshore wind farm in this study is 93%.

The potential gross energy yield of the wind farm calculated with the yield forecast described here, is transferred to the overall model described in Section 4. Since wind farms are usually built and commissioned in the summer months, the operating period, and thus the yield period, is selected as

01.06.1980 to 31.05.2010. Figure 9 indicates the mean energy yield produced by the wind farm under consideration of the availability of the wind farm in the period of 1980 to 2010.

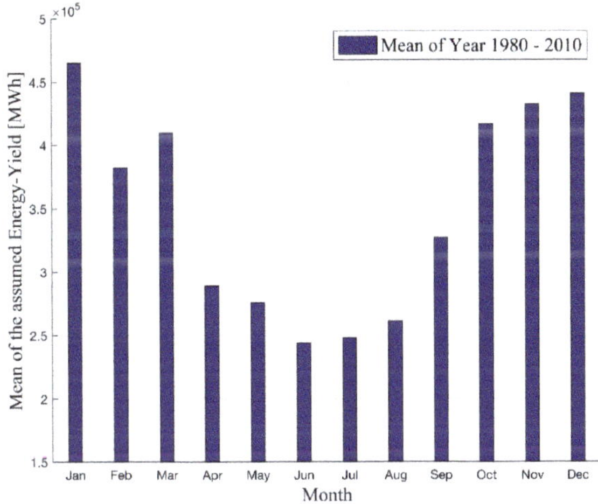

Figure 9. Monthly energy yields of the offshore wind farm as average value over the years 1980 to 2010.

6. Case Study Discussion

The model described in Section 4 and the described wind farm in Section 5 is used to analyze the possibilities of reducing the maintenance-related losses of an HVDC energy export system. The resulting energy yield forecast dataset described in Section 5 is used as the energy yield input parameter. This dataset is used, because so far, no energy yield time series for an offshore wind farm over longer periods is available.

Three different parameter variations were tested to reduce the maintenance-related losses. Table 8 gives an overview of the case study parameters and the parameters to be changed in the different variations.

Table 8. Case study parameters.

Parameter	Value
Fixed Parameters	
Operating time	30 a
Commissioning date	01.06.1980 dd.mm.yy
Maintenance date	every 01.06 dd.mm
Working time per day	12 h/d
Number of maintenance staff (AC yard 1)	2
Number of maintenance staff (AC yard 2)	2
Number of maintenance staff (DC yard)	1
Number of maintenance staff (transformer yard)	2
Modified Parameters	
Maintenance period	From 1 a to 4 a
Number of maintenance staff in den converter yard	From 2 to 10
Possible maintenance shift	From 0 d to 10 d
Cost Parameters	
Remuneration (based on [53])	100 €/MWh
Offshore staff costs per person (based on [54])	1500 €/d
Onshore staff costs per person (based on [54])	960 €/d

6.1. Maintenance Period

In the first scenario, the maintenance period of the HVDC energy export system is varied from the one year maintenance period up to the four year maintenance period. The number of maintenance staff members in the converter yard is set to two persons. The possible maintenance shift is set to 0 days. Therefore, the shift of the maintenance is not possible/allowed.

Figure 10 represents the results of the model calculation. It can be seen that the variation of the maintenance period from one year to four years influences the maintenance-related losses.

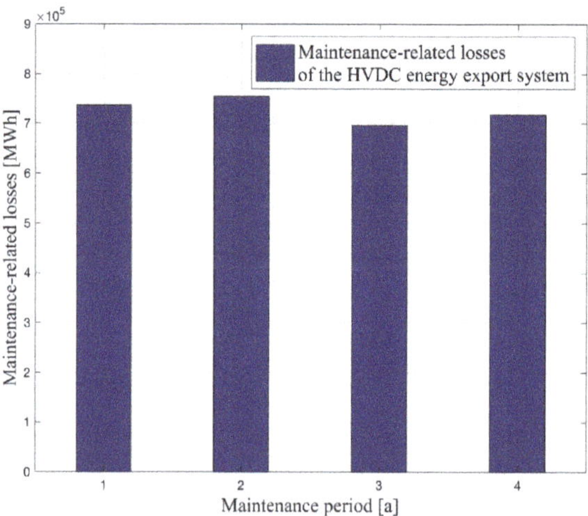

Figure 10. Maintenance-related losses for the total operation time of the HVDC system over different maintenance periods.

The x-axis shows the different maintenance periods, and the y-axis represents the maintenance-related losses. The blue bars show the maintenance-related losses over the total operation time. The result does not show a clear reduction in maintenance-related losses by only increasing the maintenance period from one year up to four years. One reason for this is that an extension of the maintenance period does not lead to a significant reduction of the maintenance time (see Figure 6a). Since maintenance always takes place in the same days of the year, it can also happen that maintenance takes place in days when the wind farm would otherwise have a high energy yield. Therefore, the maintenance-related losses can also be higher for the two year maintenance period case than for the one year maintenance period case.

6.2. Maintenace Staff

In the second scenario, the maintenance staff in the converter yard is varied from 2 to 10 persons. The maintenance period of the HVDC energy export system is set to a two-year maintenance period. Maintenance shift is set to 0 days. (i.e., no maintenance shift is not possible/allowed). The scenario with two staff members was considered the baseline case for all following calculations.

Figure 11 shows the result of the model calculation. It can be seen that the variation of the maintenance staff is reducing the maintenance-related losses.

The x-axis shows the different number of maintenance staff in the converter yard, and the y-axis represents the maintenance-related losses over the total operation time. It is noticeable that with the first doubling of the maintenance staff, maintenance-related losses can be reduced by 44.5 % (about 600 GWh maintenance-related losses were avoided). Further adding of maintenance staff only leads to a slight reduction 32% of losses (e.g., about 200 GWh could be saved by increasing staff from 4 to 6).

The magnitude of the maintenance-related losses indicates that it might be profitable to carry out the maintenance with four or more people.

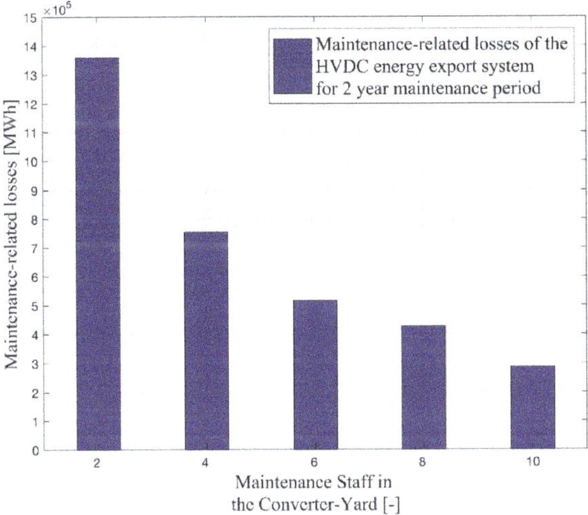

Figure 11. Maintenance-related losses for the total operation time of the HVDC system over different number of maintenance staff in the converter yard.

6.3. Maintenance Start Day

In the third scenario, the possibility to shift the maintenance start day is added. The model can now postpone the maintenance start day from 1 day up to 10 days to reduce the maintenance-related losses. Figure 12 shows the results of the second and third scenario in relation to each other.

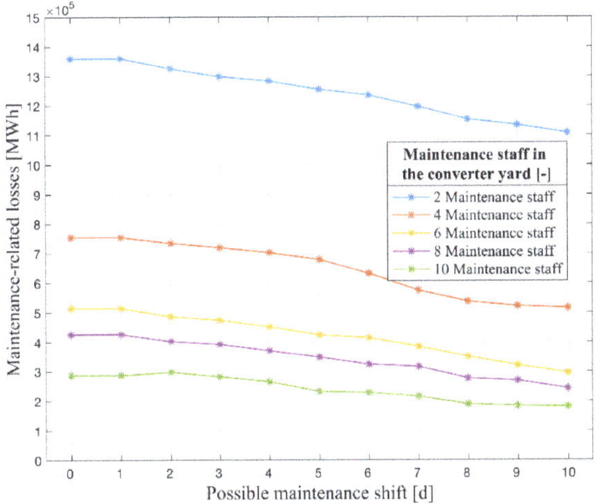

Figure 12. Maintenance-related losses for the total operation time of the HVDC system over possible maintenance shift for different number of maintenance staff in the converter-yard.

The x-axis shows different possibilities to shift the maintenance start day, and the y-axis represents the maintenance-related losses over the total operation time. The five differently colored lines represent the different numbers of maintenance staff in the converter yard.

It can be seen that the maintenance-related losses can be reduced significantly only if there is a shift margin of three days or more (up to around 50 GWh for the case of two maintenance technicians). It can also be seen that with the increasing margin to postpone the maintenance, the losses decrease. This can be explained by the fact that the model shifts the maintenance to a period in which the energy yield of the wind farm is low. The calculated maintenance-related losses should not be understood as an ultimate value, but it allows wind farm developers to get an idea of how the possibility to shift the maintenance could be profitable with respect to the losses.

The range of avoided losses indicates that it could be profitable to hire the maintenance staff for a longer period of time (multiple days). Within this period the choice of maintenance start day is optimized to minimize the maintenance-related losses.

With regard to the increase in maintenance staff, it can be seen that the curve flattens out with increasing maintenance staff. This means that postponing maintenance with increasing maintenance staff will have less effect upon maintenance-related losses.

6.4. Maintenance Costs Calculation

Based on the modeled maintenance-related losses, we calculated the maintenance-related cost change for the two parameters: the maintenance start day and the number of maintenance staff. The maintenance staff costs for the converter yard were calculated based on the modeled maintenance time for the converter yard and the maintenance staff cost parameter in Table 8, which can be seen in Figure 13a. The maintenance-related cost change was calculated by combining the direct maintenance costs (staff costs) and the indirect maintenance costs (lost remuneration), as seen in Figure 13b.

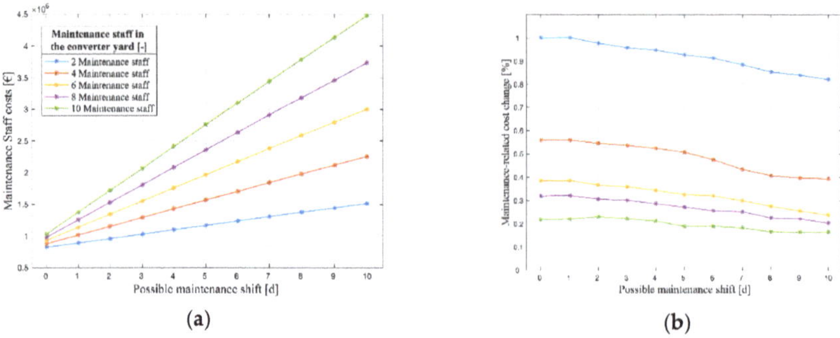

Figure 13. Maintenance staff costs (**a**) and maintenance-related costs change (**b**) for the total operation time of the HVDC system over possible maintenance shift for different number of maintenance staff in the converter-yard.

In Figure 13a,b the x-axis shows different possibilities to shift the maintenance start day. The y-axis in Figure 13a represents the maintenance staff costs for the converter yard over the total operation time. In Figure 13b the y-axis represents the maintenance-related cost change over the total operation time. The five differently-colored lines represents the different number of maintenance staff in the converter yard.

With the first doubling of the maintenance staff, the maintenance related cost can be reduced by around 44%. Additionally, it can be seen that the maintenance-related costs can be reduced by around 10%, only if there is a shift margin of six days (for the case of two maintenance technicians). It can be also seen that with an increasing margin to postpone the maintenance, the savings can be increased up to 19% (for the case of two maintenance technicians). This can be explained by the fact that the additional costs for maintenance staff are two orders of magnitude lower than the revenue losses. The range of maintenance-related cost change shows that it will be profitable to hire the maintenance staff for a longer period of time (multiple days). Within this period the choice of maintenance start day can be then optimized to minimize the maintenance-related losses, which further confirms what was concluded from Figure 12.

7. Conclusions

In this paper, we applied a bottom-up approach to analyze the maintenance-related losses for an HVDC energy export system. A maintenance model was introduced for identifying the most significant factors affecting the maintenance-related losses.

In a case study, the model was used to analyze the potential to reduce maintenance-related losses for a 1 GW offshore wind farm. Three main factors were analyzed that influence the maintenance-related losses: duration of the maintenance period, the number of maintenance staff and the shift of the maintenance. In addition, we performed a maintenance cost example to better understand the relation between the direct and indirect maintenance costs of an HVDC energy export system.

It was found that changing the length of the maintenance period has less impact on the losses than the other two factors.

Regarding the number of maintenance staff, it is noticeable that with the first doubling of the maintenance staff with respect to the baseline case, maintenance-related losses and maintenance-related cost changes can be almost halved. Further adding of maintenance staff only leads to a slight reduction of losses.

With regard to the possibility to shift the maintenance date into times with a lower energy yield from the wind farm, it was shown that the maintenance-related losses can be reduced only if the shift margin is at least three days. With increasing the margin to postpone the maintenance up to 10 days, the maintenance-related cost change decreases down to 19%. By combining the increase of maintenance staff with the shift of maintenance time, it was shown that the curve of maintenance-related losses flattens out with increasing maintenance staff. This means that postponing maintenance with increasing maintenance staff will have less effect on maintenance-related losses and costs. It was also noticed that the staff costs for the converter yard had a small effect on the maintenance-related cost change compared to the maintenance-related losses, due to the difference in the order of magnitude between the direct and indirect costs of maintenance in the converter yard.

We have shown that optimizing the maintenance of an HVDC energy export system can decrease the maintenance-related losses for an offshore wind farm to almost one half with respect to the baseline case.

It was also shown that there is an optimum number of redundant SMs in relation to the maintenance period when comparing the total costs associated with additional SMs (increasing maintenance period) with the total maintenance costs.

In future work, industry data could lead to a more accurate model result. In addition, a detailed analysis of maintenance costs, such as traveling costs and costs for maintenance staff, could also be carried out.

Author Contributions: All authors contributed to the design of the research. J.U. performed the simulations and took the lead in the analysis of the results and in writing the manuscript. All authors have read and agreed to the published version of the manuscript.

Funding: The article processing charge was funded by the German Research Foundation (DFG) and the University of Freiburg in the funding program Open Access Publishing.

Acknowledgments: We thank Mirko Schäfer and Ramiz Qussous for fruitful discussions as well as the team of Overspeed GmbH and CO. KG for supporting the work.

Conflicts of Interest: The authors declare no conflict of interest.

Nomenclature

Parameter	Description
n_{min}	Minimum number of SMs per converter-leg
n_{red}	Redundant number of SMs per converter-leg
n_{Con}	Total number of SMs in the converter
U_{SM}	SM operation voltage
U_d	DC voltage
$\hat{U}_{AC}$	AC peak voltage
R_{SM}	Reliability for one submodule
R_{CL}	Reliability for one converter-leg
R_{pos_CL}	Reliability for positive converter-leg
R_{neg_CL}	Reliability for negative converter-leg
R_{CP}	Reliability for one converter phase
R_C	Reliability for the converter
t	Time in hours
λ_{SM}	SM hazard rate
FIT_{SM}	SM failures in time
$F(k,t)$	Reliability function
n	Number of all component in a system
k	Number of components that is needed for operating a system
E_{SM}	Expected number of defective SMs
i	Time in days
$\beta(i)$	Maintenance matrix
$P_{Transf.}$	Transformer power rating
$P_{HVDC\ System}$	Power rating
$E_{yield}(i)$	Energy yield from wind farm
$E_{total\ losses}$	Maintenance-related losses

References

1. McAuliffe, F.D.; Murphy, J.; Lynch, K.; Desmond, C.; Norbeck, J.A.; Nonås, L.M.; Attari, Y.; Doherty, P.; Sorensen, J.D.; Giebhardt, J.; et al. Driving cost reductions in offshore wind–the LEANWIND project final publication. *LEANWIND Proj.* **2017**, 72.
2. Wind Europe. *Offshore Wind in Europe-Key Trends and Statistics 2018*; Wind Europe: Brussels, Belgium, 2019.
3. Wind Europe. *Offshore Wind in Europe-Key Trends and Statistics 2017*; Wind Europe: Brussels, Belgium, 2018.
4. International Energy Agency (IEA). *Offshore Wind Outlook 2019: World Energy Outlook Special Report*; International Energy Agency: Paris, France, 2019.
5. Fichtner, GGSC. *Beschleunigungs- und Kostensenkungspotenziale bei HGÜ-Offshore-Netzanbindungsprojekten – Langfassung*; Fichtner, GGSC: Stuttgart, Germany, 2016.
6. Ackermann, T. *Wind Power in Power Systems*, 2nd ed.; John Wiley & Sons, Ltd.: Hoboken, NJ, USA, 2012; ISBN 9780470974162.
7. Friedrich, K. Modern HVDC PLUS application of VSC in Modular Multilevel Converter topology. In Proceedings of the IEEE International Symposium on Industrial Electronics, Bari, Italy, 4–7 July 2010; pp. 3807–3810.
8. TenneT. Marktrelevante Informationen-Offshore. Available online: https://www.tennet.eu/de/strommarkt/transparenz/transparenz-deutschland/berichte-marktrelevante-informationen/marktrelevante-informationen/marktrelevante-informationen-offshore/ (accessed on 7 December 2019).
9. Dinwoodie, I.A.; McMillan, D. Operational strategies for offshore wind turbines to mitigate failure rate uncertainty on operational costs and revenue. *IET Renew. Power Gener.* **2014**, *8*, 359–366. [CrossRef]
10. Lakshmanan, P.; Liang, J.; Jenkins, N. Assessment of collection systems for HVDC connected offshore wind farms. *Electr. Power Syst. Res.* **2015**, *129*, 75–82. [CrossRef]
11. Martin, R.; Lazakis, I.; Barbouchi, S.; Johanning, L. Sensitivity analysis of offshore wind farm operation and maintenance cost and availability. *Renew. Energy* **2016**, *85*, 1226–1236. [CrossRef]

12. Seyr, H.; Muskulus, M. Decision Support Models for Operations and Maintenance for Offshore Wind Farms: A Review. *Appl. Sci.* **2019**, *9*, 278. [CrossRef]

13. Dalgic, Y.; Lazakis, I.; Dinwoodie, I.; McMillan, D.; Revie, M. Advanced logistics planning for offshore wind farm operation and maintenance activities. *Ocean Eng.* **2015**, *101*, 211–226. [CrossRef]

14. Gul, M.; Tai, N.; Huang, W.; Nadeem, M.H.; Ahmad, M.; Yu, M. Technical and economic assessment of VSC-HVDC transmission model: A case study of South-Western region in Pakistan. *Electron.* **2019**, *8*, 1305. [CrossRef]

15. Rienecker, M.M.; Suarez, M.J.; Gelaro, R.; Todling, R.; Bacmeister, J.; Liu, E.; Bosilovich, M.G.; Schubert, S.D.; Takacs, L.; Kim, G.K.; et al. MERRA: NASA's modern-era retrospective analysis for research and applications. *J. Clim.* **2011**, *24*, 3624–3648. [CrossRef]

16. Hussennether, V.; Rittiger, J.; Barth, A.; Worthington, D.; Dell'Anna, G.; Rapetti, M.; Hhnerbein, B.; Siebert, M. Projects BorWin2 and HelWin1–large scale multilevel voltage-sourced converter technology for bundling of offshore windpower. *Proc. CIGRE Sess. Paris Fr.* **2012**, *4*, 306.

17. Nexan. *Submarine Power Cables*; Nexan: Hannover, Germany, 2013.

18. Colla, L.; Lombardo, L.V.; Kuljaka, N.; Zaccone, E. Submarine projects in the Mediterranean Sea. Technology developments and future challenges. In Proceedings of the 1st South East European Regional CIGRÉ Conference, Portoroz, Slovenia, 7–8 June 2016; pp. 1–12.

19. Europacable. *An Introduction to High Voltage Direct Current (HVDC) Subsea Cables Systems*; Europacable: Brussels, Belgium, 2011.

20. Van der Meer, A.A.; Hendriks, R.L.; Kling, W.L. A survey of fast power reduction methods for VSC connected wind power plants consisting of different turbine types. In *EPE Wind Energy chapter 2nd seminar*; Stockholm, Sweden, 2009; pp. 23–24.

21. Chaudhary, S.K.; Teodorescu, R.; Rodriguez, P.; Kjær, P.C. Chopper Controlled Resistors in VSC-HVDC Transmission for WPP with Full-scale Converters. In Proceedings of the 2009 IEEE PES/IAS Conference on Sustainable Alternative Energy (SAE), Valencia, Spain, 28–30 September 2009; pp. 1–8.

22. Nentwig, C.; Haubrock, J.; Renner, R.H.; Van Hertem, D. Application of DC choppers in HVDC grids. In Proceedings of the 2016 IEEE International Energy Conference (ENERGYCON), Leuven, Belgium, 4–8 April 2016.

23. Feltes, C.; Wrede, H.; Koch, F.W.; Erlich, I. Enhanced fault ride-through method for wind farms connected to the grid through VSC-based HVDC transmission. *IEEE Trans. Power Syst.* **2009**, *24*, 1537–1546. [CrossRef]

24. TenneT. *Integrated Annual Report 2018*; Tennet Holding B.V; TenneT: Arnhem, The Netherlands, 2019.

25. Lesnicar, A.; Marquardt, R. A new modular voltage source inverter topology. In *Proceedings of the 10th European Conference on Power Electronics and Applications*; IEEE: Toulouse, France, 2003; pp. 1–10.

26. Konstantinou, G.; Pou, J.; Member, S.; Ceballos, S.; Agelidis, V.G.; Member, S. Active Redundant Submodule Configuration in Modular Multilevel Converters. *IEEE Trans. Power Deliv.* **2013**, *28*, 2333–2341. [CrossRef]

27. Wylie, J.; Merlin, M.C.; Green, T.C. Analysis of the effects from constant random and wear-out failures of sub-modules within a modular multi-level converter with varying maintenance periods. In Proceedings of the 2017 19th European Conference on Power Electronics and Applications (EPE'17 ECCE Europe), Warsaw, Poland, 11–14 September 2017.

28. Guo, J.; Liang, J.; Zhang, X.; Judge, P.; Wang, X.; Green, T. Reliability Analysis of MMCs Considering Sub-module Designs with Individual or Series Operated IGBTs. *IEEE Trans. Power Deliv.* **2016**, *32*, 666–677. [CrossRef]

29. Peralta, J.; Saad, H.; Dennetière, S.; Mahseredjian, J.; Nguefeu, S. Detailed and averaged models for a 401-level MMC–HVDC system. *IEEE Trans. Power Deliv.* **2012**, *27*, 1501–1508. [CrossRef]

30. Bauer, J.G.; Wissen, M.; Gutt, T.; Biermann, J.; Schäffer, C.; Schmidt, G.; Pfirsch, F. New 4.5 kV IGBT and Diode Chip Set for HVDC Transmission Applications. *PCIM Eur.* **2014**, 20–22.

31. Siemens. Fact Sheet SylWin1 HVDC Platform 2015. Available online: https://assets.new.siemens.com/siemens/assets/api/uuid:0065b6c7-7e4c-47e7-8c19-7cb754aa395c/factsheet-sylwin1-e.pdf (accessed on 7 December 2019).

32. Eberlin, S.; Hock, B. *Zuverlässigkeit Und Verfügbarkeit Technischer Systeme: Eine Einführung in Die Praxis*; Springer: Wiesbaden, Deutschland, 2014; ISBN 3658035730.

33. DIN 31051 DIN 31051: September 2012, Grundlagen der Instandhaltung 2012.

34. Ryll, F.; Freund, C. Grundlagen der Instandhaltung. In *Instandhaltung Technischer Systeme*; Springer: Berlin, Heidelberg, 2010.
35. Siemens. *Maintenance of Siemens GIS 8D 1st to 4th generation*; Siemens A.G.: Nuremberg, Germany, 2016.
36. Coccioni, R. Instandhaltung von Mittelspannungs-schaltanlagen gestern und heute. *e&i Elektrotechnik und Informationstechnik* **1998**, *115*, 559–565.
37. Shen, B.; Jiang, S.; Li, Z.; Li, G. Analysis and Treatment of Typical Problems of GIS Circuit Breaker Switching Station Installation and. *Am. J. Water Sci. Eng.* **2018**, *4*, 97–100.
38. Rajotte, C. *Guide for Transformer Maintenance*; CIGRE: Paris, France, 2011; ISBN 978-2-85873-134-3.
39. Jirutitijaroen, P.; Singh, C. The effect of transformer maintenance parameters on reliability and cost: A probabilistic model. *Electr. Power Syst. Res.* **2004**, *72*, 213–224. [CrossRef]
40. Liang, Z.; Parlikad, A. A Markovian model for power transformer maintenance. *Int. J. Electr. Power Energy Syst.* **2018**, *99*, 175–182. [CrossRef]
41. TRENCH. *Coil Products*; Trench Group: Scarborough, ON, Canada, 2016.
42. Tenzer, M.; Koch, H.; Imamovic, D. Compact Systems for High Voltage Direct Current Transmission. In *International ETG Congress*; VDE: Bonn, Germany, 2015; pp. 1–6.
43. HSP. *HVDC Wall Bushing Mounting-Operating-and Maintenance-Instructions*; HSP Hochspannungsgeräte GmbH: Troisdorf, Germany, 2015.
44. Kühn, B.; Callies, F.; Füller, A.L.; Lyding, R. *Windenergie Report Deutschland 2010*; Fraunhofer IWES: Kassel, Germany, 2010.
45. The MathWorks Inc. *MATLAB Version 9.1 (R2016b)*; The MathWorks Inc.: Natick, MA, USA, 2016.
46. Unnewehr, J.F. Data for "Reducing operational costs of offshore HVDC energy export systems through optimized maintenance.". *Zenodo* **2019**. [CrossRef]
47. Olauson, J.; Bergkvist, M. Modelling the Swedish wind power production using MERRA reanalysis data. *Renew. Energy* **2015**, *76*, 717–725. [CrossRef]
48. Carvalho, D.; Rocha, A.; Gómez-Gesteira, M.; Santos, C.S. WRF wind simulation and wind energy production estimates forced by different reanalyses: Comparison with observed data for Portugal. *Appl. Energy* **2014**, *117*, 116–126. [CrossRef]
49. Manwell, J.F.; McGowan, J.G.; Rogers, A.L. *Wind Energy Explained: Theory, Design and Application*; Wiley: Hoboke, NJ, USA, 2009; ISBN 9780470015001.
50. Mortensen, N.G.; Heathfield, D.N.; Rathmann, O.; Nielsen, M. *Wind Atlas Analysis and Application Program: WAsP 10 Help Facility*; Roskilde: DTU Wind Energy, 2011.
51. Lange, B.; Waldl, H.; Guerrero, A.G.; Heinemann, D.; Barthelmie, R.J. Modelling of offshore wind turbine wakes with the wind farm program FLaP. *Wind Energy* **2003**, *6*, 87–104. [CrossRef]
52. Janssen, K.; Faulstich, S.; Hahn, B.; Hirsch, J.; Neuschäfer, M.; Pfaffel, S.; Rohrig, K.; Sack, A.; Schuldt, L.; Stark, E. *Windenergie Report Deutschland 2014*; Fraunhofer IWES: Kassel, Germany, 2015.
53. Chrischilles, E. *EEG 2017: Eine Kostenabschätzung, Mögliche Entwicklungen der Förderkosten bis 2020 und 2025*; Institut der Deutschen Wirtschaft Köln: Köln, Germany, 2017.
54. Maples, B.; Saur, G.; Hand, M.; van de Pietermen, R.; Obdam, T. *Installation, Operation, and Maintenance Strategies to Reduce the Cost of Offshore Wind Energy*; National Renewable Energy Lab.: Golden, CO, USA, 2013.

Article

Evaluation of Anomaly Detection of an Autoencoder Based on Maintenace Information and Scada-Data

Marc-Alexander Lutz [1,*,†], Stephan Vogt [1,†], Volker Berkhout [1,†], Stefan Faulstich [1,†], Steffen Dienst [2], Urs Steinmetz [3], Christian Gück [1,†] and Andres Ortega [1,†]

[1] Fraunhofer Institute for Energy Economics and Energy System Technology, Königstor 59, 34119 Kassel, Germany; Stephan.Vogt@iee.fraunhofer.de (S.V.); volker.berkhout@iee.fraunhofer.de (V.B.); stefan.faulstich@iee.fraunhofer.de (S.F.); christian.gueck@iee.fraunhofer.de (C.G.); afortegast@gmail.com (A.O.)

[2] Trianel Windpark Borkum GmbH und Co. KG, Zirkusweg 2, 20359 Hamburg, Germany; s.dienst@trianel.com

[3] STEAG Energy Services GmbH, Rüttenscheider Str. 1-3, 45128 Essen, Germany; Urs.Steinmetz@steag.com

* Correspondence: alexander.lutz@iee.fraunhofer.de

† Current address: Fraunhofer Institute for Energy Economics and Energy System Technology, Königstor 59, 34119 Kassel, Germany.

Received: 20 December 2019; Accepted: 20 February 2020; Published: 29 February 2020

Abstract: The usage of machine learning techniques is widely spread and has also been implemented in the wind industry in the last years. Many of these techniques have shown great success but need to constantly prove the expectation of functionality. This paper describes a new method to monitor the health of a wind turbine using an undercomplete autoencoder. To evaluate the health monitoring quality of the autoencoder, the number of anomalies before an event has happened are to be considered. The results show that around 35% of all historical events that have resulted into a failure show many anomalies. Furthermore, the wind turbine subsystems which are subject to good detectability are the rotor system and the control system. If only one third of the service duties can be planned in advance, and thereby the scheduling time can be reduced, huge cost saving potentials can be seen.

Keywords: wind turbine; maintenance; autoencoder; machine learning; reliability; data driven model; service; performance

---

1. Introduction

One of the major economic impacts on the Levelized Cost of Energy (LCOE) of offshore Wind Turbines (WTs) is due to the Operation and Maintenance (O&M), which is considered to have a share between 25% and 30% according to Lei et al. [1]. Therefore, different strategies exist to reduce the LCOE by reducing the percentage of O&M-cost. An overview of those strategies is given in Wang [2]. In contradiction to that, this paper focuses on implementing and evaluating a tool for a possible predictive maintenance strategy. By detecting failures in advance, the WT downtimes can be reduced. This will have a direct impact on the LCOE. Different approaches for those tools exist. One can implement a Data Driven Model (DDM), to be precise a Normal Behaviour Model (NBM). Other approaches consider transition probabilities, while others focus on statistics to calculate values for the Mean Time to Repair (MTTR) and to calculate the failure rate. This paper focuses on a tool to model the normal behavior for each WT.

State-of-the-art WT systems monitor the condition and the performance by using Supervisory Control and Data Acquisition (SCADA)-systems, which generally record and store the data continuously. This SCADA-data forms the basis to develop the NBM.

With the model developed, a deviation from normal behavior can be considered as an anomaly. This paper aims to evaluate the anomalies generated by a NBM before a failure has occurred at the WT. These failures are documented in the form of service reports.

A literature review is given in Section 2. Hereafter, the NBM is described and trained in Section 3. In Section 3.1, the basic functionality of an Autoencoder (AE) is briefly explained. Data is needed to train the model. This data is described in Section 3.2. Adjustments to the data, filtering and fine-tuning of the parameters of the AE are to be explained in Section 3.3. Thereafter the output of the model as well its usage is presented in Section 3.4.

In Section 4, the steps in order to derive a clean failure dataset are shown. Information about downtime events are extracted out of maintenance documents and the measures undertaken at the WT and its components translated into standards (Sections 4.1 and 4.2). To ensure the correctness of the downtime events, we explain how they will be validated in Section 4.4.

With the usage of both, the historical failures of the WT (Section 4) and the trained model (Section 3), the time window before the occurrence of a failure can be investigated. This is conducted in Section 5. Subsequently, the detectability of WT failures can be evaluated (Section 5.2). The potential scheduling time and the subsystems with failures are considered to be potentially good detectable in advance. To have a holistic approach, also anomalies during expected normal behavior, are to be investigated. The results are discussed in Section 6 and concluded with Section 7.

2. Literature Review

Health monitoring of WTs is the continuous monitoring of data streams from the WT intending to generate information about its state and condition. Health monitoring of a WT comprises approaches of Condition Monitoring System (CMS) and Structural Health Monitoring System (SHM). Both approaches use a dedicated set of sensors to monitor physically relevant loads, frequencies or accelerations of a given system to predict failures and identify root causes at an early stage and increase the revenue from wind farms [3].

CMS and SHM-systems are often based on physical models. With the rise of big data analysis techniques and machine learning algorithms, DDMs based on SCADA data have been increasingly proposed and developed. The spread of DDMs has lead to early failure detection [4]. Different reviews of DDMs are conducted and focus condition monitoring [5,6], on fault prognostics [7] with the help of deep learning methods [8].

In this field, Helbing and Ritter [8] present a review of different supervised and unsupervised methods and their potential usage. Thirteen unsupervised approaches, e.g., AEs and six supervised approaches, e.g., convolutional neural networks, are listed.

Bangalore and Tjernberg [4] have implemented a non-linear autoregressive neural network to model the normal temperature of five different gearbox bearings. Zaher et al. [9] have developed a DDM using the bearing and cooling oil temperature to predict the failures within the gearbox system. Schlechtingen and Ferreira Santos [10] compare three different DDMs, the first of which is a regression model the others are Artificial Neural Network (ANN). It has been concluded that the ANN outperforms the regression model [10].

Approaches with an application of an AE on SCADA data is presented by Zhao et al. [11], Vogt et al. [12] and Deeskow and Steinmetz [13] among others. Deeskow and Steinmetz [13] have applied ensembles of Autoencoders to detect anomalies in sensor data of various types of power generating systems. They discuss these methods in comparison with supervised learning approaches. Supervised learning [3] requires a certain amount of engineering knowledge to be considered, while the unsupervised methods make anomaly detection possible with a minimum of domain knowledge.

To capture the nonlinear correlations between around ten different sensors, Jiang et al. [14] used an unsupervised learning approach to develop a Denoising Autoencoder (DAE). This concept is further developed to a multi-level DAE and presented in Wu et al. [15].

Evaluation of Application

Helbing and Ritter [8] point out that a comparison of different approaches is difficult as evaluation is often being presented as case studies with historical production data. If approaches are to be quantitatively compared, labeled data for normal production and abnormal behavior are required. This is not yet readily available. Thus, it has to be deduced from different sources that reduce the comparability of results.

Similar approaches to the AE, as described in this paper, are seen in Zhao et al. [11]. Here, findings based on three case studies are explained. Two of which deal with gearbox failures and the third of which presents a converter failure. Jiang et al. [14] evaluates the failure detectability of DAE based on SCADA data. The case studies show the detection of a generator speed anomaly and a gearbox filter blocking. In addition to that, this approach is implemented on eight of ten simulated failure scenarios put forward by Odgaard and Johnson [16]. However, there is little information available on the evaluation of health monitoring approaches with AE on a large set of failure events. This paper aims at overcoming this issue.

3. Overview and Model Implementation

In this section, the type of method for early failure detection of WT is explained as well as the implementation of the model with the data given.

3.1. Basic Functionality of Autoencoders

The AE tries to represent the input data by first encoding it, compressing it to the relevant information and decoding it again.

Within the training of the AE, the weights of the neurons are adjusted so that the loss function will yield a minimum. By doing so, the output will represent the input in the best feasible way. In Figure 1 a general set up of the architecture of the AE can be seen.

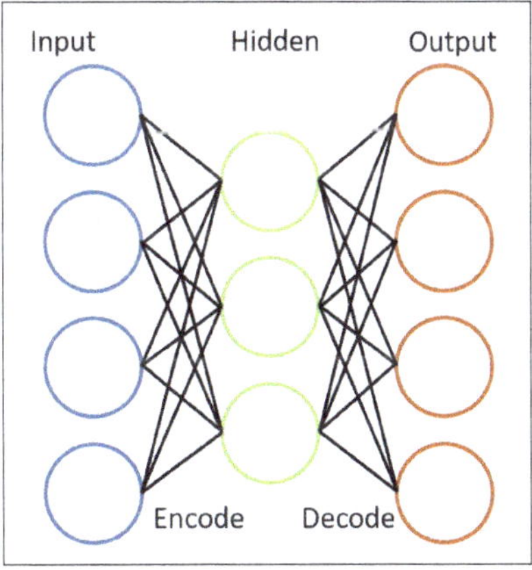

Figure 1. Autoencoder architecture.

Decoder:

$$r = g(h) \tag{1}$$

Encoder:

$$h = f(x) \tag{2}$$

Minimizing the loss function:

$$L(x, g(f(x))) = L(x, r) \tag{3}$$

As mentioned in Vogt et al. [12], different configurations of AE are possible: An Under Complete Autoencoder (UCA), an DAE or a Contractive Autoencoder (CAE). The implementation that is used within this paper is an UCA that is partly modified. This is more elaborately described in Section 3.3

3.2. Specifications of the Data Set

The dataset used consists of various offshore WTs located in the same Wind Farm (WF) in the German North Sea. All of them are of the same turbine type with a rated power 5 MW or above. The turbine type is especially designed for offshore usage. The dataset can be subdivided into two different parts: The operational data and the event data. At first, the operational data shall be discussed.

3.2.1. Operational Data

A number of around 300 sensors continuously monitor the performance of each WT. These sensors measure power, wind speed, pressure, temperature, voluminal flow and other values. The WTs are not equipped with sensors that measure the wave height or the salinity of the air. Furthermore, some of the sensors represent counters, e.g., the number of cable windings. Different sampling rates of the sensors are available. Nevertheless, all of the values are grouped into ten-minute-average values by the SCADA-system. For one timestamp with an averaging period of ten minutes, around 300 average values can be seen. A time series consisting of the timestamps is available for every WT. The data available ranges from February 2016 up to December 2019. Next to the average sensor data, an Operational Mode (OM) of each WT is given. The OM is the event triggered and described in the next section.

3.2.2. Event Data: Operational Mode

The OM indicates the state of the WT. This state could be defined as one of the following: Failure, manual stop, service, production, ready. All of the OMs are listed in Table 1. Only one OM can be valid at a time for one WT. If a new OM is given by the SCADA-systems, the old one will be discarded. Depending on the environmental conditions and the state of the WT, an OM could be present for seconds or several days. To have a complete time series, the discrete events with the information of the WT-state as given by the OM are to be linked to the equidistant operational data. This is described in the next section.

3.2.3. Preparation of the Input Data

Several OMs can appear during an averaging period; e.g., the OM of the WT can be ready, followed by production and end within the same ten minutes in a failure. Nevertheless, only one OM is linked with the equally spaced operational data time series. Therefore, if several OMs are present in one averaging period, only one is chosen depending on an hierarchical order. This order is presented in descending prioritization with the most important OM at the top in Table 1.

As a result of the concatenation of the operational data and the OMs, an equidistant time series is created where both the information of the sensor values and the information about the WT state is present. This time series is the input for the AE.

Nevertheless, further steps are necessary to have the data ready in that way that it can be used for training. Those steps consist of the following: Dropping of sensors, data imputation and data scaling, filtering for operational modes. These steps are explained subsequently.

Dropping of Sensors

It is to be expected that the time series is complete. Nevertheless, data gaps can be seen. To ensure data quality, filtering for data gaps is conducted. Only if at least 80% of all the values within the training for one sensor are integer or float numbers, this sensor is used further. If it is not the case, the sensor is dropped.

A sensor is also dropped if the value which it represents is a counter and therefore, the values are monotonous rising. Examples are the cable winding, the energy counter or the number of strokes for a specific component. Out of the around 300 sensors, roughly 25 represent counters. Those sensors are dropped.

Table 1. Hierarchy of operational modes in descending order.

Hierarchical Order	Operational Mode
1	Failure
2	Manual stop
3	Service
4	Production
5	Ready
6	Setup
7	Initialize
8	Run up
9	System test
10	Ice accumulation
11	Grid outage
12	High wind shutdown
13	Emergency supply WT
14	Emergency supply converter station
15	Shadow impact shutdown

Data Imputation and Data Scaling

A mean imputation method is chosen for this set up. If the sensor is not dropped but one or more data gaps are still present, the data gaps are substituted with the mean value of all other values of that sensor from this WT. If data gaps occur, they usually can be seen over a period of several days. Linear interpolation between the values before and after the data gap might only partly reflect the sensor behavior. E.g., the fluctuation of the wind within a day. It can be observed that the mean of data gaps regresses to the mean of the sensor with longer data gaps. Therefore with longer data gaps, an imputation with the mean is more appropriated. For data gaps of one hour or less it is expected that a linear interpolation is more appropriate. However this is not yet implemented and therefore part of future work.

Furthermore a standard scaling is implemented to prepare the data to be ready as an usable input for the AE. To standard scale the sensor values its mean value is subtracted and the result is divided by the sensors variance.

Filtering for Operational Modes

One of the reasons for the selection of an AE is that there is less amount of failure data but a high number of normal samples. In this study failure data is present in the form of service reports (see Section 4.2.1). An average number of 18 reports is issued per WT per operational year. The amount of normal samples per operational year is around 51,000, within one year. This result can be calculated as follows: Every ten minutes within a year one sample (52,560 samples) multiplied with the technical availability given in the literature to be around 96.6% (see Lutz et al. [17]). An additional reason for a high amount of normal samples can be seen due to the fact that the WT is performing according to its design specifications most of the time. This is also indicated by the time-based availability which is stated to be around 96.6% [17].

In this study, normal samples are considered to be all the timestamps of the concatenated time series where the operational mode is one of the following: Production, run-up, high-wind shutdown and ready. Furthermore if in the following a normal operational mode is mentioned, it refers to the four mentioned beforehand. Only these operational modes are used for training because they are assumed to represent normal behavior. All other are excluded from training. Purpose of the trained

AE is its ability to represent normal behavior of the WT. Any deviation from a normal state for a timestamp given in prediction period is considered to be an anomaly. The training is explained in the next section. The output of the AE in the section afterwards.

3.3. Training the Autoencoder

To train the AE, it is necessary to develop a model architecture. The initial architecture of the AE is as follows: Three hidden layers are used. The first and third of which initially consist of 1350 neurons, the second layer comprises 50 neurons. The input and output dimensions are the same. If in the following the model or an AE is mentioned it refers to the UCA. The input dimensions can vary from one implementation for one WT to another, since some sensors might be dropped. Input dimensions ranging from 243 to 263 dimensions are observed in this set up. The Keras implementation of the adam optimizer, as described in Kingma and Ba [18] is used. Its learning rate and its decay rate are the same with a value of 0.001. The number of epochs is set to 10. Furthermore, the mini-batch size during the gradient descent executed by the adam optimizer is set to 209 samples. A sample is to be understood as all sensor values for a timestamp after dropping irrelevant sensors. If the input dimension is 263, a sample comprises 263 values, one for each sensor. The selected parameters are visible in Table 2.

Table 2. Parameters of the model.

Type	Parameter
Hidden layer 1	1350
Hidden layer 2	50
Hidden layer 3	1350
Learning rate	0.001
Decay rate	0.001
Batch size	209
Epochs	10

For the activation function, a parametric rectified linear unit is chosen. Its value is set during optimization. With these specifications, the model of the AE is defined, which allows the training process to start.

First, the prepared input data is split into a training set and a validation set. This is done by iterating over the entire dataset and adding 5040 samples (35 days) to the training set and the following 1680 samples (around 12 days) to the validation set. This is repeated subsequently until all available input data out of one year is assigned. It is assumed that a training period of one year is suitable even though shorter or longer periods are possible. Additional research needs to be undertaken in the selection of the most appropriate time window for the training period. However this is part of future work. One year is assumed to be suitable because it contains seasonal dependencies that should be learned as normal behavior by the AE. As stated before four years of operational data are available. Depending on the period in which a prediction is carried out (to be explained in Section 3.4), different sizes of training periods are available. E.g., in February 2017 one year is available, in February 2018 two years. In order to enable comparable results a training period of one year is selected to train the AE.

Second, the AE is trained on the training set. The training data is mapped to the computed reconstruction and the reconstruction error is optimized. The reconstruction error is to be understood as the difference between the input data and the reconstruction. Additionally, the AE will also map the validation data, which is not part of the training data to its reconstructions, followed by another reconstruction error computation. The mean reconstruction error over the validation data then is yielded as a value, which later can be used to rate the performance.

Once the AE is trained, it is able to reconstruct the sensor values for a given input. To express if the input is abnormal or normal, we first compute the reconstruction error again. Modeling the reconstruction error as a vector of random variables leads to different methods to measure the abnormal behavior of the input based on the reconstruction error. This is described in the next paragraph.

Measure the Reconstruction Error

Using the RMSE to measure the reconstruction error, implicitly assumes that the sensor values are independently distributed. Under the assumption that some sensors are not independent of each other, the Mahalanobis distance (see Equation (4)) is used.

This measure uses the covariance matrix of the random variables to take possible dependencies into account. Furthermore, it provides a more realistic measure for this case. Since the covariance matrix of the random variables is unknown, it will be estimated from the given dataset. Using a standard covariance estimator may lead to a wrong result caused by outliers in the dataset. Therefore an outlier robust covariance estimation is used as described in Butler et al. [19]. It approximates the covariance matrix of the random variables iteratively. After estimating the covariance, it is now possible to compute the Mahalanobis distance of every input sample and use it as a score, which determines how abnormal the reconstruction error of a given sample is.

Mahalanobis distance:

$$S_x = \sqrt{(r_x - \mu)^T \Sigma^{-1} (r_x - \mu)} \tag{4}$$

where:

μ	the mean vector
Σ	the covariance matrix
r_x	the reconstruction for a timestamp x
S_x	Score for a timestamp x

By using this measure for each timestamp x, a value for the score can be calculated. If this value is higher than a chosen threshold (*thres*) the timestamp is considered to be an abnormal one. This is seen in Equation (5) and also mentioned in Vogt et al. [12].

$$Anomaly_x := \begin{cases} true, & \text{if } S_x > thres \\ false, & \text{else} \end{cases} \tag{5}$$

To calibrate the value for the threshold, one has to consider the predictions ground truth. Four different possibilities can be concluded if the predictions for an anomaly are compared to the actual OM. This is seen in Table 3.

Table 3. Potential ground truth of prediction.

Type	Description	Active OM
True negative	Score below threshold, no anomaly is detected	Normal OM
False negative	Score below threshold, no anomaly is detected	OM is not normal e.g., failure
True positive	Score above threshold, an anomaly is detected	OM is not normal e.g., failure
False positive	Score above threshold, an anomaly is detected	Normal OM

With these four possible outcomes, the false discovery rate (see Equation (6)) can be defined. This value expresses how many of the sample scores above the threshold are normal according to the OM

False discovery rate:

$$fdr := \frac{fp}{fp + tp} \tag{6}$$

where:

tp	number of true positives
fp	number of false positives
fdr	false discovery rate

This allows calibration of the threshold. At first, all scores of the training and validation dataset are calculated and sorted in ascending order. By iterating over all possible thresholds from 0 to the maximum of all calculated scores, the threshold can be calibrated. For each possible threshold the fdr (see Equation (6)) is calculated. The first threshold is chosen, where the false discovery rate is lower than the selected value. With the value for the fdr selected, it can be guaranteed that the selected percent of all detected samples are normal according to the OM. In this case, the fdr is selected to be 0.8. Thereby a higher sensitivity is given and timestamps are more easily considered to be abnormal. This is a desired behaviour since the impact of subsequent abnormal timestamps form the basis for another measure. This measure is described in Equation (7).

3.4. Using the Trained Model

The input data for the training of the AE consists out of operational data for a consecutive time of one year. Once the AE is trained and the threshold calibrated, a prediction can be made for new timestamps. It is referred to as prediction period in the following. Since the operational data (see Section 3.2.1) is available starting from February 2016, the earliest prediction can be made by February 2017. The time window for the prediction period is set to a window of seven days.

A time window of seven days is chosen out of the following reasons:

First, after performing maintenance actions at the WT it is regularly seen that some parameters of the WT differ. E.g., Oil has been refilled or some control parameters are adjusted. This leads to a changed behaviour of the WT. Yet this behaviour is within design specifications, the AE has not learned it while training and therefore it is more likely to observe an extended amount of anomalies after a maintenance action.

Second, a focus of this tool is to give decision support in a scenario where in a daily routine at several WTs preventive measures are necessary to perform. If one of the WT is more likely to fail, preventive measures need to be carried out first. With limited resources the question arises which WT should be prioritized. This tool helps in making this decision easier by identifying the WT which is more likely to fail. A prediction period of more than seven days might also be appropriate. Nevertheless the authors did not investigate this matter. It is part of future work.

For each timestamp in the prediction period, the score can be calculated (see Equation (4)). If the value for the score is above the calibrated threshold, an anomaly is detected. This boolean result is available for any timestamp in the prediction (see Equation (5)). To assess the impact of subsequently arising anomalies, a further measure is defined: The criticality. It is a counter that rises by one if an anomaly is detected and a normal operational mode is present for that timestamp. The criticality decreases by one if no anomaly is detected. It stays constant if the operational mode for that timestamp is a service and an anomaly is detected. Its lower limit is zero and its upper level is equal to the number of timestamps in the prediction time window, which is 1008 instances. This implies an anomaly for every timestamp within seven days.

The value of the criticality within the time window of the prediction period is selected as a criteria to evaluate the failure detectability of the AE. It is described in Equation (7).

$$Crit_{x_0} = 0$$

$$Crit_{x_{i+1}} = \begin{cases} Crit_{x_i}, & \text{if } S_{x_{i+1}} > thres \text{ and OM Service} \\ Crit_{x_i} + 1, & \text{if } S_{x_{i+1}} > thres \text{ and OM normal} \\ \max(0, Crit_{x_i} - 1), & \text{else} \end{cases}, \text{ for } i = 0, 1, \ldots, 1008 \tag{7}$$

With $x_1, x_2, \ldots, x_{1008}$ being the timestamps for the prediction period. The results by applying this equation are outlined in Section 5. Before doing so, a set of standardized failures has to be prepared. This is explained in the section that follows.

4. Preparation of a Clean Failure Data Set

Maintenance information about the WT is available in the form of service reports. Different maintenance engineers have described their activities at the WT-site. For each action, a text description is given, which also indicates the start and the end of the WT unavailability. Since different engineers use varying semantics for the same WT-subsystems the maintenance information needs to be standardized (see Sections 4.2.1 and 4.2.2). Before that, downtime events are explained. They serve as a tool to validate the unavailability documented in the reports. This is described in the following section.

4.1. Generation of Downtime Events

With the ten-minute average values of the sensors wind speed and power calculations and assumptions can be made. Since this calculation is rather an input to validate events it will be referred to as event data. Given the wind speed and the power for each timestamp, a decision can be made for the overall state of the WT. Wind speed and power are to be understood, as explained in the standard IEC 61400-25 [20]. The scada-events are outlined in Table 4.

Table 4. Description of scada-events for a timestamp with power and wind speed values.

Scada-Event	Power	Wind Speed
High Wind	≤ 0	>cut out wind speed
Low Wind	≤ 0	<cut in wind speed
Production	>0	>cut in & <cut out wind speed
Unspecified Downtime	≤ 0	>cut in & <cut out wind speed

With the values for wind speed and power given for each timestamp, a scada-event can be deduced. If the scada-event is the same for several sequential timestamps they will be joined together to one event. The beginning of the first timestamp indicates the start. The end of the last timestamp indicates the end of the event. This is what is to be understood if in the following a scada-event is mentioned. Within an averaging period the turbine could be both: Producing energy and consuming energy e.g., due to a cable unwinding. Hence if the values for power are averaged, this information can hardly be accessed. Nevertheless, the boundaries for wind speed and power for the different scada-events are choosen according to Table 4.

4.2. Description and Standardization of Failures

4.2.1. Event Data: Service-Reports

The scada-events provide the information if the WT is in downtime. But it does not resolve the question of why a downtime has happened. Several causes for downtime of a WT exist. It can be due to a regulation action: e.g., load curtailment, noise reduction shutdown or bird conservation actions. Furthermore, a downtime can occur if preventive or corrective measures are being performed at the WT. If so a service report will be issued by the service provider to the operator. Within this report, a detailed description of the work performed by the service team is written down. This text will describe the type of maintenance measure and which component has been the subject of the work. The service reports with a text description of the work performed are available for evaluation. A service report is available if during a maintenance activity the onboard crane system of the WT can be used for the carriage of materials for exchange, repair or enhancement of WT components or if only persons are performing measures at the WT. If the material is too heavy to be lifted with the onboard crane, no service report is available for the evaluation, as described in this paper. Different enterprises are involved and another form of documentation is used, which is not obtainable to the authors.

Since the text descriptions contained in the service report is hard to use for analysis described in Section 5 they need to be standardized. This is going to be outlined in Section 4.2.2. Next to the text description other information is also provided in the service-report. This information implies the start and end of the measure at the WT, the start and end of the unavailability, the material consumption, the technicians, the tools used, an identifier for the WT next to other details.

4.2.2. Standards

Two structures shall be introduced in the following: Reference Designation System for Power Plants® (RDS-PP®) and Zustand-Ereignis-Ursachen-Schlüssel (Engl: State Event Cause Code) (ZEUS). These structures will be used to classify the text description into standards.

RDS-PP®

The standard RDS-PP® aims at having a unique structure of WT systems and subsystems and a hierarchically dependency of those systems amongst each other. An example: On RDS-PP® level 1 all of the components related to the WT and its subsystems will be referred to as wind turbine systems. Beneath the wind turbine system on level 2, the yaw system, the drive train system and the rotor system are structured next to others. Two advantages of the standard shall be mentioned: First, the standard can be used to compare different WTs of different Original Equipment Manufacturers (OEMs) and second the components can be grouped into subsystems and systems. The first one does not apply in this paper since one WF with WTs of the same turbine type is the subject of consideration. Nevertheless, the second benefit applies and is being used and described in Section 5.

ZEUS

ZEUS is introduced by the Fördergesellschaft Windenergie und andere Dezentrale Energien (FGW) within the technical guideline TR7 [21]. ZEUS introduces several blocks that describe the state of the turbine and the state of a component. With the combination of all blocks nearly every state of the WT is defined. Each block raises a question that is answered by a certain code in the sub-block. An example: In ZEUS Block 02-08 the question is raised "Which type of maintenance is active or will be necessary to eliminate a deviation from the target state?". This answer can be given by the ZEUS-code 02-08-01, which is corrective maintenance or by the ZEUS-code 02-08-02, which indicates preventive maintenance. For a better understanding about the terminology of preventive and corrective see the British Standards Institution: Maintenance–Maintenance terminology [22].

4.3. Selection of Failures

In order to investigate only the corrective maintenance events, filtering is done according to ZEUS. Only those events are selected where a failure has happened or the measure to restore the WT to a functional state is a corrective one. Those events are further used in Section 5. The detected anomalies before such events will be evaluated. Therefore only the events with the ZEUS block shown in Table 5 are chosen.

Table 5. Selection of events by ZEUS blocks.

Zeus Block	Sub-Block	Sub-Block Description
Which type of maintenance is active or will be necessary to eliminate a deviation from the target state?	02-08-01	Corrective maintenance
How is the functional state of the element to be evaluated?	02-01-03	State after failure with need for measures
How is the functional state of the element to be evaluated?	02-01-95	Uninvestigated fault
How is the functional state of the element to be evaluated?	02-01-96	Unresolvable fault
How is the functional state of the element to be evaluated?	02-01-97	Undefined fault
How is the functional state of the element to be evaluated?	02-01-98	Other fault

If an event fulfills one of the criteria shown in Table 5, it is selected. Thereafter it has to be validated with the downtime events. This is described in the next chapter.

4.4. Validation of Failures with Downtime Events

Having created scada-events and a set of standardized failures (see Sections 4.1 and 4.2), it is necessary to validate whether a failure resulted in a downtime of the corresponding WT. Therefore, the start and the end of the unavailability as indicated in the service-reports are compared with the beginning and the end of the scada-events. It is to be expected that the downtimes of the service reports could be related to the scada events. Nevertheless, three different cases could be observed: A partial linkage, a full linkage and no linkage at all. These three different cases can be seen in Figure 2.

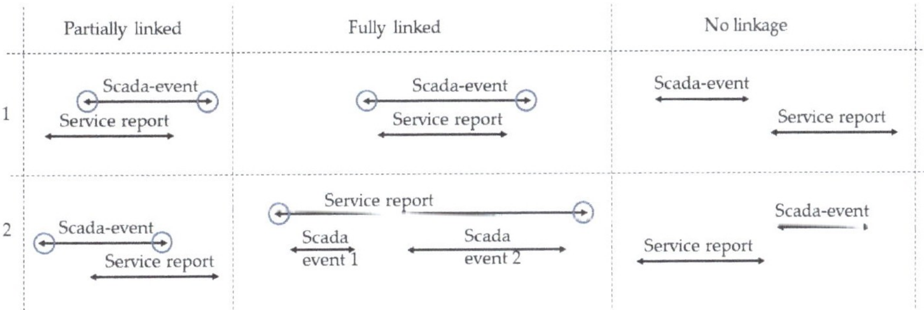

Figure 2. Validation of failures.

If the downtimes do not match, but overlap, the start and the end of the scada event, as indicated by the blue circles, are used within the clean failure dataset (e.g., partially linked Figure 2). If a service report does not overlap with the scada event, the single events are excluded from the clean failure set (e.g., no linkage in Figure 2).

As a summary, it can be stated that 35% of the events showed a full linkage, 41% showed partially linkage and 24% could not be linked and, therefore, have been removed from further analysis. No linkage is done if the WT is operating during the service. Furthermore, different entries for dates are available: The date the service report is issued and the date the service duty has happened. If the entry for the date of the service duty is not available the date of the issuing is used. This may lead to no linkage of the scada-event with the report.

Some 1495 reports have been standardized. After filtering by ZEUS, 799 of those reports remain, which are considered to indicate corrective maintenance.

5. Evaluating the Failure Detectability of the Autoencoder

In this section, the trained AE is executed on a time window of seven days: The prediction period. The maximum of the criticality, as described in Equation (7) is chosen to compare and to group the different prediction periods. Two different scenarios are considered. At first the period of expected normal operation (ENO) is discussed (see Section 5.1). At second, a period before the day of a known historical failure (see Section 5.2). The result and the comparison of the two are shown in Section 5.3.

5.1. Anomalies during Expected Normal Operation

An assumption is to be made of what is considered to be ENO. A time of three weeks in which the WT is in a normal OM the majority of the time. Therefore each day, the number of timestamps with a normal OM is counted. If the number of normal OMs is greater or equal to 138 for every single day within three weeks, a period of ENO is identified. Within those three weeks, operational modes that indicate service or error could be present, e.g., a reset of the WT but not longer than for one hour a day. Of those three weeks of ENO, the second week is selected as the period of prediction. For each WT two of those periods are identified to have the same amount of events as described in the next section.

5.2. Anomalies before Known Failures

109 reports remain after filtering for the beginning of data acquisition of one year and seven days before the failure has occurred. The evaluation is done for all of those remaining failures. The failures are described briefly: The majority was detected in the rotor system, followed by the control system and the converter system. These three systems are often ranked the highest in terms of failure occurrence. This can also be seen in Pfaffel et al. [23]. The average downtime of the failures considered is close to 2.5 days. The shortest downtime has a value of around 30 min, the most extended downtime a value of approximately 22 days. The anomalies before the 109 known failure are to be observed and the criticality calculated. The last day of the prediction period contains the downtime of the failure.

5.3. Potential Detectability of Wind Turbine Failures

Before the detectability of the failures is discussed, the criticality is to be grouped into different ranges. This is described in the next section.

5.3.1. Potential Scheduling Time

The ranges reflect the possible scope of action of the operators and are based on the authors' interpretations. The interpretations are seen in Table 6. If the value for the criticality is 0, the failure detectability is interpreted as no detection. If the criticality is greater than 432 in the prediction period of seven days, the detectability can be construed as a reliable detection. The different ranges of the criticality shown in Table 6 are further used to group the maximum of the criticality of the prediction periods of the two different scenarios (see Sections 5.1 and 5.2). The results can be seen in Figure 3.

Table 6. Potential scheduling time.

Criticality	Share of Standardized Failures	Interpretation	Interpretation in Time of Constant Anomalies [Hours]
0	0	No detection	0
1–6	8.4	Spontaneous detection	0–1
7–72	55.8	Possible short term detection	1–6
73–144	17.9	Likely mid term detection	6–24
145–432	12.6	Expected long term detection	24–72
>432	5.2	Reliable detection	>72

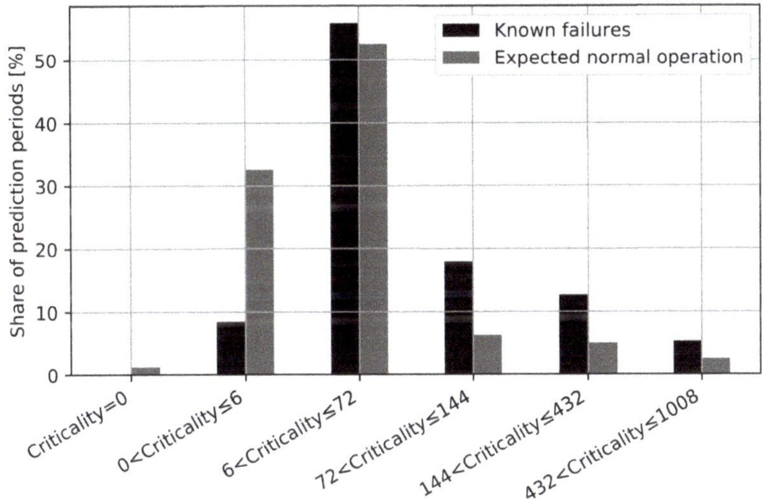

Figure 3. Comparison of prediction periods.

Figure 3 shows the maximum of the criticality in the different prediction periods. The black bars represent the prediction periods in which, on the last day, a failure has happened. The grey bars represent the prediction periods in which the WT is in ENO. The maximum of the criticality in the prediction period is grouped into different ranges (see Table 6) and divided by the total number of prediction periods in the scenario (Failure or ENO). By doing so, the share of the prediction periods can be displayed on the y-axis. In Figure 3 one can see that the black bars are more dominant at values for a higher criticality. Grey bars are more present at values for a low criticality. This underlines the capability that more anomalies are raised in periods where failure is about to happen. Furthermore, if one only considers the grouped bars for the criticality of 72 or higher it can be concluded that the black bars are almost double in their percentage value compared to the grey bars. It can be interpreted as the probability of a failure to happen is two times higher than a period of ENO to appear. Almost one-third of the prediction periods with a known failure are in this range with a criticality of 72 or higher.

It can be stated that more anomalies can be seen in periods where a failure has happened. Nevertheless, some periods of ENO show a high criticality. Several reasons could explain such a behavior: Within a reset of the WT some control parameters are changed or before the period of ENO a longer service was conducted. This leads to a new WT behavior, which has not been seen in the training, and therefore anomalies are detected.

In the next section the subsystems that are subject to expected long term detection and reliable detection (ELT&RD) (see ranges in Table 6) are to be discussed. Furthermore, some of those prediction periods are further investigated.

5.3.2. Potential Detectability of Wind Turbine Component Failures

Anomalies can be seen before failures have happened in different WT-systems. Figure 4 shows the ratio of systems share. The proportion of failures in RDS-PP® systems that are related to ELT&RD is calculated. Similar, the proportion was calculated for all RDS-PP® systems that are available for evaluation after filtering (see Section 5.2). These two proportions are divided. The result will yield a ratio for each RDS-PP® system. It is displayed in Figure 4 on the y-axis and its calculation is given in

Equation (8). The ratio indicates if the detectability of failures (ELT&RD) has increased or decreased if compared to the share of all possible detectable failures.

$$Ratio = \frac{\frac{Sys_{ELT\&RD}}{\sum_{i=1}^{n} Sys_{ELT\&RD\ i}}}{\frac{Sys}{\sum_{i=1}^{m} Sys_m}} \tag{8}$$

where:

$Sys_{ELT\&RD}$	Number of failures in specific RDS-PP® system subject to ELT&RD
Sys	Number of failures in specific RDS-PP® system. All failures available for evaluation considerd.
Ratio	Ratio of systems share
n	Number of unique RDS-PP® systems subject to ELT&RD
m	Number of unique RDS-PP® systems available for evaluation

On the *x*-axis, the systems on RDS-PP® level 2 are visible: The central lubrication system (=MDV), the generator transformer system (=MST), the rotor system (=MDA), the fire alarm system (=CKA), the control system (=MDY), the converter system (=MSE), the environmental measurement system (=CKJ) and the yaw system (=MDL).

It can be deduced that failures in the central lubrication system, the generator transformer system and the rotor system are possible better to detect. Those systems are overrepresented in failures that are considered to be ELT&RD. The ratio increases by around two. The opposite applies to failures, which happen in the converter system (=MSE), the environmental measurement system (=CKJ) and the yaw system (=MDL). Here a decrease of the ratio is seen. A ratio of one would indicate that the share of systems (ELT&RD) compared to all systems considered did not change, yet this is not recognized.

In Figure 5, the development of the criticality over the timestamps in the prediction period is visualized. On the last day of the prediction, failure is seen. The time of constant criticality indicates the OM service, which happened after a failure is detected at the WT. The failures shall be discussed briefly. An anomaly is almost present in every timestamp in the prediction period before a failure showed up in the central lubrication system (=MDV). The amount of grease was low and therefore had to be refilled. This caused a shutdown of the WT. Similar behavior of the criticality is seen before a failure appeared in the rotor system (=MDA). Some adjustments of connectors needed to be made. In the case of the failure in the meteorological measurement system (=CKJ) many anomalies are detected as well. Here both of the anemometers had to be replaced during the service after the failure.

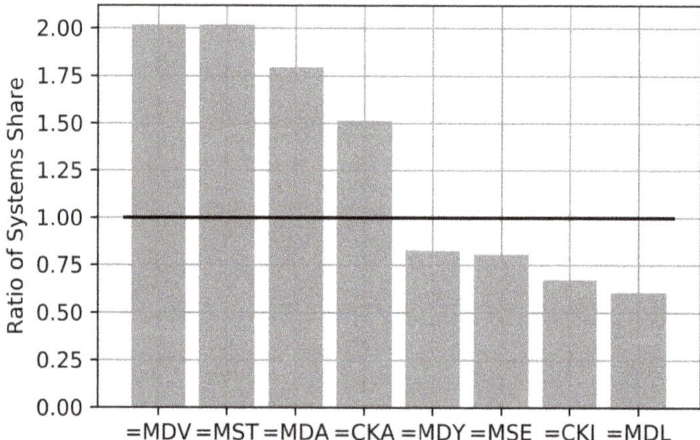

Figure 4. The ratio of Share of Systems with expected long term and reliable detection to all possible detections.

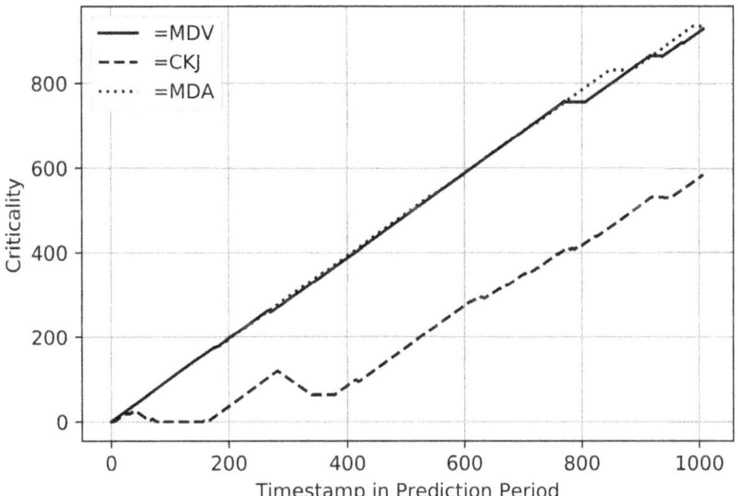

Figure 5. Development of criticality.

6. Discussion

The model of the AE was taken from Vogt et al. [12] and further developed. A hyper parameter optimization (HPO) is not implemented for every model. Rather, the best parameters of one model are selected to be valid for all other models as well. A HPO should be part and further explained in future work. As for now, the false discovery rate is chosen to select the best value for the threshold. Part of future research should probably be to apply and test different methods to calibrate the threshold. Furthermore, different time windows for the prediction period need to be investigated. Additional work should also focus on different imputation methods for different lengths of data gaps and for sensors with varying characteristics in terms of the system inertia which they are measuring. The implementation of an AE, based on the proposed methodology, on WT data is especially helpful to detect anomalies that are rather dynamic. To also investigate long term trends, a system needs to be combined with the AE developed.

By using an AE, it is possible to identify abnormal WT behaviour. Nevertheless, domain knowledge is still needed to validate if an anomaly is likely to turn into a failure. Furthermore, it is still required to identify which system, subsystem or component is the root cause for the detected anomaly. Additional research will also focus on identifying not only the WT, which is critical, but also the sensor or set of sensors that are most probably causing the anomaly. This will lead to ease of decision making.

Three assumptions are to be mentioned:

1. It is to be expected that the WT is performing according to its design specification most of the time and a training of the AE will represent normal behaviour.
2. It is expected that deviations from a normal state can be seen in the ten-minute-average data of the sensors and the operational data, and this is indicated as an anomaly by the trained AE.
3. It is expected that the information in the service reports and the descriptions of the work performed are correct and standardized correctly.

7. Summary and Outlook

The AE developed here shows a good possibility to detect historical failures in various WT-systems. This can be deduced because many anomalies are seen before a failure has happened. At first, the AE is developed, and secondly, the failures are structured and standardized according to RDS-PP$^{®}$ and ZEUS. After doing so, those two approaches were linked and the detectability of failures with the AE were validated on a set of standardized historical failures. Of all the failures, around 35.7% were subject to likely midterm detection, which can be interpreted as at least six hours of constant anomalies. About 17.8% of all failures are considered as expected long term detection, which can be understood as at least 24 h of constant anomalies. About 5.2% can be detected reliably. This can be interpreted as at least 72 h of constant anomalies. By standardizing the failures, we can state which system failures are more easily detectable. These systems were as follows: The central lubrication system, the generator transformer system and the rotor system.

The usage of an AE could help to identify failures and upcoming repair measures in various WT-systems and thereby to increase the revenue and the uptime of the WT by a significant extent.

Author Contributions: Conceptualization, M.-A.L., U.S.; methodology, M.-A.L., U.S., A.O.; software, M.-A.L., S.V., C.G., A.O.,; validation, M.-A.L.; formal analysis, M.-A.L.; investigation, V.B.; resources, M.-A.L., V.B., U.S.; data curation, S.D.; writing—original draft preparation, M.-A.L., V.B., U.S., C.G., A.O.; writing—review and editing, M.-A.L.; visualization, M.-A.L.; supervision, V.B., S.F.; project administration, V.B., S.F.; funding acquisition, V.B., S.F. All authors have read and agreed to the published version of the manuscript.

Funding: The research work presented in this paper was funded by the German Federal Ministry for Economic Affairs and Energy through the research project ModernWindABS under grant number 0324128.

Abbreviations

The following abbreviations are used in this manuscript:

AE	Autoencoder
ANN	Artificial Neural Network
CAE	Contractive Autoencoder
CMS	Condition Monitoring System
DAE	Denoising Autoencoder
DDM	Data Driven Model
ELT&RD	Expected Long Term Detection and Reliable Detection
ENO	Expected Normal Operation
FGW	Fördergesellschaft Windenergie und andere Dezentrale Energien
HPO	Hyper Parameter Optimization
LCOE	Levelized Cost of Energy
MTTR	Mean Time to Repair
NBM	Normal Behaviour Model
OEM	Original Equipment Manufacturer
OM	Operational Mode
O&M	Operation and Maintenance
ROC	Receiver Operation Characteristic
RDS-PP$^{®}$	Reference Desigantion System for Power Plants$^{®}$
SCADA	Supervisory Control and Data Acquisition
SHM	Structural Health Monitoring System
UCA	Under Complete Autoencoder
WF	Wind Farm
WT	Wind Turbine
ZEUS	Zustand-Ereignis-Ursachen-Schlüssel (Engl: State Event Cause Code)

References

1. Lei, X.; Sandborn, P.; Bakhshi, R.; Kashani-Pour, A.; Goudarzi, N. PHM based predictive maintenance optimization for offshore wind farms. In Proceedings of the IEEE/Conference on Prognostics and Health Management 2015, Austin, TX, USA, 22–25 June 2015; pp. 1–8. [CrossRef]
2. Wang, H. A survey of maintenance policies of deteriorating systems. *Eur. J. Oper. Res.* **2002**, *139*, 469–489. [CrossRef]
3. Stephan, M.; Steinmetz, U.; Schinsky, J. *Zusätzliche Einnahmen aus Windparks durch Verbesserte Datenanalyse*; VGB Powertech: Essen, Germany, 2015; pp. 62–67.
4. Bangalore, P.; Tjernberg, L.B. An Artificial Neural Network Approach for Early Fault Detection of Gearbox Bearings. *IEEE Trans. Smart Grid* **2015**, *6*, 980–987. [CrossRef]
5. Liu, W.Y.; Tang, B.P.; Han, J.G.; Lu, X.N.; Hu, N.N.; He, Z.Z. The structure healthy condition monitoring and fault diagnosis methods in wind turbines: A review. *Renew. Sustain. Energy Rev.* **2015**, *44*, 466–472. [CrossRef]
6. Martinez-Luengo, M.; Kolios, A.; Wang, L. Structural health monitoring of offshore wind turbines: A review through the Statistical Pattern Recognition Paradigm. *Renew. Sustain. Energy Rev.* **2016**, *64*, 91–105. [CrossRef]
7. Lau, B.C.P.; Ma, E.W.M.; Pecht, M. Review of offshore wind turbine failures and fault prognostic methods. In Proceedings of the IEEE 2012 Prognostics and System Health Management Conference (PHM-2012 Beijing), Beijing, China, 23–25 May 2012; pp. 1–5. [CrossRef]
8. Helbing, G.; Ritter, M. Deep Learning for fault detection in wind turbines. *Renew. Sustain. Energy Rev.* **2018**, *98*, 189–198. [CrossRef]
9. Zaher, A.; McArthur, S.; Infield, D.G.; Patel, Y. Online wind turbine fault detection through automated SCADA data analysis. *Wind Energy* **2009**, *12*, 574–593. [CrossRef]
10. Schlechtingen, M.; Ferreira Santos, I. Comparative analysis of neural network and regression based condition monitoring approaches for wind turbine fault detection. *Mech. Syst. Signal Process.* **2011**, *25*, 1849–1875. [CrossRef]
11. Zhao, H.; Liu, H.; Hu, W.; Yan, X. Anomaly detection and fault analysis of wind turbine components based on deep learning network. *Renew. Energy* **2018**, *127*, 825–834. [CrossRef]
12. Vogt, S.; Berkhout, V.; Lutz, A.; Zhou, Q. Deep Learning Based Failure Prediction in Wind Turbines Using SCADA Data. In Proceedings of the 4th Conference for Wind Power Drives 2019, Aachen, Germany, 12–13 March 2019.
13. Deeskow, P.; Steinmetz, U. Prädiktive Instandhaltung auf Basis von Big Data und Machine Learning: Monitoring mit Plus an Intelligenz, Effizienz und Wirtschaftlichkeit. *Ew Magazin für die Energiewirtschaft* **2019**, *7–8*, 46–50.
14. Jiang, G.; Xie, P.; He, H.; Yan, J. Wind Turbine Fault Detection Using a Denoising Autoencoder With Temporal Information. *IEEE/ASME Trans. Mechatron.* **2018**, *23*, 89–100. [CrossRef]
15. Wu, X.; Jiang, G.; Wang, X.; Xie, P.; Li, X. A Multi-Level-Denoising Autoencoder Approach for Wind Turbine Fault Detection. *IEEE Access* **2019**, *7*, 59376–59387. [CrossRef]
16. Odgaard, P.F.; Johnson, K.E. Wind turbine fault detection and fault tolerant control—An enhanced benchmark challenge. In Proceedings of the American Control Conference (ACC) 2013, Washington, DC, USA, 17–19 June 2013; IEEE: Piscataway, NJ, USA, 2013; pp. 4447–4452. [CrossRef]
17. Lutz, M.A.; Görg, P.; Faulstich, S.; Cernusko, R.; Pfaffel, S. Monetary–based availability: A novel approach to assess the performance of wind turbines. *Wind Energy* **2019**, *56*, 226. [CrossRef]
18. Kingma, D.P.; Ba, J. Adam: A Method for Stochastic Optimization. *arXiv* **2014**, arXiv:1412.6980.
19. Butler, R.W.; Davies, P.L.; Jhun, M. Asymptotics For The Minimum Covariance Determinant Estimator. *Ann. Stat.* **1993**, *1993*, 1385–1400. [CrossRef]
20. International Electrotechnical Commission. *61400-25-2: Communications for Monitoring and Control of Wind Power Plants—Information Models: 2016-06*; International Electrotechnical Commission: Geneva, Switzerland, 2016.

21. Fördergesellschaft Windenergie und andere Dezentrale Energien. *TG 7—Technical Guidelines for Power Generating Units—State-Event-Cause Code for Power Generating Units (ZEUS) Rubric D2—Maintenance of Power Plants for Renewable Energy, Category D2: State-Event-Cause Code for Power Generating Units (ZEUS)—Terms, Classification and Structuring of States, Events, Causes and Measures for Future Assessments and Improvements in Operation and Maintenance*; Fördergesellschaft Windenergie und andere Dezentrale Energien: Berlin, Germany, 1 October 2013.
22. British Standards Institution. *Maintenance—Maintenance Terminology*; Technical Report 13306; British Standards Institution: London, UK, 2010.
23. Pfaffel, S.; Faulstich, S.; Rohrig, K. Performance and Reliability of Wind Turbines: A Review. *Energies* **2017**, *10*, 1904. [CrossRef]

Article

Research on the Fault Characteristic of Wind Turbine Generator System Considering the Spatiotemporal Distribution of the Actual Wind Speed

Xiaoling Sheng *, Shuting Wan *, Kanru Cheng and Xuan Wang

School of Energy Power and Mechanical Engineering, North China Electric Power University, Baoding 071003, China; 2172224057@ncepu.edu.cn (K.C.); 2182224004@ncepu.edu.cn (X.W.)
* Correspondence: 52451712@ncepu.edu.cn (X.S.); 52450809@ncepu.edu.cn (S.W.);
  Tel.: +86-1365-312-3501 (X.S.); +86-1358-299-6591 (S.W.)

Received: 29 October 2019; Accepted: 7 January 2020; Published: 10 January 2020

Abstract: A reliable fault monitoring system is one of the conditions that must be considered in the design of large wind farms today. The most important factor for the fault monitoring should be the accurate diagnosis criteria with sensitive fault characteristics. Most of the current fault diagnosis criteria are obtained based on the average wind speed at the center of the hub which is not in accord with the actual wind condition in nature. So, this paper utilizes an equivalent wind speed (EWS), which can describe the actual wind speed spatiotemporal distribution on the rotor disk area considering the effects of wind shear and tower shadow, to analyze the common mechanical and electrical faults again. Firstly, the EWS model applicable to the 3-blade wind turbines is introduced; then the new fault characteristics of the wind turbine rotor aerodynamic imbalance and the stator winding asymmetry are theoretically analyzed based on the EWS model; finally, the simulation platform is built in Matlab/Simulink for comparison and the simulation result is well consistent with the theory analysis. The aim of this research is to find more accurate fault characteristics and help promoting the healthy development of wind power industry.

Keywords: wind turbines; equivalent wind speed; rotor aerodynamic imbalance; stator winding asymmetry; fault characteristics

---

1. Introduction

With the development of wind turbines (WT) in the direction of large-scale, the towers are getting higher and higher, and the blade radius is getting longer and longer, which makes the influence of wind shear and tower shadow effects on the aerodynamic load of the WT more obvious. Wind shear describes the variation in wind speed with vertical elevation, whereas tower shadow reflects the reassignment of wind speed due to the presence of a tower. Due to the factors such as wind shear and tower shadow, the actual wind speed is different everywhere on the rotor disk area, and the difference will change greatly with the WT geometry such as the blade radius and the tower outer diameter, as well as the vegetation and other geography [1]. Therefore, the wind speed in nature has been changing with time and space, which can be called the spatiotemporal distribution characteristic of wind speed.

As shown in Figure 1, an anemometer is usually installed near the hub center point O to detect the real-time wind speed, and then the average value over a period of time is taken as the average wind speed (AWS) at the center of the hub. However, the wind speed at this point does not completely reflect the wind speed on the whole rotor disk area. In particular, the rotor dimension of the modern WT is very large, and the wind speed differs across the rotor disk area and changes randomly. Thus, the wind turbine drive torque, the pitch bending moment, and the yaw moment, and so on, will change accordingly. The wind speed spatiotemporal distribution characteristics directly affect the aerodynamic load resulting in load fluctuation of the WT [2–4]. Besides, it will also affect the power output of the wind turbine generator system (WTGS) [1,5–7]. The influence of wind shear and tower shadow effects on power in terms of power fluctuation [1], power loss [5], and average output power [6] have been investigated in detail and a frequency domain approach for evaluating the impact of tower shadow and wind shear on the tie-line power oscillations was described in [7]. They concluded that both wind shear and tower shadow were sources of periodic power fluctuations and average power loss.

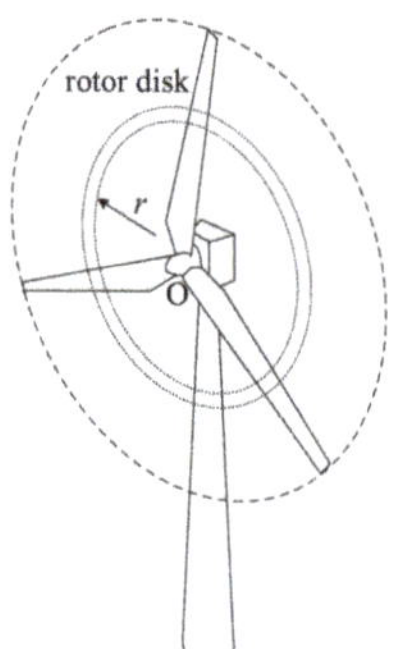

Figure 1. Wind turbine schematic diagram.

From the above researches, it can be inferred that the WTGS fault characteristic may be influenced by wind speed spatiotemporal distribution condition. However, this problem is seldom discussed in previous studies. In [8], the impact of blade mass imbalance fault on the power characteristics of a doubly-fed induction generator (DFIG) was analyzed considering the effects of wind shear and tower shadow. In [9], the influence factors causing voltage flicker of WTGS were analyzed, including wind shear, tower shadow, gearbox tooth, and blade break down. And a classifier algorithm which can detect different causes of flicker was proposed. In [10], the torque and vibration characteristics of the wind turbines drive train during voltage dips were investigated considering the wind shear and tower shadow effects. The other similar researches about this topic are truly seldom reported by now.

Currently, most of the mechanical and electrical fault diagnoses for WT with DFIG are usually based on the average wind speed at the center of the hub. Since the spatiotemporal distribution of wind speed is not considered, it may cause the analytical method or the diagnosis result to deviate from the actual situation, thus affecting the accuracy of the diagnosis. In order to better understand the fault characteristics on the condition of the actual wind speed considering the spatiotemporal distribution, two faults are selected from the common mechanical and electrical faults of the WTGS for comparative analysis. These two faults are the rotor aerodynamic imbalance of WT and the stator winding asymmetry fault of DFIG.

Aerodynamic imbalance means the aerodynamic torques of the three blades are unevenly distributed, which is different from the blade mass imbalance in [8]. The reason may be a pitch angle difference among the three blades, caused by manufacturing or control errors, or blade airfoil changes due to icing on the blade and other factors. Aerodynamic imbalance can cause the main shaft vibration and further aggravate the fatigue of the blade, bearing, gear, and other parts. At present, the aerodynamic imbalance fault study is mainly focused on two aspects. The first is to analyze the vibration characteristics of the WT to diagnose the aerodynamic imbalance fault [11,12], and the second is to study the electrical characteristics under the aerodynamic imbalance fault [13,14]. These researches on the aerodynamic imbalance fault are mostly based on the AWS at the hub center, and the spatiotemporal distribution characteristics of the actual wind speed are less considered.

It is reported that about 38% of the WTGS faults are related to the stator according to the failure statistics [15], and the stator faults is very critical because its important role in the WTGS. To avoid severe damage to stator and the WTGS, the early fault detecting such as the winding asymmetry and resistance variations may have important significance. Because the early fault characteristic is little evident, the diagnostic accuracy becomes very important. Stefani et al. [16,17] researched on the stator winding asymmetrical fault based on the frequency analysis of the rotor modulating signals. Dai [18] used the modified Hilbert–Huang Transform method to analyze the stator current in order to make the fault characteristics more obvious. Williamson et al. [19] derived simple expressions for the frequencies of the harmonic components in the steady state stator line current of a DFIG operating under various conditions of winding asymmetry. These studies have helped us understand the characteristics of stator winding asymmetry faults. However, they do not consider the spatiotemporal distribution of actual wind speed. Although Gritli et al. [20] researched the stator winding fault by using wavelet analysis under time-varying conditions, the wind shear and tower shadow were not considered.

A wind speed model named equivalent wind speed (EWS) considering the effects of wind shear and tower shadow is established in our researches [21,22] based on the relevant researches [23–25]. However, these researches only study the establishment of the wind speed model. The rear drive trains system, the generator and the control system are not taken into account in these researches. To this end, the EWS model is added to the DFIG and its control system, and a complete WTGS model with DFIG is built in this paper. Then the influence of the EWS on the mechanical and electrical fault of the WTGS is studied. Since the motor current signature analysis (MCSA) method is widely used and easier to measure, the DFIG current characteristics of the faults are analyzed mainly in the paper. The main objective of the research is to provide theoretical support for optimizing the fault monitoring system of the WTGS in the future.

2. Equivalent Wind Speed

Considering the effects of wind shear and tower shadow, the EWS of the 3-blade WT can be expressed as the sum of the average wind speed at the hub, the wind shear component and the tower shadow component:

$$V_{eq} = V_H + V_{ws} + V_{ts} \tag{1}$$

where V_{eq} is the equivalent wind speed (m/s); V_H is the wind speed at the height of hub center (m/s); V_{ws} is the wind shear fluctuation component (m/s); and V_{ts} is the tower shadow fluctuation component (m/s).

According to the former research basis, the expression corresponding to the V_{ws} and V_{ts} can be respectively expanded as [21–23]:

$$V_{ws} = V_H \left[\frac{\alpha(\alpha-1)}{8} \left(\frac{r}{H}\right)^2 + \frac{\alpha(\alpha-1)(\alpha-2)}{60} \left(\frac{r}{H}\right)^3 \cos 3\beta + \frac{\alpha(\alpha-1)(\alpha-2)(\alpha-3)}{576} \left(\frac{r}{H}\right)^4 \cos 4\beta \right] \tag{2}$$

$$V_{ts} = \frac{MV_H}{3R^2} \sum_{b=1}^{3} \left[\frac{A^2}{\sin^2 \beta_b} \ln \frac{R^2 \sin^2 \beta_b + x^2}{x^2} - \frac{2A^2 R^2}{R^2 \sin^2 \beta_b + x^2} \right]$$
$$\beta_1 = \beta \ \beta_2 = \beta_1 + \frac{2\pi}{3} \ \beta_3 = \beta_1 + \frac{4\pi}{3} \tag{3}$$

$$M = 1 + \frac{\alpha(\alpha-1)R^2}{8H^2} \tag{4}$$

where R is the blade radius (m); r is the distance from the blade element to the hub center (m), as shown in Figure 1, range from 0 to R; H is the hub height (m); α is the wind shear exponent; A is the tower radius (m); x is the distance from the rotor disk plane to the tower center line (m); β is the azimuth angle (rad); β_b (b = 1,2,3) is the azimuth angle corresponding to the three blades (rad), and $\beta_1 = \beta$, $\beta_2 = \beta_1 + 2\pi/3$, $\beta_3 = \beta_1 + 4\pi/3$.

Substitute the Equations (2) and (3) into (1) and extract V_H from the Equations, and we can get the EWS expressions as follows:

$$V_{eq} = W_{eq} V_H \tag{5}$$

$$W_{\text{eq}} = 1 + \left[\frac{\alpha(\alpha-1)}{8}\left(\frac{r}{H}\right)^2 + \frac{\alpha(\alpha-1)(\alpha-2)}{60}\left(\frac{r}{H}\right)^3 \cos 3\beta + \frac{\alpha(\alpha-1)(\alpha-2)(\alpha-3)}{576}\left(\frac{r}{H}\right)^4 \cos 4\beta\right]$$
$$+ \frac{M}{3R^2}\sum_{b=1}^{3}\left[\frac{A^2}{\sin^2\beta_b}\ln\frac{R^2\sin^2\beta_b+x^2}{x^2} - \frac{2A^2R^2}{R^2\sin^2\beta_b+x^2}\right] \tag{6}$$

W_{eq} can be called the equivalent wind speed transform coefficient. The EWS model established by Equations (5) and (6) can calculate the wind speed value at any point on the entire rotor disk area. It is an accurate and practical wind speed calculation model. Applying it to the study of the fault diagnosis, it can reflect the wind load conditions of the WT and the fault characteristics.

According to the parameters of a 3-blade 1.5 MW WT, the distribution map of the equivalent wind speed transform coefficient is plotted, as shown in Figure 2. It can be seen from the figure that due to the effects of wind shear and tower shadow, the wind speed on the rotor disk area is not uniform, and there is obvious periodic fluctuation on the time axis, and the main frequency is three times of the rotor rotating frequency.

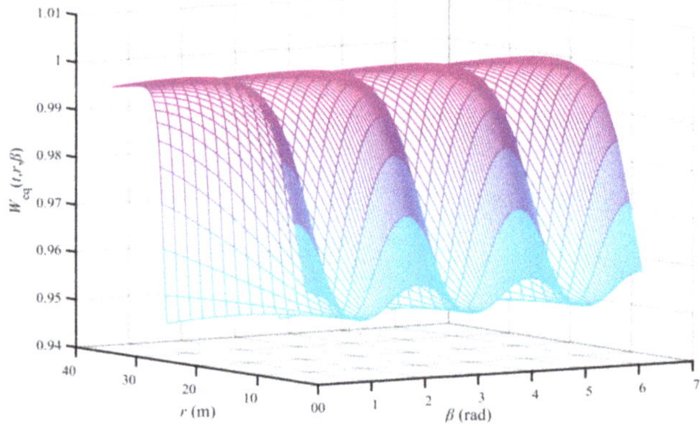

Figure 2. Equivalent wind speed transform coefficient W_{eq}.

In order to facilitate the subsequent analysis, the EWS calculation equation is simplified, and the Fourier fitting is performed on the EWS curve corresponding to the WT whose blade radius is 35 m, and V_{H} is 11 m/s. The fitting curve and the original EWS curve are shown in Figure 3.

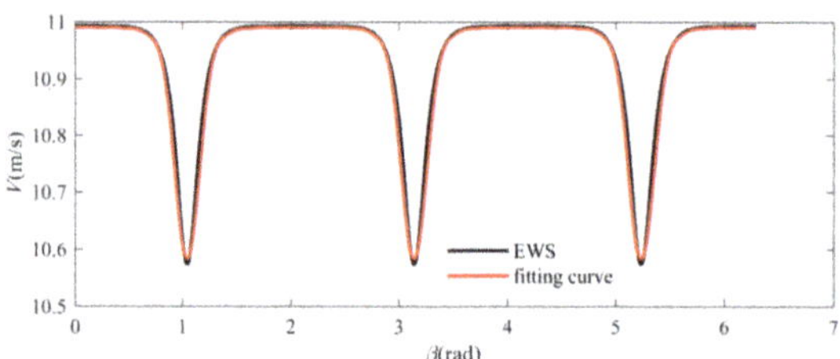

Figure 3. Original equivalent wind speed curve and the fitting curve.

The fitting curve is highly coincident with the original EWS curve, and the equation corresponding to the fitting curve is:

$$V = a_0 + a_1\cos(\omega t) + a_2\cos(2\omega t) + \ldots\ldots + a_k\cos(k\omega t) \quad k = 1,2,3\ldots \tag{7}$$

The equation of the fitting curve contains only the constant term and the trigonometric term, where the constant term a_0 is close to the hub average wind speed $V_{\text{H}} = 11$ m/s; the ω in trigonometric term has a value of 9.42, which is close to three times of the rotor rotating angular frequency ($3\omega_{\mathrm{w}}$), so the trigonometric term can be written as $\cos(3k\omega_{\mathrm{w}}t)$, and the more the number of trigonometric terms, the higher the fitting accuracy of the fitted curve and the original curve. In the above figure, when the number of trigonometric terms reaches 8, the fitting accuracy has reached 0.99. The Fourier fitting is also performed on the other EWS curve corresponding to the WT of different parameters, which is basically consistent with the above conclusion, but the number of trigonometric terms should be adjusted according to the actual situation. Therefore, V_{eq} can be approximated as:

$$V_{\mathrm{eq}} = V_{\mathrm{H}} + \sum_{k=1}^{n} V_k \cos(3k\omega_{\mathrm{w}}t + \phi_k) \tag{8}$$

where V_k and φ_k are respectively the amplitudes (m/s) and the phase angles (rad) corresponding to the k-th trigonometric term; and ω_{w} is the rotor rotating angular frequency (rad/s).

The Equation (8) is only used to facilitate the following analysis, and the final simulation adopts the Equations (1)–(6) to build the equivalent wind speed model.

According to the wind turbine aerodynamics theory, there is a proportional relationship between the mechanical torque output by the wind turbine and the quadratic wind speed. So, combining the Equation (8), the output mechanical torque T_{m} can be obtained as follows:

$$T_{\mathrm{m}} = T_{\mathrm{m}0} + \frac{2T_{\mathrm{m}0}}{V_{\mathrm{H}}} \sum_{k=1}^{n} V_k \cos(3k\omega_{\mathrm{w}}t + \phi_k) + \frac{T_{\mathrm{m}0}}{V_{\mathrm{H}}^2} \left[\sum_{k=1}^{n} V_k \cos(3k\omega_{\mathrm{w}}t + \phi_k) \right]^2 \tag{9}$$

where $T_{\mathrm{m}0}$ is the fundamental component of aerodynamic torque (N·m); $T_{\mathrm{m}0} = 0.5\rho\pi R^3 C_{\mathrm{p}} V_{\mathrm{H}}^2/\lambda$; ρ is the air density (kg/m^3); C_{p} is the optimum power coefficient; and λ is the optimum tip speed ratio.

The third term in Equation (9) is also a polynomial containing $\cos(3k\omega_{\mathrm{w}}t)$ after expansion. Then the expression of the mechanical torque obtained by combining the second and third terms is:

$$T_{\mathrm{m}} = T_{\mathrm{m}0} + \sum_{k=1}^{n} T_k \cos(3k\omega_{\mathrm{w}}t + \phi_k) \tag{10}$$

where T_k and ϕ_k are respectively the amplitudes (N·m) and the phase angles (rad) of the torque oscillation components caused by the effects of wind shear and tower shadow in the EWS.

3. Rotor Aerodynamic Imbalance and Stator Winding Asymmetry

Taking two common faults as example, this paper focuses on analyzing the different points of fault characteristics under different wind conditions of EWS and AWS, so that we can grasp the influence of EWS on the mechanical and electrical fault characteristics.

3.1. Rotor Aerodynamic Imbalance

Based on the analysis above mentioned and the Reference [13], the mechanical torque under the rotor aerodynamic imbalance and EWS can be expressed as (omitting the initial phase angle):

$$T_{\mathrm{m}} = (T_{\mathrm{m}0} - T_{\mathrm{im}}) + \sum_{k=1}^{n} T_k \cos(3k\omega_{\mathrm{w}}t) + T_{\mathrm{a}} \cos(\omega_{\mathrm{w}}t) \tag{11}$$

where T_{im} is the constant variation caused by aerodynamic imbalance (N·m); and T_{a} is the amplitude of the torque oscillation components caused by aerodynamic imbalance (N·m).

For the convenience of calculation, the above mechanical torque is simplified as follows:

$$\begin{cases} T_{\rm m} = (T_{\rm m0} - T_{\rm im}) + \sum_{k=1}^{n+1} T_k \cos(\omega_k t); \\ \omega_k = 3k\omega_{\rm w}; \ (k = 1,2,\ldots n) \\ \omega_{n+1} = \omega_{\rm w}; \\ T_{n+1} = T_{\rm a}; \end{cases} \tag{12}$$

Then calculate the DFIG rotor electrical angular speed $\omega_{\rm r}$ according to the motion equation of the DFIG [26],

$$\omega_{\rm r} = \frac{n_{\rm p}}{J} \int (T_{\rm m0} - T_{\rm im} - T_{\rm e0}) {\rm d}t + \frac{n_{\rm p}}{J} \int \sum_{k=1}^{n+1} T_k \cos(\omega_k t) {\rm d}t = \omega_{\rm r0} + \sum_{k=1}^{n+1} \frac{n_{\rm p} T_k}{J\omega_k} \sin(\omega_k t) \tag{13}$$

where $T_{\rm eo}$ is the electromagnetic torque of the DFIG (N·m); $n_{\rm p}$ is the pole pairs; and J is the equivalent moment of inertia of the WT (kg·m^2).

Let ω_1 be the angular frequency of the grid, according to the speed-frequency relationship of the DFIG, the angular frequency of the rotor current $\omega_{\rm z}$ can be obtained:

$$\omega_{\rm z} = \omega_1 - \omega_{\rm r} = \omega_{\rm z0} - \sum_{k=1}^{n+1} \frac{n_{\rm p} T_k}{J\omega_k} \sin(\omega_k t) \tag{14}$$

where $\omega_{\rm z0} = \omega_1 - \omega_{\rm r0}$, which is the fundamental frequency of rotor current (rad/s).

Then the DFIG rotor current $i_{\rm ra}$ under the EWS and aerodynamic imbalance can be obtained (the detailed computation process can be found in Appendix A):

$$i_{\rm ra} = I_{\rm r} \cos(\omega_{\rm z0} t) - \sum_{k=1}^{n} I_k \sin(\omega_{\rm z0} t + 3k\omega_{\rm w} t) - \sum_{k=1}^{n} I_k \sin(\omega_{\rm z0} t - 3k\omega_{\rm w} t) - I_{\rm r1} \sin(\omega_{\rm z0} t + \omega_{\rm w} t) - I_{\rm r1} \sin(\omega_{\rm z0} t - \omega_{\rm w} t) \tag{15}$$

where $I_{\rm r}$ is the amplitude of rotor current fundamental wave (A); $I_k = I_{\rm r} n_{\rm p} T_k / 18 J k^2 \omega_{\rm w}^2$; $I_{\rm r1} = I_{\rm r} n_{\rm p} T_{\rm a} / 2 J \omega_{\rm w}^2$.

As indicated in Equation (15), in addition to the fundamental current at frequency $\omega_{\rm z0}$, there are also modulation harmonic components at the frequencies of $\omega_{\rm z0} + 3k\omega_{\rm w}$, $\omega_{\rm z0} - 3k\omega_{\rm w}$, $\omega_{\rm z0} + \omega_{\rm w}$ and $\omega_{\rm z0} - \omega_{\rm w}$ in the rotor current. In addition, it is necessary to notice that $3k\omega_{\rm w}$ and $\omega_{\rm w}$ are the main modulation harmonic frequencies, but not only these two types of modulation frequencies in the rotor current. After these harmonics appear, frequency modulation occurs between the two kinds of harmonics, that is, new modulation frequencies appear in the current: $3k\omega_{\rm w} \pm \omega_{\rm w}$. In the Reference [27], it is studied that in the case of mass imbalance fault, many high-frequency small-amplitude components appear in the current besides the main modulation frequency $\omega_{\rm w}$. Therefore, in the case of EWS, the harmonic component at the frequency $\omega_{\rm z0} \pm (3k\omega_{\rm w} \pm \omega_{\rm w})$ in the current is composed of two parts. One part is caused by aerodynamic imbalance; the other part is caused by the modulation between $3k\omega_{\rm w}$ and $\omega_{\rm w}$.

To obtain a steady electro-mechanical energy conversion, and keep the rotating magnetic field between the stator and rotor relatively static, there should be modulation harmonic components at the frequencies of $\omega_1 \pm 3k\omega_{\rm w}$ and $\omega_1 \pm \omega_{\rm w}$ in addition to the fundamental frequency ω_1 in the stator current. The small harmonics at the frequencies of $\omega_1 \pm (3k\omega_{\rm w} \pm \omega_{\rm w})$ will also be observed on the condition of the aerodynamic imbalance fault under the EWS.

3.2. Stator Winding Asymmetry

First we analyze the stator winding asymmetry fault of the DFIG on the condition of constant AWS at the hub centre. Firstly, an inverse sequence component at the frequency $-\omega_1$ in the stator determines an inverse counter rotating magnetic field. Then the inverse sequence component produces an harmonic component in the rotor at frequency $(2 - s)\omega_1$ (s is the slip ratio) and give rise to electromagnetic and mechanical interaction between stator and rotor, which determine the further harmonic components both on stator and rotor [10]. As a consequence of this interaction, the following stator current components ω_{ss} and rotor current components ω_{sr} appear:

$$\omega_{ss} = \pm k_1 \omega_1 \quad (k_1 = 1,3,5\ldots) \tag{16}$$

$$\omega_{sr} = (2k_2 \pm s)\omega_1 \quad (k_2 = 1,2,3\ldots) \tag{17}$$

Then we will analyze the stator winding asymmetry fault of the DFIG on the condition of EWS. According to the above analysis, the inverse sequence component at the frequency $-\omega_1$ is first generated in the stator current when the stator winding asymmetry fault occurs, and then the corresponding rotor current frequency ω_{z1} can be obtained under the EWS according to the speed-frequency relationship of the DFIG and the Equations (10) and (13):

$$\omega_{z1} = \omega_1 + \omega_{r0} + \sum_{k=1}^{n} \frac{n_p T_k}{3Jk\omega_w} \sin(3k\omega_w t) = (2-s)\omega_1 + \sum_{k=1}^{n} \frac{n_p T_k}{3Jk\omega_w} \sin(3k\omega_w t) \tag{18}$$

Then, with reference to Equation (15), we can learn that, in addition to the fault frequency $(2-s)\omega_1$ caused by the stator winding asymmetry, there are also harmonic frequencies $(2-s)\omega_1 \pm 3k\omega_w$ caused by the EWS in the rotor current. However, the harmonic analysis is not over yet. The harmonic components in the rotor current will continue to induce harmonic at the frequency $3\omega_1$ and $3\omega_1 \pm 3k\omega_w$ in the stator current. Repeatedly, the harmonics in the stator and rotor currents will continue to propagate according to this law, and finally the following harmonic frequencies in the stator and rotor currents are generated:

$$\omega_{sse} = \pm k_1 \omega_1 \pm 3k\omega_w \quad (k_1 = 1,3,5\ldots) \tag{19}$$

$$\omega_{sre} = (2k_2 \pm s)\omega_1 \pm 3k\omega_w \quad (k_2 = 1,2,3\ldots) \tag{20}$$

4. Simulation Analysis

In order to verify the correctness of the theoretical analysis described above, a simulation platform of 1.5 MW WTGS with DFIG is built in the MATLAB/Simulink environment (MATLAB R2017a, MathWorks Company, Natick, MA, USA). The sketch of the simulation platform is shown in Figure 4. The simulation parameters are shown in Table 1.

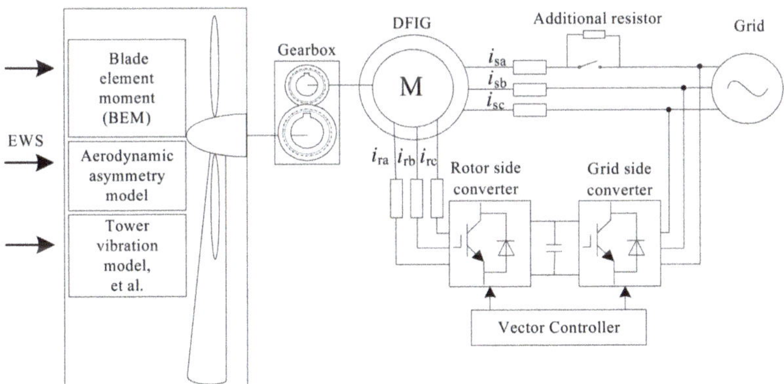

Figure 4. Simulation platform.

Table 1. Parameters of 1.5 MW doubly-fed induction generator (DFIG) wind turbine.

Parameters	Value	Parameters	Value
Rated power (MW)	1.5	Stator resistance (p.u.)	0.023
Rated wind speed (m/s)	11	Stator leakage inductance (p.u.)	0.18
Optimum tip speed ratio	10	Rotor resistance(p.u.)	0.016
Optimum power coefficient	0.5	Rotor leakage inductance (p.u.)	0.16
Blade radius (m)	35	Pairs of poles	2
Tower average radius (m)	1.7	Inertia constant (s)	0.685
Centre height of the hub (m)	70	Air density (kg/m^3)	1.225

The simulation platform mainly includes the EWS model, WT aerodynamics model, gearbox model, DFIG model, and vector control model. Among them, the WT aerodynamics model includes the blade element moment theory, the tower vibration model, and the coordinate transformation model, etc. This aerodynamics model can calculate the aerodynamic torque and other parameters of the WT output under the aerodynamic imbalance caused by the inconsistent pitch angle among three blades, and the detail can be found in our article Reference [13].

As far as the simulation of stator winding faults, two methods are usually used. One method is to simulate the stator inter-turn short circuit fault by changing the number of shorted turns, and the other is to add an additional resistor or inductor in series with the stator winding to simulate the winding asymmetry fault. Since the second method is simple and easy to implement, it is applied in many situations. For example, References [16–18,20] all adopt the second method. So, in this paper, the stator winding asymmetry fault is simulated by connecting an additional resistor in series with the stator phase A. This method is only used for the simulation of stator winding asymmetry originating from the resistance variations because of unreasonable structure design and electromagnetism thermal field, et al. The stator inter-turn short circuit fault is not considered here.

4.1. Normal Condition Simulation

Firstly, the simulation of normal running condition is performed under two different kinds of wind speed—AWS and EWS. The average constant wind speed at the hub centre is 12 m/s, and the rotor rotating speed in the normal condition is 30 r/min (the corresponding rotating frequency P is 0.5 Hz). The parameters using in the EWS model are: wind shear exponent is 0.4; the distance from the tower middle line to the blade is 4.5 m; tower radius is 1.7 m; blade radius is 35 m; the hub height is 70 m; the wind speed at the hub centre is 12 m/s. The comparison results are shown in Figures 5–8.

Figure 5 is the fast Fourier transform (FFT) spectrum of the mechanical torque output from the WT. It can be seen from the figure that there are obvious harmonic fluctuation components in the torque due to the influence of wind shear and tower shadow effects in the EWS, and the harmonic frequencies are 1.5 Hz, 3 Hz … 3*k*P (*k* takes positive number). Besides, the harmonic with frequency of 3P (1.5 Hz) is the main component. Figure 6 shows the FFT spectrum of the rotor rotating speed, which has the similar result as the torque, and also has distinct harmonic components of 3*k*P.

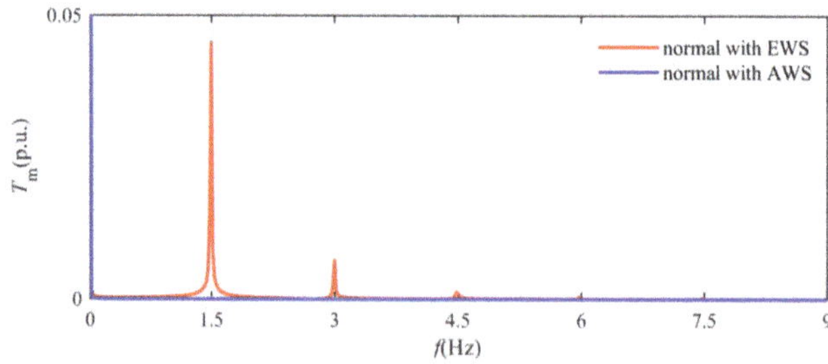

Figure 5. Fast Fourier transform (FFT) of the wind turbine (WT) mechanical torque.

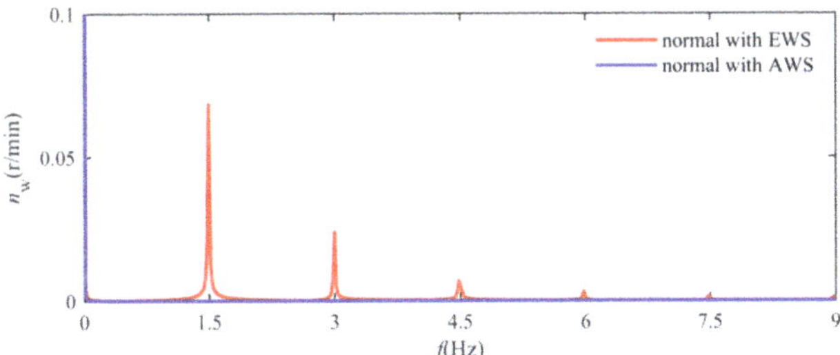

Figure 6. FFT of the WT rotor rotating speed.

Figure 7 shows the power spectral density (PSD) analysis results of the stator current. On the normal condition with EWS, there exists distinct harmonics in the stator current besides the fundamental wave. The harmonic frequencies in red line around the fundamental frequency are $50 \pm 3kP$ ($1.5k$). That is to say, the modulation frequency is $3kP$ which is three times of the rotor rotating frequency. However, these harmonics are not present on the normal condition with AWS as shown in the blue line in Figure 7.

Figure 8 shows the PSD analysis results of the rotor current. Similar to the analysis of the stator current, the harmonic with modulation frequency of $3kP$ appear on both sides of the fundamental frequency (10 Hz) of the rotor current. However, unlike the stator current results, the rotor current harmonic amplitudes are relatively high compared with the fundamental amplitude. In the stator current, the harmonic amplitude at frequency $50 + 3P$ is about 0.004 p.u., which is about 0.58% of the fundamental amplitude. However, in the rotor current, the harmonic amplitude at frequency $10 + 3P$ is about 0.0335 p.u., which is about 4.55% of the fundamental amplitude. The reason may be that the rotor side converter mainly controls the stator current, so the harmonic performance in the rotor current is relatively obvious than that of the stator current. The similar effect can be observed in the PSD spectrum. In Figure 7, the amplitude at the 51.5 Hz is about −24 dB. However, the amplitude at 11.5 Hz is about −14.75 dB in Figure 8, which is higher than that of the stator current.

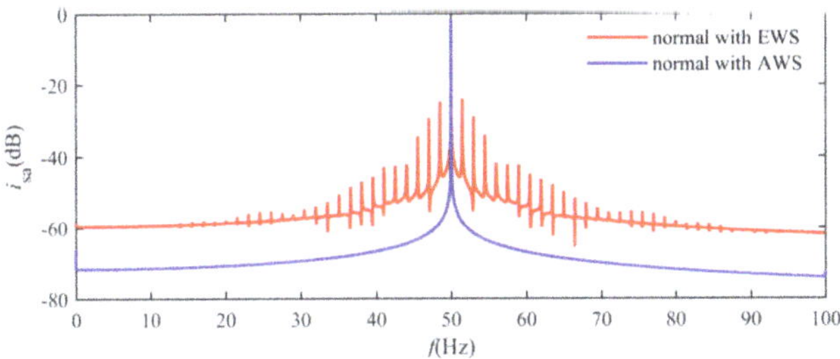

Figure 7. Stator current comparison on normal condition.

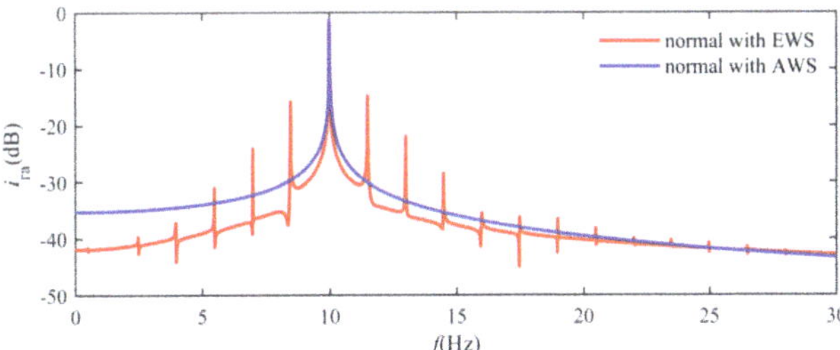

Figure 8. Rotor current comparison on normal condition.

4.2. Rotor Aerodynamic Imbalance Simulation

In this part, the rotor aerodynamic imbalance simulation is performed under AWS and EWS respectively. Under each case, three aerodynamic imbalance scenarios are simulated with the pitch angle of one blade adjusted by +1°, +2°, and +3°, respectively, while the other two blades are kept constant. The other running parameters are the same as the former mentioned in normal condition. Figures 9 and 10 show the simulation results of the stator current and the rotor current when the pitch angle is adjusted by +3°.

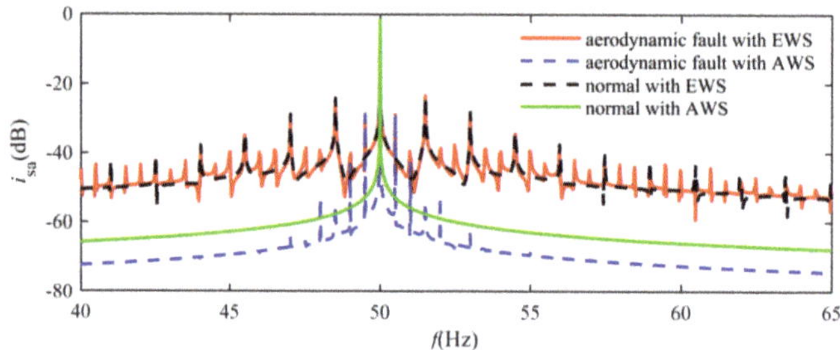

Figure 9. Stator current comparison under the aerodynamic imbalance.

Figure 9 shows the analysis results of the stator current PSD. In Figure 9, the comparison of four simulation results is given, including the aerodynamic imbalance under the EWS, the aerodynamic imbalance under the AWS, the normal operation under the EWS and the normal operation under the AWS.

It can be seen from Figure 9 that, on the condition of the aerodynamic imbalance under the AWS (in blue line), the fault harmonic frequencies in the stator current are mainly 49.5 and 50.5 Hz. That is to say, the modulation frequency is mainly the rotor rotating frequency P. The amplitudes at the modulation frequencies of 2P, 3P, 4P, 5P, and 6P are relatively small and the other frequencies are not observed in the figure. However, on the condition of aerodynamic imbalance under the EWS (in red line), the main modulation frequencies in the stator current are 3P (1.5 Hz), 6P (3 Hz) ... 3kP in addition to P. The modulation frequencies such as 2P, 4P, 5P, 7P, 8P, etc. also can be observed, however their amplitudes are relatively small compared with those of the main modulation frequencies. On the normal condition under EWS (in black line), there are only the modulation frequencies of 3kP, and on the normal condition under AWS (in green line), there is not harmonic frequency except the fundamental frequency of the stator current.

Figure 10 shows a comparison of DFIG rotor current under the four simulation conditions as mentioned above. It can be seen that, on the condition of the aerodynamic imbalance under the AWS, the harmonic frequencies of the rotor current are mainly 9.5 and 10.5 Hz, i.e., the modulation frequency is P. On the condition of aerodynamic imbalance under the EWS, the main modulation frequencies of the rotor current include 3P (1.5 Hz), 6P (3 Hz) ... 3kP in addition to P. The modulation frequencies of 2P, 4P, 5P, 7P, 8P, etc. can also be observed.

However, unlike the stator current, the main harmonic amplitudes in the rotor current are relatively high. On the condition of aerodynamic imbalance under EWS, the maximum harmonic amplitude (at 51.5 Hz) in the stator current is about 0.0044 p.u., which is about 0.66% of the fundamental amplitude. However, the maximum harmonic amplitude (at 8.5 Hz) in rotor current is about 0.036 p.u., which is about 5.7% of the fundamental amplitude. In the PSD spectrum of Figure 10, the harmonics amplitudes with the modulation frequencies of P and 3kP are higher than those of the stator current in Figure 9. The maximum harmonic at $f + 3P$ is −23.56 dB in Figure 9, and the maximum amplitude at $sf - 3P$ is −16.51 dB in Figure 10. The comparison data is shown in Table 2. Consequently, the harmonic performance in rotor current is more obvious than that of the stator current on the aerodynamic fault condition.

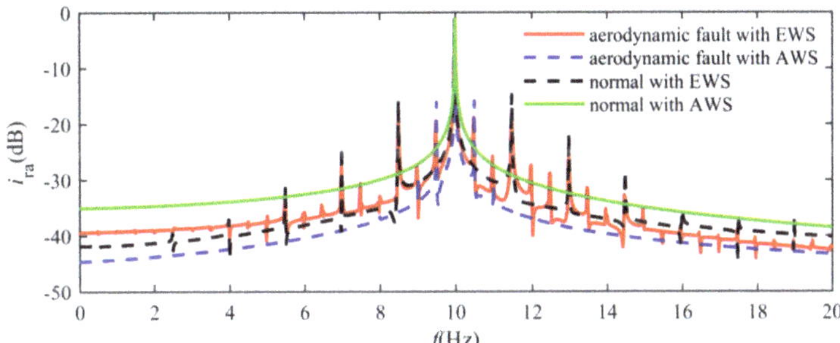

Figure 10. Rotor current comparison under the aerodynamic imbalance.

Table 2. Harmonic amplitude comparison between stator and rotor current.

Current	Ratio of the Maximum Harmonic to the Fundamental Amplitude	Maximum Harmonic Amplitude in PSD (dB)
stator current	0.66%	−23.56
rotor current	5.7%	−16.51

Figure 11 shows a comparison of the stator current PSD for three degrees of aerodynamic imbalance under EWS. It can be seen from the figure that the curves under the three imbalance degrees are basically consistent. The main difference is that the amplitude at the frequency $f \pm P$ increases as the imbalance degree increases. However, the amplitudes at the frequency $f \pm 3kP$ do not change much, mainly because they are caused by the effects of wind shear and tower shadow, and less affected by the aerodynamic imbalance. The rotor current comparison has the similar characteristic as the stator current under the three imbalance degrees.

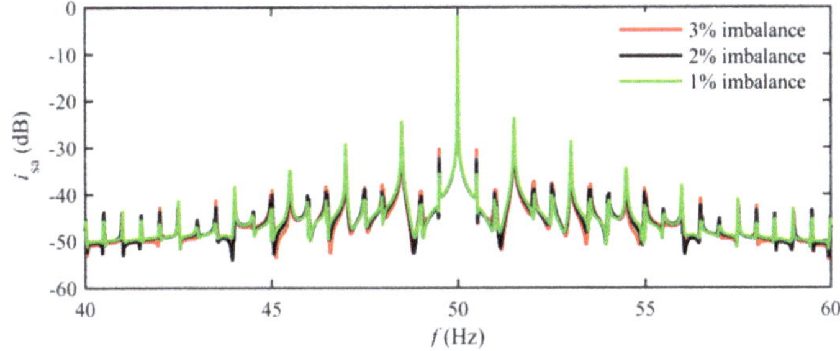

Figure 11. Stator current comparison for three aerodynamic imbalance degrees.

4.3. Stator Winding Asymmetry Simulation

Through the rotor aerodynamic imbalance fault simulation, the influence of the EWS on the mechanical fault characteristics of the WT is analyzed. Next, in order to analyze the influence on the electrical fault characteristics, the DFIG stator winding asymmetry fault is simulated.

The stator winding asymmetry has been simulated by means of an additional resistor connected in series with one stator phase winding and equal to the rated phase resistance Rs, and the resistances of the other two phases remain unchanged. Then the simulations under the two kinds of wind speed are respectively performed, which are the stator winding asymmetrical faults under the EWS and the conventional AWS. The simulation results are shown in Figures 12 and 13 below.

In Figure 12, the stator current PSD comparison of four cases is shown, including stator winding asymmetry fault under EWS, stator winding asymmetry fault under AWS, normal operation under EWS and normal operation under AWS. Figure 12a shows the result of the stator current PSD with frequency in the range of 0–300 Hz. But since the harmonic frequencies are not clear in this figure, the partial enlargement views of stator currents PSD are shown in Figure 12b,c with frequency ranges of 0–100 Hz and 120–180 Hz respectively.

There are two characteristics in the stator current PSD. Firstly, there is a clear difference between the fault and the normal condition, i.e., the odd-numbered harmonic frequencies of 150 Hz, 250 Hz, etc. exist in the stator current under the winding asymmetry fault, which is consistent with the theoretical analysis as mentioned above. Although only the fault frequencies of 150 Hz and 250 Hz are given in Figure 12a, there are other odd-numbered harmonic frequencies, which are not given because their amplitudes are relatively small. However, there are not these fault harmonics on the normal conditions. Secondly, there exists distinct difference between EWS and AWS on the condition of stator winding asymmetry fault. In the case of winding asymmetry fault with EWS, there are harmonics with modulation frequency of $3kP$ on both sides of the odd-numbered frequencies, while no such harmonics appear around the odd-numbered frequencies under stator winding asymmetry fault with AWS.

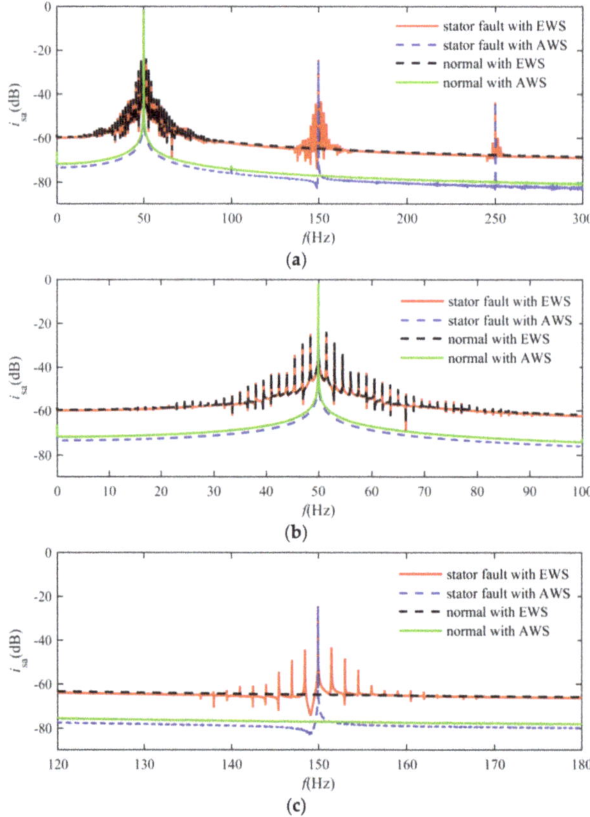

Figure 12. Stator current comparison under the stator winding asymmetry: (**a**) PSD of the stator current from 0 to 300 Hz; (**b**) PSD of the stator current from 0 to 100 Hz; (**c**) PSD of the stator current from 120 to 180 Hz.

Figure 13 shows the rotor current PSD comparison of the four cases as mentioned above. Figure 13a is the result of the rotor current PSD with the frequency in the range of 0 to 220 Hz. Figure 13b,c shows the partial enlargement views of the rotor current PSD with the frequency from 0 to 30 Hz and 80 to 120 Hz respectively.

The fault characteristic in the rotor current is basically the same as the stator current. From Figure 13a, it can be seen that the fault frequencies 90 Hz, 110 Hz, 190 Hz, 210 Hz ... $(2k \pm s)f$ (f is 50 Hz, and s is the slip ratio of -0.2) appear in the rotor current in the case of stator winding asymmetry fault except for the fundamental frequency of 10 Hz. However, there are not these fault frequencies under the normal conditions. In addition, it also can be seen from Figure 13 that there are harmonics with modulation frequency of $3kP$ on both sides of the fundamental frequency and the fault frequencies $(2k \pm s)f$ in the case of stator winding asymmetry fault with EWS. However, in the case of stator winding asymmetry with AWS, these modulation harmonics do not appear, and there are only fundamental frequency and the fault frequencies of $(2k \pm s)f$.

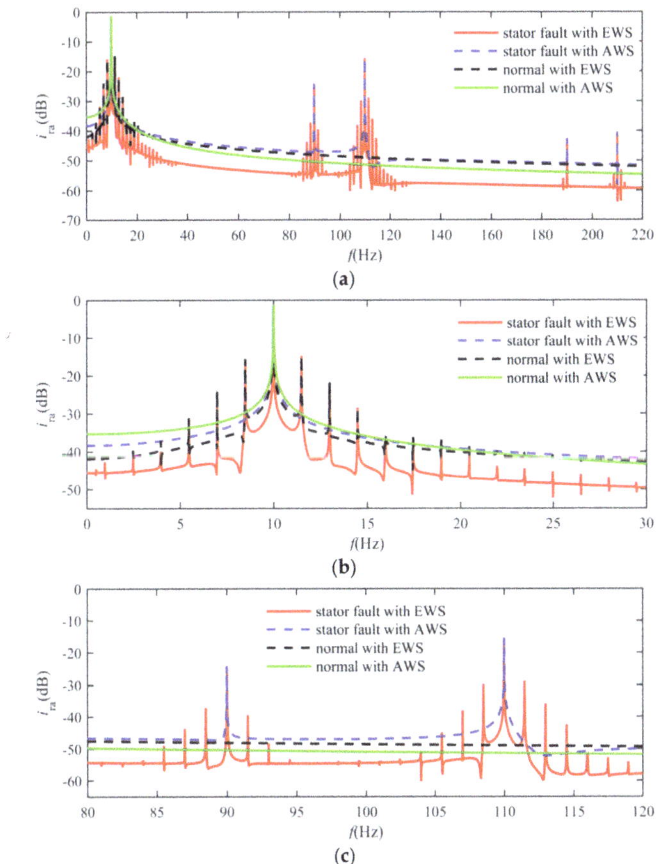

Figure 13. Rotor current comparison under the stator winding asymmetry: (**a**) PSD of the rotor current from 0 to 220 Hz; (**b**) PSD of the rotor current from 0 to 30 Hz; (**c**) PSD of the rotor current from 80 to 120 Hz.

5. Discussion

According to the analysis above mentioned, both the mechanical torque and the rotating speed of WT include the periodic fluctuation with the frequencies of $3kP$ (P is the rotor rotating frequency, k stands for positive integer) due to the effects of wind shear and tower shadow. Two typical mechanical and electrical faults under the EWS and the AWS are mainly studied, and the comparison analysis results are as follows:

Comparison results of the rotor aerodynamic imbalance simulation: in the case of AWS, the fault modulation frequency in stator and rotor currents is mainly P; and in the case of EWS, the fault modulation frequencies in stator and rotor currents are mainly P and $3kP$. In addition, there exists frequency modulation between P and $3kP$. The harmonic performance in rotor current is more obvious than that of the stator current on the aerodynamic fault condition.

Comparison results of the stator winding asymmetrical fault simulation: in the case of AWS, the fault frequency in the stator current is mainly $k_1 f$ (k_1 stands for odd number, f is the grid frequency), and the fault frequency in the rotor current is $(2k_2 \pm s)f$ (k_2 stands for positive integer); while, in the case of EWS, the fault frequencies in the stator current include $k_1 f \pm 3kP$ and $k_1 f$, and the fault frequencies in the rotor current include $(2k_2 \pm s)f \pm 3kP$ and $(2k_2 \pm s)f$. Similarly, the harmonic performance in rotor current is more obvious than that of the stator current.

6. Conclusions

As wind turbines become large-scale, even small changes in wind speed or wind direction on the rotor disk area may cause large fluctuations in aerodynamic load and mechanical torque, and further affect the operation and the fault characteristics of the WTGS. Therefore, it is important to study the distribution of the actual wind speed and then analyze the fault characteristics all over again considering the spatiotemporal distribution of actual wind speed. This paper analyzed the new characteristics of mechanical and electrical fault of WTGS with DFIG based on the EWS model, and obtained some useful conclusions. In the future, the other kinds of faults of DFIG or the faults of the other kinds of WTGS can be analyzed similarly. The research results are of great significance to improve the fault diagnosis accuracy and the fault monitoring level of WTGS.

Author Contributions: Investigation, Writing and editing, X.S.; Data curation, K.C.; Software, data analysis, X.W.; Visualization and supervision, S.W. All authors have read and agreed to the published version of the manuscript.

Funding: This research was funded by National Natural Science Foundation of China, No.51777075, Natural Science Foundation of Hebei Province, No.E2019502064, and the Fundamental Research Funds for the Central Universities, No.2018MS121.

Conflicts of Interest: The authors declare no conflict of interest.

Acronyms

WT	wind turbines
DFIG	doubly-fed induction generator
WTGS	wind turbine generator system
AWS	average wind speed at the hub center
EWS	equivalent wind speed
MCSA	motor current signature analysis
PSD	power spectral density
FFT	fast Fourier transform

Appendix A

The computation process of the rotor current in Equation (15) is shown as follows:

$$
\begin{aligned}
i_{ra} &= I_r \cos\left(\int_0^t \omega_z d\tau\right) \\
&= I_r \cos\left\{\int_0^t \left[\omega_{z0} - \frac{n_p}{J} \sum_{k=1}^{n+1} \frac{T_k}{\omega_k} \sin(\omega_k \tau)\right] d\tau\right\} \\
&= I_r \cos\left[\omega_{z0} t + \sum_{k=1}^{n+1} \frac{n_p T_k}{J \omega_k^2} \cos(\omega_k t)\right] \\
&= I_r \left\{\cos(\omega_{z0} t) \cos\left[\sum_{k=1}^{n+1} W_k \cos(\omega_k t)\right] - \sin(\omega_{z0} t) \sin\left[\sum_{k=1}^{n+1} W_k \cos(\omega_k t)\right]\right\} \\
&\approx I_r \left\{\cos(\omega_{z0} t) - \sin(\omega_{z0} t)\left[\sum_{k=1}^{n+1} W_k \cos(\omega_k t)\right]\right\} \\
&= I_r \cos(\omega_{z0} t) - \sum_{k=1}^{n+1} I_r W_k \sin(\omega_{z0} t) \cos(\omega_k t) \\
&= I_r \cos(\omega_{z0} t) - \sum_{k=1}^{n+1} \frac{I_r W_k}{2}\left[\sin(\omega_{z0} t + \omega_k t) + \sin(\omega_{z0} t - \omega_k t)\right] \\
&= I_r \cos(\omega_{z0} t) - \sum_{k=1}^{n} I_k \sin(\omega_{z0} t + 3k\omega_w t) - \sum_{k=1}^{n} I_k \sin(\omega_{z0} t - 3k\omega_w t) - I_{r1} \sin(\omega_{z0} t + \omega_w t) - I_{r1} \sin(\omega_{z0} t - \omega_w t)
\end{aligned}
$$

where $W_k = n_p T_k / J(\omega_k)^2$.

References

1. Cem, E.; Tugce, C. In case study: Investigation of tower shadow disturbance and wind shear variations effects on energy production, wind speed and power characteristics. *Sustain. Energy Technol. Assess* **2019**, *35*, 148–159.
2. Wang, H.; Zhang, B.; Qiu, Q. Numerical study of the effects of wind shear coefficients on the flow characteristics of the near wake of a wind turbine blade. *Proc. Inst. Mech. Eng. Part A* **2016**, *230*, 86–98. [CrossRef]
3. Nilay, S.U.; Oguz, U. Effect of steady and transient wind shear on the wake structure and performance of a horizontal axis wind turbine rotor. *Wind Energy* **2013**, *16*, 1–17.
4. Houtzager, I.; van Wingerden, J.W.; Verhaegen, M. Wind turbine load reduction by rejecting the periodic load disturbances. *Wind Energy* **2013**, *16*, 235–256. [CrossRef]
5. Wen, B.; Wei, S.; Wei, K.; Yang, W.; Peng, Z.; Chu, F. Power fluctuation and power loss of wind turbines due to wind shear and tower shadow. *Front. Mech. Eng.* **2017**, *12*, 321–332. [CrossRef]
6. Wen, B.; Wei, S.; Wei, K.; Yang, W.; Peng, Z.; Chu, F. Influences of wind shear and tower shadow on the power output of wind turbine. *J. Mech. Eng.* **2018**, *54*, 124–132. [CrossRef]
7. Tan, J.; Hu, W.H.; Wang, X.R.; Chen, Z. Effect of tower shadow and wind shear in a wind farm on AC tie-line power oscillations of interconnected power systems. *Energies* **2013**, *6*, 6352–6372. [CrossRef]
8. Wan, S.T.; Cheng, K.R.; Sheng, X.L.; Wang, X. Characteristic analysis of DFIG wind turbine under blade mass imbalance fault in view of wind speed spatiotemporal distribution. *Energies* **2019**, *12*, 3178. [CrossRef]
9. Milad, F.; Asghar, A.F. Recognition and assessment of different factors which affect flicker in wind turbines. *IET Renew. Power Gen.* **2016**, *10*, 250–259.
10. Miao, F.L.; Shi, H.S.; Zhang, X.Q. Impact of the converter control strategies on the drive train of wind turbine during voltage dips. *Energies* **2015**, *8*, 11452–11469. [CrossRef]
11. Zhao, M.H.; Jiang, D.X.; Li, S.H. Research on fault mechanism of icing of wind turbine blades. In Proceedings of the World Non-Grid-Connected Wind Power and Energy Conference, Nanjing, China, 24–26 September 2009; pp. 356–359.
12. Niebsch, J.; Ramlau, R.; Thien, T.N. Mass and aerodynamic imbalance estimates of wind turbines. *Energies* **2010**, *3*, 696–710. [CrossRef]
13. Sheng, X.L.; Wan, S.T.; Cheng, L.F.; Li, Y.G. Blade aerodynamic asymmetry fault analysis and diagnosis of wind turbines with doubly fed induction generator. *J. Mech. Sci. Technol.* **2017**, *31*, 5011–5020. [CrossRef]
14. Xiang, G.; Wei, Q. Imbalance fault detection of direct-drive wind turbines using generator current signals. *IEEE Trans. Energy Convers.* **2012**, *27*, 468–476.

15. Estefania, A.; Andres, H.E.; Emilio, G.L. Current signature analysis to monitor DFIG wind turbine generators: A case study. *Renew. Energy* **2018**, *116*, 5–14.
16. Stefani, A.; Yazidi, A.; Rossi, C.; Filippetti, F.; Casadei, D.; Capolion, G.A. Doubly fed induction machines diagnosis based on signature analysis of rotor modulating signals. *IEEE Trans. Ind. Appl.* **2008**, *44*, 1711–1721. [CrossRef]
17. Casadei, D.; Filippetti, F.; Rossi, C.; Stefani, A. Closed loop bandwidth impact on doubly fed induction machine asymmetries detection based on rotor voltage signature analysis. In Proceedings of the 2008 43rd International Universities Power Engineering Conference, Padova, Italy, 1–4 September 2008; pp. 1–5.
18. Dai, Z.J. Research and Implementation of Fault Diagnosis for Stator and Rotor of Doubly-Fed Induction Generator Based on Improved HHT. Master's Thesis, Shanghai Electric Institute, Shanghai, China, 2015.
19. Williamson, S.; Djurović, S. Origins of stator current spectra in DFIGs with winding faults and excitation asymmetries. In Proceedings of the 2009 IEEE International Electric Machines and Drives Conference, Miami, FL, USA, 3–6 May 2009; pp. 563–570.
20. Gritli, Y.; Zarri, L.; Rossi, C.; Filippetti, F.; Capolino, G.A.; Casadei, D. Advanced diagnosis of electrical faults in wound-rotor induction machines. *IEEE Trans. Ind. Electron.* **2013**, *60*, 4012–4024. [CrossRef]
21. Wan, S.T.; Cheng, L.F.; Sheng, X.L. Effects of yaw error on wind turbine running characteristics based on the equivalent wind speed model. *Energies* **2015**, *8*, 6286–6301. [CrossRef]
22. Wan, S.T.; Cheng, L.F.; Sheng, X.L. Numerical analysis of the spatial distribution of equivalent wind speed in large-scale wind turbines. *J. Mech. Sci. Technol.* **2017**, *31*, 965–974. [CrossRef]
23. Dolan, D.S.L.; Lehn, P.W. Simulation model of wind turbine 3p torque oscillations due to wind shear and tower shadow. *IEEE Trans. Energy Convers.* **2006**, *21*, 717–724. [CrossRef]
24. De Kooning, J.D.M.; Vandoorn, T.L.; De Vyver, V.J.; Meersman, B.; Vandevelde, L. Shaft speed ripples in wind turbines caused by tower shadow and wind shear. *IET Renew. Power Gen.* **2014**, *8*, 195–202. [CrossRef]
25. Kong, Y.G.; Wang, J.; Gu, H.; Wang, Z.X.; Xu, D.L. Dynamic modeling of wind turbine wind speed based on wind shear and tower shadow effect. *Acta Energ. Sol. Sin.* **2011**, *32*, 1237–1244.
26. He, Y.K.; Hu, J.B.; Xu, L. *Operation Control of Grid Doubly-Fed Induction Generator*, 2nd ed.; Electric Power Press: Beijing, China, 2014; pp. 52–54.
27. Sheng, X.L.; Wan, S.T.; Cheng, L.F.; Li, Y.G. Characteristics analysis of DFIG under the blade mass imbalance fault. *Acta Energ. Sol. Sin.* **2017**, *38*, 1324–1332.

Article

Assessment of Early Stopping through Statistical Health Prognostic Models for Empirical RUL Estimation in Wind Turbine Main Bearing Failure Monitoring

Jürgen Herp [1,*], Niels L. Pedersen [2] and Esmaeil S. Nadimi [1]

[1] The Maersk Mc-Kinney Moller Institute, University of Southern Denmark, 5230 Odense, Denmark; esi@mmmi.sdu.dk

[2] Diagnostics, Siemens Gamesa Renewable Energy, 7330 Brande, Denmark; Niels.L.Pedersen@siemensgamesa.com

* Correspondence: herp@mmmi.sdu.dk

Received: 2 December 2019; Accepted: 20 December 2019; Published: 23 December 2019

Abstract: Details about a fault's progression, including the remaining-useful-lifetime (RUL), are key features in monitoring, industrial operation and maintenance (O&M) planning. In order to avoid increases in O&M costs through subjective human involvement and over-conservative control strategies, this work presents models to estimate the RUL for wind turbine main bearing failures. The prediction of the RUL is estimated from a likelihood function based on concepts from prognostics and health management, and survival analysis. The RUL is estimated by training the model on run-to-failure wind turbines, extracting a parametrization of a probability density function. In order to ensure analytical moments, a Weibull distribution is assumed. Alongside the RUL model, the fault's progression is abstracted as discrete states following the bearing stages from damage detection, through overtemperature warnings, to over overtemperature alarms and failure, and are integrated in a separate assessment model. Assuming a naïve O&M plan (wind turbines are run as close to failure as possible without regards for infrastructure or supply chain constrains), 67 non run-to-failure wind turbines are assessed with respect to their early stopping, revealing the potential RUL lost. These are turbines that have been stopped by the operator prior to their failure. On average it was found that wind turbines are stopped 13 days prior to their failure, accumulating 786 days of potentially lost operations across the 67 wind turbines.

Keywords: wind turbines; condition monitoring; inference; neural networks; remaining-useful-lifetime; main bearing

---

1. Introduction

Health prognostics in asset assessment, including remaining-useful-lifetime (RUL) estimations, are key elements in operation, and maintenance (O&M) strategies. This can help to increase production and/or reduce O&M costs. As wind power is a leading renewable energy source, availability, reliability, and lifetimes are taken incrementally into account by investors. In this work, we focus on slow developing faults, which, when not addressed, can cause unwanted or unnecessary costly downtime [1–8]. SPecifically, we will focus on (i) investigate main bearing monitoring, (ii) give a full account of the underlying neural network (NN) approach presented by Herp et al. [9] on different timescales, (iii) present how this ties into O&M efforts, and (iv) compare expected main bearing RUL with model predictions. The later will give raise to a discussion on early stopping of wind turbines and potential waste of RUL (when O&M planning requires to stop the wind turbine ahead of time).

Models for assessing a systems current and future condition are often data-driven, and thus different approaches can be used depending on the type of data at hand. Trappey at al. [10] uses a combined statistical and NN approach, in which principal component analysis is used to extract relevant features. A NN was trained with back-propagation on these features to learn and predict the condition of power transformers. You et al. [11] addresses the use of NNs, by including temporal dependencies, they propose a diagnostic approach for electrical car batteries through the utilization of recurrent neural networks (RNN). On the other hand, Herp et al. [12] proposes a pure statistical approach based on the history and descriptive statistics of the observed data, using a Gaussian process to predict the future of bearing failure time-series in wind turbines. Hong et al. [13] similarly addresses the problem of bearing monitoring by extracting features and using an approach combining NN with self-organizing map approach to estimate the confidence of the bearing health states. In contrast to the aforementioned, Si [14] proposed an approach based on the time-series stochastic properties. Assuming an underlying driving Brownian motion, a closed-form predictive distribution for the time it takes the signal to reach a threshold can be calculated. Medical research uses survival analysis to understand the relations between patients' clinical features and the effectiveness of treatment options. Khan et al. [15] and others like Katzman et al. [16] have recently shown that prognostic and health management, and NNs can be combined to outperform existing state-of-the-art survival analysis models. A comprehensive overview of advances in RUL prediction can be found in Si et al. [17], Chapter 1. For a summary on wind turbine monitoring techniques, we refer to Márquez [18].

Catering to the need of health prognostics, the details of the model proposed by Herp et al. [9] are inspired by concepts of the above mentioned work. It adopts the closed-form approach by Si et al. [17] by assuming the RUL can be quantified by a closed-form expression and embeds it in a NN, to make confidence calculations of the predictions easier. A Weibull distribution has been chosen as a lifetime distribution, since literature already has shown its use in connection with NNs [19–22]. In contrast to Ranganath et al. [19] deep exponential families model, this work will limit itself to Weibull distribution's only. Yang et al. [23] and Aggarwal et al. [24] developed similar Weibull based NNs and RNNs, respectively, but focus amongst others on different disciplines than wind turbine monitoring.

For the sake of completeness, we refer the interested reader to Herp et al. [9] for a comparative study between different RUL estimation frameworks in empirical bearing fault prediction, as this study will not be concerned with the topic.

This paper is organized as follows: In Section 2, the methods for the RUL estimation are described in detail. The presented methodology is applied in Section 3 showing the potential gain from re-evaluating early stopping. A discussion on the results and the underlying assumption are provided in Section 4. The paper is closed by a final remark in Section 5.

2. Methodology

As this study is concerned with both the empirical estimation of the remaining-useful-lifetime of main bearings and the assessment of early stopping, both topics will have a dedicated section of their own. Even if treated separately here, we will later show that these methods are very much intertwined when it comes to monitoring and operation of one or more wind turbines. In Figure 1 we present how a monitoring and operation work flow might look like in order to make educated decisions based on estimations of the RUL. Fault monitoring is embedded in a back-end operations monitoring framework, which facilitates fault detection. We will not spend more thoughts on fault detection in this work and assume that one of the many fault detection approaches described in the literature can be applied. On a fault monitoring level, upon having collected data regarding the fault-to-be, we can distinguish between three main strategies for later decision making. (1) Assessing the situation at the time of detection, draw a conclusion and decide on an O&M action. Options (2) and (3) are assessments over time, where the aforementioned option derives a new assessment at given condition monitoring (CM) points spaced out over the time until failure, and the latter provides an assessment at any given time. For the remainder of this work we refer to, (1) *Initial Assessment* , (2) *Discrete Assessment*, and (3) *Continuous Assessment•*, collectively, as *Empirical Remaining-Useful-Lifetime Estimation*. The conclusion of the assessments is then facilitated again in the Operations Monitoring, while actions are left outside the monitoring framework as they are a matter of O&M management. In the conclusion step, one would estimate the benefit of continuing or stopping operations. We refer to this as *Assessment of Early Stopping*.

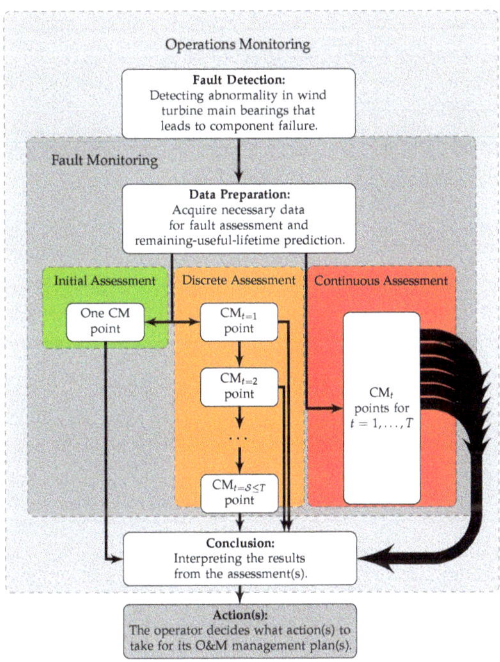

Figure 1. High-level workflow for remaining-useful-lifetime O&M management.

2.1. Data and Notation

This study is concerned with two types of data, time-series data and event data. The time-series data are composed out of a subset of SCADA data and other health indicators, while the event data contains an event description together with its start and stop time. Since a main bearing failure will be investigated as a case study, the SCADA and event data are associated with the failure (including events for the initial damage detection, failure warnings, and failure alarms) and stems from 132 wind turbines of the same type, of which 35 are operated until failure.

Regarding the time-series data, we let $\{x_1,\ldots,x_T\}$ be a process of measurement vectors, $x_t \forall t = 1,\ldots,T$, containing measurements for m variables, i.e., $x_t = [x_1,\ldots,x_m]^\top$. Successive samples on an interval a to b are denoted $x_{[a,b]}$. The feature space available in this study is the same as used by Bach-Andersen et al. [25] and Herp et al. [9], namely, active power, generator rpm, gear box oil temperature, ambient temperature, wind speed, nacelle temperature, and a bearing health indicator. Apart from the bearing health indicator, all features are sampled each 10 min as averages of the past 10 min interval. The bearing health indicator is based on energy bands in vibration spectra and is provided as an event-based measurement. For the remainder of this work, the SCADA and health indicator data are re-sampled to hourly timestamps. The time-series data are the foundation for the predictive models described in Section 2.2 and employed in Section 3.

The event data are a set of discrete data points where we let $\mathcal{E} = \{E_1,\ldots,E_K\}$ be the set of all events $E_k \forall k = 1,\ldots,K$, and we let $\mathcal{T} = \{1,\ldots,t,\ldots,T\}$ be the time at which an event can occur in the reference frame of the time-series data. An outtake of these data is shown in Table 1. Starting with the initial detection of the damage, hundreds of events are recorded until failure. The failure of the wind turbine of Table 1 is indicated by the red vertical line (see graph), and the event associated with the failure is *bearing overtemperature alarm*. Wind turbines for which this event is recorded are referred to as run-to-failure wind turbines. These wind turbines also contain the event *bearing temperature warning*, shown by the orange vertical line in Table 1. Other combinations of events, different from *bearing overtemperature warning* and *bearing overtemperature alarm*, might likewise carry valuable information between the initial detection of the damage and the wind turbine failure. Considering a chain of successive and/or simultaneous events, the dependency between two sets of events $\mathcal{E}_l \Rightarrow \mathcal{E}_{l'}$ will be defined by:

$$\mathcal{E}_l \Rightarrow \mathcal{E}_{l'} \qquad\qquad l \neq l' \tag{1a}$$

$$\mathcal{E}_l, \mathcal{E}_{l'} \subseteq \mathcal{E} \tag{1b}$$

$$\mathcal{E}_l \cap \mathcal{E}_{l'} \neq \varnothing. \tag{1c}$$

Table 1. Binary Event Mapping: Example of mapping from recorded event data into a binary array for identifying states, including bearing temperature warning (-) and bearing temperature alarm (-).

Event	Start Time	Stop Time
E_{21}	12 October 2014, 09:59:01	13 October 2014, 10:12:43
E_1	12 October 2014, 15:39:06	13 October 2014, 10:22:43
$\vdots$	$\vdots$	$\vdots$
E_3	19 October 2014, 02:22:00	19 October 2014, 13:00:00
E_{10}	19 October 2014, 02:42:59	19 October 2014, 13:00:00

$\Downarrow$

				Events				
Sample	E_1	E_2	$\cdots$	E_{k-1}	E_k	E_{k+1}	$\cdots$	E_K
$t-2$	0	0	$\cdots$	0	1	0	$\cdots$	1
$t-1$	0	0	$\cdots$	0	1	0	$\cdots$	1
t	0	0	$\cdots$	0	1	1	$\cdots$	0
$t+1$	0	0	$\cdots$	0	0	1	$\cdots$	0
$t+2$	1	0	$\cdots$	0	0	1	$\cdots$	1

$\Downarrow$

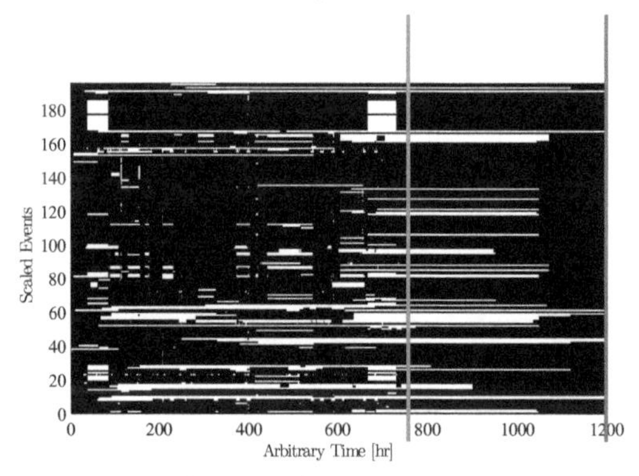

The found sets $\mathcal{E}_k$ can then be used to establish a state as defined later on in Equation (15).

2.2. Empirical Remaining-Useful-Lifetime Estimation

As we are concerned with the time between detecting a failure and the failure itself, we let the RUL be defined to be positive bound, i.e., RUL $\in [0, \infty)$. Following the lead of probabilistic estimations of the remaining-useful-lifetime in literature, this section provides the missing detailed account for the underlying framework used in Herp et al. [9] for wind turbine bearing failure.

In this study, $\eta(t)$ is referred to as the hazard function and $H(t)$ is referred as to the cumulative hazard function. Following textbooks on survival analysis [26], $H(t)$ leads to a cumulative distribution function for a positive random variable RUL:

$$\mathbb{P}(\text{RUL} \leq t) = 1 - \exp\left(-\int_0^t \eta(w)dw\right) = 1 - e^{-H(t)}, \tag{2}$$

and a probability density function:

$$\mathbb{P}(t) = \frac{\partial}{\partial t}\mathbb{P}(\text{RUL} \leq t) = \frac{\partial}{\partial t}\left(1 - e^{-H(t)}\right) = e^{-H(t)}\eta(t). \tag{3}$$

As a consequence, for each cumulative distribution function, $\mathbb{P}(\text{RUL} \leq t)$, there exists $H(t)$ such that $H(t) = -\log(1 - \mathbb{P}(\text{RUL} \leq t))$. In terms of remaining-useful-lifetime, large values of the hazard function indicate a higher chance of an failure to occur up to the current time, i.e., higher failure rate in terms of fault prediction. For the remainder of the study, the focus will be on the right tail of the distribution under consideration, which will define the RUL distribution [26]:

$$L(t) = 1 - \mathbb{P}(\text{RUL} \leq t) = e^{-H(t)}. \tag{4}$$

As the time it takes to observe the remaining-useful-lifetime is the remaining lifetime of the wind turbine itself, the true remaining-useful-lifetime, $\hat{\text{RUL}}$, cannot be observed until it is too late. This premise is known as censored observations in survival analysis [26]. Let RUL be some positive random variable indicating the remaining-useful-lifetime. An observation of RUL is said to be censored whenever it has been observed point wise. In this framework one might distinguish between true $\hat{\text{RUL}}$ and observed RUL. The true $\hat{\text{RUL}}$ is always contained in the observation. When $\hat{\text{RUL}}$ equals the observed value, the process is referred to as non-censored. More specifically we distinguish between (i): right censored data $\Delta = 1$, such that RUL is know to be above a threshold t (i.e., RUL $\in (t, \infty)$), and (ii) non-censored $\Delta = 0$, when $\hat{\text{RUL}}$ equals the observed value (i.e., RUL $= t$). Censoring the remaining-useful-lifetime of wind turbine prohibits simply using the mean value of the already failed wind turbines, as this will lead to underestimating the RUL.

A full mathematical proof of the following description can be found in Patti et al. [27]. In Figure 2 the censoring idea is illustrated: the parametrized probability density function tightens when the failure is observed or is pushed beyond the current observed point, if the failure is censored.

Under the assumption of non-informative censoring [26], the likelihood reads:

$$\mathcal{L} = \mathbb{P}(t)^{\Delta}\mathbb{P}(\text{RUL} > t)^{1-\Delta} \tag{5}$$
$$= \mathbb{P}(t)^{\Delta}L(t)^{1-\Delta} \tag{6}$$
$$= e^{-\Delta H(t)}\eta(t)^{\Delta}e^{-(1-\Delta)H(t)} \tag{7}$$
$$= \eta(t)^{\Delta}e^{-H(t)}. \tag{8}$$

Thus, the log-likelihood can be written as:

$$\log \mathcal{L} = \Delta \log \eta(t) - H(t). \tag{9}$$

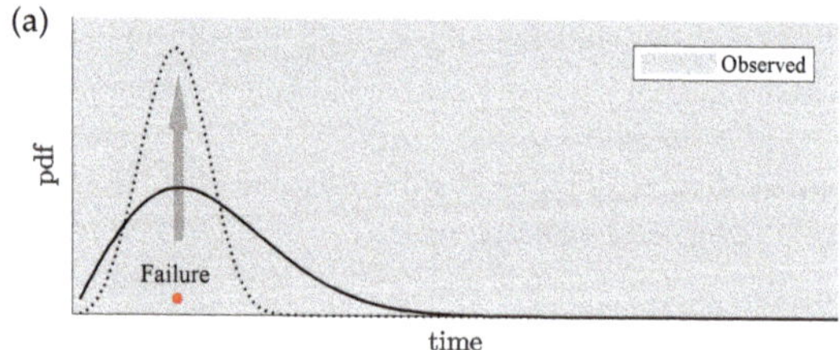

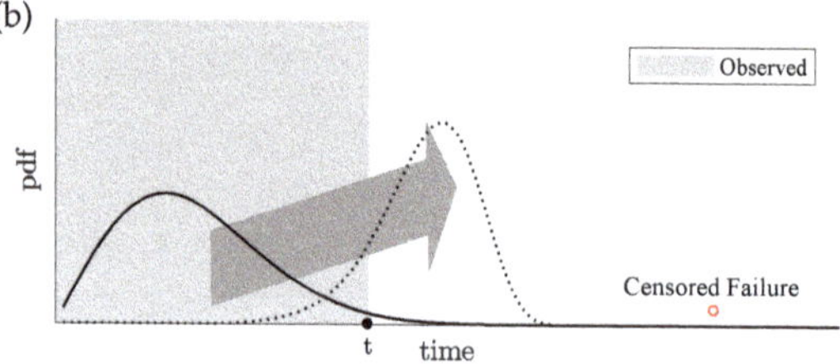

Figure 2. (**a**) Given the failure is observed (non-censored) it is desired to harden a distribution at the time of failure. (**b**) A distribution is estimated based on the measurements up to time t. As the failure is not observed *yet* (censored) the distribution is pushed beyond the point t into the unobserved future.

For non-censored and censored failures, the log-likelihood will always have a negative contribution with increasing t, penalizing as time progresses. Following Herp et al. [9] we consider a Weibull distribution, for its tractable properties, intuitive parametrization and use in other related studies [19–22]:

$$\mathbb{P}(t) = \frac{\alpha}{\beta} \left(\frac{t}{\beta}\right)^{\alpha-1} e^{-\left(\frac{t}{\beta}\right)^{\alpha}} \tag{10}$$

$$\mathbb{P}(\text{RUL} \leq t) = 1 - e^{\left(\frac{t}{\beta}\right)^{\alpha}}. \tag{11}$$

Here, $t \in [0, \infty)$, $\beta \in (0, \infty)$, and $\alpha \in (0, \infty)$, where α and β are referred to as the scale and shape of the distribution. Beside other features, such as analytic moments, the Weibull distribution has an analytic hazard and cumulative hazard function, which can be obtained by Equations (2) and (3):

$$\eta(t) = \left(\frac{t}{\beta}\right)^{\alpha-1} \frac{\alpha}{\beta}. \tag{12}$$

$$H(t) = \left(\frac{t}{\beta}\right)^{\alpha} \tag{13}$$

The cost associated with the Weibull function is then given by Equation (9) and reads:

$$\log(\mathcal{L}) = \Delta \left(\alpha \log\left(\frac{t}{\beta}\right) + \log(\alpha) - \log(t)\right) - \left(\frac{t}{\beta}\right)^{\alpha}. \tag{14}$$

2.2.1. Initial Assessment

The initial assessment contains only one estimation of the remaining-useful-lifetime, based on the descriptive statistics and maximizing the likelihood as described by Equation (14). Figure 3 shows the Weibull model for run-to-failure wind turbines (bearing overtemperature *alarm*), wind turbines running into bearing overtemperature *warning*, wind turbines excluded from the model construction (as they do not run into any aforementioned event), are illustrated by their cumulative distribution as a scatter plot (*non-failing*). For later comparison, a *stopped* by operator model is obtained for the non-failing wind turbines. Furthermore, the Kaplan-Meier [28] estimation (*KM*) is provided for reference, as it makes no assumption with regard to the shape of the probability distribution.

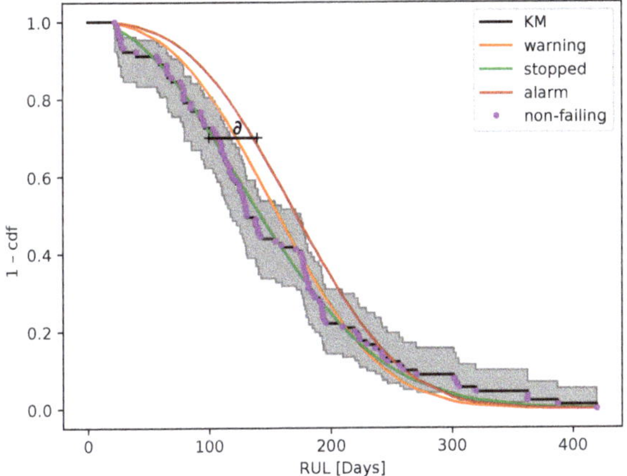

Figure 3. Initial Assessment: Comparison of different models up to different events including the cumulative distribution over wind turbines not running into an event. For comparison the Kaplan–Meier (KM) estimation is provided for the non-failing wind turbines, including confidence bound. δ is the mismatch between a non run-to-failure (non-failing) wind turbine and the bearing overtemperature alarm (alarm).

2.2.2. State Abstraction and Discrete Assessment

The same principle as in Figure 3 applies to the discrete assessment. However, in order to perform the discrete assessment, condition monitoring (CM) points will need to be taken into account. How these CM points are extracted is described in this section.

Between initial detection and failure of a damaged bearing, the bearing undergoes transitions from a minor fatigue to advanced fatigue and damage. In combination with Equation (1), this section addresses how to identify individual stages of the bearing fatigue to lay the foundation of the *discrete assessment* later on.

Consider an event E_k or set of events $\mathcal{E}_k$, from a library of events $\mathcal{E} = \{E_1, \ldots, E_K\}$ (abstracted by Equation (1)), linked to the operation of a wind turbine at time t, to be dependent on the data collected up to time t and a hidden state variable $s^{(t)}$, $s^{(t)}$ representing the current state of the wind turbine. The probability for $\mathcal{E}_k$ can be given by:

$$\mathbb{P}(\mathcal{E}_k \mid s^{(t)}, \mathbf{x}_{[1,t]}) = \frac{\mathbb{P}(\mathcal{E}_k, s^{(t)}, \mathbf{x}_{[1,t]})}{\mathbb{P}(s^{(t)} \mid \mathbf{x}_{[1,t]})\mathbb{P}(\mathbf{x}_{[1,t]})}, \tag{15}$$

further, S_m is referred to as the hidden states, defined by the separability of the process $\{\mathbf{x}_1, \ldots, \mathbf{x}_T\}$ into $S \leq T$ states. The transition between those states is called state transition, where $s^{(t)}$ is the length of the current state with samples $\mathbf{x}_{[S_S,t]}$, with S_S being the time of the last state transition. Following Herp et al. [29,30] and Prescott Adam et al. [31], a state in a set of time-series can be abstracted by considering only the maximum likelihood of $\mathbb{P}(s^{(t)}, \mathbf{x}_{[1,t]})$, where

$$\mathbb{P}(s^{(t)}, \mathbf{x}_{[1,t]}) = \sum_{(s^{t-1})} \underbrace{\mathbb{P}(s^{(t)} \mid s^{(t-1)})}_{\text{conditional prior}} \underbrace{\mathbb{P}(\mathbf{x}_t \mid s^{(t-1)}, \mathbf{x}_{[1,t-1]})}_{\text{sample model}} \mathbb{P}(s^{(t-1)}, \mathbf{x}_{[1,t-1]}). \tag{16}$$

Here the conditional prior and sample model are implicit depending on known hyper-parameters $\beta = [\beta_c, \beta_m]$, where β_c and β_m are the parametrization of the conditional prior and the sample model. Thus, $\mathbb{P}(s^{(t)} \mid s^{(t-1)}) \equiv \mathbb{P}(s^{(t)} \mid s^{(t-1)}, \beta_c)$ and $\mathbb{P}(\mathbf{x}_t \mid s^{(t-1)}, \mathbf{x}_{[1,t-1]}) \equiv \mathbb{P}(\mathbf{x}_t \mid s^{(t-1)}, \mathbf{x}_{[1,t-1]}, \beta_m)$.

Given the nature of the data we are seeking a sample model that addresses changes in the first and second moment of the time-series. This is achieved by classifying each $s^{(t)}$ by the two moments $\mathbb{E}[\mathbf{x}]$ and $\mathbb{E}[\mathbf{x}^2]$. Given sufficient statistics for each $s^{(t)}$, a posterior distribution for $\mathbf{x}_t$ is given for the iterative update [32,33]:

$$\mu_t = \frac{\mu_0 \kappa_0 + \mathbb{E}[\mathbf{x}_{[1,t]}]}{\kappa_0 + t}, \tag{17a}$$

$$\kappa_t = \kappa_0 + 1, \tag{17b}$$

$$\alpha_t = \alpha_0 + t, \tag{17c}$$

$$\zeta_t = \sum_{j=1}^{t} (\mathbf{x}_j - \mathbb{E}[\mathbf{x}])^2, \tag{17d}$$

$$\gamma_t = \gamma_0 + \frac{1}{2}\zeta_t + \frac{\kappa_0 t (\mathbb{E}[\mathbf{x}_{[1,t]}] - \mu_0)^2}{2(\kappa_0 + t)}, \tag{17e}$$

with $\mu_0, \kappa_0, \alpha_0, \gamma_0$ being the previous statistics. A Student's t-distribution can then be used for the posterior distribution [34] with ν degrees of freedom and μ and σ as mean and variance, it follows that:

$$\mathbb{P}(\mathbf{x}_t \mid s^{(t-1)}, \mathbf{x}_{(1,t-1)}) = \mathrm{St}_{2\alpha_t}\left(\mu_t, \frac{\gamma_t}{\alpha_t}\frac{\kappa_t + 1}{\kappa_t}\right). \tag{18}$$

The conditional prior is chosen as a hazard function, as motivated in Section 2.2, such that

$$\mathbb{P}(s^{(t)} \mid s^{(t-1)}) = \begin{cases} \eta & \text{if } s^{(t)} = 1 \\ 1 - \eta & \text{if } s^{(t)} = s^{(t-1)} + 1 \\ 0 & \text{else} \end{cases}, \tag{19}$$

here η is a hazard function of a geometric sequence and given by the elapsed time since the last state transition:

$$\eta = \frac{1}{\mathbb{E}[s^{(t)}]}. \tag{20}$$

For the remainder of this work, the algorithm is implemented with a supervised probability update as proposed by Herp et al. [29].

In order to quantify states, a piecewise constant regression for finite state numbers is performed on the cumulative probability state transitions:

$$\hat{P}(t) = \sum_{s=2}^{S+1} \alpha_s \chi_{A_s}(t), \tag{21}$$

$$\text{s.t.}$$

$$\underset{\alpha_s, \chi_{A_s}}{\arg\max} \ (P(t) - \hat{P}(t))^2$$

$$\text{length}(A_s) \geq 5$$

where α_s is a real number, the states cumulative probability, and A_s the interval of the sth sate, i.e., $[S_{m-1}, S_m)$, χ_{A_s} is an indicator function that is 1 if $t \in A$, and 0 otherwise. The constraint on A_s comes from the delay of detecting a new state.

In detail, the state abstraction for the health indicator can be seen in Figure 4. From top to bottom the health indicator for a wind turbine is shown in (a), (b) shows the log-likelihood for the state transitions (gray-scale) and its maximum likelihood, (c) shows the state transition probability density function and cumulative density function, in addition the abstracted states in accordance with Equation (21) are provided.

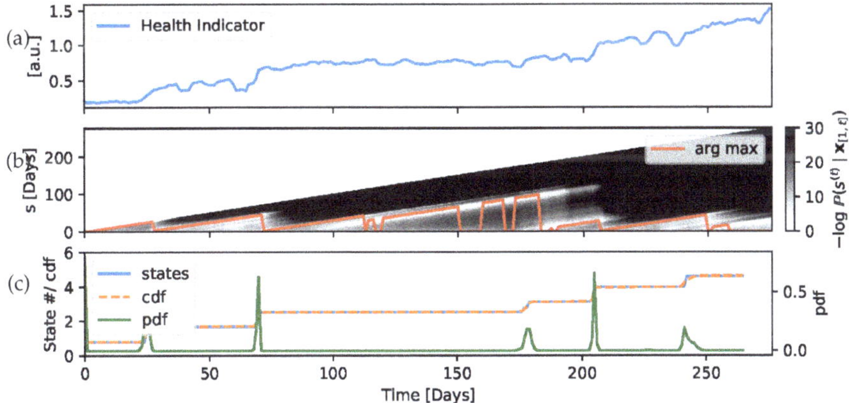

Figure 4. State Abstraction: (**a**) Wind turbine health indicator. (**b**) log-likelihood for state transition. (**c**) Transition probability and states.

The state abstraction give us a measure for the length of each state. Given the time for each turbine to any state transition, the RUL probabilities can be calculated equivalent to the initial assessment for any combinations of state transitions. The implementation and results of this approach are discussed in the case study of Section 3, providing the RUL probabilities from initial detection to any of the state transition. Figure 10).

2.2.3. Continuous Assessment: RUL Recurrent Neural Network

We employ a recurrent neural network (RNN) for the continuous assessment for maximizing the cost function given by Equation (14) at each t. In general, this is an optimization problem of finding a function or distribution in the space of hazard function, $F \in \mathbf{H}$, that maximizes the model's likelihood given historical data, $\mathbf{x}_{[1,t]}$. The space of all possible closed-from distributions in $\mathbf{H}$ is too large to obtain solutions in an easy manner. Furthermore, as $\mathbf{x}_{[1,t]}$ can contain any information at time t, including information of the previous states [30] the optimization for all $\mathbf{x}$ is computational not feasible. However, constraining the problem to a Weibull distribution and letting the Weibull distribution be dependent on $\mathbf{x}_{[1,t]}$, it then follows from Equations (12) and (13), as well as Equation (14) that:

$$\underset{\alpha>0,\beta>0}{\arg\max} \log \mathcal{L}(\alpha,\beta,\mathrm{RUL},\Delta,\mathbf{x}_{[1,t]}) = \sum_{i=1}^{t} \left(\Delta_i \alpha(\mathbf{x}_{[1,i]}) \log\left(\frac{\mathrm{RUL}}{\beta(\mathbf{x}_{[1,i]})}\right) \right.$$

$$\left. + \Delta_i \left[\log\left(\alpha(\mathbf{x}_{[1,i]})\right) - \log(\mathrm{RUL}) \right] - \left(\frac{\mathrm{RUL}}{\beta(\mathbf{x}_{[1,i]})}\right)^{\alpha(\mathbf{x}_{[1,i]})} \right) \tag{22}$$

This optimization is still not feasible, as for a large dataset, each parameter needs to be obtained through consideration of all historic data. In the context of Figure 5, $\boldsymbol{\theta}$ can be written as a vector $\boldsymbol{\theta} = [\alpha,\beta]^{\top}$. Omitting mathematical details on node level, which can be found in a wide range of textbooks such as Goodfellow et al. [35], each node in Figure 5 outputs $o_i = g(\mathbf{x}_i, o_{i-1}, \boldsymbol{\omega}_0)$, for $i = 1, \ldots, t$, a function of the data $\mathbf{x}_t$, previous output o_{i-1}, and the networks topology $\boldsymbol{\omega}_0$. The network's output at each t will be of the form $\boldsymbol{\theta} = m(\mathbf{x}_t, o_{t-1}, \boldsymbol{\omega}) = [\alpha,\beta]^{\top}$, where $m(\cdot)$ is the mapping for the topology of the chosen RNN. Equation (22) is thus reduced to solving the gradient by back-propagation for a set of optimal or suboptimal $\boldsymbol{\omega}$:

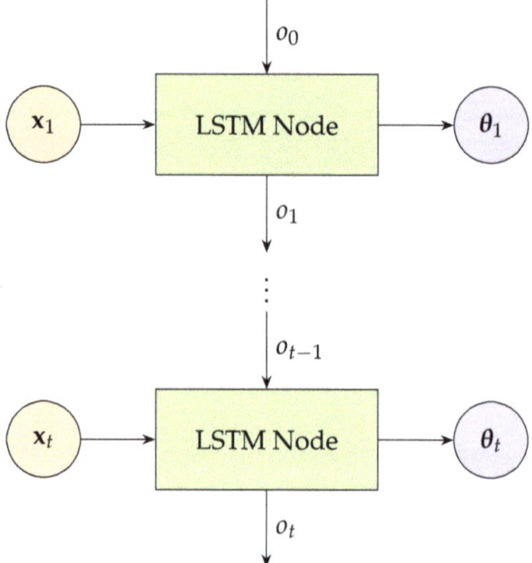

Figure 5. Schematic drawing the recurrent neural network sequence: input feature, current, previous, and future hidden states o_t.

$$\arg\max_{\omega} \log \mathcal{L}(\omega, \text{RUL}, \Delta, \mathbf{x}_{[1,t]}) = \sum_{i=1}^{t} \left(\Delta_i \left[\alpha_i \log\left(\frac{\text{RUL}}{\beta_i}\right) + \log(\alpha_i) - \log(\text{RUL}) \right] - \left(\frac{\text{RUL}}{\beta_i}\right)^{\alpha_i} \right) \qquad (23)$$

For the remainder of this study, we use a Long Short Term Memory (LSTM) RNN [35] and maintain the topology, feature space, and optimization as in Herp et al. [9] (See Figure 6). The sequence length is set to 7 days. An example on how predictions look like can be found in Figure 7, showing the assumable RÛL and model prediction. The prediction is shown in terms of the first *moment*, *median*, and *arg max* of the Weibull distribution at each time t. The distribution itself is illustrated as shaded areas.

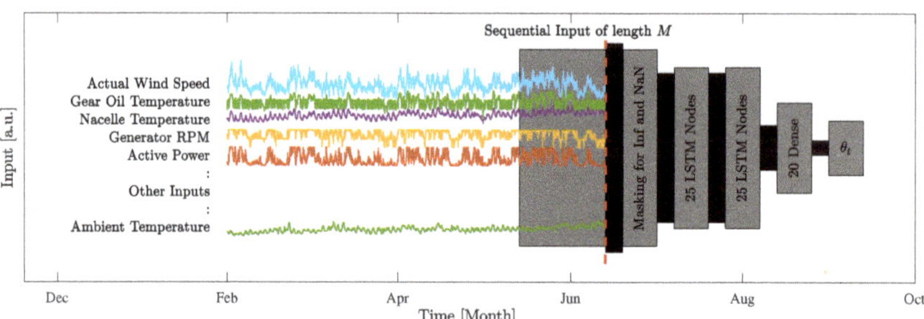

Figure 6. Illustration of the RNN used to predict the RUL. Selected SCADA time-series are shown as input [9].

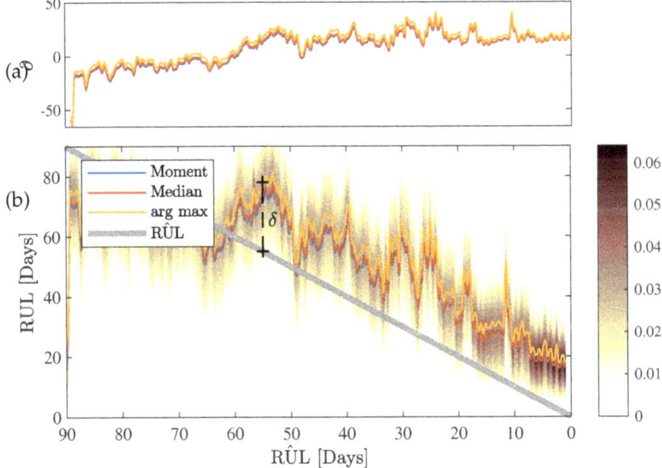

Figure 7. (**a**) Continuous prediction of RUL of a non run-to-failure wind turbine, compared to the expected run-to-failure RÛL. (**b**) Mismatch between RÛL and the predictive distributions first moment, median, and arg max.

3. Assessment of Early Stopping—A Bearing Failure Study

This section combines the afore mentioned methodology and applies it to 67 non run-to-failure wind turbines. For each assessment, all turbines are investigated with respect to early stopping, showing the lost remaining RUL after turbine operations were stopped. The remaining turbines are used for model construction.

Consider the example of Figure 3 where models for the *bearing overtemperature warning* and *bearing overtemperature alarm* are illustrated. Comparing each non run-to-failure wind turbine to any of the event models shows a mismatch δ, which is defined as the distance between any given wind turbine and the predictive models. This mismatch is interpreted as the lost remaining RUL.

Under continuous assessment, Figure 7 shows the mismatch between the presumably true RÛL and the model prediction for a non run-to-failure wind turbine. In this case we consider an average measure for each wind turbine:

$$D^{(\text{turbine})} \equiv \frac{1}{T} \sum_{t=1}^{T} \delta_t^{(\text{turbine})}, \tag{24}$$

where

$$\delta_t^{(\text{turbine})} \equiv \text{RUL}_t^{(\text{turbine})} - \text{RÛL}_t^{(\text{turbine})}. \tag{25}$$

$D^{(\text{turbine})}$ will then be a measure of the specific turbine's potential remaining RUL. In the following we employ the methodology of this section in a case study of mean bearing failures.

When addressing whether or not to stop a wind turbine and perform maintenance is a topic on its own. Many factors, e.g., weather condition, availability of equipment, spare parts, and manpower, have to be taken into consideration when deciding and planing O&M tasks. In the following, we simplify towards a pure time-based approach. Many wind turbines are stopped before they run into failure; based on the proposed RUL estimations, the potential remaining time until failure can be estimated.

3.1. Initial Assessment

Model construction can be facilitated up to different events, including *bearing overtemperature warning* and *bearing overtemperature alarm*. For the wind turbines in this study, the respective models are shown in Figure 3. Assuming that all turbines experience the same fault, the difference between the two models (warning and alarm) should be an offset in the predictive horizon. However, variability in environmental features and the bearing failure's development can lead to changes of the time it takes to undergo a state transition from the *bearing overtemperature warning* to the *bearing overtemperature alarm* event. This is illustrated in Figure 8, labeled warning vs. alarm. The low curvature of the graph indicates that the difference between the two models is relatively small, and relatively constant throughout the RUL space. Given the limited number of wind turbines, that are operated beyond RUL > 250 days and turbines that run into any event before RUL < 100 days, broader confidence bounds can be observed in these RUL ranges. Besides the time between warning and alarm, we can consider a model for the 67 non run-to-failure wind turbines, referred to as the *stopped* model in Figure 3. A mismatch δ for this model with respect to the warning and alarm model is shown in Figure 8. Comparing the models, turbines that have run 100 days, and were stopped too early would have had 30 days left of operations until running into failure, i.e., the *bearing overtemperature alarm*. Or if a more conservative stopping criterion is desired, 20 days until the *bearing ovetemperature warning*. Variability in wind turbine operations leads to a point where the expected mismatch becomes negative; in this regime, the wind turbine is operated at the right tail of the RUL probability curve of Figure 3. An operator might interpret this as the time where the wind turbine has overextended the confidence of continuing operations, when comparing to the ensemble of already failed turbines.

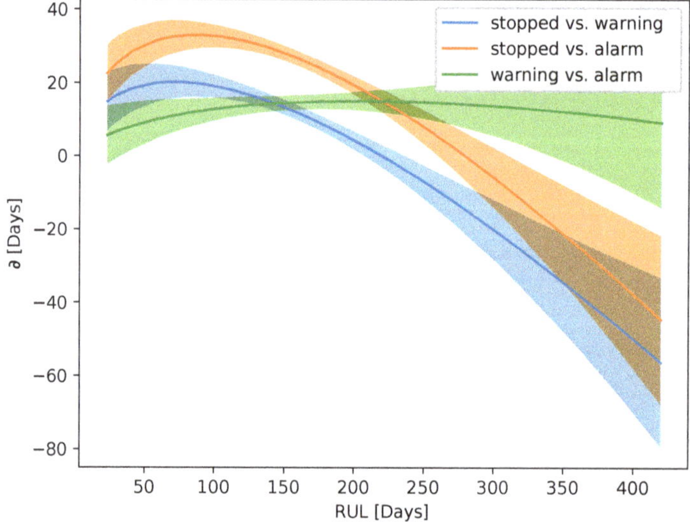

Figure 8. Comparison of initial assessment of RUL with the 67 wind turbines that did not run into either *bearing overtemperature* events, *bearing overtemperature warning*, and *bearing overtemperature alarm* models. 0.95 confidence bounds are provided.

The mismatch to failure over all individual wind turbines can be seen in Figure 9a as a histogram. The largest mismatch is 57 days, while the smallest is −40 days. The mean value and median are 11 and 15 days, respectively. The bulk of turbines is concentrated in the quantiles between 3 and 23 days. The combined mismatch accumulates to 782 days.

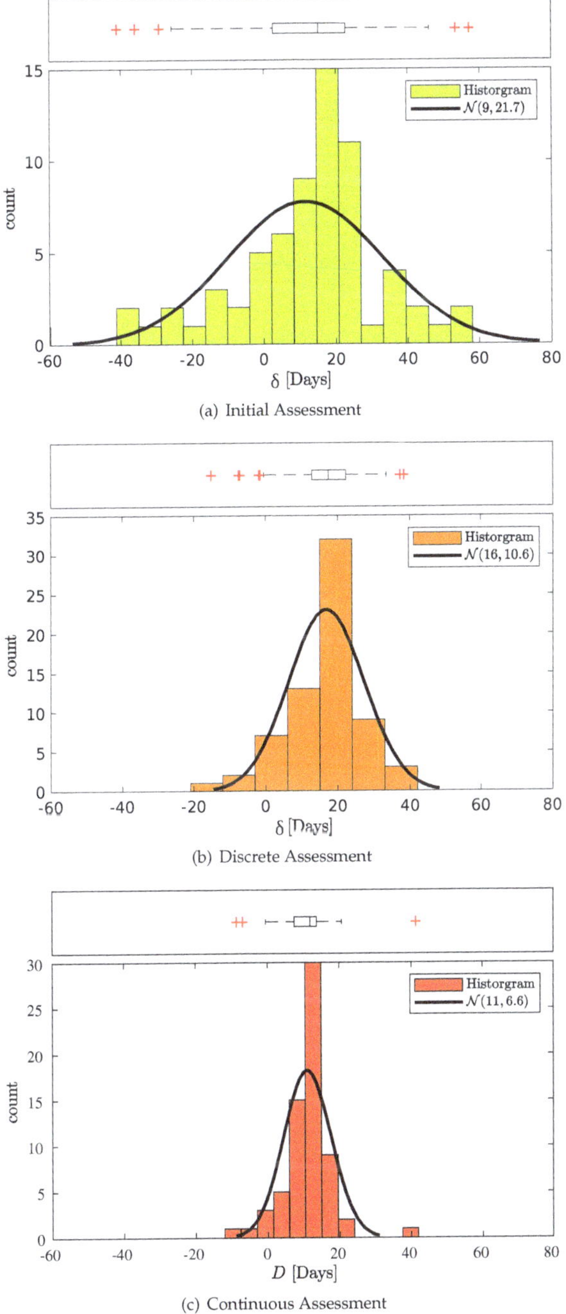

(a) Initial Assessment

(b) Discrete Assessment

(c) Continuous Assessment

Figure 9. Distribution over mismatch for the 67 wind turbines that did not run into either *bearing overtemperature* events. (**a**) initial, (**b**) discrete, and (**c**) continuous assessment.

3.2. Discrete Assessment

In its core, the discrete assessment with CM points at each state transition, is the same as the initial assessment but applied at each CM point. As failure of a component can go through different stages, the operator might want to re-evaluate when the component undergoes the next state transition.

The state detection and abstraction is performed as outlined in Section 2.2.2. As the bearing failures in this case study are slowly developing faults, we impose a minimum state length of $A_s \geq 5$ days, as seen in the constraint of Equation (21). This is done in order to prevent fast switching states in case of random outliers.

In Figure 10, the the cumulative state lengths are shown for each turbine. For each state transition, the RUL for that state is provided. This gives the operator information on when to expect the next state transition of a damaged component. Remark: as the number of states increases, the models become less defined as confidence bounds increase. This is caused by the limited sample size of turbines that experience three or more states transitions.

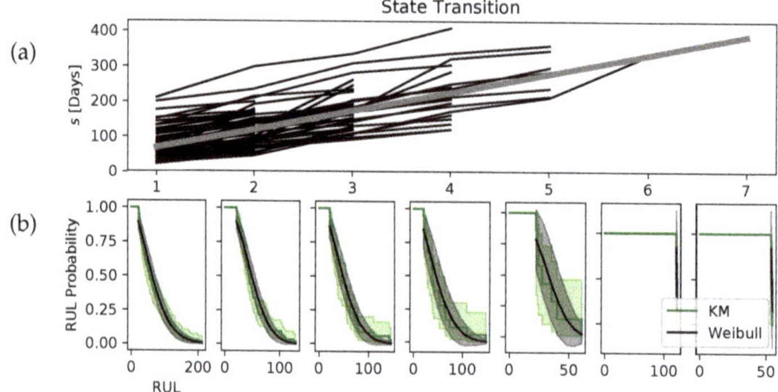

Figure 10. (**a**) Cumulative state length of wind turbines, including best linear fit. (**b**) RUL probability based on the length between each state transition.

The expected RUL is obtained by comparing the wind turbine's cumulative state length with the state length of states for wind turbines abstracted with the event pattern $\mathcal{E}_l \Rightarrow \mathcal{E}_{l'}$, where $\mathcal{E}_{l'}$ contains the *bearing overtemperature alarm*. The mismatch to failure over all individual wind turbine can be seen in Figure 9b as a histogram. The largest mismatch is 38 days, while the smallest is -15 days, the mean value and median are 16 and 17 days, respectively. The bulk of turbines is concentrated in the quantiles between 10 and 20 days. The combined mismatch accumulates to 836 days.

3.3. Continuous Assessment

As shown in Figure 7 the continuous assessment is concerned with the prediction of the RUL at each time step. Focusing on either the first moment, median, or arg max of the Weibull distribution, it becomes apparent that there is a discrepancy between the model's prediction and expected RÛL. This is the aforementioned mismatch $\delta_t^{(\mathrm{turbine})}$ and may differ for each wind turbine. Similar graphs can be drawn for other turbines, with varying degrees of $\delta_t^{(\mathrm{turbine})}$, containing also negative values of $\delta_t^{(\mathrm{turbine})}$. The average mismatch to failure over all individual wind turbines can be seen in Figure 9c as a histogram. The largest mismatch is 41 days, while the smallest is -8 days; the mean value and median are 11 and 12 days, respectively. The bulk of turbines are concentrated in the quantiles between 10 and 15 days. The combined mismatch accumulates to 739 days.

4. Discussion

Common for all assessments is a high count of wind turbine mismatches at the respective mean values. These peaks can contain more than twice the amount of wind turbines than their neighboring bins, indicating not only the most likely mismatch, but also a firm consistency indicating a tendency to stop too early. However, two points are worth mentioning, (i) the different assessments do not yield the same results, and (ii) the accumulative mismatch is based on naïve O&M assumption.

4.1. Discrepancy Between Assessments

The discrepancy between the different assessments stem from the underlying nature of the descriptive statistics. As the initial assessment applies for the total length of failure, no iterative adjustment occurs in the monitoring effort, thus assessing wind turbines that are further into a fault process are not presented by the initial assessment. On the other hand, as the name implies, it gives an initial assessment of the expected length of the failure state and their relative probability based on an ensemble of wind turbines.

A step closer to a real representation of the RUL is provided by the discrete assessment. While the failure state can be split further into shorter states, the RUL and probability of failure is obtained by propagating non run-to-failure wind turbines through the RUL probabilities for each state transition. Besides, the discrete assessment can provide a measure of RUL until the next state transition, aiding an operator to known when to stop a wind turbine's operation for more conservative O&M strategies. For both the initial assessment and discrete assessment, RUL of individual wind turbines and, from there, the mismatch δ, are ensemble measures, i.e., when addressing the RUL of a single wind turbine with respect to a cumulative distribution obtained by all other turbines that were undergoing the same failure.

The initial and discrete assessments are in their core curve fitting problems for one or more datasets of wind turbine RULs. When the number of CM points goes towards the number of sample vectors, $S \rightarrow T$, one can consider the discrete and continuous assessment as equivalent. While the aforementioned assessment was an ensemble measure, the continuous assessment offers a real-time monitoring approach for individual wind turbines, where the RUL is updated at each successive time step. This makes the prediction of the RUL more and more reliable the closer a wind turbine is to failure, but provides poor or misleading assessments early in the fault process, where predictions are wide [9]. The further a wind turbine is in a fault state, a continuous assessment becomes more and more desirable.

All together, when combined as proposed in Figure 1, the proposed assessment offers an operator information at desired stages of a main bearing fault.

4.2. The Naïve O&M Assumption

The cumulative, potential lost RUL has to be looked upon with caution. Firstly, naïve O&M assumption takes away any other consideration that might play a role in stopping a turbine, and secondly, wind turbines might continue to operate reliably even if models suggest otherwise. Apart from the predictive error due to model constraints, it becomes apparent that RUL, i.e., a measure of time, is a difficult measure that is not easily related to the physical parameters of a wind turbine's operation. If a wind turbine is not operated due to lack of wind or other external factors, no matter how good a model, predictions will underestimate the RUL in any of the suggested frameworks.

5. Conclusions

We have proposed three fault monitoring and assessment concepts, namely, *initial assessment*, *discrete assessment*, and *continuous assessment*, showing that in any of the frameworks, wind turbines are stopped earlier than necessary compared to a global ensemble. Based on the descriptive statistics of 67 non-run-to-failure wind turbines, wind turbines are stopped 13 days too early, prior to their failure, accumulating, on average, 786 of potentially lost days of operations under the naïve O&M assumption made in this study. Future work will probably focus on the inclusion of supply chain and weather constraints to represent a more realistic model mismatch behavior.

Author Contributions: Conceptualization:, J.H., N.L.P. and E.S.N.; Methodology: J.H.; Resources: N.L.P.; Data curation: J.H.; Writing—Original draft preparation: J.H.; Writing—review and editing, J.H.; Supervision, E.S.N. All authors have read and agreed to the published version of the manuscript.

Funding: This project was funded by Siemens Gamesa Renewable Energy and the University of Southern Denmark.

Acknowledgments: The authors would like to take the chance to thank Siemens Gamesa Renewable Energy for providing us with the necessary data.

Conflicts of Interest: The authors declare no conflict of interest. The funders had no role in the design of the study; in the collection, analyses, or interpretation of data; in the writing of the manuscript, or in the decision to publish the results.

Abbreviations

cdf	cumulative distribution function
CM	Condition Monitoring
KM	Kaplan-Meier
NN	Neural Network
O&M	Operation and Maintenance
pdf	probability density function
RNN	Recurrent Neural Network
RUL	Remaining-Useful-Lifetime
SCADA	Supervisory Control and Data Acquisition

Nomenclature

a	scalar
$\mathbf{a}$	vector
$\mathbf{A}$	matrix
$\mathbf{x}_t$	sample vector
$\mathbf{x}_{[a,b]}$	samples on the interval a to b
$\mathcal{E}$	set of all possible events
E_k	k^{th} event
$\mathcal{E}_l$	subset of $\mathcal{E}$
RUL	random variable for the remaining-useful-lifetime
RÛL	real remaining-useful-lifetime
η	hazard function
H	cumulative hazard function
$\Delta \in \{0,1\}$	censoring variable
$\mathcal{L}$	likelihood function
$\beta = [\beta_c, \beta_m]$	hyper-parameter for conditional prior and sample model
$\mathbb{P}(s^{(t)} \mid s^{(t-1)}) \equiv \mathbb{P}(s^{(t)} \mid s^{(t-1)}, \beta_c)$	independent conditional prior
$\mathbb{P}(s^{(t)} \mid \mathbf{x}_{[1,t]})$	probability distribution over the current state
$\mathbb{P}(\mathbf{x}_t \mid s^{(t-1)}, \mathbf{x}_{[1,t-1]}) \equiv \mathbb{P}(\mathbf{x}_t \mid s^{(t-1)}, \mathbf{x}_{[1,t-1]}, \beta_m)$	sample model
δ	mismatch between RUL and RÛL
D	average mismatch

References

1. Petersen, K.R.; Madsen, E.S.; Bilberg, A. Offshore Wind Power at Rough Sea: The need for new Maintenance Models. In Proceedings of the 20th EurOMA Conference, Dublin, Ireland, 9–12 June 2013.
2. Nielsen, J.; Sørensen, J. On risk-based operation and maintenance of offshore wind turbine components. *Reliab. Eng. Syst. Saf.* **2011**, *96*, 218–229, doi:10.1016/j.ress.2010.07.007. [CrossRef]

3. Blanco, M.I. The economics of wind energy. *Renew. Sustain. Energy Rev.* **2009**, *13*, 1372–1382, doi:10.1016/j.rser.2008.09.004. [CrossRef]

4. van Bussel, G.J.W.; Zaaijer, M.B. Reliability, Availability and Maintenance Aspects of Large-Scale Offshore Wind Farms, a Concepts Study. In Proceedings of the MAREC 2001 Marine Renewable Energies Conference, Newcastle, UK, 27 March 2001; Volume 113, pp. 119–126.

5. Wilkinson, M.; Spinato, F.; Knowles, M. Towards the Zero Maintenance Wind Turbine. In Proceedings of the 41st International Universities Power Engineering Conference (UPEC '06), Newcastle, UK, 6–8 Septemeber 2006; Volume 1, pp. 74–78, doi:10.1109/UPEC.2006.367718. [CrossRef]

6. Kim, N.; An, D.; Choi, J. *Prognostics and Health Management of Engineering Systems: An Introduction*; Springer: Berlin, Germany, 2016.

7. Sheppard, J.W.; Kaufman, M.A.; Wilmer, T.J. IEEE Standards for Prognostics and Health Management. *IEEE Aerosp. Electron. Syst. Mag.* **2009**, *24*, 34–41, doi:10.1109/MAES.2009.5282287. [CrossRef]

8. Walford, C.A. *Wind Turbine Reliability: Understanding and Minimizing Wind Turbine Operation and Maintenance Costs*; Technical Report SAND2006-1100; Sandia National Laboratories: Albuquerque, NM, USA, 2006. Available Online: https://energy.sandia.gov/wp-content/gallery/uploads/SAND-2006-1100.pdf (accessed on 12 December 2019).

9. Herp, J.; Pedersen, N.L.; Nadimi, E.S. A Novel Probabilistic Long-Term Fault Prediction Framework Beyond SCADA Data—With Applications in Main Bearing Failure. *J. Phys. Conf. Ser.* **2019**, *1222*, 012043, doi:10.1088/1742-6596/1222/1/012043. [CrossRef]

10. Trappey, A.J.C.; Trappey, C.V.; Ma, L.; Chang, J.C. Integrating Real-Time Monitoring and Asset Health Prediction for Power Transformer Intelligent Maintenance and Decision Support. In *Engineering Asset Management—Systems, Professional Practices and Certification*; Tse, P.W., Mathew, J., Wong, K., Lam, R., Ko, C., Eds.; Springer: Cham, Switzerland, 2015; pp. 533–543.

11. You, G.W.; Park, S.; Oh, D. Diagnosis of Electric Vehicle Batteries Using Recurrent Neural Networks. *IEEE Trans. Ind. Electron.* **2017**, *64*, 4885–4893, doi:10.1109/TIE.2017.2674593. [CrossRef]

12. Herp, J.; Ramezani, M.H.; Bach-Andersen, M.; Pedersen, N.L.; Nadimi, E.S. Bayesian state prediction of wind turbine bearing failure. *Renew. Energy* **2018**, *116*, 164–172, doi:10.1016/j.renene.2017.02.069. [CrossRef]

13. Hong, S.; Zhou, Z.; Zio, E.; Wang, W. An adaptive method for health trend prediction of rotating bearings. *Dig. Signal Process.* **2014**, *35*, 117–123, doi:10.1016/j.dsp.2014.08.006. [CrossRef]

14. Si, X.S. An Adaptive Prognostic Approach via Nonlinear Degradation Modeling: Application to Battery Data. *IEEE Trans. Ind. Electron.* **2015**, *62*, 5082–5096, doi:10.1109/TIE.2015.2393840. [CrossRef]

15. Khan, S.; Yairi, T A review on the application of deep learning in system health management. *Mech. Syst. Signal Process.* **2018**, *107*, 241–265, doi:10.1016/j.ymssp.2017.11.024. [CrossRef]

16. Katzman, J.; Shaham, U.; Bates, J.; Cloninger, A.; Jiang, T.; Kluger, Y. DeepSurv: Personalized Treatment Recommender System Using A Cox Proportional Hazards Deep Neural Network. *BMC Med. Res. Methodol.* **2018**, *18*, 24. [CrossRef]

17. Si, X.; Zhang, Z.; Hu, C. *Data-Driven Remaining Useful Life Prognosis Techniques: Stochastic Models, Methods and Applications*; Springer Series in Reliability Engineering; Springer: Berlin/Heidelberg, Germany, 2017.

18. Márquez, F.P.G.; Tobias, A.M.; Pérez, J.M.P.; Papaelias, M. Condition monitoring of wind turbines: Techniques and methods. *Renew. Energy* **2012**, *46*, 169–178, doi:10.1016/j.renene.2012.03.003. [CrossRef]

19. Ranganath, R.; Tang, L.; Charlin, L.; Blei, D.M. Deep Exponential Families. *arXiv* **2014**, arxiv:1411.2581.

20. Ranganath, R.; Perotte, A.; Elhadad, N.; Blei, D. Deep Survival Analysis. *arXiv* **2016**, arXiv:1608.02158.

21. Ali, J.B.; Chebel-Morello, B.; Saidi, L.; Malinowski, S.; Fnaiech, F. Accurate bearing remaining useful life prediction based on Weibull distribution and artificial neural network. *Mech. Syst. Signal Process.* **2015**, *56–57*, 150–172, doi:10.1016/j.ymssp.2014.10.014. [CrossRef]

22. Mazhar, M.; Kara, S.; Kaebernick, H. Remaining life estimation of used components in consumer products: Life cycle data analysis by Weibull and artificial neural networks. *J. Op. Manag.* **2007**, *25*, 1184–1193, doi:10.1016/j.jom.2007.01.021. [CrossRef]

23. Yang, Z.; Chen, C.; Wang, J.; Li, G. Reliability Assessment of CNC Machining Center Based on Weibull Neural Network. *Math. Probl. Eng.* **2015**, *2015*, 8. [CrossRef]

24. Aggarwal, K.; Atan, O.; Farahat, A.K.; Zhang, C.; Ristovski, K.; Gupta, C. Two Birds with One Network: Unifying Failure Event Prediction and Time-to-failure Modeling. In Proceedings of the 2018 IEEE International Conference on Big Data (Big Data), Seattle, WA, USA, 10–13 December 2018; pp. 1308–1317.

25. Bach-Andersen, M.; Rømer-Odgaard, B.; Winther, O. Flexible non-linear predictive models for large-scale wind turbine diagnostics. *Wind Energy* **2016**, *20*, 753–764, doi:10.1002/we.2057. [CrossRef]

26. Kleinbaum, D.; Klein, M. *Survival Analysis: A Self-Learning Text*, 3rd ed.; Statistics for Biology and Health; Springer: New York, NY, USA, 2011.

27. Patti, S.; Biganzoli, E.; Boracchi, P. Review of the Maximum Likelihood Functions for Right Censored Data. A New Elementary Derivation. In *COBRA Preprint Series*; bepress: Berkeley, CA, USA, 2007.

28. Kaplan, E.L.; Meier, P. Nonparametric Estimation from Incomplete Observations. *J. Am. Stat. Assoc.* **1958**, *53*, 457–481, doi:10.1080/01621459.1958.10501452. [CrossRef]

29. Herp, J.; Ramezani, M.H. State transition of wind turbines based on empirical inference on model residuals. In Proceedings of the 2015 Conference on Research in Adaptive and Convergent Systems, RACS 2015, Prague, Czech Republic, 9–12 October 2015; pp. 32–37, doi:10.1145/2811411.2811509. [CrossRef]

30. Herp, J.; Ramezani, M.H.; Nadimi, E. Dependency in State Transitions of Wind Turbines-Inference on model Residuals for State Abstractions. *IEEE Trans. Ind. Electron.* **2017**, doi:10.1109/TIE.2017.2674580. [CrossRef]

31. Prescott Adams, R.; MacKay, D.J.C. Bayesian Online Changepoint Detection. *arXiv* **2007**, arXiv:0710.3742.

32. Fisher, R.A. On the Mathematical Foundations of Theoretical Statistics. *Philos. Trans. R. Soc. Lond. A Math. Phys. Eng. Sci.* **1922**, *222*, 309–368, doi:10.1098/rsta.1922.0009. [CrossRef]

33. Hipp, C. Sufficient Statistics and Exponential Families. *Ann. Stat.* **1974**, *2*, 1283–1292, doi:10.1214/aos/1176342879. [CrossRef]

34. Fisher, R.A. Applications of "Student's" distribution. *Metron* **1925**, *5*, 90–104.

35. Goodfellow, I.; Bengio, Y.; Courville, A. *Deep Learning*; MIT Press: Cambridge, MA, USA, 2006. Available online: http://www.deeplearningbook.org (accessed on 12 December 2019)

Article

Gearbox Fault Prediction of Wind Turbines Based on a Stacking Model and Change-Point Detection

Tongke Yuan [1,*], Zhifeng Sun [1] and Shihao Ma [2]

[1] College of Electrical Engineering, Zhejiang University, Hangzhou 310027, China; eeszf@zju.edu.cn
[2] Wuzhong Baita Wind Power Corporation Limited, Wuzhong 751100, China; ma_sh@ecidi.com
* Correspondence: yuantongke@zju.edu.cn; Tel.: +86-15306539411

Received: 12 October 2019; Accepted: 4 November 2019; Published: 6 November 2019

Abstract: The fault diagnosis and prediction technology of wind turbines are of great significance for increasing the power generation and reducing the downtime of wind turbines. However, most of the current fault detection approaches are realized by setting a single alarm threshold. Considering the complicated working conditions of wind farms, such methods are prone to ignore the fault, send out a false alarm, or leave insufficient troubleshooting time. In this work, we propose a gearbox fault prediction approach of wind turbines based on the supervisory control and data acquisition (SCADA) data. A stacking model composed of Random Forest (RF), Gradient Boosting Decision Tree (GBDT), and Extreme Gradient Boosting (XGBOOST) was constructed as the normal behavior model to describe the normal conditions of the wind turbines. We used the Mahalanobis distance (MD) instead of the residual to measure the deviation of the current state from the normal conditions of the turbines. By inputting the MD series into the proposed change-point detection algorithm, we can obtain the change point at which the fault symptom begins to appear, and thus achieving the fault prediction of the gearbox. The proposed approach is validated on the historical data of 5 wind turbines in a wind farm, which proves its effectiveness to detect the fault in advance.

Keywords: wind turbines; fault prediction; stacking model; normal behavior model; change-point detection; SCADA

1. Introduction

In recent years, due to the increasingly serious energy crisis and environmental pollution, many countries of the world have vigorously developed new energy sources. As green renewable energy, wind energy has received great attention from countries all over the world. According to the 2018 annual report of the Global Wind Energy Council (GWEC) [1], the global offshore and onshore installed capacity has reached 23,140 MW and 568,409 MW, respectively. In 2018, the new installed capacity of onshore wind turbines reached 46,820 MW. GWEC expects the onshore market to install upwards of 50 GW per year until 2023. With the increase of the installed capacity of wind power, more urgent requirements are put forward for the daily fault detection and operation and maintenance (O&M) of wind turbines. Due to the remote location of the wind farm, the WTs are exposed to highly variable and harsh weather conditions [2], which makes the O&M cost of wind turbines higher than other traditional power generation methods. Especially when the wind farm is built at sea, desert, etc., its O&M costs will be much higher. Therefore, it is necessary to vigorously develop wind turbine fault diagnosis technology and to provide early warning of faults and achieve certain accuracy and practicability. Currently, wind turbine manufacturers and research institutes are trying to make breakthroughs in the fault diagnosis and O&M technology of wind turbines.

There are many sensors installed on the turbines to record the working conditions of each component. According to the utility of the data recorded by sensors, the parameter system of the wind turbine can be divided into the supervisory control and data acquisition (SCADA) system, which mainly consists of the performance parameters and the condition-monitoring system (CMS) based on the vibration parameters [3]. For different fault modes, the parameters such as temperature, power, speed, and blade angle of the turbine will not be the same. Therefore, the relationship between these data and the fault can be explored through machine learning algorithms. SCADA and CMS systems form a big data system for wind turbines. The advantage of big data is that as long as the amount of data is large enough, by selecting appropriate data mining methods and optimizing the data mining process, it is possible to predict unknown parameters and its regularity.

For the fault diagnosis of wind turbines, Professor Andrew Kusiak of the Intelligent Systems Laboratory of the University of Iowa has made great achievements. Kusiak et al. [4] developed virtual models of a wind turbine which were extracted by six machine learning algorithms. The power output and rotor speed were selected as performance parameters to test the proposed methodology. In [5], a three-level fault prediction methodology based on the SCADA data and status data of 4 wind turbines was proposed. In order to predict the blade angle asymmetry and implausibility faults, an association rule mining algorithm was applied and five classifiers were used to identify the best model [6]. Also, the bearing faults were analyzed by building a normal behavior model based on the neural network algorithms in [7]. The test results on five over-temperature events validated its effectiveness on bearing faults prediction.

Generally speaking, the approaches for wind turbine fault detection/prediction through data mining algorithms are divided into two major ways. One is through normal behavior modeling (NBM) [8]. By selecting the historical SCADA data under normal conditions as the training set, the normal behavior model of the key components of the turbines can be constructed. Support Vector Machine (SVM) [9,10], neural network [11], and deep belief networks [12] are commonly used algorithms in NBM. Then, according to the statistical characteristics of the normal conditions of the wind turbine, an empirical threshold is set as an alarm line so that any fault deviates from the normal conditions can be detected. Bangalore et al. [13] constructed an artificial neural network (ANN) normal behavior model to predict the gearbox fault. During the anomaly detection stage, a Mahalanobis Distance (MHD) method was presented to describe the deviation between the ANN model and the operating condition. Finally, the case studies of four wind turbines illustrated the effectiveness of the proposed approach. For other subcomponents of the turbines, such as the generator bearings and slip rings, etc., the evolution of their faults develops faster than that of gearbox and thus it is difficult to make timely intervention before the fault occurs. Therefore, it is particularly important to achieve the fault prediction of these subcomponents through advanced analysis techniques. In [14], a nonlinear state estimate technique (NSET) is proposed to construct the normal behavior model of the generator temperature. In [15], the selected model is principal component regression and can identify the incoming generator slip ring fault approximately one day in advance. In [16], the authors proposed a model-based approach to predict generator bearing temperature and thus achieving the early detection of bearing and gearbox faults.

The other way is a classification-based approach. The classification-based approach is to divide the operating state of the wind turbine into different categories (label), such as normal operation, shutdown, or a specific type of fault. Each sample in the training set consisting of SCADA historical data has a label. Then the classifier after the training process can be used to determine the type of fault of the turbines. Association rule is a commonly used algorithm. The authors in [6] used the Apriori algorithm to identify the correlation from the database composed of fault categories, various operating parameters, and history records and then identified the specific fault mode. Similarly, in [17], the Apriori algorithm was applied to find the key event codes of wind turbines.

For the NBM approach, no matter which algorithm is used to establish the prediction model, most of the approaches during the fault prediction stage are achieved by analyzing the residual and setting a single alarm threshold. The authors in [18] used the Mahalanobis distance instead of the residual and applied the Weibull distribution to obtain the threshold of the fault alarm. Wu et al. [19] developed a normal behavior model for wind turbines gearboxes using echo state networks. The potential faults were detected through residual evaluation and dynamic threshold setting. However, using a single alarm threshold or an adaptive threshold method often has a time uncertainty problem, that is, an accurate time point at which the fault symptom begins to appear cannot be given. Moreover, an inappropriate threshold setting may cause a false alarm or miss detection, which makes the fault prediction based on SCADA data has the problems of low prediction accuracy and low computational efficiency. In order to improve the reliability and accuracy of fault prediction, we propose a gearbox fault prediction approach based on a stacking model and CUSUM (Cumulative Sum) control chart. The framework of the proposed method can be divided into four parts, which are the data preprocessing, feature selection, the normal behavior modeling process, and the fault prediction process, where several algorithms or their combinations were applied. The main contributions of our work and the functionality of these algorithms are summarized as follows:

1. During the data preprocessing process, we select the SCADA historical data under the normal working conditions and apply the Density Based Spatial Clustering of Applications with Noise (DBSCAN) algorithm and the quartile method to recognize and filter out the three types of outliers and abnormal data. The DBSCAN algorithm plays an important part in filtering out most of the outliers and then the quartile method is applied for further preprocessing to improve the data quality.
2. During the feature selection process, three data mining algorithms are used to select the features most relevant to the gearbox oil temperature as the input variables of the model.
3. Based on the Stacking idea, a normal behavior model including Random Forest (RF), GBDT, and XGBOOST is established to describe the temperature characteristics of the gearbox under normal operation state. By comparing the performance of seven single models on three metrics, we finally select three models, RF, GBDT, XGBOOST, as the basic models of our proposed stacking model. The proposed stacking model trains three basic models in parallel which can avoid the defect of a single machine learning model like overfitting and low prediction accuracy. The case study illustrates the superiority and effectiveness of the proposed model.
4. The change-point detection algorithm based on CUSUM control chart is applied to detect the change points in the MD series. Compared with the fault detection algorithm by setting a single threshold, the proposed approach can accurately locate the time at which the fault symptom begins to appear. A case study on five faulty turbines shows the proposed approach can predict the faults 13 to 22 days in advance.

The rest of this paper is organized as follows: Section 2 briefly introduces three types of abnormal data and illustrates the necessity of data preprocessing. Section 3 clarifies the structure of the proposed approach. Section 4 introduces the data set used in this work and the entire process of normal behavior modeling. In Section 5, we use the data of five turbines to test the performance of our proposed approach and list the test results. Section 6 summarizes the full paper and points out the future research direction.

2. Background Review

The complete wind turbine consists of the gearbox system, generator system, pitch system, yaw system and other key components [20]. The gearbox is an important component of the wind turbine transmission system. Its main function is to transmit the power generated by the blade under the action of the wind to the generator and achieve a higher speed. At present, the gearbox manufacturing technology is relatively mature and has high reliability, however, due to its frequent high-speed, heavy-duty, and high-impact working environment, it is prone to suffer various faults within an operational period of 5 years, and has to be replaced [21]. Moreover, the system shutdown caused by gearbox failure takes the longest time and a gearbox replacement can cost up to 14.5 percent of the wind turbine maintenance cost [20]. Therefore, gearbox fault prediction is necessary.

Generally, gearbox faults can be reflected by the change of gearbox oil temperature. In [22], the authors illustrated that the rise of oil temperature could indicate the development of the gearbox faults. And an artificial neural network model was constructed to predict the oil temperature. In this work, the gearbox oil temperature is modeled to describe the working status of the wind turbine.

The normal behavior modeling approach has high requirements for the quality of training data: fault-free SCADA data must be accurately selected [23]. However, due to sensors' accuracy degradation, communication failure, and curtailment, the turbine can operate in different modes even under normal conditions [24]. Figure 1 shows three common types of anomaly data in the power curve:

- The lower stacked abnormal data is represented as a horizontally dense data cluster in the scatter plot, as shown in Type 1 in Figure 1. Type 1 indicates the points where the wind power output from the wind turbine is 0 regardless of the wind speed. Such abnormal data appears due to damage to the power measuring instrument or abnormal communication equipment, fault of the wind turbine, etc. [25].
- Wind curtailment data. The "Wind curtailment" data appears as a horizontally dense cluster of data in the scatter plot, typically in the middle of the curve, as shown in Type 2 in Figure 1. Such anomaly data shows that the output power of the wind turbine does not change with the change of wind speed. Wind curtailment is the reduction in electricity generation below what a system of well-functioning wind turbines are capable of producing [26], which is usually enforced by wind farms. The following are the main causes of "Wind curtailment". First, China's energy structure is still dominated by thermal power generation, supplemented by new energy, so wind power generation is only used as regulating power. Second, because of the excess capacity of wind power, it is difficult to store large capacity wind power resources, so it is necessary to curtail wind power. Finally, due to the instability of wind speed, the electric energy generated by the wind turbine is also unstable. When the wind speed has a huge fluctuation, the grid-connected cannot be performed forcibly, so the wind curtailment operation has to be carried out at this time.
- The dispersive anomaly data is randomly distributed around the curve, as shown in Type 3 in Figure 1. The distribution of such anomaly data is irregular and is usually caused by sensor accuracy degradation, instrument fault, or noise during the signal propagation.

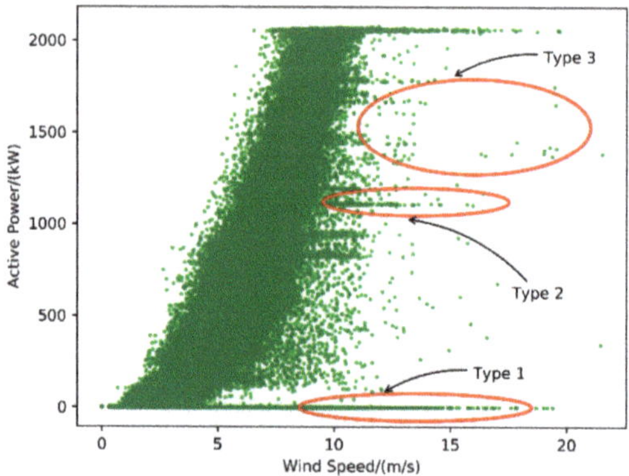

Figure 1. Three common types of anomaly data in the power curve.

If these raw SCADA data are directly used for analysis and modeling, it will inevitably lead to a decrease in prediction accuracy and difficulty in achieving research goals. Therefore, it is necessary to preprocess the raw data collected by the SCADA system before establishing the normal behavior model.

3. Framework for Proposed Fault Prediction Approach

The main idea of our proposed fault prediction approach is to fit the normal operating behavior of wind turbines by establishing a normal behavior model of the gearbox oil temperature, and then use the Mahalanobis distance, instead of the residual, to describe the deviation between the actual measured value and the model output value. Finally, the Change-point detection based on CUSUM control charts is applied to implement the gearbox fault prediction. The overall flowchart of the fault prediction model proposed in this study can be divided into the training process and application process, as shown in Figure 2. The training process deals with the necessary data preprocessing and feature selection, which are performed based on the historical data collected by the SCADA system to establish the normal behavior model of the gearbox. During the application process, the real-time SCADA data is input into the normal behavior model after the same data preprocessing and feature selection. There is a certain deviation between the predicted values of the model and the actual values recorded by the SCADA system. The larger the deviation is, the higher the possibility of a fault happens. In this work, we apply the Mahalanobis distance to describe this deviation. By inputting the Mahalanobis distance series into the proposed Change-point detection algorithm, the time at which the fault occurs and the time at which the fault symptom begins to appear can be obtained, and thus the fault prediction can be realized.

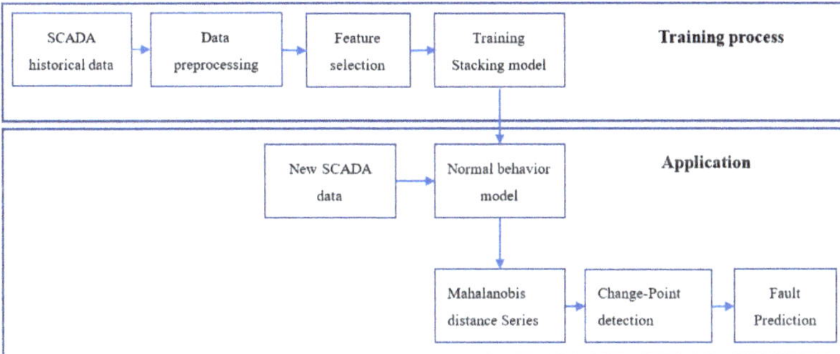

Figure 2. Overview of the overall framework.

3.1. Data Preprocessing

From Figure 1, we can see intuitively that the three types of anomaly data points show obvious outlier characteristics in the power curve. In this work, we apply the combination of DBSCAN and the quartile method to identify the anomaly points and outliers. The case study of wind turbines using SCADA data has validated this data cleaning method, which demonstrates its effectiveness and generality.

DBSCAN is a typical density-based clustering algorithm, which is designed to discover the clusters and the noise in a spatial database [27]. According to two input parameters, Eps and MinPts, the algorithm divides the data to be clustered into three categories: core points, border points, and noise points. A core point refers to the point in a cluster that there are at least a minimum number of points in an Eps-neighborhood of that point. The Eps-neighborhood of border points contains fewer points than core points. And the points not belong to core points or border points are defined as outliers (noise), which are the anomaly data to be filtered in this work. Compared with the K-Means clustering algorithm, the DBSCAN does not need to determine the number of cluster centers in advance and can identify clusters of arbitrary shapes and has strong anti-noise ability.

The quartiles are a type of quantile which can be applied to check for outliers [28]. In this work, we used the first quartile (Q_1) and the third quartile (Q_3) to achieve the further preprocessing of the data after the DBSCAN method. The first quartile refers to the middle number between the smallest number and the median of the data set. The third quartile is defined as the middle value between the median and the highest value of the data set. By calculating the Q_1 and Q_3, we can get the interquartile range (*IQR*) which defined as follows:

$$IQR = Q_3 - Q_1 \tag{1}$$

According to the *IQR*, we can get the normal data bounds expressed as follows:

$$[F_l, F_u] = [Q_1 - 1.5IQR, Q_3 + 1.5IQR] \tag{2}$$

where F_l and F_u are the lower limit and upper limit of normal data, respectively. Any data lying outside the defined bound can be considered an outlier. In this work, the quartile method is adopted by dividing the wind speed data into different intervals and then to filter the active power data in each interval correspondingly. The interquartile range is the same as the variance and standard deviation, which can represent the dispersion of statistical data of a variable. But the interquartile range is relatively robust statistics since the value of *IQR* does not change significantly with individual abnormal data. However, this method has its limitations, that is, the method can correctly and effectively identify these abnormal data only when the proportion of abnormal data to the total amount of data is small.

Hence, in this study, the combination of the DBSCAN and quartile method is applied for data preprocessing. It can be seen from Figure 1 that the proportion of three kinds of abnormal data is relatively large. If we user the quartile method directly, the abnormal data is bound to affect the value of the *IQR*. Moreover, the density of these abnormal data is significantly less than normal data. Therefore, in the data preprocessing stage, the DBSCAN is firstly used to identify and remove most of the abnormal data, the SCADA data is further processed by the quartile method to improve the quality of data cleaning.

3.2. Feature Selection

Feature selection is a significant part of the data mining process. As the input of the model, features determine the quality of the model. Generally, feature selection is executed by selecting a subset from many features or applying a suitable method to reduce the dimension. The purpose of feature selection is to remove redundancy, reduce the number of features, improve model accuracy and reduce run time. In this work, to establish the normal behavior model of the wind turbine gearbox, the gearbox oil temperature is selected as the output variable. Three feature selection algorithms, namely, filter model [29], wrapper model with recursive feature elimination (RFE) algorithm [30,31], and RF model [32,33], are applied to select the features most relevant to the gearbox oil temperature as the input variables of the model. These three algorithms are typical feature selection methods based on how to generate feature subsets [34]. They will calculate the most relevant features according to different rules, which can avoid the defect of a single method. By inputting the SCADA data with whole 39 parameters as input variables, we obtained the top 10 most relevant parameters ranked by each algorithm and finally choose 15 parameters as the input variables.

3.3. The Stacking Model

The oil temperature of the gearbox is affected by many factors including the working condition and the inherent properties of wind turbines, thus making the status information complicated. To avoid the defects of the single model, we propose a model based on the Stacking strategy by combining three regressors for temperature predictions. At present, the machine learning model based on the stacking strategy has been widely applied in Kaggle competition, sentiment classification [35], speech recognition [36], and many other application scenarios. As far as we know, there are still no researchers who have applied stacking strategy to wind turbine fault prediction.

In this work, a two-layer stacking model is proposed to construct the normal behavior model of gearbox oil temperature as shown in Figure 3.

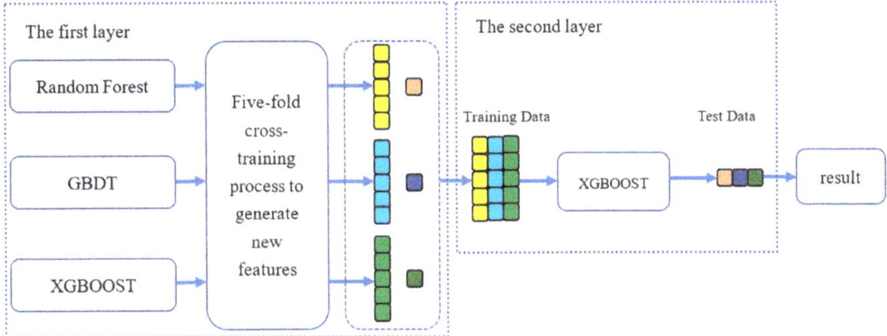

Figure 3. Structure of the two-layer stacking model.

The first layer consists of three basic models, including RF [32], GBDT [37], and XGBOOST [38]. The idea of stacking is to train these basic models in parallel and combine them by training a meta-model to output predictions based on the multiple predictions returned by these basic models [39]. For each basic model, we use five-fold cross-training, which is similar to k-fold cross-validation, to generate training data and test data for the meta-model in the second layer. Figure 4 describes the details of this process.

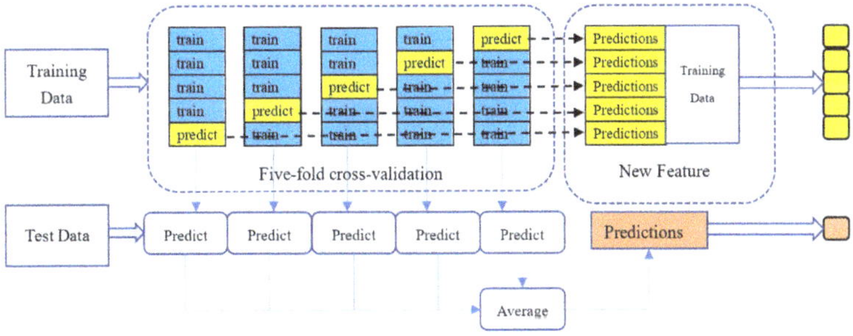

Figure 4. Structure of the five-fold cross-validation process.

The training data is divided into five folds. In each training process, four folds are used for training and the remaining one fold is used for making predictions for observations. In other words, the five-fold cross-training consists in training on four folds in order to make predictions on the remaining fold and that iteratively so that to obtain predictions for observations in any folds. By combining these predictions of three basic models as new features, we can obtain the training data of the meta-model in the second layer. Other than this, each basic model can produce five predictions of the original test data. We take the average of these predictions as the test data for the meta-model. In the second layer, with the training data and the test data produced by the cross-training process, the XGBOOST model is trained as the meta-model and output the final predictions based on the test data.

The proposed stacking model acts as a normal behavior model to generate a prediction of the gearbox oil temperature based on the given input features. In this work, by inputting the 15 features we selected, the stacking model will output the prediction of gearbox oil temperature at this time. Its superiority lies in the part that it can train three single machine learning models in parallel, which greatly accelerates the training process. The performance of our stacking model on the test data has validated its effectiveness.

3.4. Fault Prediction Approach

3.4.1. Mahalanobis Distance Series

The Mahalanobis distance is an effective method for calculating the similarity of two unknown sample sets. Compared to the Euclidean distance, the MD is scale-invariant and takes into account the correlations of the data set. At present, the MD has been introduced into the field of wind turbine fault detection to determine the fault alarm threshold [12,13]. In this work, we use the MD to quantify the deviation between the test data (real-time SCADA data) and the normal behavior model. The MD for the normal behavior model can be expressed as follows:

$$MD_{nor_i} = \sqrt{\left(X_{nor_i} - \mu_{nor}\right)C_{nor}^{-1}\left(X_{nor_i} - \mu_{nor}\right)^{T}} \tag{3}$$

where $X_{nor_i} = [err_i, MV_i]$. MV_i represents the i-th measured value recorded by the SCADA system during the five-fold cross-validation process and err_i is the validation error; C_{nor} and μ_{nor} are the covariance matrix and the mean value vector of X_{nor}; MD_{nor_i} is the Mahalanobis distance of the vector X_{nor_i}.

During the fault prediction stage, we input the real-time SCADA data into the normal behavior model and calculate the corresponding MD series as the input of the subsequent change-point detection algorithm. The MD in this stage is defined as follows:

$$MD_{test_i} = \sqrt{\left(X_{test_i} - \mu_{nor}\right)C_{nor}^{-1}\left(X_{test_i} - \mu_{nor}\right)^{T}} \tag{4}$$

where $X_{test_i} = [err_i, MV_i]$ is consists of the test error err_i and the measured value MV_i during the fault prediction stage; C_{nor} and μ_{nor} are the covariance matrix and the mean value vector obtained from the normal behavior model. 1; $MD_{test} = \left[MD_{test_1}, MD_{test_2}, \ldots, MD_{test_N}\right]$ is the final MD series as the input of the change-point detection algorithm.

3.4.2. Change-Point Detection

At present, most of the fault detection approaches of wind turbines analyze the residual series by setting the single threshold or adaptive alarm threshold, which has the problem of time uncertainty. In this work, we applied the change-point detection algorithm based on the CUSUM control chart to analyze the MD series.

The CUSUM control chart is drawn by the cumulative sum of the difference between the observed value and the target value. It is widely used in the field of economics because of its simple and intuitive characteristics. If the statistical value is higher than the average value over a period of time, this amount of data will continue to accumulate, and the CUSUM control chart will show a steady growth trend and vice versa. Therefore, the CUSUM control chart can detect whether there is a change in the series, but the detailed information of the change point, such as confidence level and confidence interval, cannot be obtained only through the control chart [40]. Therefore, in this work, we propose a fault prediction algorithm combining the CUSUM control chart and change-point detection, which can accurately capture the small abnormal changes in the MD series. Algorithm 1 below describes the calculation process of the algorithm.

Algorithm 1: Change-Point Detection.

Data: Mahalanobis Distance Series MD and its length M; Confidence Level C; Number of bootstrap samples performed N

Result: index of the change point

1 Calculate the average $\overline{MD} = \frac{MD_1 + MD_2 + \ldots + MD_M}{M}$

2 Calculate the cumulative sums by adding the difference between the current value and the average to the previous sum. $S_i = S_{i-1} + \left(X_i - \overline{X}\right)$ for $i = 1, 2, \ldots, M$ and $S_0 = 0$

3 Calculate the original S_{diff} which defined as: $S_{diff} = S_{\max} - S_{\min}$—where $S_{\max} = \max\limits_{i=1,2,\ldots,M} S_i$, $S_{\min} = \min\limits_{i=1,2,\ldots,M} S_i$

4 for j in *range* (1, N) do

5 Generate a bootstrap sample of the original MD series, denoted $MD_1^j, MD_2^j, \ldots, MD_M^j$

6 Based on the bootstrap sample, calculate the bootstrap CUSUM, denoted $S_0^j, S_1^j, \ldots, S_M^j$

7 Calculate the difference of the bootstrap CUSUM, denoted S_{diff}^j

8 Compare each S_{diff}^j with the original S_{diff}, let X be the number of bootstraps for which $S_{diff}^j < S_{diff}$ for $j = 1, 2, \ldots, N$

9 Calculate the confidence level P that a change occurred as a percentage: $P = 100\frac{X}{N}\%$ and compare with the input confidence level C

10 If $P > C$, a change has occurred with a confidence level of C

11 Output the index of the change point m, let m be such that: $|S_m| = \max\limits_{i-0,1,\ldots,M} |S_i|$

The input parameters of the algorithms are the MD series, its length and set a confidence level C and the number of bootstrap samples performed N. The functionality of the proposed approach is to find the possible change point in the MD series obtained from our normal behavior model and then output the index of the change point with the given confidence level. In other words, the change point output by our algorithm is calculated by a large number of repeated sampling and obtained based on the probability. Our fault detection approach will perform bootstrap sampling N times on the original MD series and will generate a new random sorted MD series each time. If the majority of the new series are recognized as having change points, we are confident about the conclusion that there exist change points in the original MD series. For example, if the algorithm finds that there are 900 new MD series containing change points in the total 1000 times of bootstrap sampling, we can draw a conclusion that the original MD series have change points with a confidence level of 90%. The final number of change points depends on the confidence level we set. If we set a relatively low confidence level, there is a high possibility that we will get many change points, which including the points at which the real fault happened as well as the points turn out to be unreliable. To get more reliable results, we can set a higher confidence level so that we will much more confident about the change points obtained from our algorithms and they will be closer to the real conditions.

The principle of using the CUSUM control chart to judge whether a change has taken place in the series is to generate new MD series by multiple sampling without replacement. If the maximum difference of the cumulative sum of the new series is mostly smaller than that of the original series, we have reason to believe that a change must have occurred in the original MD series. The more this happens in N times sampling, the higher the confidence level of a change happens. Once a change has been detected, it is also necessary to determine when the change occurred. In this work, we use the binary search algorithm to specify the index of the change point. The original MD series is divided into two parts based on the change point found by the change-point detection algorithm, similar to the binary search algorithm, we iteratively execute the Algorithm 1 on each subseries until all the change points that satisfy the confidence level are accurately found. Figure 5 is the flowchart of the entire change-point detection process.

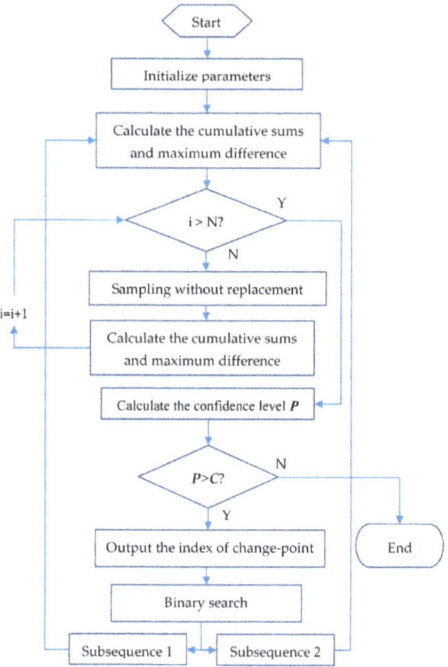

Figure 5. The flowchart of the change-point detection process.

4. Modeling Normal Gearbox Behavior

4.1. Data Description

The SCADA data used in this work comes from 173 wind turbines in a wind farm located in Ningxia, China. In this work, the SCADA data contains 37 parameters for each wind turbine with a one-minute interval, which can be divided into four categories, including the condition parameters, the health parameters, the performance parameters, and the controlling parameters [41]. We use two types of wind turbines with rated powers of 1500 kW and 2000 kW as the research object, and correspondingly establish two normal behavior models to fit the gearbox oil temperature under normal working conditions of turbines. Table 1 shows the specifications of two types of wind turbines. The SCADA data are re-sampled using the 5-min interval.

Table 1. Specification of two types of wind turbines.

Wind Turbine Parameters	Type	
	A	B
Rated power	1500 kW	2000 kW
Cut-in, rated wind speed	3 m/s, 11 m/s	2.5 m/s, 10 m/s
Rotor diameter	88 m	115 m
Blade length	43 m	56.5 m
Hub height	80 m	90 m

Table 2 lists the detailed descriptions of the modeling datasets. A totally of 16 turbines were used for modeling analysis. To depict the normal behavior of the turbines as completely as possible, we used the SCADA data of five turbines to train the normal behavior model, and there were no gearbox faults during the corresponding time interval. In addition, the historical SCADA data of turbines #36 and #149 in a month period is selected to test the effectiveness of the normal behavior model we established. Finally, we use five turbines (#38, #58 for type A and #159, #173, #162 for type B) to test our proposed fault prediction algorithm. According to the wind farm fault alarm record, these five turbines experienced gearbox oil temperature exceeding limit faults during the selected time interval.

Table 2. Description of modeling datasets.

Type	Dataset	Turbines Considered	Timestamps
A	Training	#34, #35, #42, #44, #52	1 July 2018–1 January 2019
	Testing normal behavior	#36	1 July 2018–31 July 2018
	Testing fault prediction	#38	1 March 2018–4 April 2018
		#58	1 April 2018–1 May 2018
	Training	#150, #152, #153, #156, #169	1 July 2018–1 January 2019
B	Testing normal behavior	#149	1 December 2018–30 December 2018
	Testing fault prediction	#159	30 March 2018–30 April 2018
		#173	1 April 2018–30 April 2018
		#162	10 February 2018–18 March 2018

4.2. Data Preprocessing

After removing the duplicate and missing values in the original SCADA data, in order to accurately describe the normal conditions of the wind turbines, a power curve is built from SCADA data where the outliers are filtered out through the method we proposed above. In this section, we use the historical data of turbine #159 during the half-year period to validate the performance of the proposed data cleaning method. The result is provided in Figure 6.

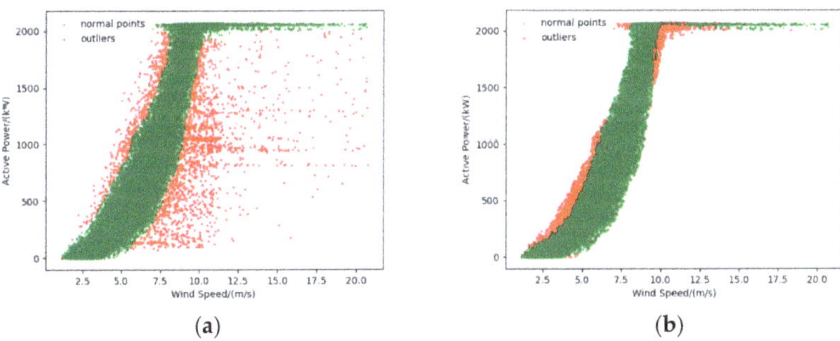

Figure 6. Performance of the proposed data preprocessing method: (**a**) result of the DBSCAN and (**b**) result of the quartile method after filtered out by the DBSCAN.

The DBSCAN method is firstly applied to identify and filter out most of the outliers as shown in Figure 6a, and we additionally use the quartile method to further filter the SCADA data. Through the combination of the DBSCAN and quartile method, we obtain the power curve seen in Figure 7. It is not difficult to see from the figure that the data preprocessing method can filter out the outliers in the power curve, that is, corresponding to the abnormal working conditions of the wind turbine.

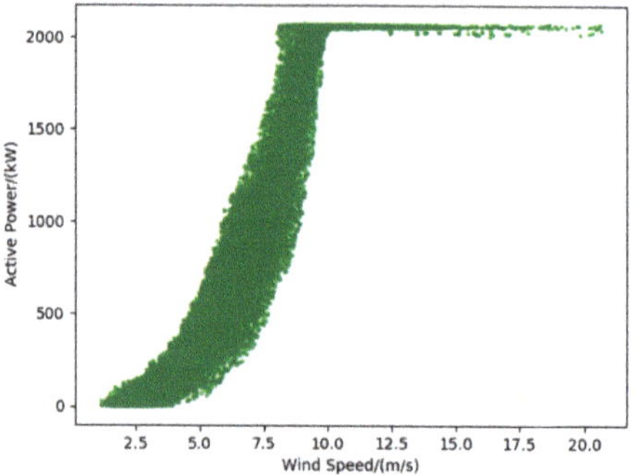

Figure 7. The power curve after data preprocessing.

4.3. Feature Selection

After deletion of the abnormal data of the turbines, three data-mining algorithms (filter model, wrapper model with recursive feature elimination algorithm, and Random Forest) are used to select the most relevant parameters for predicting the gearbox oil temperature. Considering that the rise of oil temperature is a gradual process that cannot change in a sudden, the input parameters measured at past intervals may influence the future state of the turbines [4]. In this work, we introduce the gearbox oil temperature parameter at two past intervals, (t-1) and (t-2), into input parameters to predict the oil temperature at time t. The time interval is 5 min. Table 3 presents the 10 most relevant parameters of all 39 parameters. Based on the result of three feature selection algorithms, we finally choose the following 15 parameters as the input variables for gearbox oil temperature prediction, namely gearbox oil temperature(t-1), gearbox oil temperature(t-2), active power, gearbox bearing A temperature, gearbox bearing B temperature, wind speed, generator speed, rotor speed, generator stator L3 temperature, voltage phase C, generator torque, active consumption, generator bearing A temperature, pitch angle, and rotor temperature.

Table 3. Result of the three feature selection algorithms.

No.	Filter	Wrapper	RF
1	Gearbox oil Temperature(t-1)	Active power	Gearbox oil Temperature (t-1)
2	Gearbox oil Temperature(t-2)	Active consumption	Gearbox bearing A temperature
3	Gearbox bearing A temperature	Gearbox bearing A temperature	Gearbox bearing B temperature
4	Gearbox bearing B temperature	Gearbox bearing B temperature	Gearbox oil Temperature (t-2)
5	Wind speed	Generator Stator L1 temperature	Wind speed
6	Active power	Generator Stator L3 temperature	Active power
7	Generator speed	Wind speed	Rotor temperature
8	Rotor speed	Gearbox oil Temperature (t-1)	Voltage phase b, c
9	Generator torque	Gearbox oil Temperature (t-2)	Rotor speed
10	Generator Stator L3 temperature	Generator bearing A temperature	Pitch angle

4.4. Model Construction

In this section, the normal behavior model of gearbox oil temperature is constructed based on the stacking strategy. We use the SCADA data of ten turbines to establish the stacking models of two types of wind turbines, five for type A and the remaining five for type B, as shown in Table 2. To evaluate the performance of the proposed model, three metrics, mean absolute error (MAE), root mean square error (RMSE), and coefficient of determination (R^2), are applied, which are defined as follows:

$$\text{MAE} = \frac{1}{N}\sum_{i=1}^{N}\left|\hat{y}_i - y_i\right| \tag{5}$$

$$\text{RMSE} = \sqrt{\frac{1}{N}\sum_{i=1}^{N}(\hat{y}_i - y_i)^2} \tag{6}$$

$$R^2 = 1 - \frac{\sum\limits_{i=1}^{N}(\hat{y}_i - y_i)^2}{\sum\limits_{t=1}^{N}(y_i - \bar{y})^2} \tag{7}$$

where y_i is the i-th measured value of oil temperature, $\hat{y}_i$ represents the i-th prediction of the model and $\bar{y}$ refers to the mean value of the measured values.

Furthermore, seven single machine learning models, including k-Nearest Neighbor (KNN), Support Vector Machine (SVM), Decision Tree (DT), Adaboost, RF, GBDT, and XGBOOST, are applied for comparison with the proposed stacking model. In particular, the latter three models are the basic models of our proposed stacking model and the XGBOOST model is also selected as the meta-model in the second layer. We set aside 15% of the training data as the test set. Table 4 lists the performance of the models. We can observe that the three models, RF, GBDT, and XGBOOST, achieve better performance than other single models, and the stacking model based on these three models achieves the best performance with the highest R^2 scores and the lowest MAE and RMSE.

Table 4. Performance of the models.

Model	A			B		
	MAE	**RMSE**	**R^2**	**MAE**	**RMSE**	**R^2**
KNN	1.586	2.064	0.935	1.383	1.765	0.965
SVM	0.512	0.714	0.959	0.657	1.033	0.988
DT	0.644	0.948	0.928	0.416	0.662	0.985
Adaboost	0.689	0.845	0.943	0.675	0.893	0.991
RF	0.289	0.389	0.996	0.262	0.407	0.995
GBDT	0.266	0.381	0.997	0.279	0.408	0.995
XGBOOST	0.265	0.380	0.997	0.282	0.409	0.995
Stacking	0.219	0.352	0.998	0.254	0.376	0.997

4.5. Testing Normal Behavior Model

In this section, in order to evaluate whether the proposed model can accurately describe the normal behavior of wind turbines, we select the historical SCADA data of turbine #36 (type A) and #149 (type B) for a period of a month to test the two models constructed above. Table 5 shows the metrics of the two normal behavior models on the test set. Figures 8 and 9 show the performance of the two models on type A and B respectively, including the comparison of the measured value and the model predictions, the Mahalanobis distance and the CUSUM control chart.

Table 5. Performance of the two normal behavior models.

Type	Turbine	MAE	RMSE	R^2
A	#36	0.440	0.589	0.972
B	#149	0.343	0.435	0.998

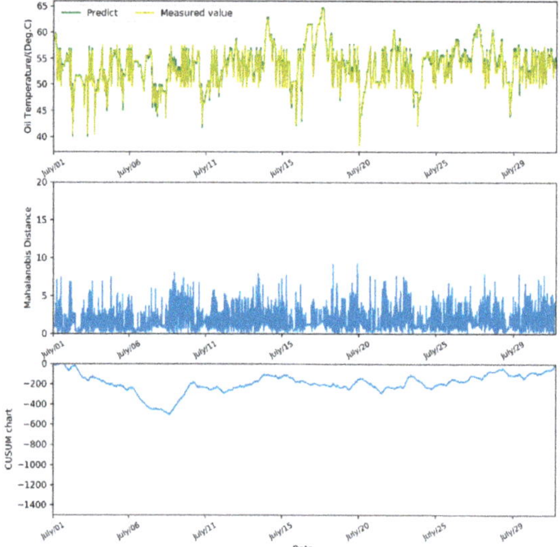

Figure 8. Testing normal behavior model results for Turbine #36.

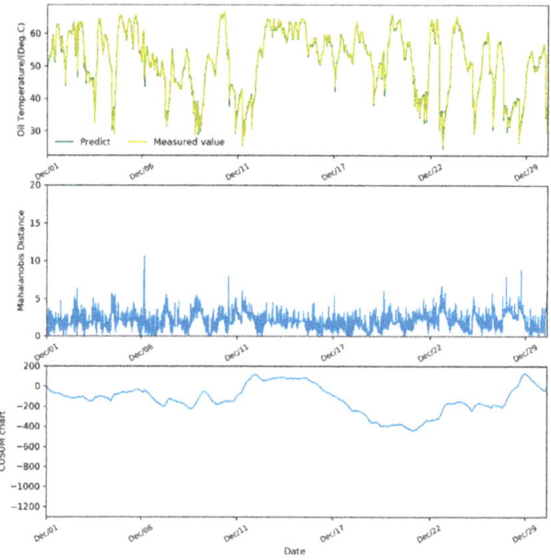

Figure 9. Testing normal behavior model results for Turbine #149.

As can be seen from Figures 8 and 9, the predicted value of the model for the gearbox oil temperature can accurately fit the measured value, and the gearbox oil temperature values fall within the normal working range of (20, 75). The Mahalanobis distance values of the two turbines are relatively stable, and correspondingly, there is no sharp change occurred in the CUSUM control chart. This proves that the proposed model can accurately describe the temperature change of the gearbox during normal operating conditions. Hence, it can be concluded that the normal behavior model we built can be used for the modeling analysis of any turbines of the same type in the wind farm, which means there is no need to establish a normal behavior model for each turbine.

5. Testing Fault Prediction Approach

In this section, the proposed approach has been applied in five wind turbines (two turbines of type A and three turbines of type B) to predict the gearbox fault. According to the fault alarm record of the wind farm, the five turbines have suffered gearbox oil temperature over-limit fault during the selected time period. First, the test data is input into the normal behavior model we established to obtain the MD series on the test data set. Then, the change-point detection algorithm is applied to analysis the MD series, and the time point at which the fault occurs and the time point at which the fault symptom begins to appear can be obtained to achieve the fault prediction. In this work, the time interval of the MD series is 5 min, the number of sampling is set to 1000, and the confidence level is 0.99 and 0.90 respectively. The test results of the fault prediction algorithm are as follows.

5.1. Test Results for Turbine #38

Figure 10 displays the fault prediction results of turbine #38, which consists of the gearbox oil temperature curve, MD series, and CUSUM control chart in turn. There are two change points detected by the change-point detection algorithm which were marked in the figure. The change point 1 indicates that the fault symptom begins to appear at 11:42 on 15 March 2018, which is the 2148th sample point. The change point 2 indicates the actual fault that happened at 11:24 on 29 March 2018, which is the 4015th sample point calculated by our algorithm. It can be seen from the temperature curve the measured value has exceeded the upper limit at the corresponding time of change point 2, at which the MD value also has a drastic increase.

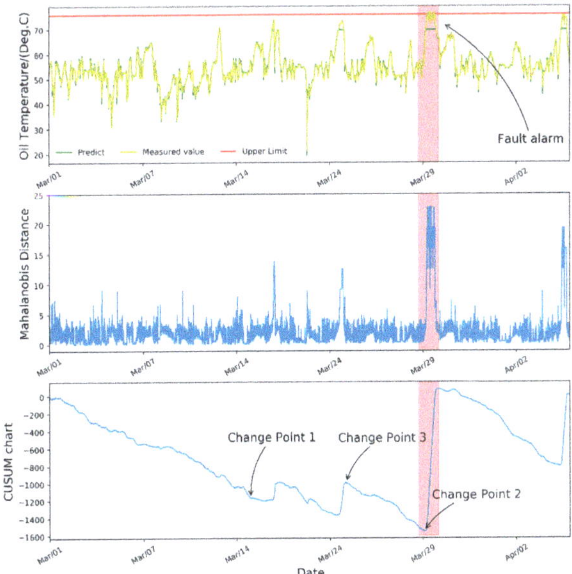

Figure 10. Testing the fault prediction approach results for Turbine #38.

From Table 6, it is worth noting that the confidence level of change point 3 is 0.90, which means we only get two change points under the confidence level of 0.99. The confidence level indicates how confident the result is that the fault actually occurs or the fault symptom begins to appear. The change point 1 occurred with 99% confidence. While the change point 3 occurred with 90% confidence. That means in the 1000 times of bootstrap sampling, the change point 3 was detected at least 900 times while the change point 1 and 2 were detected at least 990 times. Therefore, we are much more confident about the change point 1 at which the fault symptoms begin to appear.

Table 6. The information of the change points for Turbine #38.

Change Point	Confidence Level	Timestamp
1	0.99	15 March 2018 11:42
2	0.99	29 March 2018 11:24
3	0.90	25 March 2018 11:35

According to the fault alarm record of the wind farm, the gearbox oil temperature over-limit fault did happen at 15:10 on 29 March 2018, which is almost consistent with the time point 2 calculated by the change point detection algorithm. Although the gearbox oil temperature at change point 3 is close to the upper alarm limit, no gearbox fault has occurred. Thus, change point 3 is only a potential change point with a low confidence level. Therefore, we can achieve the fault prediction by sending out an alarm signal at the time of change point 1 to arrange the operation and maintenance work. The case study preliminarily verifies the effectiveness of our fault prediction algorithm to recognize the fault symptoms, and thus send out the alarm signal 14 days in advance. Other than this, the proposed change point algorithm can generate different results based on the confidence level we set. Generally speaking, the lower the confidence level we set, the more change points been recognized. But quite low confidence levels will lead to an unreliable result like a false alarm.

5.2. Test results for Turbine #173

Similarly, Figure 11 shows the fault prediction results of turbine #173. The fault symptom begins to appear at 03:24 on 6 April 2018, with corresponding to the change point 1. The change point 2 indicates the fault finally happened at 08:32 on 19 April 2018. Table 7 lists the information about the four change points. When the confidence level is reduced from 0.99 to 0.90, in addition to the original change point 1 and 2, the algorithm also detects the change point 3 and 4. However, there is no fault happened at the timestamp corresponding to change point 3 and 4.

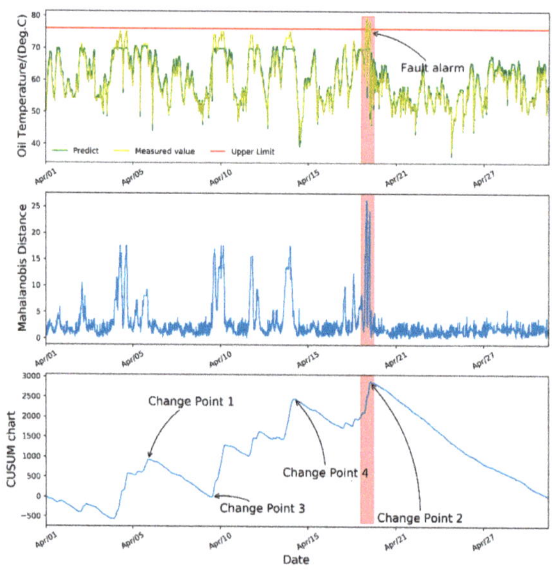

Figure 11. Testing the fault prediction approach results for Turbine #173.

Table 7. The information of the change points for Turbine #173.

Change Point	Confidence Level	Timestamp
1	0.99	6 April 2018 03:24
2	0.99	19 April 2018 08:32
3	0.90	9 April 2018 16:23
4	0.90	13 April 2018 21:05

According to the fault alarm record, turbine #173 had a temperature over-limit fault at 06:52 on 19 April 2018. We are much more confident about the change point 1 at which the fault symptoms beginning to appear. Therefore, the fault prediction algorithm can detect the fault 13 days in advance. Through the analysis of the above two turbines, we can find that the proposed algorithm can not only identify the time of the fault happens but also accurately detect the time point at which the fault symptom begins to appear with a certain confidence level so as to realize the fault prediction.

5.3. Test Results for Other Turbines

The proposed algorithm was applied to the other three wind turbines to prediction the gearbox fault. All these three turbines have suffered gearbox oil temperature faults during the selected time period. Table 8 lists the test results. It can be seen from the figure that the faults can be detected 16 to 22 days in advance. The difference in prediction results may be related to the type of wind turbines and the severity of the fault.

Table 8. Prediction results for the other three wind turbines.

Wind Turbine	Fault Prediction Date	Fault Record Date	Length of Prediction
#58	3 April 2018	19 April 2018	16 days
#159	30 March 2018	18 April 2018	19 days
#162	21 February 2018	14 March 2018	22 days

6. Conclusions

In this work, a gearbox fault prediction approach has been proposed by building a normal behavior model and analyzing the MD series through the change-point detection algorithm. During the modeling process, a data cleaning method composed of the DBSCAN and quartile was applied to identify and drop out the anomaly data. Then, two stacking models were constructed for each type of wind turbines by inputting 15 features to describe the gearbox oil temperature under normal conditions. Compared with other single models, the stacking model can achieve better performance on three metrics and can also overcome the defects of the single model, such as the over-fitting problem.

The effectiveness of the proposed approach was verified using the SCADA data of a wind farm in Ningxia, China. Considering that there are a large number of wind turbines in the wind farm, we use the SCADA data of five turbines for half a year period to establish the normal behavior model of the turbines, so that the model can be used for fault prediction of all other turbines of the same type without separately building models for each turbine. Finally, we verified the fault prediction algorithm on two normal turbines and five fault turbines. The case study shows that the algorithm can detect the fault symptoms 13 to 22 days in advance.

The proposed approach in this paper mainly has two aspects. The Mahalanobis distance, on the one hand, is applied to quantify the discrepancy between the model output value and the actual value of oil temperature. It can be seen from the MD figures that the series of turbines under normal conditions is relatively stable and there is no large fluctuation. Contrary to this, the MD value of the faulty turbines will have a drastic increase when the fault happens, which proves its rationality of using MD to measure the deviation between the actual and normal conditions of turbines. On the other hand, the change-point detection algorithm can not only identify the time point at which the fault occurs, but more importantly, it can detect the time point at which the fault symptom begins to appear, thereby scheduling the operation and maintenance work in advance, and realizing the fault prediction.

In conclusion, we have proposed a fault prediction approach based on a stacking model and change-point detection. The stacking model, as the normal behavior model to describe the normal conditions of wind turbines, its prediction accuracy is obviously higher than that of other single models and is especially suitable for dealing with the occasions with a large amount of data and complex parameter relationships in wind farm. The stacking model can also be applied to power forecasting and the power optimization of wind farms. In addition, compared to the traditional fault detection approach by setting an alarm threshold, our approach is capable of providing more information on the incoming fault. It can not only determine the time at which the fault symptoms beginning to appear, for each change point, it also provides further information including a confidence level representing the possibility that a fault symptom is about to appear. The higher the confidence level is, the more we are confident about the fault prediction result. During the actual application stage, by setting a proper confidence level in advance, the personnel of the wind farm can arrange the operation and maintenance work at the corresponding time of the change point detected by the algorithm, repairing the gearbox in advance, and achieving the goal of fault prediction.

It is worth noting that, unlike setting a single threshold for fault detection, we use the method of change-point detection for the MD series to predict the fault. However, due to the inherent limitations of the change-point detection algorithm itself and the complicated state of the turbines during actual operation, it is also possible to detect the existence of change points in the MD series of a normal turbine, which may lead to a false alarm. In this case, it is of vital importance to select the proper test interval during the actual application stage. For example, in this paper, we use the SCADA data for a period of a month for fault prediction. That is to say, the length of the MD series can significantly affect the accuracy of the final detection result. Take Figures 8 and 10 as examples. They display the test results on normal turbines and fault turbines respectively. There is an obvious difference between their CUSUM control chart. The CUSUM chart of normal turbines (Figure 8) may fluctuate slightly in the short term, but the trend in the one-month period is relatively stable, with its absolute value fluctuating from 0 to 500. While for the fault turbines in Figure 10, the CUSUM chart has a steeper trend and especially when the fault happens, there is a sudden change in direction of CUSUM with their absolute values changing from 0 to 1600. In future work, the appropriate prediction interval should be selected in the application process to reduce the false alarm rate. In addition, the proposed approach will be applied for fault prediction of other components of the wind turbine.

Author Contributions: T.Y. proposed the main idea for the paper and prepared the manuscript. Z.S. supervised the paper and reviewed the manuscript. S.M. contributed the data set and supervised the research.

Funding: This work was supported by Commonweal Technology Research Project of Zhejiang Province, China (grant No. LGG18F030005).

Conflicts of Interest: The authors declare no conflict of interest.

Abbreviations

CUSUM	Cumulative Sum
NBM	Normal Behavior Modeling
DNSCAN	Density-Based Spatial Clustering of Applications with Noise
IQR	Interquartile Range
RF	Random Forest
GBDT	Gradient Boosting Decision Tree
XGBOOST	Extreme Gradient Boosting
MD	Mahalanobis Distance
RFE	Recursive Feature Elimination
KNN	k-Nearest Neighbor

SVM   Support Vector Machine
DT   Decision Tree
Adaboost Adaptive Boosting

References

1. Brussels, B. GWEC Global Wind Report 2018. 2019. Available online: https://gwec.net/global-wind-report-2018/ (accessed on 21 June 2019).
2. Tchakoua, P.; Wamkeue, R.; Ouhrouche, M.; Slaoui-Hasnaoui, F.; Tameghe, T.; Ekemb, G. Wind Turbine Condition Monitoring: State-of-the-Art Review, New Trends, and Future Challenges. *Energies* **2014**, *7*, 2595–2630. [CrossRef]
3. Leahy, K.; Gallagher, C.; O'Donovan, P.; Bruton, K.; O'Sullivan, D. A Robust Prescriptive Framework and Performance Metric for Diagnosing and Predicting Wind Turbine Faults Based on SCADA and Alarms Data with Case Study. *Energies* **2018**, *11*, 1738. [CrossRef]
4. Kusiak, A.; Li, W. Virtual Models for Prediction of Wind Turbine Parameters. *IEEE Trans. Energy Convers.* **2010**, *25*, 245–252. [CrossRef]
5. Kusiak, A.; Li, W. The prediction and diagnosis of wind turbine faults. *Renew. Energy* **2011**, *36*, 16–23. [CrossRef]
6. Kusiak, A.; Verma, A. A Data-Driven Approach for Monitoring Blade Pitch Faults in Wind Turbines. *IEEE Trans. Energy Convers.* **2010**, *2*, 87–96. [CrossRef]
7. Kusiak, A.; Verma, A. Analyzing bearing faults in wind turbines: A data-mining approach. *Renew. Energy* **2012**, *48*, 110–116. [CrossRef]
8. Tautz-Weinert, J.; Watson, S.J. Using SCADA data for wind turbine condition monitoring—A review. *IET Renew. Power Gener.* **2017**, *11*, 382–394. [CrossRef]
9. Santos, P.; Villa, L.F.; Renones, A.; Bustillo, A.; Maudes, J. An SVM-based solution for fault detection in wind turbines. *Sensors* **2015**, *15*, 5627–5648. [CrossRef]
10. Wenyi, L.; Zhenfeng, W.; Jiguang, H.; Guangfeng, W. Wind turbine fault diagnosis method based on diagonal spectrum and clustering binary tree SVM. *Renew. Energy* **2013**, *50*, 1–6. [CrossRef]
11. Bangalore, P.; Tjernberg, L.B. An Approach for Self Evolving Neural Network Based Algorithm for Fault Prognosis in Wind Turbine. In Proceedings of the 2013 IEEE Grenoble Conference, Grenoble, France, 16–20 June 2013; pp. 1–6.
12. Wang, H.; Wang, H.; Jiang, G.; Li, J.; Wang, Y. Early Fault Detection of Wind Turbines Based on Operational Condition Clustering and Optimized Deep Belief Network Modeling. *Energies* **2019**, *12*, 984. [CrossRef]
13. Bangalore, P.; Letzgus, S.; Karlsson, D.; Patriksson, M. An artificial neural network-based condition monitoring method for wind turbines, with application to the monitoring of the gearbox. *Wind Energy* **2017**, *20*, 1421–1438. [CrossRef]
14. Guo, P.; Infield, D.; Yang, X. Wind Turbine Generator Condition-Monitoring Using Temperature Trend Analysis. *IEEE Trans. Sustain. Energy* **2012**, *3*, 124–133. [CrossRef]
15. Astolfi, D.; Castellani, F.; Natili, F. Wind turbine generator slip ring damage detection through temperature data analysis. *Diagnostyka* **2019**, *20*, 3–9. [CrossRef]
16. Garlick, W.G.; Dixon, R.; Watson, S.J. A model-based approach to wind turbine condition monitoring using SCADA data. *ICSE* **2009**, *38*, 536–548.
17. Yeh, C.H.; Lin, M.H.; Lin, C.H.; Yu, C.E.; Chen, M.J. Machine Learning for Long Cycle Maintenance Prediction of Wind Turbine. *Sensors* **2019**, *19*, 1671. [CrossRef]
18. Bangalore, P.; Tjernberg, L.B. An Artificial Neural Network Approach for Early Fault Detection of Gearbox Bearings. *IEEE Trans. Smart Grid* **2015**, *6*, 980–987. [CrossRef]
19. Wu, X.; Wang, H.; Jiang, G.; Xie, P.; Li, X. Monitoring Wind Turbine Gearbox with Echo State Network Modeling and Dynamic Threshold Using SCADA Vibration Data. *Energies* **2019**, *12*, 982. [CrossRef]
20. Pinar Pérez, J.M.; García Márquez, F.P.; Tobias, A.; Papaelias, M. Wind turbine reliability analysis. *Renew. Sustain. Energy Rev.* **2013**, *23*, 463–472. [CrossRef]
21. Ragheb, A.; Ragheb, M. Wind turbine gearbox technologies. In Proceedings of the 1st International Nuclear & Renewable Energy Conference (INREC), Amman, Jordan, 21–24 March 2010; pp. 1–8.

22. Zaher, A.; Mcarthur, S.D.J.; Infield, D.G.; Patel, Y. Online Wind Turbine Fault Detection through Automated SCADA Data Analysis. *Wind Energy* **2010**, *12*, 574–593. [CrossRef]
23. Leahy, K.; Gallagher, C.; O'Donovan, P.; O'Sullivan, D.T.J. Issues with Data Quality for Wind Turbine Condition Monitoring and Reliability Analyses. *Energies* **2019**, *12*, 201. [CrossRef]
24. Butler, S.; Ringwood, J.; O'Connor, F. Exploiting SCADA system data for wind turbine performance monitoring. In Proceedings of the Conference on Control & Fault-tolerant Systems, Nice, France, 9–11 October 2013.
25. Xi, Y.; Lu, Z.; Ying, Q.; Yong, M.; O'Malley, M. Identification and Correction of Outliers in Wind Farm Time Series Power Data. *IEEE Trans. Power Syst.* **2016**, *31*, 4197–4205.
26. Bird, L.; Lew, D.; Milligan, M.; Carlini, E.M.; Estanqueiro, A.; Flynn, D.; Gomez-Lazaro, E.; Holttinen, H.; Menemenlis, N.; Orths, A.; et al. Wind and solar energy curtailment: A review of international experience. *Renew. Sustain. Energy Rev.* **2016**, *65*, 577–586. [CrossRef]
27. Ester, M.; Kriegel, H.P.; Sander, J.; Xu, X. A density-based algorithm for discovering clusters a density-based algorithm for discovering clusters in large spatial databases with noise. In Proceedings of the Second International Conference on Knowledge Discovery and Data Mining, Portland, OR, USA, 2–4 August 1996; AAAI Press: Portland, OR, USA, 1996; pp. 226–231.
28. Hyndman, R.J.; Fan, Y. Sample Quantiles in Statistical Packages. *Am. Stat.* **1996**, *50*, 361–365.
29. Brown, G. A New Perspective for Information Theoretic Feature Selection. In Proceedings of the 12th International Conference on Artificial Intelligence and Statistics (AISTATS) 2009, Clearwater Beach, FL, USA, 16–19 April 2009; Volume 5, pp. 49–56.
30. Guyon, I.; Weston, J.; Barnhill, S.; Vapnik, V. Gene Selection for Cancer Classification using Support Vector Machines. *Mach. Learn.* **2002**, *46*, 389–422. [CrossRef]
31. Dy, J.G.; Brodley, C.E. Feature Selection for Unsupervised Learning. *J. Mach. Learn. Res.* **2004**, *5*, 845–889.
32. Breiman, L. Random Forests. *Mach. Learn.* **2001**, *45*, 5–32. [CrossRef]
33. Zhu, R.; Zeng, D.; Kosorok, M.R. Reinforcement Learning Trees. *J. Am. Stat. Assoc.* **2015**, *110*, 1770–1784. [CrossRef]
34. Deng, X.; Li, Y.; Weng, J.; Zhang, J. Feature selection for text classification: A review. *Multimed. Tools Appl.* **2018**, *78*, 3797–3816. [CrossRef]
35. Vishwanath, D.; Gupta, S. Adding CNNs to the Mix: Stacking models for sentiment classification. In Proceedings of the IEEE Annual India Conference (INDICON), Bangalore, India, 16–18 December 2016; pp. 1–4.
36. Wang, Z.; Wang, D. Recurrent deep stacking networks for supervised speech separation. In Proceedings of the IEEE International Conference on Acoustics, Speech and Signal Processing (ICASSP), New Orleans, LA, USA, 5–9 March 2017; pp. 71–75.
37. Friedman, J.H. Greedy function approximation: A gradient boosting machine. *Ann. Stat.* **2001**, *29*, 1189–1232. [CrossRef]
38. Chen, T.; Guestrin, C. XGBoost: A Scalable Tree Boosting System. In Proceedings of the 22nd ACM SIGKDD International Conference on Knowledge Discovery and Data Mining, San Francisco, CA, USA, 13–17 August 2016; ACM: San Francisco, CA, USA, 2016; pp. 785–794.
39. Li, Y.; Yang, Z.; Chen, X.; Yuan, H.; Liu, W. A stacking model using URL and HTML features for phishing webpage detection. *Future Gener. Comput. Syst.* **2019**, *94*, 27–39. [CrossRef]
40. Taylor, W.A. *Change-Point Analysis: A Powerful New Tool for Detecting Changes*; Baxter Healthcare Corporation: Deerfield, IL, USA, 2000; Available online: https://variation.com/wp-content/uploads/change-point-analyzer/change-point-analysis-a-powerful-new-tool-for-detecting-changes.pdf (accessed on 21 June 2019).
41. Zhao, Y.; Li, D.; Dong, A.; Kang, D.; Lv, Q.; Shang, L. Fault Prediction and Diagnosis of Wind Turbine Generators Using SCADA Data. *Energies* **2017**, *10*, 1210. [CrossRef]

Article

Optimal Preventive Maintenance of Wind Turbine Components with Imperfect Continuous Condition Monitoring

Ahmed Raza [1] and Vladimir Ulansky [2,3,*]

[1] Projects and Maintenance Section, The Private Department of the President of the United Arab Emirates, Abu Dhabi 000372, UAE; ahmed.awan786@gmail.com

[2] Research and Development Department, Mathematical Modelling & Research Holding Limited, London W1W 7LT, UK

[3] Department of Electronics, Robotics, Monitoring Technology, and IoT, National Aviation University, 03058 Kyiv, Ukraine

* Correspondence: vulanskyi@mmrholding.org; Tel.: +44(0)-2038-236-006 or +38(0)-6327-549-82

Received: 17 June 2019; Accepted: 15 September 2019; Published: 8 October 2019

Abstract: Among the different maintenance techniques applied to wind turbine (WT) components, online condition monitoring is probably the most promising technique. The maintenance models based on online condition monitoring have been examined in many studies. However, no study has considered preventive maintenance models with incorporated probabilities of correct and incorrect decisions made during continuous condition monitoring. This article presents a mathematical model of preventive maintenance, with imperfect continuous condition monitoring of the WT components. For the first time, the article introduces generalized expressions for calculating the interval probabilities of false positive, true positive, false negative, and true negative when continuously monitoring the condition of a WT component. Mathematical equations that allow for calculating the expected cost of maintenance per unit of time and the average lifetime maintenance cost are derived for an arbitrary distribution of time to degradation failure. A numerical example of WT blades maintenance illustrates that preventive maintenance with online condition monitoring reduces the average lifetime maintenance cost by 11.8 times, as compared to corrective maintenance, and by at least 4.2 and 2.6 times, compared with predetermined preventive maintenance for low and high crack initiation rates, respectively.

Keywords: condition monitoring; preventive maintenance; sensor; uncertainty quantification; false positive; false negative; maintenance cost

---

1. Introduction

Currently, there is a global transformation of energy production toward renewable energy sources, and wind power is a source that is developing at the highest rate. According to the authors of [1], the overall capacity of all wind turbines (WTs) installed worldwide, by the end of 2018, was 597 GW, of which 50.1 GW was added in 2018 alone. Thus, the increase in wind power amounted to 8.4% in 2018. Wind power developed most rapidly in China and the United States, adding 21 GW and 7.6 GW in 2018, reaching 217 GW and 96 GW, respectively [1]. With such rapid growth of wind energy production, special attention is given to the lowered cost of 1 kWh of produced energy. Thus, according to IRENA [2], 1 kWh of energy produced by an offshore WT has decreased from 0.17 USD/kWh in 2010 to 0.14 USD/kWh in 2017. A significant reduction in the cost of produced wind energy can be achieved by reducing the cost of operation and maintenance (O&M). Indeed, O&M costs for offshore wind farms account for 25% of the total cost [3]. Therefore, many scientific publications have been devoted to optimizing the maintenance of WT [4–10].

When analyzing possible maintenance strategies for WT components, we use the terminology of the standard EN 13306: 2017 [11]. By this standard, the following basic maintenance types can be applied to WT components: preventive, predetermined, condition-based, corrective, and predictive.

Traditionally, preventive maintenance has relied on statistics data calculating such reliability indicators as mean time to failure, and the cumulative distribution function of time to failure to determine the optimum periodicity of component replacement. Preventive maintenance without previous condition investigation is called predetermined maintenance [11]. The main effect achieved when using predetermined preventive maintenance is reducing the probability of failure in the interval between services. Today, however, preventive maintenance activities can be effectively improved by using built-in sensors that can directly monitor the component condition. When using information from sensors, the operator observes an approaching failure, which allows for planning the necessary actions to replace the component and minimize potential losses.

Condition-based maintenance (CBM) is also preventive maintenance, but it includes a combination of condition monitoring and corresponding maintenance actions. Condition monitoring can be either scheduled or continuous. A distinctive feature of condition-based maintenance is the introduction of a preventive or replacement threshold, which is different from the degradation failure threshold.

Corrective maintenance is carried out after failure occurrence and intended to restore the component operability. Corrective maintenance is typically used for components which have failures that are not safety-related and do not lead to significant economic losses. Corrective maintenance may also use periodic inspections or continuous condition monitoring on the base of sensors; however, maintenance actions are performed only after failure identification.

Predictive maintenance comprises condition monitoring and forecasting the future component condition, where maintenance decision-making depends on the results of the prediction, so that the component can be replaced or restored before its failure. Predictive maintenance can also be based on periodic or continuous condition monitoring. However, the information obtained during monitoring is used to estimate the remaining useful life and to plan maintenance actions.

Thus, all of the types of maintenance considered, except for predetermined, are based on varying degrees of condition monitoring. The literature [12–14] gives a comprehensive analysis and discussion of condition monitoring methods used for WT components. The authors of [15] present an in-depth analysis of CBM strategies for offshore wind energy equipment. Currently, the most promising type of condition monitoring is the use of sensors that provide reliable continuous measurements and targeted evaluation of critical data. The use of sensors allows for avoiding manual inspections, reducing the probability of failure occurrence, reducing the downtime, and helping to plan maintenance actions [16]. Therefore, we will further highlight the studies on the optimization of the maintenance of deteriorating WT components using continuous condition monitoring.

Byon et al. [4] considered optimal repair strategies for WT that operate in different weather conditions. It is assumed that the sensors are error-free, and instantaneously reveal the system condition. Yildirim et al. [6] considered an integrated framework for wind farm maintenance that uses real-time sensor data to improve wind farm maintenance and operational decisions. A dynamic maintenance cost associated with preventive and corrective maintenance decisions is proposed. The study performed by Canizo et al. [17] presented a big data analytical approach for predictive maintenance of WT using a big data processing framework to generate predictive models based on historical data. Interactive fault-tolerant monitoring predicts the state of WT every 10 minutes. Fu et al. [18] considered a method for analyzing the reliability of a network monitoring wind turbine blades based on wireless sensor networks. Krishna [19] considered a remote instrument monitoring system, which is a dispatching control and data acquisition system based on a wireless sensor network, with a set of sensors distributed throughout the turbines of wind power plants. Nilsson and Bertling [20] presented a lifecycle cost analysis of different preventive and corrective maintenance strategies of WT, where a condition monitoring system is used to improve maintenance planning for a single WT onshore and a wind farm offshore. Kerres et al. [21] considered a model that is capable of estimating O&M cost for different maintenance strategies over the WT lifecycle. Three simulated types of maintenance include run-to-failure maintenance, maintenance with annual inspections of the gearbox and generator, as well as the installation of a vibration condition monitoring system that detects 90% of the gearbox and generator defects. Van Horenbeek et al. [22] considered a new concept of modeling various failure modes based on the P–F curve. The constructed stochastic simulation model allows for quantifying the economic benefits of introducing an imperfect condition monitoring system in the WT gearbox. Research proves that a condition monitoring system is beneficial, in comparison to the current maintenance strategy. Kaiser and Gebraeel [23] considered a sensory-updated predictive maintenance policy, which uses degradation models and sensor signals to predict and update the residual life distribution. Bryant and John [24] considered an asset management model of WT based on the method of modeling Petri nets. The condition monitoring system continuously monitors WT components to capture early indication of component failure, so that preventive maintenance can be planned. Besnard and Bertling [25] considered three maintenance strategies of WT blades, including visual inspections, inspection with a condition-monitoring technique, and online condition monitoring. The expected lifecycle maintenance costs associated with different approaches are calculated and compared. Pattison et al. [26] considered a new architecture for the implementation of reliability-centered maintenance of offshore WT. The architecture supposes three integrated modules for intelligent condition monitoring, reliability, and maintenance modeling, as well as maintenance schedules that provide cost-effective preventive maintenance management of offshore WTs. Ghamlouch et al. [27] considered a preventive maintenance model for a system subject to uncertainty due to the stochastic nature of the deterioration and production processes. Shafiee and Finkelstein [28] considered an optimal proactive maintenance policy for continuous monitoring systems prone to stochastic degradation. The task of proactive group maintenance is performed when the degradation level of the subsystem exceeds the "alert" threshold (less than the failure threshold), which reduces the probability of system failures. Shafiee et al. [29] considered an optimal CBM strategy for a multi-blade offshore wind turbine system subjected to stress corrosion cracking and environmental shocks. Wang et al. [30] considered a new model for diagnosing faults and predicting the remaining useful life of WT components with limited degradation data. The model can be used for predictive maintenance optimization of WT components. Marugan et al. [31] considered a novel approach for false alarm detection and prioritization applied to a real dataset from a vibration monitoring system of a WT. The studies [32–34] considered a preventive maintenance model where the replacement can be corrective, due to a failure, or preventive, after a predetermined time τ, depending on what occurs first. The model assumes that condition monitoring is perfect and immediately detects any fault, and each faulty component is replaced with a new one.

Thus, to date, a significant number of articles [4,6,17–28,32–34] have been published in which continuous condition monitoring is used to reduce maintenance cost or increase the availability of WT components or other degrading systems. A common assumption in these publications is that condition monitoring is perfect. However, the information received by the operator from the sensors is usually distorted by noise and measurement errors [35]. Therefore, decisions made based on information received from sensors may be erroneous, which may ultimately lead to significant economic losses. Indeed, a false alarm can be associated with an unscheduled interruption in the operation of a WT for a while, and result in carrying out unreasonable work. As shown in [22], the added economic effect from the use of a perfect condition-monitoring system is €99,844 for one turbine. However, this effect can be significantly reduced by false alarms and low detectability. The level of detectability depends on the probability of a false negative that is associated with undetected failure, which may result in significant economic losses, sometimes connected even to the destruction of WT equipment. Therefore, when optimizing the preventive maintenance of the WT components, it is necessary to consider the validity of the information received from the built-in sensors.

Table 1 shows the results of the analysis of maintenance models using continuous condition monitoring. From Table 1, it follows that in almost all studies on maintenance optimization, the continuous condition monitoring is assumed to be perfect. Only the research performed by Van Horenbeek et al. [22] indicates that the effect of using sensors is highly dependent on the performance of the condition monitoring system, including the impact of false alarms.

This study proposes a new mathematical model of preventive maintenance of WT components on the basis of imperfect continuous condition monitoring. For the first time, equations for calculating the interval probabilities of a false positive, true positive, false negative, and true negative, are derived. A numerical example illustrates that the probabilities of decisions made based on the sensor information depend on both the noise and the degradation process parameters. Mathematical modeling of the WT blades preventive maintenance is given. Numerical calculations show that preventive maintenance with online condition monitoring of WT blades is much more effective than predetermined preventive maintenance and corrective maintenance. The proposed mathematical model of preventive maintenance based on imperfect condition monitoring can be used, not only for WT components, but also for some other deteriorating systems.

Table 1. Classification of maintenance models on the basis of continuous condition monitoring.

Type of Maintenance	Continuous Condition Monitoring	Studies
Corrective	perfect	[4,6,18,20,21]
	imperfect	-
Preventive	perfect	[4,6,19–21,24–27,32–34]
	imperfect	[22]
Condition-based	perfect	[28,29]
	imperfect	-
Predictive	perfect	[17,23]
	imperfect	-

2. Quantification of the Uncertainty of Continuous Condition Monitoring

Let the state of the WT component be continuously monitored with the help of built-in sensors in the interval $(0, \tau)$, where τ is the periodicity of preventive maintenance. Further, we suppose that condition monitoring is imperfect, i.e., an operable WT component can be judged as inoperable at any instant of the interval $(0, \tau)$, and vice versa, an inoperable WT component can be judged as operable. We assume that a WT component is replaced at the time τ if the sensors do not indicate a failure during the interval $(0, \tau)$ or at any instant of the interval $(0, \tau)$, if at that instant the sensors indicate failure. Decisions made based on monitoring the condition of the WT component during the interval $(0, \tau)$ are reduced to conducting a preventive or corrective maintenance. After any type of maintenance, the WT component becomes as good as new. The trustworthiness of continuous condition monitoring results has a significant impact on these decisions.

Let us consider the graph of decision-making when monitoring the condition of a WT component in the interval $(0, \tau)$, which is shown in Figure 1. A priori, the WT component can be in the operable state in the interval $(0, \tau)$ with probability $P(\tau)$ or the inoperable state with probability $1 - P(\tau)$. According to the results of continuous condition monitoring of the WT component in the interval $(0, \tau)$ one of the following events may occur. The component will not fail, and the sensors will not give an alarm (true positive—TP), the component will not fail, but the sensors will provide a warning (false positive—FP), the component will fail, and the sensors will alarm (true negative—TN), and finally the component will fail, but the sensors will not trigger an alarm (false negative—FN). The probabilities of these events are marked as $P_{TP}(\tau)$, $P_{FP}(\tau)$, $P_{TN}(\tau)$, and $P_{FN}(\tau)$ in the graph of Figure 1.

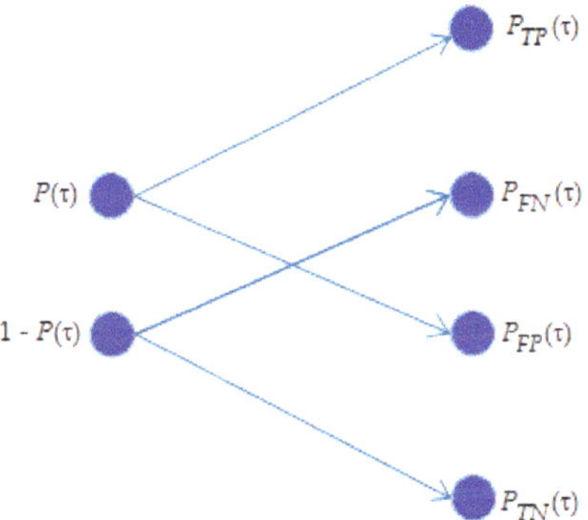

Figure 1. Graph of decision-making when monitoring the condition of the wind turbine (WT) component in the interval (0, τ), where $P(\tau)$ is the probability that the WT component will be in the operable state in the interval (0, τ), $P_{TP}(\tau)$ is the probability that the component will not fail during the interval (0,τ), and the sensors will not give an alarm (true positive), $P_{FP}(\tau)$ is the probability that the component will not fail during the interval (0,τ), and the sensors will give an alarm (false positive), $P_{TN}(\tau)$ is the probability that the component will fail, and the sensors will alarm (true negative), and $P_{FN}(\tau)$ is the probability that the component will fail, but the sensors will not trigger an alarm (false negative).

It should be noted that the probabilities indicated in Figure 1 are determined on a time interval (0, τ) of continuous condition monitoring, i.e., these probabilities are interval probabilities, and do not correspond to some point in time.

Let us determine the probabilities of a TP, FP, TN, and FN in the interval (0, τ). Assume that the random variable H (H ≥ 0) denotes the time to failure of the WT component with a failure probability density function (PDF) ω(η). The random variable H is the smallest root of the stochastic equation

$$X(t) - FT = 0, \tag{1}$$

where $X(t)$ is the stochastic process of the WT component degradation and FT is the degradation failure threshold.

Since the sensors are assumed to be error-prone, the observed process $Y(t)$ can be represented as the sum of the degradation processes $X(t)$ and noise $e(t)$.

$$Y(t) = X(t) + e(t). \tag{2}$$

Let us consider the random variable H*, which is the smallest root of the stochastic equation

$$Y(t) - FT = 0. \tag{3}$$

The random variable H* is the occasional time until the stochastic process $Y(t)$ crosses the failure threshold FT. Figure 2 illustrates the difference between the realizations $x(t)$ and $y(t)$ of the stochastic processes $X(t)$ and $Y(t)$, and realizations η and η* of the random variables H and H*. Equation (3) shows that random variable H* is a function of random variable H and noise e(t). That is why the realizations of time to failure η* and η do not match in Figure 2.

The presence of noise $e(t)$ in Equation (3) leads to a random error in evaluation of time to failure, which is defined as follows:

$$\Delta = H^* - H. \tag{4}$$

Random variables H ($0 < H < \infty$) and Δ ($-\infty < \Delta < \infty$) have an additive relationship. Therefore, the random variable H* is defined in a continuous range of values from $-\infty$ to ∞.

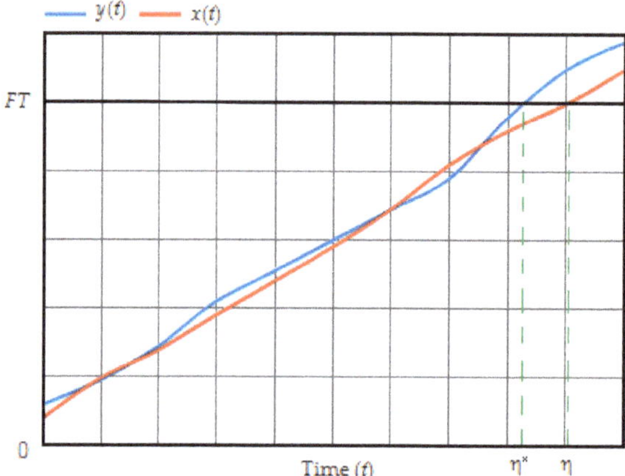

Figure 2. Possible realizations of the stochastic processes $X(t)$ and $Y(t)$, and the realizations η and η^* of the random variables H and H*.

Using the definitions of random variables H and H*, we formulate events related to FP, TP, FN, and TN during the interval $(0, \tau)$ as follows:

$$FP(0,\tau) = [(H \geq \tau) \cap (H^* \leq \tau)], \tag{5}$$

$$TP(0,\tau) = [(H \geq \tau) \cap (H^* \geq \tau)], \tag{6}$$

$$FN(0,\tau) = [(H \leq \tau) \cap (H^* \geq \tau)], \tag{7}$$

$$TN(0,\tau) = [(H \leq \tau) \cap (H^* \leq \tau)]. \tag{8}$$

The probabilities of events (5)–(8) can be found by calculating the probability of hitting a random point {H, H*} in a 2D domain created by the boundaries of variation of each random variable, and is equal to 2-fold integral over this area. Let us denote the joint PDF of the random variables {H, H*} as $\omega_0(\eta, \eta^*)$. The event $FP(0, \tau)$ corresponds to the 2D domain with the following limits: $\tau \leq H < \infty$ and $-\infty < H^* \leq \tau$. By integrating the PDF $\omega_0(\eta, \eta^*)$ within the specified region, we determine the probability of FP.

$$P_{FP}(0, \tau) = P[FP(0, \tau)] = \int_{\tau}^{\infty} \int_{-\infty}^{\tau} \omega_0(\eta, \eta^*) d\eta^* d\eta. \tag{9}$$

We derive the probabilities of events $TP(0, \tau)$, $FN(0, \tau)$, and $TN(0, \tau)$ analogically to the probability $P_{FP}(0, \tau)$. Conducting some mathematical manipulations we obtain

$$P_{TP}(0, \tau) = P[TP(0, \tau)] = \int_{\tau}^{\infty} \int_{\tau}^{\infty} \omega_0(\eta, \eta^*) d\eta^* d\eta, \tag{10}$$

$$P_{FN}(0, \tau) = P[FN(0, \tau)] = \int_{0}^{\tau} \int_{\tau}^{\infty} \omega_0(\eta, \eta^*) d\eta^* d\eta, \tag{11}$$

$$P_{TN}(0, \tau) = P[TN(0, \tau)] = \int_{0}^{\tau} \int_{-\infty}^{\tau} \omega_0(\eta, \eta^*) d\eta^* d\eta. \tag{12}$$

As can be seen from (9)–(12), in order to find the probabilities FP, TP, FN, and TN, the joint PDF $\omega_0(\eta, \eta^*)$ should be known. Let us present the joint PDF $\omega_0(\eta, \eta^*)$ as follows:

$$\omega_0(\eta, \eta^*) = \omega(\eta)\theta(\eta^* | \eta), \tag{13}$$

where $\theta(\eta^* | \eta)$ is the conditional PDF of random variable H^* provided that $H = \eta$.

From (4), when $H = \eta$, the random variable H^* reduces to $H^* = \eta + \Delta$. Since between the random variables H and Δ exists the additive relationship, the following equality thus holds:

$$\theta(\eta^* | \eta) = f(\eta^* - \eta | \eta), \tag{14}$$

where $f(\eta^* - \eta | \eta)$ is the conditional PDF of the random error in evaluation of time to failure, provided that $H = \eta$.

Substituting (14) into (13) gives

$$\omega_0(\eta, \eta^*) = \omega(\eta)f(\eta^* - \eta | \eta). \tag{15}$$

The relation (15) makes it possible to simplify (9)–(12). By substitution of (15) into (9), we get

$$P_{FP}(0, \ \tau) = \int_{\tau}^{\infty} \omega(\eta) \int_{-\infty}^{\tau} f(\eta^* - \eta | \eta)\, d\eta^* d\eta. \tag{16}$$

Making the substitution $\delta = \eta^* - \eta$ in the internal integral of (16) results in the following expression for FP:

$$P_{FP}(0, \ \tau) = \int_{\tau}^{\infty} \omega(\eta) \int_{-\infty}^{\tau - \eta} f(\delta | \eta)\, d\delta d\eta. \tag{17}$$

Accomplishing a similar change of variable in expressions (10)–(12), we get

$$P_{TP}(0, \ \iota) - \int_{\tau}^{\infty} \omega(\eta) \int_{\tau - \eta}^{\infty} f(\delta | \eta)\, d\delta d\eta. \tag{18}$$

$$P_{FN}(0, \ \tau) = \int_{0}^{\tau} \omega(\eta) \int_{\tau - \eta}^{\infty} f(\delta | \eta)\, d\delta d\eta. \tag{19}$$

$$P_{TN}(0, \ \tau) = \int_{0}^{\tau} \omega(\eta) \int_{-\infty}^{\tau - \eta} f(\delta | \eta)\, d\delta d\eta. \tag{20}$$

As seen from (17)–(20), to calculate the probabilities of FP, TP, FN, and TN when performing continuous condition monitoring of a WT component, the PDF $\omega(\eta)$ and $f(\delta | \eta)$ are supposed to be known. We should also note that formulas (17)–(20) are generalized, i.e., they can be used for any stochastic degradation process $X(t)$ and any noise $e(t)$.

3. Maintenance Model with Imperfect Continuous Condition Monitoring

Let us consider the mathematical maintenance model of a WT component in the interval $(0, \tau)$ with imperfect continuous condition monitoring. Further, we consider the problem of minimizing the expected maintenance cost per unit of time of a WT component for an infinite time horizon. The decision to conduct a maintenance action on the WT component is based on information received from the sensors. The decision to conduct corrective maintenance within the interval $(0, \tau)$ is made in the case of failure detection, which corresponds to the occurrence of the $TN(0, \tau)$ event. In the case of the $FP(0, \tau)$

event occurring, the preventive maintenance is carried out. If the alarm signal does not appear in the interval $(0, \tau)$, it means that a $TP(0, \tau)$ event occurred, or that the component failed, and an $FN(0, \tau)$ event appeared. Since the time horizon is infinite, it is appropriate to choose the expected maintenance cost per unit of time as the maintenance effectiveness indicator of the WT component. Since after any type of maintenance, the WT component becomes as good as new, then according to the authors of [34], the expected cost of maintenance per unit of time can be determined as the ratio of the average maintenance cost for the regeneration cycle to the average duration of this cycle.

$$E[C_u(\tau)] = \frac{E[C_a(\tau)]}{E[T_{RC}(\tau)]}. \tag{21}$$

The optimization problem is as follows:

$$\tau_{opt} \Rightarrow \min_{\tau} E[C_u(\tau)], \tag{22}$$

where τ_{opt} is the optimal periodicity of preventive maintenance.

The average maintenance cost for one regeneration cycle consists of two parts. The first part determines the cost of maintenance for the case when a failure occurs in the interval $(0, \tau)$, i.e., $H \leq \tau$. Let the failure occur at the moment $H = \eta \leq \tau$. Then the cost of maintenance includes the cost of corrective repair if the sensors detect the failure, and the cost of losses due to unrevealed failure state of the WT component if the sensors do not identify the failure.

$$E[C_a(\tau)|H = \eta \leq \tau] = C_{CM}P_{TN}(\eta, \tau|H = \eta) + C_{UF}(\tau - \eta)P_{FN}(\eta, \tau|H = \eta), \tag{23}$$

where C_{CM} is the cost of corrective maintenance, C_{UF} is the loss cost per unit of time due to unrevealed failure, $P_{TN}(\eta, \tau|H = \eta)$ is the conditional probability of TN in the interval (η, τ) provided that $H = \eta$, and $P_{FN}(\eta, \tau|H = \eta)$ is the conditional probability of FN in the interval (η, τ) provided that $H = \eta$.

Using (19) and (20) we determine the conditional probabilities $P_{FN}(\eta, \tau|H = \eta)$ and $P_{TN}(\eta, \tau|H = \eta)$ as follows:

$$P_{FN}(\eta, \tau|H = \eta) = \int_{\tau-\eta}^{\infty} f(z|\eta)dz, \tag{24}$$

$$P_{TN}(\eta, \tau|H = \eta) = \int_{-\infty}^{\tau-\eta} f(z|\eta)dz. \tag{25}$$

The second part determines the cost of maintenance for the case when no failure occurs in the interval $(0, \tau)$, i.e., $H > \tau$. In this case, the cost of maintenance includes the cost of preventive maintenance, either if the sensors incorrectly detect a failure in the interval $(0, \tau)$, or if the sensors do not send an alarm signal in the interval $(0, \tau)$.

$$E[C_a(\tau)|H = \eta > \tau] = C_{PM}^{FP}P_{FP}(0, \tau|H = \eta) + C_{PM}^{TP}P_{TP}(0, \tau|H = \eta), \tag{26}$$

where C_{PM}^{FP} and C_{PM}^{TP} are, respectively, the cost of preventive maintenance due to the FP and TP events, $P_{FP}(0, \tau|H = \eta)$ is the conditional probability of FP in the interval $(0, \tau)$ provided that $H = \eta$, and $P_{TP}(0, \tau|H = \eta)$ is the conditional probability of TP in the interval $(0, \tau)$ provided that $H = \eta$.

The values of C_{PM}^{FP} and C_{PM}^{TP} may not match, because in the case of occurrence of FP, the preventive maintenance is unscheduled, and in the case of TP, the preventive maintenance is scheduled.

On the base of (17) and (18), we determine the conditional probabilities $P_{FP}(0, \tau|H = \eta)$ and $P_{TP}(0, \tau|H = \eta)$

$$P_{FP}(0, \tau|H = \eta) = \int_{-\infty}^{\tau-\eta} f(\delta|\eta)d\delta, \tag{27}$$

$$P_{TP}(0, \tau | H = \eta) = \int_{\tau - \eta}^{\infty} f(\delta | \eta) d\delta. \tag{28}$$

Applying the law of total expectation, we determine the average maintenance cost for one regeneration cycle:

$$E[C_a(\tau)] = \int_0^{\tau} E[C_a(\tau) | H = \eta \leq \tau] w(\eta) d\eta + \int_{\tau}^{\infty} E[C_a(\tau) | H = \eta > \tau] w(\eta) d\eta. \tag{29}$$

Substituting (23) and (26) into (29) gives

$$E[C_a(\tau)] = C_{CM} P_{TN}(0, \tau) + C_{UF} \int_0^{\tau} (\tau - \eta) P_{FN}(\eta, \tau | H = \eta) w(\eta) d\eta + C_{PM}^{FP} P_{FP}(0, \tau) + C_{PM}^{TP} P_{TP}(0, \tau). \tag{30}$$

Let us now determine the mean time of the regeneration cycle. In deriving the expression for $E[T_{RC}(\tau)]$, we will consider that within the interval $(0, \tau)$, the maintenance is possible only due to the occurrence of either an FP event or a TN event. Assume that $H = \eta \leq \tau$. Then, the conditional mathematical expectation of the regeneration cycle is

$$E[T_{RC}(\tau) | H = \eta \leq \tau] = \eta P_{TN}(\eta, \tau | H = \eta) + (\tau - \eta) P_{FN}(\eta, \tau | H = \eta). \tag{31}$$

Now suppose that $H = \eta > \tau$. In this case, we have

$$E[T_{RC}(\tau) | H = \eta > \tau] = \int_0^{\tau} z \omega_{FP}(z | \eta) dz + \tau P_{TP}(0, \tau | H = \eta), \tag{32}$$

where $\omega_{FP}(z | \eta)$ is the conditional PDF of an FP in the interval $(0, z)$, $0 < z \leq \tau$, provided that $H = \eta$.

The conditional PDF $\omega_{FP}(z | \eta)$ is a derivative of the cumulative distribution function of the time to a false alarm under condition that $H = \eta$

$$\omega_{FP}(z | \eta) = \frac{d}{dz} \left\{ \int_{-\infty}^{z - \eta} f(\delta | \eta) d\delta \right\}. \tag{33}$$

Using the theorem on the derivative of an integral by a variable upper limit, we find

$$\omega_{FP}(z | \eta) = f(z - \eta | \eta). \tag{34}$$

Substituting (34) into (32) gives

$$E[T_{RC}(\tau) | H = \eta > \tau] = \int_0^{\tau} z f(z - \eta | \eta) dz + \tau P_{TP}(0, \tau | H = \eta), \tag{35}$$

Using the law of total expectation, we determine the mean time of the regeneration cycle:

$$E[T_{RC}(\tau)] = \int_0^{\tau} E[T_{RC}(\tau) | H = \eta \leq \tau] w(\eta) d\eta + \int_{\tau}^{\infty} E[T_{RC}(\tau) | H = \eta > \tau] w(\eta) d\eta. \tag{36}$$

By substitution of (31) and (35) into (36), we have

$$E[T_{RC}(\tau)] = \int_0^\tau P_{TN}(\eta, \tau | H = \eta)\omega(\eta)d\eta + \int_0^\tau (\tau - \eta)P_{FN}(\eta, \tau | H = \eta)\omega(\eta)d\eta +$$

$$\int_\tau^\infty \int_0^\tau f(z - \eta | \eta)\omega(\eta)dzd\eta + \tau \int_\tau^\infty P_{TP}(0, \tau | H = \eta)\omega(\eta)d\eta. \tag{37}$$

For the case when the condition monitoring is perfect, Equation (37) reduces to

$$E[T_{RC}(\tau)] = \int_0^\tau \omega(\eta)d\eta + \tau \int_\tau^\infty \omega(\eta)d\eta, \tag{38}$$

which coincides with the well-known expression [34].

4. Example: Preventive Maintenance of Wind Turbine Blades

4.1. Methods of Condition Monitoring of Wind Turbine Blades

Modern WT rotor blades are usually made of composite materials. Different types of damage may occur in WT blades. The most common damages to the WT rotor blades are as follows [36]: cracks along adhesive joints, cracks parallel to the fiber direction, cracks along the plane between plies, etc. The causes of the defects that appear in WT rotor blades are analyzed in many studies, for example [36–39]. Blade cracks can be detected by periodical visual inspections, periodical inspections on the base of nondestructive testing, or online condition monitoring using sensors. A detailed analysis of different testing, inspecting, and online monitoring technologies for WT blades and other components is described in [14,38,40,41]. As mentioned in many studies, online condition monitoring is a promising technique for reducing the maintenance cost and downtime of WT [42–45]. The online monitoring methods include the acoustic emission detection method, the thermal imaging method, the ultrasonic method, the fiber optics method, and some others [38]. According to the authors of [38], ultrasonic methods allow for the identification of cracks, several mm in size. Nowadays, optical fiber sensors are widely used in different structural health monitoring systems [46–51]. The use of an online monitoring system based on Brillouin optical time-domain analysis allows for detecting a crack as small as 1.5 cm, when monitoring a wind turbine blade [52].

Further, we assume that one of the methods, as mentioned above, is used for online condition monitoring of WT blade damages. Some forms of damage, once they have appeared in the composite blade structure, will quickly spread and may overload other structural components that finally may lead to a failure of the WT. The destruction of a WT blade usually leads to significant economic losses.

4.2. Crack Degradation Model

An increase in the size of the crack leads to a rise in the probability of failure of the WT blade. The dependence of the length of a growing crack on time is a monotonic function. In the general case, such a function is convex. The following random function can be used to approximate the monotonic stochastic process of crack degradation:

$$X(t) = At^\gamma, \tag{39}$$

where A is the random degradation rate of crack [cm/unit of time] and γ is the exponent of time.

If $\gamma > 1$, then the realizations of the random function $X(t)$ have a convex shape.

For calculating the average maintenance cost per unit of time we need to know the conditional PDF of random error in evaluating the time to failure $f(\delta|\eta)$. Following the method described in [53], and assuming that the noise $e(t)$ is a white noise, we derive the following equation for the PDF $f(\delta|\eta)$:

$$f(\delta|\eta) = \gamma\left(\frac{FT}{\eta}\right)\left|\left(\frac{\delta+\eta}{\eta}\right)^{\gamma-1}\right|\Omega\left\{FT\left[1-\left(\frac{\delta+\eta}{\eta}\right)^{\gamma}\right]\right\},\tag{40}$$

where $\Omega(e)$ is the PDF of the random measurement error.

For the sake of simplicity, assume that the crack degradation function is linear, i.e., $\gamma = 1$. In this case, from (40) we obtain

$$f(\delta|\eta) = \left(\frac{FT}{\eta}\right)\Omega\left(-\frac{FT}{\eta}\delta\right).\tag{41}$$

In the case when the measurement noise is Gaussian white noise, (41) transforms into the following equation:

$$f(\delta|\eta) = \left(\frac{1}{\sigma_e^2\sqrt{2\pi}}\right)\left(\frac{FT}{\eta}\right)\exp\left\{-\frac{1}{2\sigma_e^2}\left[\frac{(-FT)\delta}{\eta}\right]^2\right\},\tag{42}$$

where σ_e is the standard deviation of the measurement noise.

For a linear stochastic process of crack degradation, the PDF of the time to failure is given by [54]:

$$\omega(\eta) = \frac{m_1\sigma_1^2\eta^2 + \sigma_1^2\eta(FT - m_1\eta)}{\sqrt{2\pi}\sigma_1^3\eta^3}\exp\left\{-\frac{(FT - m_1\eta)^2}{2\sigma_1^2\eta^2}\right\}\left[\int_0^\infty \varphi(a)\,da\right]^{-1},\tag{43}$$

where m_1 and σ_1 are the mathematical expectation and standard deviation of the random degradation rate of crack, and $\varphi(a)$ is the Gaussian PDF of the random degradation rate of crack $A\left[\frac{cm}{unit\ of\ time}\right]$. The distribution of A is truncated because negative rates are impossible.

The crack can begin its growth at any time during the lifetime of the blade. Thus, for determining the average lifetime maintenance cost, the expected maintenance cost during the regeneration cycle of the blade should be multiplied by the crack initiation rate (θ), and the lifetime of a WT (T_{LT}).

$$E[C_{LT}(\tau)] = \theta T_{LT} E[C_a(\tau)].\tag{44}$$

4.3. Model Parameters

In determining the optimal periodicity of preventive maintenance, we mostly use the initial data given in [25] for an offshore WT. As indicated in [25], even a small crack size of 20–30 mm for the delamination may reduce the strength of the blades, and may finally be a reason for failure. Further, we assume that $FT = 60$ cm. The cost of corrective maintenance due to catastrophic failure is $C_{CM}^{CF} = \text{€}440,400$, which includes the cost of a new blade, the cost of production losses, the cost of ordering a boat and crane, and the cost of installation. The cost of ordering an offshore boat and inspection C_M is assumed to be €12,500. The cost of corrective maintenance due to degradation failure is usually significantly less than C_{CM}^{CF} because the damaged blade can be repaired in situ. In calculations, we assume that $C_{CM} = \text{€}100,000$. Furthermore, it will be shown that, for a degrading component, a false alarm usually occurs when the value of the degradation parameter approaches the threshold FT. Based on this we assume that losses due to an FP will include the cost of ordering an offshore boat, the inspection cost, and the cost of in situ repair by patch technique because of advanced degradation. Therefore, we assume that $C_{PM}^{FP} = \text{€}20,000$. The cost of preventive maintenance due to TP event at the end of the interval $(0, \tau)$, we assume to be equal C_{PM}^{FP}, i.e., $C_{PM}^{TP} = \text{€}20,000$. The loss cost per unit of time due to unrevealed failure, we calculate assuming that, as in [25], the average power production is 2 MW and the price for electricity is 50 €/MWh. In this case, we calculate that $C_{UF} = 72,000\ \frac{\text{€}}{\text{month}}$. It should be noted that in 2018, the cost of electricity in Europe was significantly higher varying from

100.5 €/MWh in Bulgaria to 312.3 €/MWh in Denmark [55]. Following study [25], we assume that the crack initiation rate θ is in the range of 0.05–0.2 [1/year] and the lifetime T_{LT} for one WT is 25 years.

Table 2 shows a summary of the model parameters.

Table 2. Model parameters.

Parameter	Symbol	Value
Cost of corrective maintenance due to degradation failure	$C_{CM}\ [\text{€}]$	100,000
Cost of corrective maintenance due to catastrophic failure	$C_{CM}^{CF}\ [\text{€}]$	440,000
Cost of preventive maintenance due to false positive	$C_{PM}^{FP}\ [\text{€}]$	20,000
Cost of preventive maintenance due to true positive	$C_{PM}^{TP}\ [\text{€}]$	20,000
Loss cost per unit of time due to unrevealed failure	$C_{UF}\ \left[\frac{\text{€}}{\text{month}}\right]$	72,000
Cost of ordering an offshore boat and inspection	$C_M\ [\text{€}]$	12,500
Crack initiative rate	$\theta\ [1/\text{year}]$	0.05–0.2
WT lifetime	$T_{LT}\ [\text{year}]$	25

5. Results and Discussion

This section of the article presents the results and discussion of the preventive maintenance optimization of WT blades on the basis of the mathematical model developed in Sections 2 and 3.

5.1. Investigation of Probabilities of Correct and Incorrect Decisions of Online Condition Monitoring

The probabilities of correct and incorrect decisions at the online condition monitoring of the WT blades are calculated using formulas (17)–(20). As can be seen from (17)–(20), to calculate these probabilities, PDF $f(\delta|\eta)$ and $\omega(\eta)$ should be known. Using expressions (42) and (43), let us investigate the behavior of PDF $f(\delta|\eta)$ and $\omega(\eta)$.

Figure 3 shows a 3D presentation of the conditional PDF $f(\delta|\eta)$, plotted by 3D Surface Plotter when $FT = 60$ cm and $\sigma_e = 6$ cm.

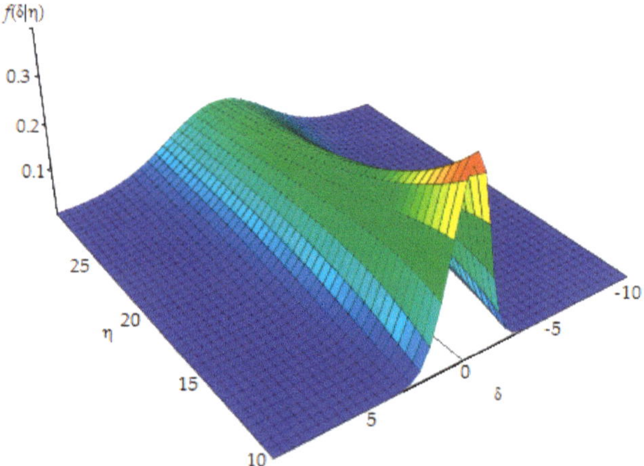

Figure 3. A 3D presentation of the conditional PDF of error when evaluating the time to failure, depending on the arguments δ and η.

As can be seen in Figure 3, the conditional PDF $f(\delta|\eta)$ flattens with an increase in the failure time η, which indicates an increase in the variance of the error in evaluating the time to failure.

Figure 4a shows a plot for the conditional PDF of the measurement error in evaluating the time to failure when $\eta = 6$ months. As can be seen in Figure 4a, the PDF $f(\delta|\eta)$ has a symmetric Gaussian distribution. For comparison, Figure 4b shows a plot for the PDF of the measurement error $\Omega(e)$. By comparing the plots presented in Figure 4a and 4b, we can see that both PDFs are of the same form, though the abscissa axis denotes the error in evaluating the time to failure for the PDF $f(\delta|\eta)$ and the measurement error of sensors for the PDF $\Omega(e)$. Thus, the conditional PDF $f(\delta|\eta)$ allows the conversion of sensor errors into errors in evaluating the time to failure of a WT component. Essentially, this is the conversion of spatial characteristics (errors of sensors) into temporary characteristics (errors in evaluating the time to failure). Knowledge of PDF $f(\delta|\eta)$ allows for calculating the probabilities of the events FP, TP, FN, and TN during any interval of condition monitoring.

Figure 5a shows the plot of the PDF $w(\eta)$ when $m_1 = 5 \frac{cm}{month}$, $\sigma_1 = 2.5 \frac{cm}{month}$, and $FT = 60$ cm. As can be seen in Figure 5, PDF $w(\eta)$ begins to grow sharply at $\eta = 4.4$ months and reaches a maximum at $\eta = 8.9$ months.

Figure 5b shows the dependence of the probability of failure $Q(\eta)$ as a function of time to failure. As can be seen from Figure 5b, the probability of failure reaches 10% at $\eta = 5.6$ months, 25% at $\eta = 8.9$ months, and 50% at $\eta = 12$ months. From the graph in Figure 5b, we can see that the highest increase in the probability of failure falls on the time interval of 5–12 months.

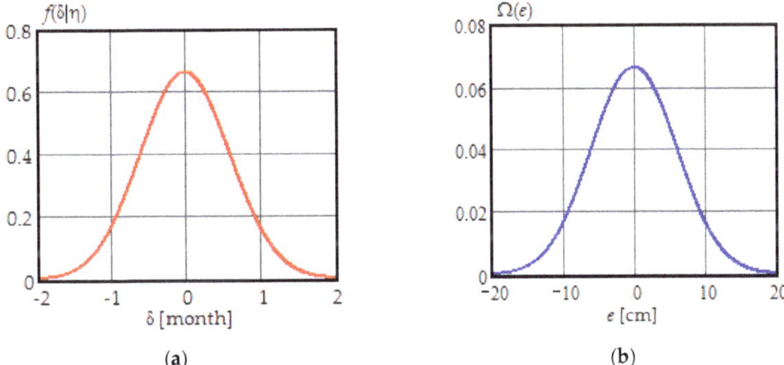

Figure 4. (**a**) The plot of the conditional probability density function (PDF) of the error in evaluating the time to failure when $\eta = 6$ months; (**b**) The plot of the PDF of the measurement error of sensors.

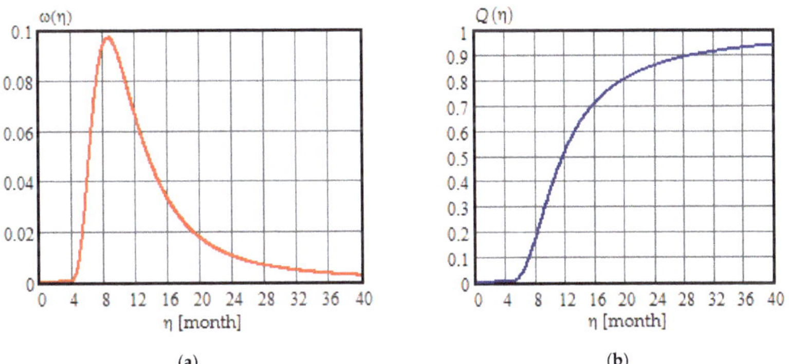

Figure 5. (**a**) Probability density function of time to failure; (**b**) Probability of failure.

Figures 6–9 show the dependence of the probabilities of a FP, TP, FN, and TN on the length of the monitoring interval τ when $\sigma_e = 6$ cm and the rest of data are the same as for Figure 5.

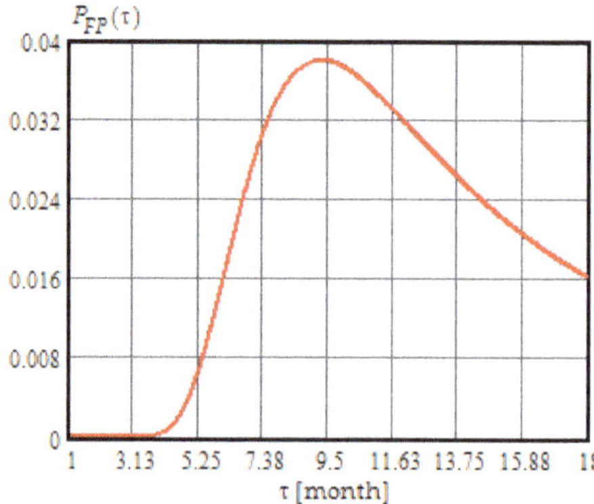

Figure 6. The dependence of the probability of false positive on the length of the interval τ.

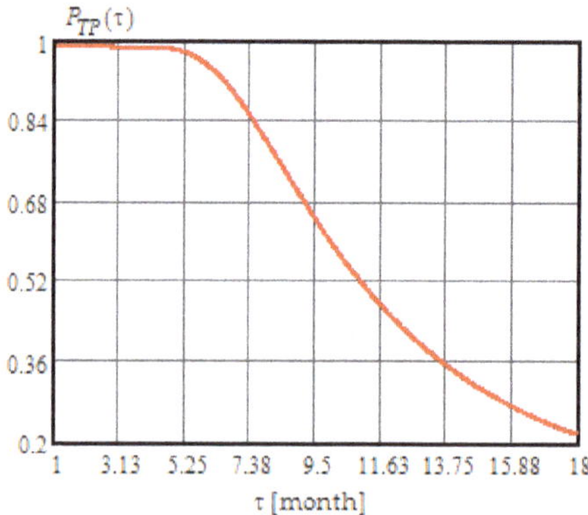

Figure 7. The dependence of the probability of true positive on the length of the interval τ.

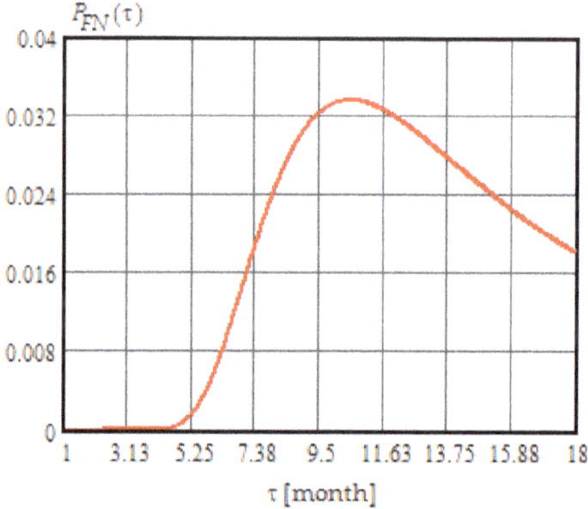

Figure 8. The dependence of the probability of false negative on the length of the interval τ.

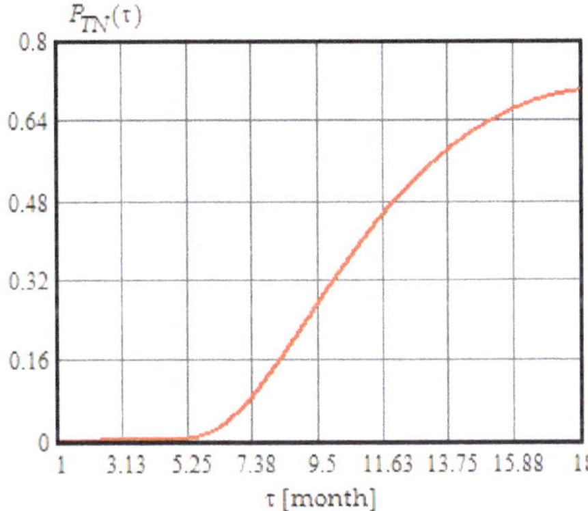

Figure 9. The dependence of the probability of true negative on the length of the interval τ.

From the analysis of dependencies in Figures 6–9, we can draw the following conclusions:

1. All probabilities depend on the length of the monitoring interval τ;
2. The probability of an FP begins to go up remarkably at $\tau = 4$ months and gets to 3.8% at $\tau = 9.5$ months, and then slowly decreases to 1.6 % at $\tau = 18$ months;
3. The probability of TP decreases slowly, reaching 98.5% at $\tau = 4.75$ months, and then begins to decrease rapidly, reaching 22.4% at $\tau = 18$ months;
4. The probability of FN begins to increase remarkably at $\tau = 5.25$ months reaching the maximum value of 3.4% in the vicinity of the maximum of PDF of time to failure when $\tau = 10.6$ months, and then decreases reaching 1.8% at $\tau = 18$ months;
5. The probability of TN begins to increase significantly at $\tau = 5.5$ months when its value is 0.7%, and then increases sharply reaching 74.2% at $\tau = 18$ months.

The behavior of the curves in Figures 6–9 requires some explanation. The dependence of $P_{FP}(\tau)$ in Figure 6 is explained by the behavior of the sum of the crack size and sensor noise according to Equation (2). When the crack size is small and far from the FT threshold, the probability that the sum of the crack size and sensor noise exceeds the threshold FT is low. That is why for small crack size, the probability of a false alarm is negligible. However, as the mathematical expectation of the crack size approaches the FT threshold, the probability that the sum of the crack size and noise exceeds the threshold FT increases. Therefore, the probability of a false alarm also increases, reaching a maximum at $\tau = 9.5$ months. When the mathematical expectation of the crack size exceeds the FT threshold, the probability of a false alarm decreases, because most probably, the unit failed, and as is well-known, the false alarms occur only for operable units.

Let us explain the dependence of $P_{TP}(\tau)$ in Figure 7. When the crack size is small and far from the FT threshold, the probability that the sum of the crack size and sensor noise is less than the threshold FT is high. That is why, for small crack size, the probability of a TP event is high. However, as the crack size rises toward FT threshold, the probability that the sum of the crack size and noise is less than the threshold FT decreases. Therefore, the probability of a TP also decreases, reaching 22.4% at $\tau = 18$ months.

The dependence of $P_{FN}(\tau)$ in Figure 8 is also explained by the behavior of the sum of the crack size and sensor noise in respect to the threshold FT. When the crack size is small and far from the FT threshold, the probability of failure is also small, according to Figure 5b. That is why the probability that the crack size exceeds FT threshold, and that the sum of the crack size and sensor noise is less than the threshold FT, is shallow. Therefore, for small crack size, the probability of an FN is negligible. Beginning from $\tau = 5.25$ months the probability of failure increases remarkably, which means that an increasing number of realizations of the stochastic process $X(t)$ exceed the threshold FT. However, for some of these realizations, the sum of the crack size and noise is less than the threshold *FT*, which leads to false-negative events. The probability of an FN reaches the maximum at $\tau = 10.6$ months where the increase in the probability of failure is the maximum. When the mathematical expectation of the crack size moves up from the *FT* threshold, the probability of an FN decreases because it is unlikely that the sum of the crack size and sensor noise will be less than the threshold *FT*.

The dependence of $P_{TN}(\tau)$ in Figure 9 is opposite to that in Figure 7. When the crack size is small and far from the *FT* threshold, the probability that the sum of the crack size and sensor noise exceeds the threshold *FT* is low. That is why, for small crack size, the probability of a TN event is also low. However, as the crack size rises, the probability that the sum of the crack size and noise is higher than the threshold *FT* increases. Therefore, the probability of a TN also increases, reaching 74.2% at $\tau = 18$ months.

Thus, from the conducted numerical analysis of the behavior of the probabilities of correct and incorrect decisions when continuously monitoring the condition of the blades, it follows that these probabilities are not constant in time and should be considered when optimizing the periodicity of preventive maintenance.

It should be noted that in the maintenance models with periodic condition monitoring, the probabilities of an FP, TP, FN, and TN are considered in many studies, for example [56–59]. However, in these mathematical models, the authors assumed that the probabilities of correct and incorrect decisions had constant values and do not depend on time and parameters of the degradation process.

5.2. Investigation of the Expected Cost of Preventive Maintenance with Online Condition Monitoring

Figure 10 shows the dependence of the expected maintenance cost per unit of time on the periodicity of preventive maintenance for the data given in Table 2.

As can be seen in Figure 10, the expected cost of maintenance per unit of time is a convex function of the periodicity of preventive maintenance τ. When τ is small, the expected maintenance cost is too high due to frequently conducted preventive maintenance. When τ is large, the expected maintenance cost is also high due to dominated corrective maintenance cost C_{CM}, which is significantly higher than preventive maintenance cost C_{PM}^{TP}. Optimal periodicity of preventive maintenance ensures the minimal expected cost of maintenance per unit of time, providing a trade-off between the weighted cost of preventive and corrective maintenance. Indeed, at the optimal periodicity of 6 months, the expected maintenance cost per unit of time is minimal and equal to 3600 €/month.

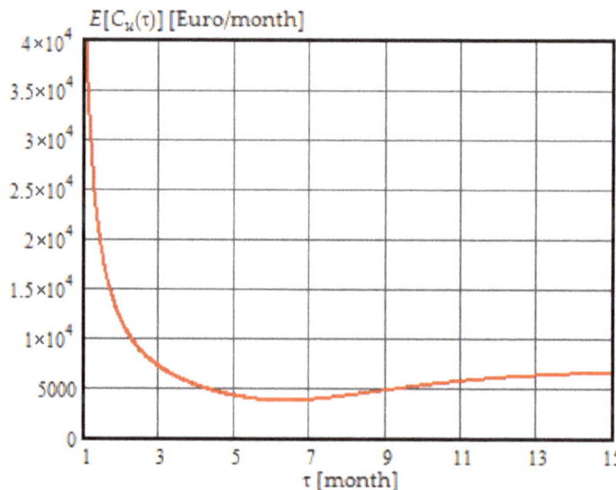

Figure 10. The dependence of the expected cost of maintenance per unit of time on the periodicity of preventive maintenance.

The obtained result requires clarification. Modern WT are usually designed to work for 20–25 years [41,60]. Therefore, the optimal periodicity of preventive maintenance refers to the case of a crack occurrence and its further growth in the WT blade. Calculating (30) for the data given in Table 2, we obtain that the average maintenance cost $E[C_a(\tau)]$ for one regeneration cycle is €21,230. Substituting θ and T_{LT} from Table 2, and $E[C_a(\tau)]$ into (44) we calculate that the range of the average lifetime maintenance cost for one WT blade is €26,540 to €106,160 depending on the crack initiation rate.

Let us compare preventive maintenance with online condition monitoring and predetermined preventive maintenance without condition monitoring. With online condition monitoring, preventive maintenance is only performed if a crack in the WT blade occurs and develops. If there is no crack, then only online condition monitoring is performed. In the absence of an online condition monitoring system, preventive maintenance should be carried out after a specified time interval, regardless of whether there is a crack or not. Independently of the type of preventive maintenance, corrective maintenance is accomplished after the failure occurrence.

A mathematical model of the predetermined preventive maintenance of WT blades is given in Appendix A. Figure 11 shows the dependence of the lifetime maintenance cost on the periodicity of preventive maintenance.

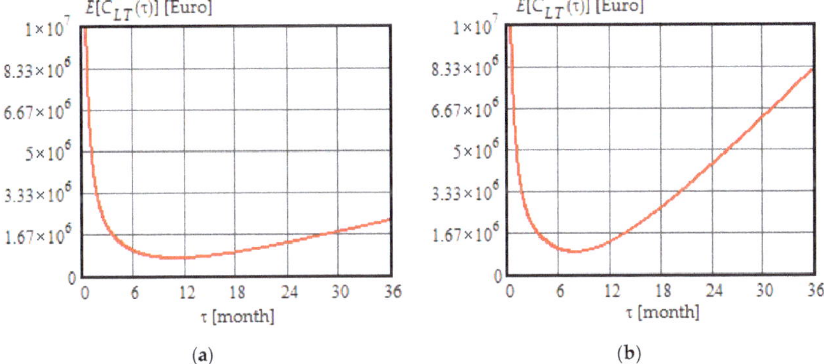

Figure 11. The dependence of the average lifetime maintenance cost on the periodicity of predetermined preventive maintenance when $C_{UF} = 72,000 \frac{€}{month}$: (**a**) Initiation crack rate is 0.05 [1/year]; (**b**) Initiation crack rate is 0.2 [1/year].

From Figure 11, we can draw the following conclusions:

1. The average lifetime maintenance cost is a convex function of the periodicity of predetermined preventive maintenance;
2. The optimal periodicity of predetermined preventive maintenance decreases from 11 months to 8 months, when the crack initiation rate increases from 0.05 [1/year] to 0.2 [1/year];
3. The minimal value of the average lifetime maintenance cost increases with an increase in the crack initiation rate. Indeed, when $\theta = 0.05$ [1/year], then $E[C_{LT}(\tau)] = €687,200$; when $\theta = 0.2$ [1/year], then $E[C_{LT}(\tau)] = €878,500$.

Comparing the results of calculations for preventive maintenance with online condition monitoring and without it, we note that when using condition monitoring, the maintenance cost is many times less (25.9 and 8.3, depending on θ). The reason for this is the fact that with online monitoring, preventive maintenance of the blades is carried out in the presence of a crack. Without condition monitoring, preventive maintenance is carried out after each interval τ, regardless of whether there is a crack or not.

The plots in Figure 11a,b match the case when the failure is unrevealed and $C_{UF} > 0$. Let us evaluate the average lifetime maintenance cost assuming that the failure is revealed, i.e., $C_{UF} = 0$. Figure 12a,b show the corresponding plots.

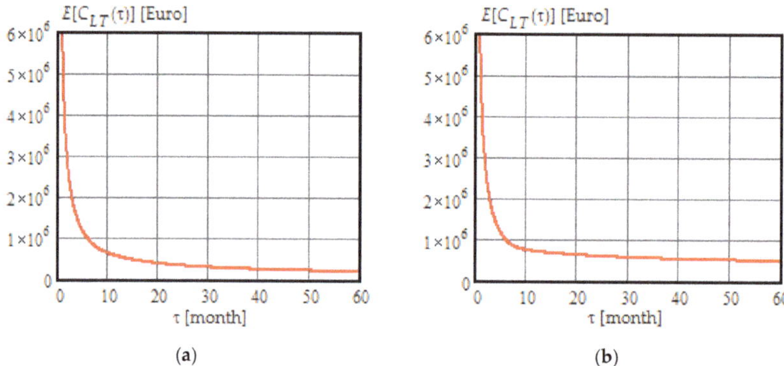

Figure 12. The dependence of the average lifetime maintenance cost on the periodicity of predetermined preventive maintenance when $C_{UF} = 0$: (**a**) Initiation crack rate is 0.05 [1/year]; (**b**) Initiation crack rate is 0.2 [1/year].

As can be seen in Figure 12a,b, over the interval (0, 8) months, the cost function decreases sharply, and with $\tau > 8$ months, the average lifetime maintenance cost slowly decreases reaching at $\tau = 60$ months the values of €196,700 and €486,800 with $\theta = 0.05$ [1/year] and $\theta = 0.2$ [1/year], respectively.

It should be noted that when using corrective maintenance without condition monitoring and $C_{CM}^{CF} = €440,000$ the average lifetime maintenance cost is €550,000 for $\theta = 0.05$ [1/year] and €2,200,000 when $\theta = 0.2$ [1/year].

Table 3 shows a summary of the average lifetime cost calculations for different maintenance types for WT blades.

Table 3. Summary of the average lifetime maintenance cost for different maintenance types.

Crack Initiation Rate, θ [1/year]	Average Lifetime Maintenance Cost [€/blade]			
	Preventive Maintenance with Online Condition Monitoring	Corrective Maintenance	Predetermined Preventive Maintenance	
			$C_{UF} = 72,000\ [\frac{\text{€}}{\text{month}}]$	$C_{UF} = 0$
0.05	46,700	550,000	687,200	196,700
0.2	187,500	2,200,000	878,500	486,800

From the analysis of Table 3, it follows that for any crack initiation rate, when using corrective or predetermined preventive maintenance, the average lifetime maintenance cost is many times higher than when using preventive maintenance with online condition monitoring.

In the numerical example, for simplicity, we assumed that the crack propagation model is linear. If the crack propagation function is not linear, in this case, Equations (1)–(38) are valid for any stochastic process of crack degradation and any measurement noise. The conditional PDF (40) is valid for any stochastic degradation process that can be represented by the model (39). The PDF (42) and (43) can be used only for the linear degradation model. The PDF of time to failure corresponding to the nonlinear model of degradation (39), i.e., in the case when $\gamma > 1$, is given in [54]. It should be noted once again that the proposed approach to optimizing preventive maintenance based on continuous condition monitoring is not limited to the model of the degradation process (39). Equations (1)–(38) can be used for any model of a stochastic degradation process. For example, there is a variety of stochastic processes suitable for the description of crack growth. We can mention here the stochastic gamma process for modeling gradual damage monotonically accumulating over time [61,62] and a class of increasing Markov processes [63,64].

6. Conclusions

This article has proposed a new approach to optimizing preventive maintenance with online condition monitoring, which assumes that the built-in sensors are error-prone and based on the information they provide, incorrect decisions when determining the condition of the WT component are possible. A mathematical model has been proposed to evaluate the probabilities of correct and incorrect decisions made on the results of continuous condition monitoring the deteriorating components of WT. Significantly, the article first has solved the problem of interval evaluation of probabilities of such events as false positive, true positive, false negative, and true negative for the case of continuous condition monitoring. It has been shown that probabilities of correct and incorrect decisions depend on the interval of continuous condition monitoring and are functions of the parameters of the degradation model. Furthermore, the probabilities of a false positive, true positive, false negative, and true negative have been incorporated into the mathematical model of preventive maintenance with continuous condition monitoring allowing to determine the average maintenance cost for one regeneration cycle, the mean time of the regeneration cycle, the expected maintenance cost per unit of time, and the average lifetime maintenance cost. The proposed mathematical equations for the expected cost of preventive maintenance with online condition monitoring are valid for any stochastic deterioration process and any measurement noise. The developed mathematical model has been examined on an example of optimizing the preventive maintenance of WT blades. By numerical calculations, it has been shown that with the optimal periodicity of preventive maintenance of 6 months the expected maintenance cost per unit of time is around 3600 Euro/month; the average maintenance cost for one regeneration cycle is €21,230; and the average lifetime maintenance cost for one WT blade is €26,540 to €106,160 for the lifetime of 25 years and at the crack initiation rate lying in the range of 0.05–0.2 [1/year]. It has also been shown that optimal preventative maintenance with online condition monitoring reduces the average lifetime maintenance cost by 11.8 times compared with corrective maintenance without condition monitoring, and by at least 4.2 and 2.6 times compared with predetermined preventive maintenance for low and high crack initiation rates, respectively.

The proposed mathematical model of preventive maintenance based on imperfect online condition monitoring can be used for a wide range of deteriorating systems since the basic equations do not limit the type of stochastic degradation process and the type of measurement noise.

Our future work will include an investigation of preventive maintenance of WT blades for the case of the nonlinear stochastic process of crack propagation; application of the developed mathematical model to some other components of WT; modification of the proposed maintenance model for case of a minimal repair policy; and development of a predictive maintenance model of WT components based on imperfect continuous monitoring.

Author Contributions: This article presents the collective work of two authors. The authors (A.R. and V.U.) jointly participated in the conceptualization of the problem, development of mathematical models, numerical calculations, and writing the article.

Funding: This research received no external funding.

Conflicts of Interest: The authors declare no conflict of interest.

Abbreviations

The following abbreviations exist in the manuscript:

CBM	Condition-based maintenance
FN	False negative
FP	False positive
GW	Gigawatt
kwh	Kilowatt hour
MW	Megawatt
MWh	Megawatt hour
O&M	Operation and maintenance
PDF	Probability density function

TN	True negative
TP	True positive
WT	Wind turbine

Nomenclature

H	random time to failure of a WT component
$\omega(\eta)$	probability density function of time to failure
$Q(\eta)$	probability of failure
τ	periodicity of preventive maintenance
$X(t)$	stochastic process of a WT component degradation
$e(t)$	stochastic process of noise
$Y(t)$	sum of stochastic processes $X(t)$ and $Y(t)$
FT	degradation failure threshold
H^*	random time until stochastic process $Y(t)$ crosses failure threshold FT
Δ	random error in evaluation of time to failure
$\omega_0(\eta, \eta^*)$	joint probability density function of random variables H and H^*
$P(\tau)$	probability of failure-free operation of the WT component in the interval $(0, \tau)$
$FP(0, \tau)$	a false positive event in the interval $(0, \tau)$
$TP(0, \tau)$	a true positive event in the interval $(0, \tau)$
$FN(0, \tau)$	a false negative event in the interval $(0, \tau)$
$TN(0, \tau)$	a true negative event in the interval $(0, \tau)$
$P_{FP}(0, \tau)$	probability of a false positive in the interval $(0, \tau)$
$P_{TP}(0, \tau)$	probability of a true positive in the interval $(0, \tau)$
$P_{FN}(0, \tau)$	probability of a false negative in the interval $(0, \tau)$
$P_{TN}(0, \tau)$	probability of a true negative in the interval $(0, \tau)$
$\theta(\eta^*|\eta)$	conditional probability density function of random variable H^*
$f(\delta|\eta)$	conditional probability density function of the random error in evaluation of time to failure
$E[C_a(\tau)]$	average maintenance cost for the time between renewals
$E[T_{RC}(\tau)]$	average duration of the time between renewals (regeneration cycle)
$E[C_u(\tau)]$	expected cost of maintenance per unit of time
τ_{opt}	optimal periodicity of preventive maintenance
C_{CM}	cost of corrective maintenance due to degradation failure
C_{CM}^{CF}	cost of corrective maintenance due to catastrophic failure
C_{UF}	loss cost per unit of time due to unrevealed failure
C_{PM}^{FP}	cost of preventive maintenance due to false positive
C_{PM}^{TP}	cost of preventive maintenance due to true positive
$\Omega(e)$	probability density function of the sensor random measurement error
σ_e	standard deviation of the measurement noise
m_1	mathematical expectation of the random degradation rate of crack
σ_1	standard deviation of the random degradation rate of crack
A	random degradation rate of crack
γ	exponent of time
$\varphi(a)$	Gaussian probability density function of the random degradation rate of crack
θ	crack initiation rate
T_{LT}	lifetime of a WT
C_M	cost of ordering an offshore boat and inspection

Appendix A

In the case of predetermined preventive maintenance, the average maintenance cost for the time between renewals includes the expected cost due to corrective maintenance, preventive maintenance and the losses due to unrevealed failure, i.e.,

$$E[C_a(\tau)] = C_{CM} \int_0^\tau \omega(\eta)d\eta + C_{UF} \int_0^\tau (\tau - \eta)\omega(\eta)d\eta + C_{PM}^{TP} \int_\tau^\infty \omega(\eta)d\eta, \tag{A1}$$

where $0 < \tau \leq \theta^{-1}$.

If $C_{UF} = 0$ in (A1), we obtain well-known formula in age replacement model [34].

If the predetermined preventive maintenance is carried out with periodicity τ, then the total number of preventive maintenance cycles over the WT lifetime is

$$N = \frac{T_{LT}}{\tau}. \tag{A2}$$

The average number of cracks in the WT blade during the service life is determined as the product of the crack initiation rate and the lifetime, i.e.

$$n_{crack} = \theta T_{LT}. \tag{A3}$$

Based on (A1)–(A3), the average lifetime maintenance cost is given by

$$E[C_{LT}(\tau)] = n_{crack}E[C_a(\tau)] + (N - n_{crack})C_M, \tag{A4}$$

where C_M is the cost of maintenance actions including ordering an offshore boat and inspection.

As seen from (A4), the average lifetime maintenance cost includes the cost of maintenance in the presence of a crack and the cost of maintenance in its absence. If a crack appeared in each maintenance interval τ, then $\tau = \theta^{-1}$, $N = n_{crack}$ and the second term in (A4) would be zero.

References

1. Wind Power Capacity Worldwide Reaches 597 GW, 50.1 GW Added in 2018. Available online: https://wwindea.org/blog/2019/02/25/wind-power-capacity-worldwide-reaches-600-gw-539-gw-added-in-2018/ (accessed on 25 February 2019).
2. Renewable Power Generation Costs in 2017. Available online: https://www.irena.org/-/media/Files/IRENA/Agency/Publication/2018/Jan/IRENA_2017_Power_Costs_2018.pdf (accessed on 8 January 2018).
3. Wind Turbine Maintenance Costs to Almost Double by 2020. Available online: https://www.edie.net/news/6/Win-turbine-maintenance-costs-to-nearly-doubl/ (accessed on 19 March 2015).
4. Byon, E.; Ntaimo, L.; Ding, Y. Optimal maintenance strategies for wind turbine systems under stochastic weather conditions. *IEEE Trans. Reliab.* **2010**, *59*, 393–404. [CrossRef]
5. Wind Farms: Preventive to Predictive Maintenance with Remote Monitoring. Available online: https://innovate.ieee.org/innovation-spotlight/wind-farm-preventive-maintenance-local-control-unit/ (accessed on 30 March 2017).
6. Yildirim, M.; Gebraeel, N.Z.; Sun, X.A. Integrated predictive analytics and optimization for opportunistic maintenance and operations in wind farms. *IEEE Trans. Power Syst.* **2017**, *32*, 4319–4328. [CrossRef]
7. Ouyang, T.; Yusen He, Y.; Huang, H. Monitoring wind turbines' unhealthy status: A data-driven approach. *IEEE Trans. Emerg. Top. Comput. Intell.* **2019**, *3*, 163–172. [CrossRef]
8. Huang, L.L.; Fu, Y.; Mi, Y.; Cao, J.L.; Wang, P. A Markov-chain-based availability model of offshore wind turbine considering accessibility problems. *IEEE Trans. Sustain. Energy* **2017**, *8*, 1592–1600. [CrossRef]
9. Leonardi, F.; Messina, F.; Santoro, C. A risk-based approach to automate preventive maintenance tasks generation by exploiting autonomous robot inspections in wind farms. *IEEE Access* **2019**, *7*, 49568–49579. [CrossRef]
10. Besnard, F.; Fischer, K.; Tjernberg, L.B. A model for the optimization of the maintenance support organization for offshore wind farms. *IEEE Trans. Sustain. Energy* **2013**, *4*, 443–450. [CrossRef]
11. BS EN 13306: 2017. Maintenance—Maintenance Terminology. Available online: https://shop.bsigroup.com/ProductDetail/?pid=000000000030324472 (accessed on 1 December 2017).
12. Artigao, E.; Martín-Martíneza, S.; Honrubia-Escribano, A.; Gómez-Lázaro, E. Wind turbine reliability: A comprehensive review towards effective condition monitoring development. *Appl. Energy* **2018**, *228*, 1569–1583. [CrossRef]
13. Yang, B.; Dongbai, S. Testing, inspecting and monitoring technologies for wind turbine blades: A survey. *Renew. Sustain. Energy Rev.* **2013**, *22*, 515–526. [CrossRef]
14. Stetco, A.; Dinmohammadi, F.; Zhao, X.; Robu, V.; Flynn, D.; Barnes, M.; Keane, J.; Nenadic, G. Machine learning methods for wind turbine condition monitoring: A review. *Renew. Energy* **2019**, *133*, 620–635. [CrossRef]

15. Kang, J.; Jose Sobral, J.; Soares, C.G. Review of Condition-Based Maintenance Strategies for Offshore Wind Energy. *J. Mar. Sci. Appl.* **2019**, *18*, 1–16. [CrossRef]

16. Sensor Solutions for Predictive Maintenance of Wind Turbines and Generators. Available online: https://www.micro-epsilon.co.uk/news/2019/2019-04-09-ME334-Sensor-solutions-for-predictive-maintenance-of-wind-turbines-and-generators/ (accessed on 9 May 2019).

17. Canizo, M.; Onieva, E.; Conde, A.; Charramendieta, S.; Trujillo, S. Real-time predictive maintenance for wind turbines using big data frameworks. In Proceedings of the 2017 IEEE International Conference on Prognostics and Health Management (ICPHM), Dallas, TX, USA, 19–21 June 2017; pp. 1–8.

18. Fu, Z.; Luo, Y.; Gu, C.; Li, F.; Yue, Y. Reliability analysis of condition monitoring network of wind turbine blade based on wireless sensor networks. *IEEE Trans. Sustain. Energy* **2019**, *10*, 549–557. [CrossRef]

19. Krishna, D.G. Preventive maintenance of wind turbines using remote instrument monitoring system. In Proceedings of the 2012 IEEE Fifth Power India Conference, Murthal, India, 19–22 December 2012; pp. 1–5.

20. Nilsson, J.; Bertling, L. Maintenance management of wind power systems using condition monitoring systems-life cycle cost analysis for two case studies. *IEEE Trans. Energy Convers.* **2007**, *22*, 223–229. [CrossRef]

21. Kerres, B.; Fischer, K.; Madlener, R. Economic Evaluation of maintenance strategies for wind turbines: A stochastic analysis. *IET Renew. Power Gener.* **2015**, *9*, 766–774. [CrossRef]

22. Van Horenbeek, A.; Van Ostaeyen, J.; Duflou, J.R.; Pintelon, L. Quantifying the added value of an imperfectly performing condition monitoring system—application to a wind turbine gearbox. *Reliab. Eng. Syst. Saf.* **2013**, *111*, 45–57. [CrossRef]

23. Kaiser, K.A.; Gebraeel, N.Z. Predictive maintenance management using sensor-based degradation models. *IEEE Trans. Syst. Man Cybern. Part A Syst. Hum.* **2009**, *39*, 840–849. [CrossRef]

24. Bryant, L.; John, A. Modelling wind turbine degradation and maintenance. *Wind Energy* **2016**, *19*, 571–591.

25. Besnard, F.; Bertling, L. An approach for condition-based maintenance optimization applied to wind turbine blades. *IEEE Trans. Sustain. Energy* **2010**, *1*, 77–83. [CrossRef]

26. Pattison, D.; Segovia Garcia, M.; Xie, W.; Quail, F.; Revie, M.; Whitfield, R.I.; Irvine, I. Intelligent integrated maintenance for wind power generation. *Wind Energy* **2016**, *19*, 547–562. [CrossRef]

27. Ghamlouch, H.; Fouladirad, M.; Grall, A. The use of real option in condition-based maintenance scheduling for wind turbines with production and deterioration uncertainties. *Reliab. Eng. Syst. Saf.* **2019**, *188*, 614–623. [CrossRef]

28. Shafiee, M.; Finkelstein, M. A proactive group maintenance policy for continuously monitored deteriorating systems: Application to offshore wind turbines. *Proc. Inst. Mech. Eng. Part O J. Risk Reliab.* **2015**, *229*, 373–384. [CrossRef]

29. Shafiee, M.; Finkelstein, M.; Bérenguer, C. An opportunistic condition-based maintenance policy for offshore wind turbine blades subjected to degradation and environmental shocks. *Reliab. Eng. Syst. Saf.* **2015**, *142*, 463–471. [CrossRef]

30. Wang, J.; Liang, Y.; Zheng, Y.; Gao, R.X.; Zhang, F. An integrated fault diagnosis and prognosis approach for predictive maintenance of wind turbine bearing with limited samples. *Renew. Energy* **2019**, *142*, 642–650. [CrossRef]

31. Marugán, A.P.; Chacon, A.M.; Marquez, F.P. Reliability analysis of detecting false alarms that employ neural networks: A real case study on wind turbines. *Reliab. Eng. Syst. Saf.* **2019**, *191*, 106574. [CrossRef]

32. Gertsbakh, I. *Reliability Theory: With Applications to Preventive Maintenance*; Springer: New York, NY, USA, 2001; pp. 84–89.

33. Nakagawa, T. Replacement and preventive maintenance models. In *Handbook of Performability Engineering*; Misra, K.B., Ed.; Springer: London, UK, 2008; pp. 807–823.

34. Nakagawa, T. *Maintenance Theory of Reliability*; Springer: London, UK, 2005; pp. 220–224.

35. Kavaz, A.G.; Barutcu, B. Fault detection of wind turbine sensors using artificial neural networks. *J. Sens.* **2018**, *2018*, 5628429. [CrossRef]

36. McGugan, M.; Pereira, G.; Sørensen, B.F.; Toftegaard, H.; Branner, K. Damage tolerance and structural monitoring for wind turbine blades. *Philos. Trans. Royal Soc. A. Math. Phys. Eng. Sci.* **2015**, *373*, 1–16. [CrossRef]

37. Shokrieh, M.M.; Rafiee, R. Simulation of fatigue failure in a full composite wind turbine blade. *Compos. Struct.* **2006**, *74*, 332–342. [CrossRef]

38. Cian, C.C.; Lee, J.R.; Bang, H.J. Structural health monitoring for a wind turbine system: A review of damage detection method. *Meas. Sci. Technol.* **2008**, *19*, 1–20.

39. Wedel-Heinen, J.; Hayman, B.; Bronsted, P. Materials challenges in present and future wind energy. *Mater. Res. Soc. Bull.* **2008**, *33*, 343–354.

40. Li, D.; Ho SC, M.; Song, G.; Ren, L.; Li, H. A review of damage detection methods for wind turbine blades. *Smart Mater. Struct.* **2015**, *24*, 1–24. [CrossRef]

41. Hameed, Z.; Hong, Y.S.; Cho, Y.M.; Ahn, S.H.; Song, C.K. Condition monitoring and fault detection of wind turbines and related algorithms: A review. *Renew. Sustain. Energy Rev.* **2009**, *13*, 1–39. [CrossRef]

42. Hyers, R.W.; McGowan, J.G.; Sullivan, K.L.; Manwell, J.F.; Syrett, B.C. Condition monitoring and prognosis of utility scale wind turbines. *Energy Mater.* **2006**, *1*, 187–203. [CrossRef]

43. García, F.P.; Tobias, A.M.; Pinar, J.M.; Papaelias, M. Condition monitoring of wind turbines: Techniques and methods. *Renew. Energy* **2012**, *46*, 169–178. [CrossRef]

44. Lu, B.; Li, Y.; Xin Wu, X. A Review of Recent Advances in Wind Turbine Condition Monitoring and Fault Diagnosis. In Proceedings of the 2009 IEEE Power Electronics and Machines in Wind Applications, Lincoln, NE, USA, 24–26 June 2009; pp. 1–7.

45. Wang, K.; Sharma, V.S.; Zhang, Z.Y. SCADA data based condition monitoring of wind turbines. *Adv. Manuf.* **2014**, *2*, 61–69. [CrossRef]

46. Lopez-Higuera, J.M.; Rodriguez, L.; Quintela, A.; Cobo, A.; Madruga, F.J.; Conde, O.M.; Lomer, M.; Quintela, M.A.; Mirapeix, J. Fiber optics in structural health monitoring. In *Advanced Sensor Systems and Applications IV*; Brian, G., Ed.; Curran Associates, Inc.: Red Hook, NY, USA, 2010.

47. Güemes, A.; Fernández-López, A.; Díaz-Maroto, P.F. Structural health monitoring in composite structures by fiber-optic sensors. *Sensors* **2018**, *18*, 1094. [CrossRef] [PubMed]

48. Murayama, H.; Wada, D.; Igawa, H. Structural health monitoring by using fiber-optic distributed strain sensors with high spatial resolution. *Photonic Sens.* **2013**, *3*, 355–376. [CrossRef]

49. Jaaskelainen, M. *Fiber Optic Distributed Sensing Applications in Defense, Security, and Energy*; Udd, E., Du, H.H., Wang, A., Eds.; SPIE: Orlando, FL, USA, 2009; p. 731606.

50. Mishnaevsky, L.; Branner, K.; Petersen, H.; Beauson, J.; McGugan, M.; Sørensen, B. Materials for wind turbine blades: An overview. *Materials* **2017**, *10*, 1285. [CrossRef]

51. Oh, K.Y.; Park, J.Y.; Lee, J.S.; Epureanu, B.I.; Lee, J.K. A novel method and its field tests for monitoring and diagnosing blade health for wind turbines. *IEEE Trans. Instrum. Meas.* **2015**, *64*, 1726–1733. [CrossRef]

52. Zhang, F.; Li, Y.; Yang, Z.; Zhang, L. Investigation of wind turbine blade monitoring based on optical fiber Brillouin sensor. In Proceedings of the 2009 International Conference on Sustainable Power Generation and Supply, Nanjing, China, 6–7 April 2009; pp. 1–4.

53. Raza, A.; Ulansky, V. Modelling of predictive maintenance for a periodically inspected system. *Procedia CIRP* **2017**, *59*, 95–101. [CrossRef]

54. Ignatov, V.A.; Ulansky, V.V.; Taisir, T. *Prediction of Optimal Maintenance of Technical Systems*; Znanie: Kiev, Ukraine, 1981; pp. 9–10. (In Russian)

55. Electricity Price Statistics. Available online: https://ec.europa.eu/eurostat/statistics-explained/index.php/Electricity_price_statistics (accessed on 21 May 2019).

56. Berrade, M.; Cavalcante, A.; Scarf, P. Maintenance scheduling of a protection system subject to imperfect inspection and replacement. *Eur. J. Oper. Res.* **2012**, *218*, 716–725. [CrossRef]

57. Berrade, M.D.; Scarf, P.A.; Cavalcante, C.A.V.; Dwight, R.A. Imperfect inspection and replacement of a system with a defective state. A cost and reliability analysis. *Reliab. Eng. Syst. Saf.* **2013**, *120*, 80–87. [CrossRef]

58. Zequeira, R.I.; Bérenguer, C. Optimal scheduling of non-perfect inspections. *IMA J. Manag. Math.* **2006**, *17*, 187–207. [CrossRef]

59. He, K.; Maillart, L.M.; Prokopyev, O.A. Scheduling preventive maintenance as a function of an imperfect inspection interval. *IEEE Trans. Reliab.* **2015**, *64*, 983–997. [CrossRef]

60. Nachimuthu, S.; Zuo, M.J.; Ding, Y. A decision-making model for corrective maintenance of offshore wind turbines considering uncertainties. *Energies* **2019**, *12*, 1408. [CrossRef]

61. Van Noortwijk, J.M.; Pandey, M.D. A stochastic deterioration process for time-dependent reliability analysis. In Proceedings of the Eleventh IFIP WG 7.5 Working Conference on Reliability and Optimization of Structural Systems, Banff, AB, Canada, 2–5 November 2003; pp. 259–265.

62. Van Noortwijk, J.M.; Cooke, R.; Kok, M. A Bayesian failure model based on isotropic deterioration. *Eur. J. Oper. Res.* **1995**, *82*, 270–282. [CrossRef]
63. Abdel-Hameed, M. Inspection and maintenance policies of devices subject to deterioration. *Adv. Appl. Probab.* **1987**, *19*, 917–931. [CrossRef]
64. Abdel-Hameed, M. Optimal predictive maintenance policies for a deteriorating system: The total discounted cost and the long-run average cost cases. *Commun. Stat. Theory Methods* **2004**, *33*, 735–745. [CrossRef]

 *energies*

Article

A Context-Aware Oil Debris-Based Health Indicator for Wind Turbine Gearbox Condition Monitoring

Kerman López de Calle [1,2,*], Susana Ferreiro [1], Constantino Roldán-Paraponiaris [1] and Alain Ulazia [3]

[1] Intelligent Information Systems unit, IK4-TEKNIKER, Iñaki Goenaga street 5, 20600 Eibar, Spain
[2] Department of Computer Science and Artificial Intelligence, University of the Basque Country (UPV/EHU), Faculty of Informatics 649, 20080 Donostia, Spain
[3] Department of NE and Fluid Mechanics, University of the Basque Country (UPV/EHU), Otaola 29, 20600 Eibar, Spain
* Correspondence: kerman.lopezdecalle@tekniker.es

Received: 16 July 2019; Accepted: 27 August 2019; Published: 2 September 2019

Abstract: One of the greatest challenges of optimising the correct operation of wind turbines is detecting the health status of their core components, such as gearboxes in particular. Gearbox monitoring is a widely studied topic in the literature, nevertheless, studies showing data of in-service wind turbines are less frequent and tend to present difficulties that are otherwise overlooked in test rig based works. This work presents the data of three wind turbines that have gearboxes in different damage stages. Besides including the data of the SCADA (Supervisory Control And Signal Acquisition) system, additional measurements of online optical oil debris sensors are also included. In addition to an analysis of the behaviour of particle generation in the turbines, a methodology to identify regimes of operation with lower variation is presented. These regimes are later utilised to develop a health index that considers operation states and provides valuable information regarding the state of the gearboxes. The proposed health index allows distinguishing damage severity between wind turbines as well as tracking the evolution of the damage over time.

Keywords: condition monitoring; condition based maintenance; wind turbine; oil debris monitoring; gearbox

1. Introduction

In a world with an ever-increasing electric energy demand, wind energy is getting attention, and has become the fastest growing renewable energy source because of its availability and abundance [1,2]. In this way, the global wind turbine installed power capacity is an increasing trend [3], and two phenomenons are arising: wind turbine (WT) hub size is increasing; and, the fast expansion of wind farms (WF) requires finding better settlements. Consequently, industries are expanding to inhospitable locations, such as offshore hard-to-reach places [4] in a pursuit of better wind resources, as typically, these locations provide higher wind power resources with less turbulence [5].

One of the biggest burdens wind farms face are the Operation & Maintenance (O&M) costs that, as the authors of [1] state, can comprise 10–20% of the total cost of energy (COE) for wind project, and reach up to 35% for a WT at the end of life, a figure that goes up to 30% in the case of big offshore wind farms [6]. Furthermore, various works have related higher failure rates in bigger WTs as compared to smaller ones [7–9]. Finding a positive correlation between average wind speed and failure rate that is reinforced in offshore sites [8,10]. Therefore, it is necessary to assess the health state of the systems and subsystems of WTs, in order to organise maintenance actions and reduce downtimes and losses due to unforeseen stops. In this field, condition monitoring systems (CMS) are well known technology with proven success for health status detection, fault identification and prediction [1,2,9]. Such technology allows identifying the state of the assets remotely, reducing considerably the need of visual inspections on site, which is a costly matter mostly for offshore farms [11].

Particularly, gearboxes represent a delicate component of WTs [10]. The various failure statistic analysis carried out lead to some controversy on its tendency to failure [12], as some of them have less failures reported [8,12], whereas others find high failure rates related to gearboxes [13]. Anyway, most of the studies associate the longest WT downtimes to this component [3,12] and emphasise the need of proper monitoring techniques to avoid them. Additionally, it is considered one of the costliest parts of the WTs [10] and although there are attempts to replace them with direct drives to reduce costs, studies question this assumption, and still the majority of offshore wind turbines rely on gearboxes [14]. Consequently, their monitoring is of vital importance.

Gearbox and gearbox related subsystem condition monitoring (CM) has been broadly addressed on the literature. For that purpose a wide variety of approaches have been tested, such as vibration, oil debris monitoring (ODM), acoustic emissions and current signature among others. Vibration-based CMS prevail over other kinds of sensors [2,9], with much research carried out on the field of signal processing of vibration signals in time, frequency, time/frequency and order domains [3]. However, some works find ODM techniques also interesting, for their higher correlation with wear creation [15], or for the added value they have for monitoring both the oil quality and the state of the gearbox parts [1,2].

Nevertheless, even if the advantages of CM are proven [2,11], its transfer from experimental tests to real WT use cases is less known in the literature [16]. This is because the variability of operation conditions of WTs affects the extraction of indicators while it specially damages the systems of WTs [2]. Most of the works presenting real in-service WT data are based on the use of SCADA (Supervisory Control and Data Acquisition) data, which is readily available in general. Typically, it is used to compare performances among WTs using power curves [17]. These benchmarking procedures are usual for other O&M issues such as pitch misalignment correction [18,19] and the identification of defective anemometers [20]. Additionally, temperatures from the SCADA have been modelled and compared over time to use differences as alarms as the different works reviewed by the authors of [21] show. However, the success of these techniques is limited [3]. Partly, because of external influences (such as the outside temperature) that require the alarms to be manually supervised by operators [21]. Consequently, the inclusion of additional CM sensors in operating WTs is flourishing [9], and an increasing number of works present findings from real cases of use:

- In the work by the authors of [22], two case studies are analysed: in the first one, physical principles that relate the difference in temperatures with the efficiency, rotational speed and power output are used, and the approach is validated by using the deviation of the temperature with respect to power in order to foresee a failure; in the second one, vibration and particle counter sensors are used and the evolution of the signals is studied before and after the replacement of a bearing. They suggest using cumulative particle counts to better detect failures instead of direct particle creation measurements and to combine various sensors in order to improve confidence in the diagnosis.

- In the work by the authors of [23], they create a health indicator based on the centroids proposed by a Self Organising Map in order to group WTs according to health status using SCADA data. This way operators are given additional information regarding the health state of the WTs, and can plan consequently.

- In the work by the authors of [16], current and vibration analyses are used to diagnose a turbine drive train, they emphasise on the difficulty of using signal processing techniques that are only proven at laboratory scale, and recognise the complexity of calculating the remaining useful life (RUL) and establishing damage/healthy thresholds, especially with a lack of available historical data.

- In the work by the authors of [24],the vibration signature of a sample of healthy wind turbines is shown, most relevant indicators are identified on averaged power spectra and the dependence of amplitudes on the operation is studied. They conclude that the high impact of wind speed on vibration amplitudes has to be taken into account to develop CMS.

- In the work by the authors of [14], the data gathered before and after a planetary gear was changed due to spalling is examined. They integrate temperature, vibration and particle counter signals in order to reduce false alarms, and prove the ability to distinguish healthy and warning states.
- In the work by the authors of [25], they suggest the use of moving averages (of both short and long term trends) of ODM to generate a count rate propagation model. Then, they establish an acceptance threshold based on the equivalent maximum angle of spall which is related to bearing geometry; and, lastly, they estimate remaining useful life (RUL).

Most of the works related to on service WT utilise vibration and/or oil debris sensors [3,9]. The works based on ODM from the previously mentioned ones [14,22,25] agree on the same difficulties for the development of ODM systems: the need of averaging or using cumulative values instead of using directly particle generation rates; and the tendency of particle creation rate to vary with operation. These findings are supported by the extensive work of the authors of [26], in which a full-scale WT gearbox of 750 kW is tested with in-line and online sensors and samples taken along the time. In their findings, the need of filtering influences caused by operational conditions is remarked; they recommend to focus in trends instead of in absolute values, and suggest considering big particle size (>14 µm) indicators in particular; also, they identify that damaged gearboxes have much higher debris generation rates than healthy ones.

Taking into consideration the interest of having real on-service WT operation data analysed, and that some of the limitations of ODM of WT are already identified on the literature, this work aims to provide a better insight for the development of ODMs. For that purpose, the data obtained in three WTs monitored with oil debris sensors are studied for a period of six months; the readings of the sensor are compared to other traditional SCADA based monitoring techniques; and, lastly, a study of the different operation states is carried out to determine which filtering criteria is better to develop an health index that considers operating conditions.

2. Data and Methodology

2.1. Wind Farm and Turbines

This study analyses the data produced by 3 WTs which are located in the wind farms at Bayo and Monteros, in Zaragoza (Spain). Both wind farms are close one another and undergo similar influences of the wind. The natural barriers of the Iberian System mountain ranges in the south and the Pyrenees mountain ranges in the north constitute a funnel effect that creates the meteorological occurrence known as cierzo; a dry, usually cold and accelerated flux of air intensified by the natural funnel going through the Ebro valley. Cierzo is more frequent during winter and the beginning of spring, and is compensated by the antagonistic phenomenon known as bochorno, that goes in the opposite direction to cierzo and tends to be softer. Additionally, these oposing phenomena provide the wind with copious kinetic energy and make the region an interesting location for the exploitation of wind energy [27].

The WTs have a 58 m diameter rotor and three blades. Their rated power is 850 kW and cut-in and cut-out wind speeds are 3 m/s and 20 m/s, respectively. They have planetary gearboxes with 1/62 transmission ratios coupled with asynchronous generators. The mineral lubricant is cleaned by offline oil filters and the online oil debris optical sensors is installed in a bypass of the lubrication system.

Regarding the health status of the gearboxes, visual and endoscopic inspections carried out on-site reveal different levels of damage. Two of the gearboxes show medium wear levels (WT 1 and WT 2) with micropitting present in most of the gears, whereas the last one is diagnosed with medium–high wear level showing greater surfaces damaged by micropitting in some gears and pitting in the sun gear. However, no corrective actions have been recommended yet.

2.2. Optical Oil Debris Sensor

Oil samples can be taken and analysed offline in laboratories, however, this procedure delays the decision making process and requires to access the WTs. Therefore, online oil debris sensors are an attractive way of determining the quality of the lubricant and safeguard the components of the gearbox.

In particular, this work uses a optical oil debris sensor. This kind of sensors monitor the fluid condition and contamination using optical technology by capturing high-resolution images of the moving fluid, and later applying advanced processes of image digitisation and spectral analysis. They detect, quantify and classify the particles bigger than 4 microns by size and/or shape, in addition of distinguishing these particles from air bubbles [28]. Besides wind turbine lubrication system monitoring, this kind of technology is well-suited for other industrial applications such as automotive, steel sector, wastewater treatment or cement industries [29] as all of the previous use lubrication systems.

2.3. Dataset

The study is based on a dataset consisting of six months long records of 3 WTs. The data records are taken with one minute frequency from the SCADA. At the same time, additional measurements provided by online optical oil debris sensors are taken. Variables from the SCADA represent the operation of the WTs, whereas the ones provided by the sensor indicate the amount of particles of size greater than 4, 6 and 14 micrometers (ISO.4, ISO.6 and ISO.14, respectively) present on the lubricating oil according to the ISO 4406 standard [30]. These values of the oil sensor represent the particle generation rate, as the oil is being continuously filtered. Details of the variables of the SCADA and the oil debris sensor with the units of measurement are presented in the Table 1.

Table 1. Variables available in the dataset and measurement units by data source.

Source	System	Units of Measurement
SCADA	Pitch angle	°
	Gearbox temperature	°C
	Wind speed	m/s
	Generator speed	rpm
	Active power	kW
Oil debris sensor	ISO.4	scale
	ISO.6	scale
	ISO.14	scale

For privacy reasons the data is shown in a normalised way along this work within a 0 to 1 range corresponding to minimum and maximum values of each of the variables in the dataset.

2.4. Methodology

In order to gain better insight on the use of oil debris sensors to obtain health indicators, the study has two parts: an exploration and correlation analysis stage, in which an overview of the data is presented and some methods of the literature contrasted; and the comparison of operation regions and health index (HI) development, where different operating regimes are compared and the most appropriate one is chosen as the basis to develop a HI. The methods used in each of the parts are presented below.

2.4.1. Exploration and Correlation Analysis

In an initial stage, various visualisation and correlation techniques are used:

- Pearson correlation: Coefficient used to measure the degree of linear association between two variables, presented in the work by the authors of [31].
- Spearman's correlation: Nonparametric coefficient that reflects the degree of monotonicity between two variables explained in the work by the authors of [32].
- Principal component analysis (PCA): PCA is a orthogonal transformation that turns a set of variables into a set of linearly uncorrelated variables. This technique is widely used for visualisation purposes in order to reduce multidimensional spaces to lower dimensional representations with the minimum information loss [33].
- Exponential moving average: In contrast to regular moving average where the average of a window of values in taken, this type of moving average is used when latest values are need to have more importance. The implementation used in this work can be found on the work by the authors of [34].
- Local regression (LOESS): This nonparametric method is used on local subsets of data. The implementation is based on the work by the authors of [35].
- Decision trees: Decision trees are machine learning (ML) algorithms used for classification. Their goal is to predict values of a target variable based on the inputs. Trees are built by splitting input variables with criteria that maximise the probability of having instances of certain group in each partition. The trees used in this work used Gini impurity index for partitioning and are implemented in the work by the authors of [36], which is based on the work by the authors of [37].

2.4.2. Comparison of Operation Regions and Health Index (HI) Development

During the initial exploration, the influence of the operation in the particle creation rates is detected; nevertheless, as there is no clear correlation identified between operation variables and particle creation, it is decided to consider only the measurements that are taken under the same operation conditions. Furthermore, a methodology is used to define which operating conditions are the most appropriate for monitoring purposes. The following techniques were used.

- Operation region (OR) definition: In order to find the optimal instants for taking measurements, several operating regions (OR) are explored. Each operating region is defined by a set of rules/criteria, such as wind speed in range (x m/s, y m/s), active power equals nominal power, etc. The ORs analysed in this work are suggested by experts in the domain, and are depicted in the following Figure 1 with a short explanation added in Table 2.
- Operation states (OS): Instead of considering each time instant individually as a data point with certain associated variable values (pitch angle, active power, ISO.4... etc.) groupings of data points have been studied. These groups or operation states are generated considering the different operation regions previously presented, and correspond to a set of chronologically continuous points over time fulfilling the rules proposed by the OR. Every time the machine works under the criteria of an OR we say it has entered in a new OS related to that OR, that lasts as long as the WT keeps working under the constraints of the OR.

Table 2. Descriptions of the different operating regions.

Operating Region	Characteristics
Nominal	Stable power generation. Varying pitch
N. & low pitch	Similar to nominal, but more restrictive and not including high wind speeds, delimited using pitch values.
Ramp to nominal	Ranges from about the middle of the power curve to beginning of nominal operation.
Ramp	Values taken only during the power ramp.
Pre-ramp	Values taken before the generator speed ramp starts.

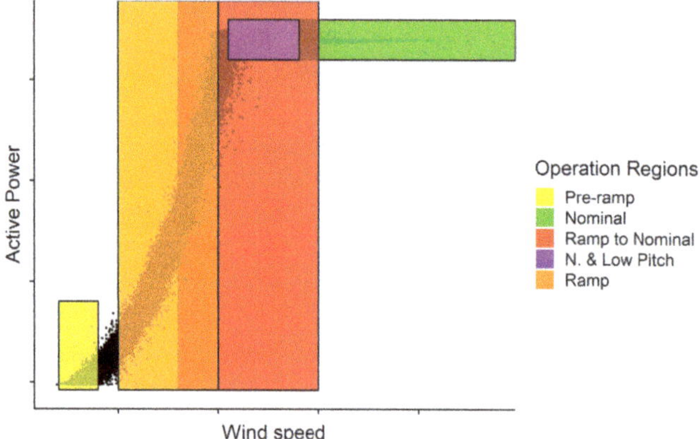

Figure 1. Operation regions over active power against wind speed plot.

- Operation clustering: With the purpose of analysing the steadiness of the different OR the following procedure was used in order to generate data cluster representing the variability of the operation in each OS according to the different ORs.

1. Scale all the variables between 0 and 1 corresponding to the maximum and minimum values of each variable.
2. Taking an OR (Example:Nominal) find the respective number of OS occurrences in the dataset $\{OS\}_{i=1}^{m}$, where m is the number of occurrences.
3. Create a matrix for each OS_i where $i = 1, 2, \ldots, m$:

$$OS_i = \begin{pmatrix} a_{11}^i & a_{12}^i & \cdots & a_{1p}^i \\ a_{21}^i & a_{21}^i & \cdots & a_{2p}^i \\ \vdots & & \ddots & \vdots \\ a_{n_i1}^i & \cdots & \cdots & a_{n_ip}^i \end{pmatrix} \tag{1}$$

where p is equal to the number of sensors considered and n_i is the length of the i-th OS; therefore, these matrices contain the values of the p operation variables along the OS.

4. Then, the difference vector of each variable is calculated by OS. This vectors represent the variability of the operation during the OS and give as a result the new matrix D:

$$D_i = \begin{pmatrix} d_{11}^i & d_{12}^i & \cdots & d_{1p}^i \\ d_{21}^i & d_{22}^i & \cdots & d_{2p}^i \\ \vdots & \vdots & \ddots & \vdots \\ d_{n_i-1,1}^i & d_{n_i-1,2}^i & \cdots & d_{n-1,p}^i \end{pmatrix}, \quad i = 1, 2, \ldots, m \tag{2}$$

where $d_{jk} = a_{j+1,k}^i - a_{jk}^i$,that is, each element of the difference vector is the difference between the measurement in that instant (j) and the following measurement $(j + 1)$, for each $j = 1, 2, \ldots, (n-1), k = 1, 2, \ldots, p$.

5. Then matrix R is computed.

$$R = \begin{pmatrix} r_{11} & r_{12} & \cdots & r_{1p} \\ r_{21} & r_{22} & \cdots & r_{2p} \\ \vdots & \vdots & \ddots & \vdots \\ r_{m1} & r_{m2} & \cdots & r_{mp} \end{pmatrix} \tag{3}$$

R is the result of computing the columnwise quadratic mean of the D_i matrices, and represents the average values of the variability considering both negative and positive values. They are computed in the following way.

$$r_{ik} = \sqrt{\frac{\sum_{j=1}^{n_i-1} d_{jk}}{n_i - 1}} \tag{4}$$

6. From the R matrix, two metrics are obtained:

   (a) Centroid: The average position of the points contained in W. Computed as follows.

$$Centroid = (\mu_1, \ldots, \mu_p) \tag{5}$$

   where for all $k = 1, \ldots, p$ the average μ_k is calculated as follows.

$$\mu_k = \frac{1}{m} \sum_{i=1}^{m} r_{ik} \tag{6}$$

   (b) Cluster dispersion: Mean of the variable variance value that represents how disperse the cluster is; it is calculated as follows.

$$Dispersion - \sum_{k=1}^{p} \sigma_k \tag{7}$$

   Each σ_k for $k = 1, 2, \ldots, p$ is the standard deviation computed in the following way.

$$\sigma_k = \frac{1}{m} \sum_{i=1}^{m} (r_{ik} - \mu_k)^2$$

This procedure is repeated separately for each WT and OR. Therefore, there are five ORs by three WTs, a total of 15 data clusters.

- Operation state and cluster metrics: From the clusters of data and the OS some metrics are calculated that help identifying the most interesting OR. These metrics are as follows.

  - Weekly occurrence ratio: Average number of times per week the WT enters in an OS as defined in the OR.
  - Steadiness: The euclidean distance from the centroid (or mean point) of a cluster to the total steadiness (no variation) point.
  - Dispersion: Indicates how spread the data points within a cluster are. Defined previously in Cluster dispersion.

3. Results

As in Section 2.4, results chapter is divided in two parts. The first part, Section 3.1, explains the exploratory analysis that is carried out over the dataset and the relations found between the variables. The second part, Section 3.2, shows the steps that were taken in order to identify the best conditions for obtaining measurements along the time in order to obtain a health index of the gearboxes.

3.1. Exploration and Correlation Analysis

Taken a sample of the whole dataset, Pearsons and Spearmans correlations are studied. In order to identify any possible difference between power generation and during no generation, correlation is also measured in separate samples. However, no significant correlations (neither Pearsons nor Spearmans) are found between operation variables and particle generation data. Regarding the operational variables, some show high degree of association because of the control system. Furthermore, the association of the same variables among WTs over overlapped time spans yields high correlation which means they face similar environmental conditions (wind). However, this is not the case of ODSs, that do not correlate from one WT to another. Nevertheless, in light of the strong correlations between particle indicators (ISO.4, ISO.6 and ISO.14), and following the advice provided by the authors of [26], it is decided to follow the study using ISO.14 indicator as only indicator for particles in order to simplify the study.

After the brute correlation study, the variables are visually studied against the wind speed in Figure 2. The different variables of the SCADA data plotted against the wind speed show the typical patterns that can be found in wind turbines, and are extremely similar one to another. The greatest (but yet small) differences are found in gearbox temperature, suggesting there could be some differences in the cooling system or on the efficiency of the gearboxes.

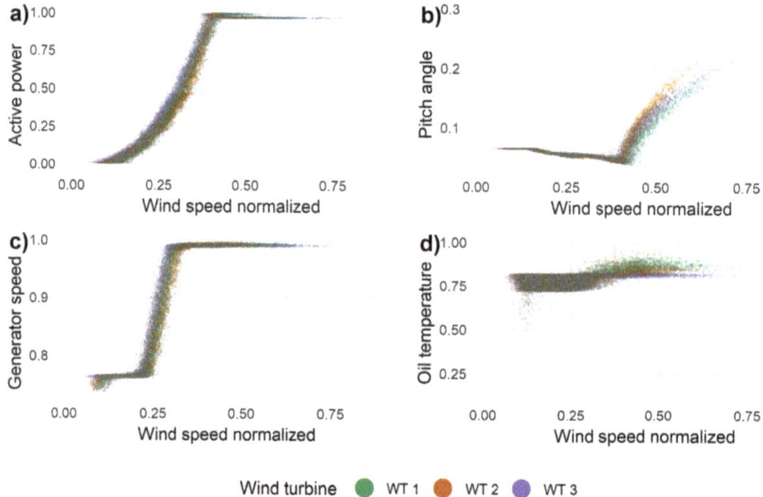

Figure 2. SCADA variables against wind speed by WT. (**a**) Active power against wind speed, (**b**) pitch angle against wind speed, (**c**) generator speed against wind speed and (**d**) oil temperature against wind speed.

As the signals of the ODSs are discrete, much noisier and is almost impossible to visualise anything in the raw measurements, the measurements are given some pretreatments by averaging the values in 0.33 m/s wind speed bins, which creates the pattern visible in Figure 3c.

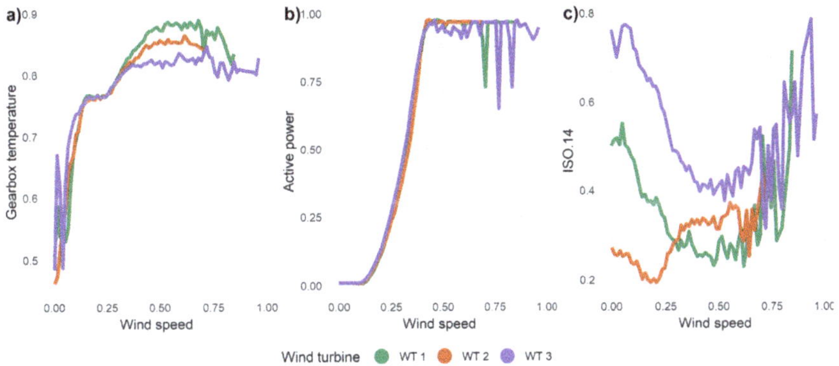

Figure 3. Averaged values of gearbox temperature, active power and ISO.14 in 0.33 m/s wind speed bins. (**a**) Gearbox temperature. (**b**) Active power. (**c**) ISO.14.

Averaged values show great differences between the particle creation rates among WTs. At the same time, the influence of operation over particle generation is visible. Interestingly, the behaviours do not coincide exactly between WTs: WT 1 and WT 3 show big similarities, with high wear creation with low wind speeds and lower wear creation at medium speed or nominal operation; meanwhile, WT 2 shows a different pattern, as its wear creation increases proportionally with wind speed. These differences in the behaviours of the WTs regarding particle creation and operation are also present in the averaged values of wear generation during power production and no power production (including: idling, generator turning without active power generation and idling because of overload) as Figure 4 demonstrates.

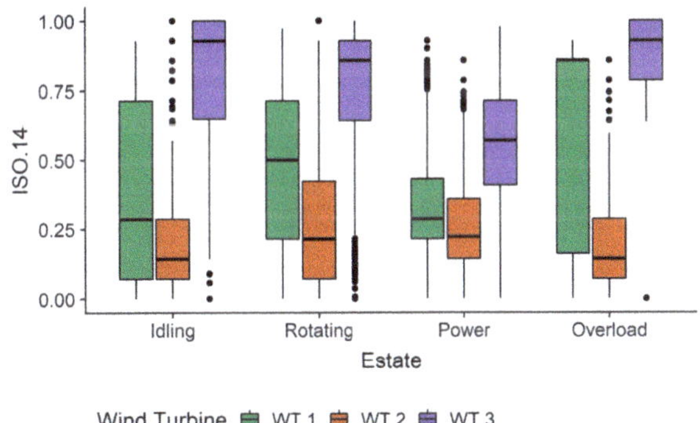

Figure 4. Boxplots of ISO.14 particle creation rates during different operations.

Again, WT 2 does not act as the other WTs. In any case, the previous figures suggest the particle creation is greater during no power generation, meaning braking and acceleration could be causing higher particle creations. Furthermore, there is a clear distinction in the mean level of particle creation rates that match the visual diagnostics of the gearboxes, showing higher values in WT 3, and lower values for WT 1 and WT 2. This variation of particle levels that indicates disparate damage severity, is complex to detect by just paying attention to the the SCADA variables. As Figure 3a,b shows, the same binned in variables typically used for condition monitoring by benchmarking (Active power and gearbox temperature) are not sufficiently different in order to make comparisons between turbines and determine whether WTs could be damaged. Whereas these differences are clearly visible in the binned ISO.14 values (Figure 3c).

This fact is clearer when cumulative particle creations are used. Instead of using raw signals, using cumulative particle rates provides a better insight of the degradation process, as it allows us distinguishing changes in the slopes. In Figure 5, we see clear and increasing differences between WTs in the trends generated when plotting cumulative particle creations against cumulative power generation. If cumulative temperature is observed, the differences among WTs, even if existent, are quite small which reduces the possibility to correctly diagnose failures using only SCADA data. Additionally, the presence of similar shapes for the three turbines in both temperature and particle creation and considering the same periods of time are studied indicate some common factor could be causing the sharp increase in the middle of the curves, which is visible in both variables.

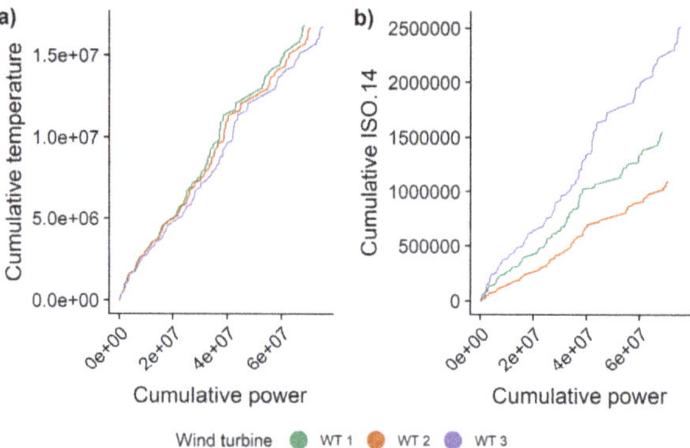

Figure 5. Cumulative values against cumulative active power. (**a**) Temperature. (**b**) ISO.14.

In order to reproduce previous findings in the literature, it is decided to analyse braking and acceleration registered in the SCADA data. For doing so, the Generator speed of five days of operation is taken and it is manually labelled adding "braking", "boosting" or "other" labels. Then, five lagged variables of the generator speed, a exponentially smoothed generator speed (using a bin size of 15) and a difference vector are created. With this data a decision tree is trained using the default parameters for classification cases and it is used to segment the remaining generator speed data in brake/boost/other. With the data split in these groups, it is possible to study the sequences occurring in the data. Boosting and braking sequences are studied by measuring the spearmans correlations of the ISO.14 variables with the smoothed generator speed. In Figure 6 the distribution of the correlations obtained by WT is presented.

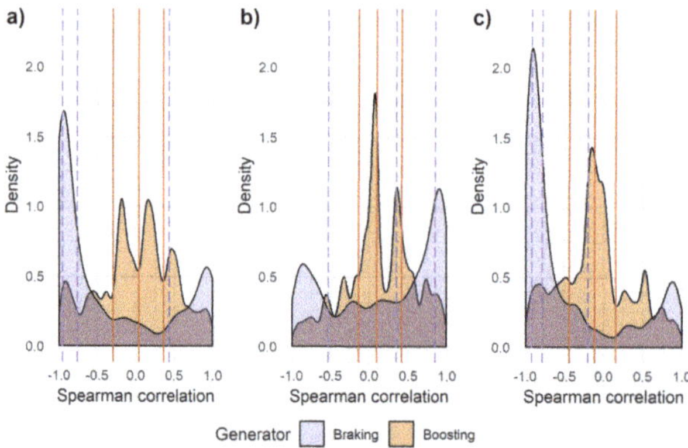

Figure 6. Density plot of the spearman correlation of ISO.14 in respect to the exponentially smoothed generator speed by wind turbine, vertical lines represent quartiles. (**a**) WT 1, (**b**) WT 2 and (**c**) WT 3.

The distribution of the correlation shows here different behaviours. In the braking sequences, WT 1 and WT 3 (Figure 7a,c) have bimodal distributions with a minor mode in strong positive correlation values and the major mode in very strong negative correlation values. This implies that there is a predominant tendency to create more particles during stops (speed decreases and particle generation increases), but is not always the case, as in some cases the correlation is positive (speed decrease with particle decrease). In WT 2 the opposite behaviour is identified, even if the correlation distribution is also bimodal, the major mode is on positively correlated values, meaning in this WT there is a tendency to decrease particle generation when the generator is stopping. Regarding the boosting sequences, the overall correlation values are quite low, which implies there is no clear relation between the increasing speed and the particle creation. The predominance of the major mode in very negative correlation values together with the quartile lines so far from the 0 value indicates braking generate an increase in particle creation, at least for turbines WT 1 and WT 3.

Taking WT 3, boosting and braking sequences were analysed in depth. The following Figure 7 presents two examples of sequences with strong spearman correlation values with negative and positive correlation.

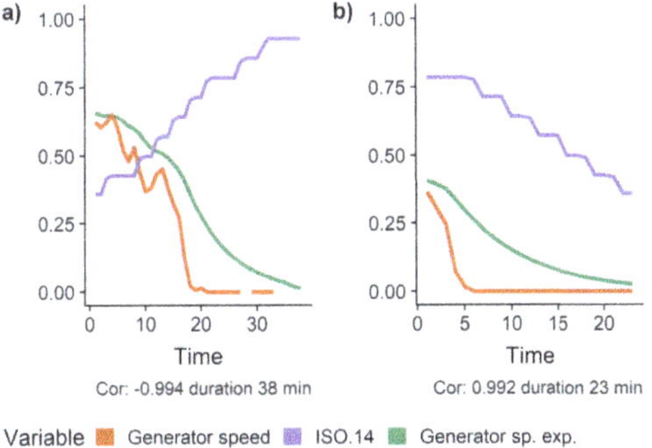

Figure 7. Examples of OSs showing high spearman correlation between ISO.14 and smoothed generator speed. (**a**) Positive correlation. (**b**) Negative correlation.

Interestingly, despite the there is a clear unbalance in the number occurrences, stopping can lead to both an increase or a decrease in particle generation. Note that the generator speed decreases faster than the ISO.14 level, and the nature of the exponentially smoothed speed is more similar to the one of the ISO.14 variable that is more influenced by the inertia of the system than generator speed.

The same procedure is followed for boosting, in this case, considering predominant correlation is near 0 (meaning there is no monotonicity) examples with low correlation are also studied. Figure 8 displays occurrences with high positive correlations (a), highly negative correlations (b) and no correlation (c).

With the uniform distribution of correlation for boosting cases and the different cases shown in Figure 8 there is no way of identifying an expected behaviour of the particle generation during boosting sequences. Furthermore, Figure 8c reveals an unexpected behaviour during idling. As the sensor is giving high ISO.14 particle levels. This fact occurs mostly in WT 3 but is also reported in WT 1, but with a lower frequency. Off-line oil filters should operate continuously regardless of the operation of the machine, but this finding suggest the filter could be stopping in certain situations, which explains also the big difference of particle generation found in Figure 4.

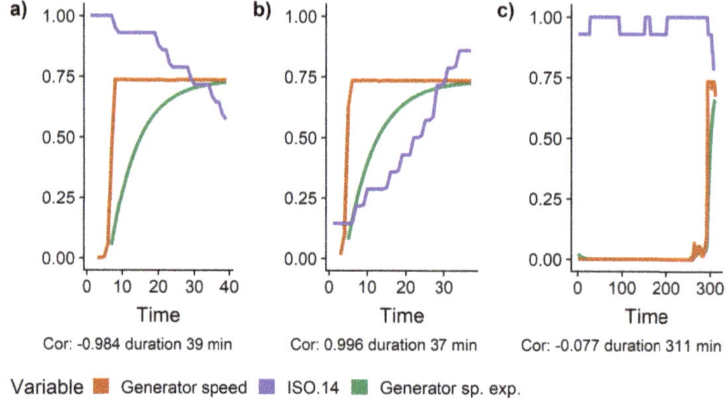

Figure 8. Boosting examples with disparate spearman correlation values between ISO.14 and smoothed generator speed. (**a**) Positive correlation. (**b**) Negative correlation. (**c**) No correlation.

3.2. Comparison of Operation Regions and Health Index (HI) Development

At this point the influence of operation over the particle creation is evident; therefore, it is decided to isolate measurements taken under similar conditions to compare them along the time and use these filtered measures to build a HT. For this purpose, the procedure explained in Section 2.4.2 is carried out. The operational data is taken, different operation regions are defined one by one as explained in Figure 1 , the operation states produced in each turbine are generated and once all WT have been processed, the OSs are studied. Considering that there could be a delay between operation conditions and the effect of those conditions on the oil debris content, it is decided to remove occurrences (Operation States) shorter than a minimum length. In order to define the most appropriate minimum required duration, the following table, Table 3, is created, where the effect of filtering with different duration is presented.

Table 3. Weekly OS frequency of different minimum times by each wind turbine (WT) and operation region (OR).

Turbine	OR	>5	>10	>15	>30	>45
WT 1	N. & pitch	179	64	30	8	3
WT 1	Nominal	1073	532	357	189	113
WT 1	Ramp-to-nominal	2204	1159	794	417	294
WT 1	Pre-ramp	1119	367	155	25	9
WT 1	Ramp	3451	1906	1380	757	481
WT 2	N. & pitch	44	15	6	2	2
WT 2	Nominal	985	507	339	183	128
WT 2	Ramp-to-nominal	1833	1057	758	427	297
WT 2	Pre-ramp	1040	325	140	29	5
WT 2	Ramp	3316	1893	1340	744	502
WT 3	N. & pitch	262	103	50	21	11
WT 3	Nominal	1144	600	410	206	127
WT 3	Ramp-to-nominal	2191	1159	820	453	320
WT 3	Pre-ramp	1299	403	174	19	3
WT 3	Ramp	3282	1840	1339	754	511

The time filter reveals that most occurrences have very short duration, as moving the filter from 5 to 10 min reduces the number of occurrences to a half in most of the ORs. Furthermore, very restrictive ORs, such as N. & pitch, are less present in the database, and the ones with wider limits (that also coincide with the most frequent wind speeds) are more present in the dataset. Considering longer OSs should reduce the amount of noise created by previous operation regimes, while there should be a sufficient week rate in order to obtain enough indicators over time, it is decided to keep OSs that last longer than 10 min.

Following with the procedure, once data matrix R is obtained for each WT, it is possible to see the different clusters that are created and represent the variability of the measurements. Figure 9 represents the two principal components of the operational variables in the dataset (pitch angle, gearbox temperature, wind speed, generator speed and active power) that are generated in WT 1 using PCA algorithm. The variability retained by each dimension is displayed in the axes. It is interesting to see the representations that the different ORs take. The first principal component (Dim 1) mostly contains pitch difference, generator speed and gearbox temperature. Most of the clusters have great part of the variation related to this feature, whereas in the second component (Dim 2), wind speed, active power and generator speed are causing most of the variation, and this dimension affects mainly Ramp and Ramp-to-nominal clusters that show very high dispersion.

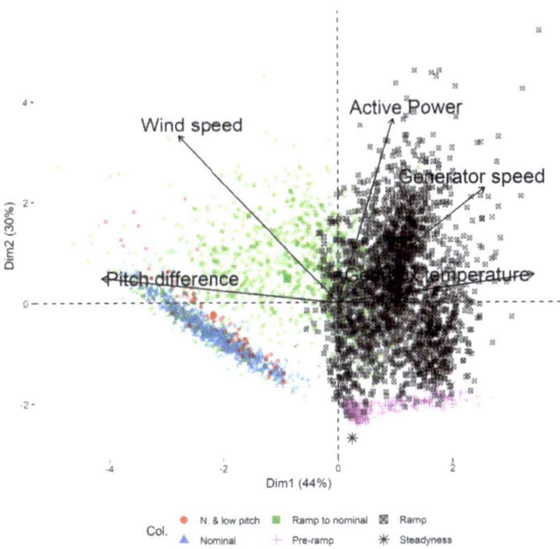

Figure 9. Example of the PCA obtained from the different OR and RMS (root mean square) values of the OS generated in WT 1.

The steadiness point, that is, the point with no variation is also represented on the graphs as a star. The euclidean distance to that point (the steadiness) is measured and the results presented in the following Table 4:

Table 4. Steadiness of the different OR by WT.

	N. & Pitch	Nominal	Ramp to Nominal	Pre-Ramp	Ramp
WT 1	2.00×10^{-3}	1.50×10^{-3}	1.50×10^{-3}	3.00×10^{-4}	1.10×10^{-3}
WT 2	1.40×10^{-3}	1.00×10^{-3}	1.10×10^{-3}	4.00×10^{-4}	1.00×10^{-3}
WT 3	2.00×10^{-3}	1.50×10^{-3}	1.50×10^{-3}	4.00×10^{-4}	1.10×10^{-3}
Average	1.80×10^{-3}	1.30×10^{-3}	1.40×10^{-3}	4.00×10^{-4}	1.10×10^{-3}

There are extreme differences regarding how much operation variables vary during the OSs. From the least steady OR (Ramp-to-nominal) to the steadiest one (Pre-ramp) the distance is ten times bigger, meaning the Pre-ramp operation regime is much steadier than the Ramp-to-nominal OR.

Regarding the variation of the clusters, that is, how close from the centroid the data-points are, the results displayed in Table 5 are obtained.

Table 5. Variation of the ORs by WT.

	N. & Pitch	Nominal	Ramp to Nominal	Pre-Ramp	Ramp
WT 1	7.56×10^{-5}	4.64×10^{-5}	7.58×10^{-5}	6.96×10^{-6}	5.76×10^{-4}
WT 2	1.98×10^{-5}	2.78×10^{-5}	6.19×10^{-4}	9.03×10^{-6}	5.36×10^{-4}
WT 3	7.31×10^{-5}	5.51×10^{-5}	6.80×10^{-4}	9.32×10^{-6}	6.34×10^{-4}
Average	5.62×10^{-5}	4.31×10^{-5}	6.86×10^{-4}	8.44×10^{-6}	5.82×10^{-4}

The smallest variation values are obtained by Pre-ramp OR, with clear difference to with the rest of the ORs.

Considering the information provided by the frequency study, the PCA visualisation, distance to steadiness and centroid variation, which OR should be chosen for HI purposes is determined. N. & pitch is discarded because of it low occurrence frequency, Ramp and Ramp-to-nominal show high dispersion, which means there are operational fluctuations in the ORs they delimit. Between Nominal and Pre-ramp ORs, according to steadiness and dispersion criteria Pre-ramp should be chosen, therefore, it is decided to consider the measurements taken under Pre-ramp OR.

Taking the Pre-ramp OR, the ODM variables as well as the other SCADA variables are averaged with the RMS (root mean square) value of each OS. Figures 10 and 11 display the RMS values obtained from the OSs generated using these points. Over the points, the coloured curves represent the fitting provided by the LOESS algorithm using a high span fraction (0.75) in order to retain the trend instead of local variations. The grey shade represents the 95% confidence interval of the fitted curve.

Note that power generation stays under very strict limits, shows almost no variation along the time and, besides the latest trend values (that have less data points), almost no difference between turbines. Similarly, temperature shows higher variation but the trend keeps very stable along the time. Interestingly, the differences in temperature that are visible do not match the expectations: WT 1 and WT 2 (diagnosed with medium wear wear level) have higher temperature values than WT 3 (diagnosed with medium–high wear level). Regarding particle generation, it is possible to see the sudden increase WT 2 and WT 3 have could be related to the same increase of OSs with higher Active power production (Figure 10b). The differences in trend values do correspond to the damage levels of the gearboxes, showing that WT 3 is in a worse condition than the rest of the turbines.

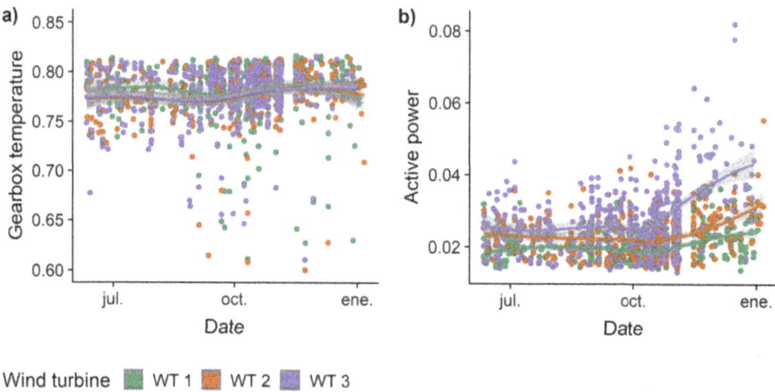

Figure 10. RMS values of different variable OSs throughout the time, generated using type Nominal OR. Curve is obtained by local regression (LOESS) smoothing. (**a**) Gearbox temperature. (**b**) Active power.

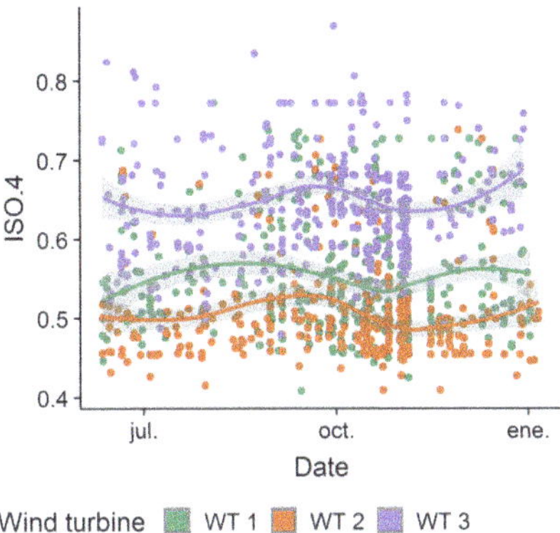

Figure 11. RMS values of ISO.14 variable and smoothed trend fitted with LOESS.

Lastly, the comparison of the evolution of the particles in different ISO particle size (Figure 12) demonstrates the correlations among ISOs that are found in the correlation analysis. ISO.14 shows the greatest difference between WTs, which means it could reflect the damage status in a more accurate way, and therefore it is better suited for comparison purposes. However, the contrast between WTs is clearly visible also in the rest of the ISOs. The scale of the figure is ranged between 0 and 25 ISO values and thresholds proposed by laboratory experience are included for both warnings (20/18/15) and danger (21/19/16) for ISO 4/6/14, respectively. Note that, even if the trends are far from reaching the thresholds, the real-time measurements surpass the thresholds more than once. This means that operation variation can cause great spikes in the particle generation rate and, in order to obtain an overview of the condition of the gearbox, it is required to focus on the trends instead of in instantaneous values. Considering ISO.14 keeps under very low values, the state of the gearbox could be considered yet to be healthy. Either if the smoothed trend would reach values close to the thresholds or it would increase sharply, gearbox should be considered in danger.

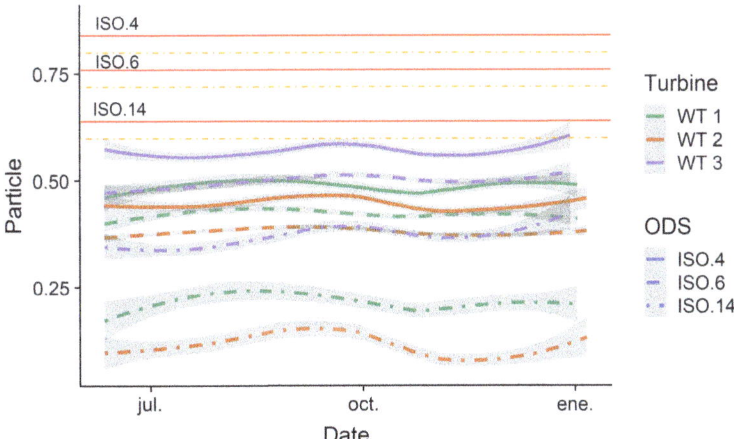

Figure 12. LOESS smoothed particle generation rates (ISO 4, 6 and 14) of Pre-ramp ORs by WT. The horizontal lines at the top represent warning and danger thresholds for the different particle sizes.

4. Discussion and Future Outlook

This study presents the data obtained through the monitoring of three WTs with oil debris optical sensors during six months. In this way, turbines with gearboxes in different stages of deterioration are presented and compared. Initially, an attempt to correlate the operation of the WTs and ODSs has been carried out. After that, a way to identify repeatable and steady operation regimes has been used as a basis for developing a health indicator.

Different works have shown in-service oil debris monitoring in the literature [22,25] or have studied the relation of the operation in behaviour of ODSs in full scale test rigs [26]. Nevertheless, this work is particular as it presents both: first an study of the influence of the operation on debris generation; and then, it proposes a method for the identification of the optimal instants to obtain measurements considering operation.

The number of studied turbines is reduced and the installation of the lubrication systems is equal according to our knowledge. However, two clear behaviours have been discovered during the exploratory phase: two of the turbines show similar trends whereas another one seems to behave in a completely different way. Therefore, the results here displayed should be understood in this context.

We have faced difficulties when working with the noise of raw ODS's measurements that are also reported in other works [22,25,26], and using the techniques already present in the literature (cumulative particle rates [22,25]) has been useful to reduce the noise. In comparison to the sole use of variables that are directly obtained from the SCADA (active power and generator temperature), significant improvement has been detected when using ODSs, as they show greater differences between the health status of the gearbox. This fact validates the thesis of the need of including additional sensors for defining with higher accuracy the damage levels of the systems [22].

Due to the varying operation and the difference behaviours of WTs, it has been difficult to find clear correlations between operation and particle creation. ISO measurements are highly correlated among themselves, but it is difficult to find association to other operation variables. Only in a detailed inspection of braking and acceleration periods, contrarily to what is reported in the literature [26], a general tendency to increase particle generation has been detected when generator is braking, whereas no increase has been detected during acceleration. Furthermore, some periods of high particle creation have been identified when the generator is idling, this phenomenon could be caused by different behaviours of the oil filtering system, but this assumption has not been proved.

In concordance to the findings of the authors of [26], gearboxes that are more deteriorated (the case of WT 3) have shown a tendency to generate more particles rates than gearboxes in better condition (WT 1 and WT 2). This fact is clearly visible in the cumulative particle creation or the binned ISO against wind speed plots. Also, the pattern of particle generation that is visible in the binned wind speed of WT 1 and 3 reminds the Stribeck curve, which could explain the high particle creation rates at low speeds and the high rates at higher speeds. Furthermore, the differences in particle creation rates are more evident when considering bigger particle size (ISO.14), as the authors of [26] stated. Nevertheless, the sensor in WT 2 provides patterns that differ from the sensors in WT 1 and WT 2. Considering how close the patterns are in WT 1 and WT 3, two hypothesis could be possible: either sensor is not working correctly; or the lubrication system is affected by factors not included in the SCADA.

Regarding the development of the HI, in contrast to the proposals of other works in the literature that make trends over the whole data [25], our method considers the operation regime in which the measurements are taken and uses only measurements that are obtained under the same circumstances. On the one hand, this leads to have periods of time without indicators, but this issue has been considered by choosing ORs with high number of occurrences. On the other hand, an analysis of the operation has been carried out to identify instants with lesser variation in the operation, which should provide more stable measurements and less influenced by the operation. However, using smoothing techniques has still been necessary in order to make trends visible.

According to the analysis of the operation, the WTs tend to move fastly from one OR to another, as the high number of short OSs reveals. Regarding the steadiness of the different ORs analysed, in the pre-ramp zone the operational variables remain more stable than in the rest of ORs that are in the power ramp, as they show a bigger distance to steadiness point.

As the authors of [16] recognise, establishing limits for admissible and nonadmissible damage is one of the biggest challenges in in-service machinery. The limits proposed in this work are based on laboratory experience but might need to be readjusted by interacting with bigger WT databases, as there is no total failure record in the dataset under study.

Lastly, even if ODSs have been demonstrated capable of detecting diverse levels of damage in gearboxes, with the current analysis it is not possible to determine which component of the gearbox is really damaged, which is possibly to do with other sensors such as vibration sensors. In order to detect the root cause of the damage with ODS, a characterisation of the kind of particles would be needed, including shape and elemental composition in addition to the particle count and size. Without these requirements, visual inspection will be needed to determine which component is exactly damaged.

The findings of this work suggest a promising future for optical oil debris sensors in the field of WT monitoring. At the same time, the need of being aware of all the details of the case study is also concluded, as there are some inconsistencies that are not explainable by the sole analysis of the SCADA and ODS data, but might be explainable if more details of the installation of the sensors and the WT itself were made available. In this regard, improving the cooperation and trust between WT owner and researchers would be a key factor for doing better analyses.

The addition of more monitored turbines to the study as well as prolonging the studied period of time would validate the results and determine whether one of the groups is just anomalous for external reasons (such as unreported differences in the systems), or there are really other turbines in which the oil debris follows the same behaviours. In this same line, it would be interesting to keep observing the differences in the HI that are proposed while visual inspections are done periodically in order to prove the validity of the approach for diagnosing the state of the turbine and also to learn to adjust the limits of admissible/nonadmissible thresholds of particle generation before having severe damages.

Furthermore, using the characterisation of the generated debris in order to related the visual inspections with the results provided by online sensor would give and additional value to the monitoring, as the root cause of failure could be identified. Also, adding vibration sensors in order to determine which source of data could provide better insight, more valuable information or identifying possibilities of synergy would be of interest.

Author Contributions: Conceptualisation, all the authors; Methodology, all the authors; Original draft preparation, K.L.d.C.; Writing—Review & Editing, C.R.-P., A.U. and K.L.d.C.; Software, K.L.d.C., C.R.-P. and S.F.; Supervision, S.F. and A.U.

Funding: This work was performed with the financial support of the FRONTIERS IV (ELKARTEK KK-2018/00096) Project financed by Eusko Jaurlaritza.

Conflicts of Interest: The authors declare no conflicts of interest. The funders had no role in the design of the study; in the collection, analyses, or interpretation of data; in the writing of the manuscript, or in the decision to publish the results.

References

1. Tchakoua, P.; Wamkeue, R.; Ouhrouche, M.; Slaoui-Hasnaoui, F.; Tameghe, T.A.; Ekemb, G. Wind turbine condition monitoring: State-of-the-art review, new trends, and future challenges. *Energies* **2014**, *7*, 2595–2630. [CrossRef]

2. Hameed, Z.; Hong, Y.S.; Cho, Y.M.; Ahn, S.H.; Song, C.K. Condition monitoring and fault detection of wind turbines and related algorithms: A review. *Renew. Sustain. Energy Rev.* **2009**, *13*, 1–39. [CrossRef]

3. Nie, M.; Wang, L. Review of condition monitoring and fault diagnosis technologies for wind turbine gearbox. *Procedia CIRP* **2013**, *11*, 287–290. [CrossRef]

4. Marti-Puig, P.; Blanco, A.M.; Cárdenas, J.J.; Cusidó, J.; Solé-Casals, J. Feature selection algorithms for wind turbine failure prediction. *Energies* **2019**, *12*, 453. [CrossRef]

5. Manwell, J.F.; McGowan, J.G.; Rogers, A.L. *Wind Energy Explained: Theory, Design and Application*; John Wiley & Sons, Ltd.: Hoboken, NJ, USA, 2010. [CrossRef]

6. Crabtree, C.J.; Zappalá, D.; Hogg, S.I. Wind energy: UK experiences and offshore operational challenges. *Proc. Inst. Mech. Eng. Part A J. Power Energy* **2015**, *229*, 727–746. [CrossRef]

7. Echavarria, E.; Hahn, B.; van Bussel, G.J.W.; Tomiyama, T. Reliability of Wind Turbine Technology Through Time. *J. Sol. Energy Eng.* **2008**, *130*, 031005. [CrossRef]

8. Su, C.; Yang, Y.; Wang, X.; Hu, Z. Failures analysis of wind turbines: Case study of a Chinese wind farm. In Proceedings of the 2016 Prognostics and System Health Management Conference (PHM-Chengdu 2016), Chengdu, China, 19–21 October 2016. [CrossRef]

9. García Márquez, F.P.; Tobias, A.M.; Pinar Pérez, J.M.; Papaelias, M. Condition monitoring of wind turbines: Techniques and methods. *Renew. Energy* **2012**, *46*, 169–178. [CrossRef]

10. Carroll, J.; McDonald, A.; McMillan, D. Failure rate, repair time and unscheduled O&M cost analysis of offshore wind turbines. *Wind Energy* **2016**. [CrossRef]

11. Nilsson, J.; Bertling, L. Maintenance management of wind power systems using condition monitoring systems—Life cycle cost analysis for two case studies. *IEEE Trans. Energy Convers.* **2007**, *22*, 223–229. [CrossRef]

12. Pfaffel, S.; Faulstich, S.; Rohrig, K. Performance and reliability of wind turbines: A review. *Energies* **2017**, *10*, 1904. [CrossRef]

13. Hahn, B.; Durstewitz, M.; Rohrig, K. Reliability of Wind Turbine–Experiences of 15 years with 1500 WTs. In *Wind Energy*; Peinke, J., Schaumann, P., Barth, S., Eds.; Springer: Berlin/Heidelberg, Germany, 2006. [CrossRef]

14. Koltsidopoulos Papatzimos, A.; Dawood, T.; Thies, P.R. Data Insights from an Offshore Wind Turbine Gearbox Replacement. *J. Phys. Conf. Ser.* **2018**, *1104*. [CrossRef]

15. Kattelus, J.; Miettinen, J.; Lehtovaara, A. Detection of gear pitting failure progression with on-line particle monitoring. *Tribol. Int.* **2018**, *118*, 458–464. [CrossRef]

16. Artigao, E.; Koukoura, S.; Honrubia-Escribano, A.; Carroll, J.; McDonald, A.; Gómez-Lázaro, E. Current signature and vibration analyses to diagnose an in-service wind turbine drive train. *Energies* **2018**, *11*, 960. [CrossRef]

17. Gonzalez, E.; Stephen, B.; Infield, D.; Melero, J.J. Using high-frequency SCADA data for wind turbine performance monitoring: A sensitivity study. *Renew. Energy* **2019**, *131*, 841–853. [CrossRef]

18. Elosegui, U.; Egana, I.; Ulazia, A.; Ibarra-Berastegi, G. Pitch angle misalignment correction based on benchmarking and laser scanner measurement in wind farms. *Energies* **2018**, *11*, 3357. [CrossRef]

19. Astolfi, D. A Study of the Impact of Pitch Misalignment on Wind Turbine Performance. *Machines* **2019**, *7*, 8. [CrossRef]

20. Rabanal, A.; Ulazia, A.; Ibarra-Berastegi, G.; Sáenz, J.; Elosegui, U. MIDAS: A Benchmarking Multi-Criteria Method for the Identification of Defective Anemometers in Wind Farms. *Energies* **2019**, *12*, 28. [CrossRef]

21. Tautz-Weinert, J.; Watson, S.J. Using SCADA data for wind turbine condition monitoring—A review. *IET Renew. Power Gener.* **2017**, *11*, 382–394. [CrossRef]

22. Feng, Y.; Qiu, Y.; Crabtree, C.J.; Long, H.; Tavner, P.J. Monitoring wind turbine gearboxes. *Wind Energy* **2013**, *16*, 728–740. [CrossRef]

23. Blanco, M.A.; Gibert, K.; Marti-Puig, P.; Cusidó, J.; Solé-Casals, J. Identifying health status of wind turbines by using self organizing maps and interpretation-oriented post-processing tools. *Energies* **2018**, *11*, 723. [CrossRef]

24. Escaler, X.; Mebarki, T. Full-Scale Wind Turbine Vibration Signature Analysis. *Machines* **2018**, *6*, 63. [CrossRef]

25. Dupuis, R. Application of Oil Debris Monitoring For Wind Turbine Gearbox Prognostics and Health Management. In Proceedings of the Annual Conference of the Prognostics and Health Management Society, Portland, OR, USA, 10–16 October 2010.

26. Sheng, S. Monitoring of Wind Turbine Gearbox Condition through Oil and Wear Debris Analysis: A Full-Scale Testing Perspective. *Tribol. Trans.* **2016**. [CrossRef]

27. Cuadrat Prats, J. El clima de Aragón. *Geografía Física de Aragón. Aspectos Generales y Temáticos*; Asociación de Geógrafos Español: Murcia, Spain, 2004; pp. 15–26.

28. Mabe, J.; Zubia, J.; Gorritxategi, E. Photonic low cost micro-sensor for in-line wear particle detection in flowing lube oils. *Sensors* **2017**, *17*, 586. [CrossRef] [PubMed]

29. Lopez, P.; Mabe, J.; Miró, G.; Etxeberria, L. Low cost photonic sensor for in-line oil quality monitoring: Methodological development process towards uncertainty mitigation. *Sensors* **2018**, *18*, 2015. [CrossRef] [PubMed]

30. British Standard. *BS ISO 4406:1999 Hydraulic Fluid Power—Fluids—Method for Coding the Level of Contamination by Solid Particles*; International Organization for Standardization: Geneva, Switzerland, 1999. [CrossRef]

31. Pearson, K. VII. Note on regression and inheritance in the case of two parents. *Proc. R. Soc. Lond.* **1895**, *58*, 240–242.

32. Spearman, C. The proof and measurement of association between two things. *Am. J. Psychol.* **1904**, *15*, 72–101. [CrossRef]

33. Hotelling, H. Analysis of a complex of statistical variables into principal components. *J. Educ. Psychol.* **1933**, *24*, 417–441. [CrossRef]

34. Ulrich, J. TTR: Technical Trading Rules. R Package Version 0.23-4. 2018. Available online: https://cran.r-project.org/web/packages/TTR/TTR.pdf (accessed on 15 July 2019).

35. Cleveland, W.; Grosse, E.; Shyu, W. Local regression models. In *Statistical Models in S*; Chambers, J.M., Hastie, T.J., Eds.; Software Pacific Grove: Wadsworth, OH, USA, 1992; pp. 309–376.

36. Therneau, T.; Atkinson, B. rpart: Recursive Partitioning and Regression Trees. R Package Version 4.1-13. 2018. Available online: https://cran.r-project.org/web/packages/rpart/vignettes/longintro.pdf (accessed on 15 July 2019).

37. Breiman, L.; Friedman, J.; Olshen, R.; Stone, C. *Classification and Regression Trees*; Wadsworth Int. Group: Wadsworth, OH, USA, 1984; Volume 37, pp. 237–251.

Article

Dynamic Fault Monitoring of Pitch System in Wind Turbines using Selective Ensemble Small-World Neural Networks

Meng Li and Shuangxin Wang *

School of Mechanical, Electronic and Control Engineering, Beijing Jiaotong University, Beijing 100044, China
* Correspondence: shxwang1@bjtu.edu.cn; Tel.: +86-010-5168-7021

Received: 10 May 2019; Accepted: 20 August 2019; Published: 23 August 2019

Abstract: Pitch system failures occur primarily because wind turbines typically work in dynamic and variable environments. Conventional monitoring strategies show limitations of continuously identifying faults in most cases, especially when rapidly changing winds occur. A novel selective-ensemble monitoring strategy is presented to diagnose the most pitch failures using Supervisory Control and Data Acquisition (SCADA) data. The proposed strategy consists of five steps. During the first step, the SCADA data are partitioned according to the turbine's four working states. Correlation Information Entropy (CIE) and 10 indicators are used to select correlation signals and extract features of the partition data, respectively. During the second step, multiple Small-World Neural Networks (SWNNs) are established as the ensemble members. Regarding the third step, all the features are randomly sampled to train the SWNN members. The fourth step involves using an improved global correlation method to select appropriate ensemble members while in the fifth step, the selected members are fused to obtain the final classification result based on the weighted integration approach. Compared with the conventional methods, the proposed ensemble strategy shows an effective accuracy rate of over 93.8% within a short delay time.

Keywords: pitch system; dynamic fault monitoring; selective ensemble learning; small-world neural network (SWNN); reliability

---

1. Introduction

Doubtless, safe operation of the pitch system is a key to ensure power stability and reliable braking of wind turbines [1]. The dynamic turbulence or unsteadiness gusts not only provide power for the pitch system, but also produces the most stress or dynamic loading on the blades [2]. Historically, pitch system faults are largely caused by dynamic loading situations due to uncertainty in the wind resource intensity and duration [3,4]. Such situations have frequently led to tragic accidents, as well as casualties and asset losses. The pitch system, therefore, has to frequently change blade angles and adjust blade speed to avoid being destroyed [5]. It is worth mentioning that a wind turbine has four working states: start-up, wind speed regulation, power regulation, and cut-out. The level of the wind speed is the decision-maker for state transitions, which leads to the coordinated action of the pitch system. When the wind speed varies over a wide range, the state switches and the pitch system will have an increased ability to ensure the safety of the turbine. Conversely, when the wind speed fluctuates within a small range, the state is locked, and the pitch system also performs small movements frequently to capture the maximum wind energy. Whether it is a large movement or a small one, the pitch movement often lags behind the wind speed [6]. Once a failure occurs, it is difficult to find it timely and accurately. The existing Supervisory Control and Data Acquisition (SCADA) system can send an alarm after the faults, but it has no intelligent monitoring function to provide an early warning and accurate location information before the fault. The current way relies on operators to detect abnormal situations and make corrective decisions based on enough safety intelligence.

Generally, fault monitoring approaches are divided into model-based methods and data-driven methods. The model-based methods use explicit system dynamic models and control theories to generate residuals for fault monitoring. Alternatively, the data-driven methods use data mining techniques to capture discrepancies between observed data and that predicted by a model. Such discrepancies will reflect whether the machine is in normal or failure mode, which requires a classifier to judge. Recently, some Artificial Intelligent (AI) classifiers, such as neural networks [7–12], machine learning methods [13–15], and deep learning methods [16,17], have been widely applied in classifying the incipient faults of wind turbines. These methods are really very effective for some faults within a certain working state, but it seems impossible for them to diagnose other faults under other working states. Considering actual demand, it is necessary to establish a systematic fault monitoring system that covers multiple dynamic working states. Moreover, a targeted analysis of what types of faults will occur in different states is critical to build the monitoring system.

Fortunately, an ensemble learning technique is a better method in solving the above problems and Boosting and Bagging are two common approaches of ensemble learning. Boosting [18] is a cascade training method that uses the same data to train the ensemble members one-by-one. It requires a strong dependency from a series of ensemble members. Specifically, if the former members are not well trained, the latter members will be affected and show bad performance. Moreover, Boosting is easy to be interrupted during training when there is a small interference, leading to the overall failure of the training [19]. Bagging is a separate training approach [20] that requires multiple single ensemble members to perform the same task [21,22]. Using this training mode, the ensemble members are homogeneous or heterogeneous, and their alternative algorithms, such as support vector machine (SVM), artificial neural networks (ANN) and naive Bayes [23], should be as simple and effective as possible. More importantly, the final result of the ensemble learning is a comprehensive decision output which is obtained by fusing the results of the multiple ensemble members based on a certain combination method [24]. The use of an ensemble learning technique in monitoring pitch failures is still rare, however. Pashazadeh fused Multi-Layer Perceptron (MLP), Radial Basis Function (RBF), Decision Tree (DT), and K-Nearest Neighbor (KNN) classifiers can be used together to detect early faults in wind turbines [25]. Dey compared three cascade fault diagnosis schemes to address the issue of fault detection and isolation for wind turbines [26]. Concluded from the above-limited applications, the ensemble learning technique is indeed an efficient strategy to identify failures and improve classification performance. Additionally, the neural networks are often used as the alternative algorithms of the ensemble members, because the neural network is a "universal approximator" and has better capabilities in processing multidimensional nonlinear data [27–29].

To achieve the higher fault diagnosis performance, the ensemble learning should have three basic principles. First, there must be enough data to train the ensemble members. The training data in this paper are recorded from a wind farm SCADA system for one year, and the Bootstrap sampling method is used to create samples by varying the data to solve the shortage problems in some data. Second, ensemble members should have different classification characteristics, which are not only diverse but also complementary. The small world neural networks (SWNN) [30] are suitable to be the ensemble members because they are semi-random neural networks and easily can achieve excellent performance [9,31]. The probability p is used to describe the degree of random reconnection of the SWNN's structures. Most noteworthy is that the SWNNs randomly can produce diverse networks with different structures when probability p is a deterministic value. Additionally, the SWNNs are rather easy to be trained by forwarding propagation and error feedback when using the same initial values. Third, a wise ensemble strategy also is required, depending on the types of ensemble members [32]. Usually, for the ensemble members based on neural networks, the ensemble strategies use voting, weighted voting or meta-learning methods to obtain the final ensemble outputs [33]. An advantage is the ensemble members are independent and irrelevant, which is helpful to improve the classification efficiency and accuracy.

Consequently, a five-step selective ensemble strategy for dynamic fault monitoring of a pitch system is proposed. Taking the first step, the fault-causing data are partitioned according to the working states of the wind turbines, the Correlation Information Entropy (CIE) method is used to select correlation signals from the SCADA system and 10 indicators are designed to extract features of the partitioned data. Multiple SWNNs are established as ensemble members in the second step. During the third step, the features are randomly sampled to train the ensemble members. Regarding the fourth step, an improved global correlation method is used to select appropriate ensemble members. The selected members are fused to obtain a final result based on weighted integration approach in the fifth step. The final result is called the ensemble output. Considering testing and validation purposes, two case comparisons are used to verify the effectiveness of the proposed ensemble strategy.

The remaining Sections are organized as follows: Section 2 gives the fault analysis of the pitch system under different working states; Section 3 describes the novel selective-ensemble monitoring strategy and the entire process of its five steps; Sections 4 and 5 give comparison examples to demonstrate the effectiveness of the proposed ensemble model. Finally, Section 6 concludes this paper.

2. Dynamic Fault Analysis of the Pitch System

To gain a proper grasp of the failure regularity, estimates from statistics provide all pitch fault information to support the establishment of the fault monitoring strategy. The pitch fault information includes fault types, the number of faults, and the working state of the wind turbine when a fault occurs. The structure of the pitch system is first given, then the pitch fault information is recorded from a real wind farm. More details about the information analysis can be seen in the following.

2.1. Pitch System

Figure 1 shows an example structure of the pitch system, where a pitch system is installed in the hub of a wind turbine. The pitch system consists of one central controller and three pitch devices (the #1, #2 and #3 pitch device). The central controller is the command center of the pitch system, which is used to control three pitch devices, respectively. Normally, three pitch devices are relatively independent and each one has its own individual actuator. The three pitch devices, in practice, usually are working synchronously, therefore, the #1 pitch device is chosen as an example to describe typical pitch control operations.

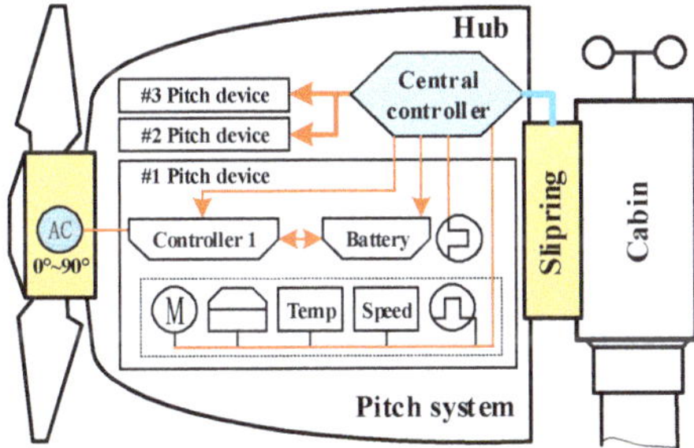

Figure 1. The structure of the pitch system in a wind turbine.

Regarding Figure 1, the #1 pitch device consists of a shaft controller, AC motor, battery, redundant encoder, and two limit switches. The AC motor executes the pitch actions after instructions are sent from the shaft controller. These actions occur in conjunction with other subsystems, such as the servo motor driver, brake resistor, gearbox, rotary photoelectric encoder, blade angle encoder, and limit switch. Electrical switch failures are the common type of faults in the pitch system due to the frequency response needed by this complex system, however, only a few electrical switch faults can be indicated by the SCADA system, most of them are unrecognizable and without any alerts.

2.2. Fault Rules

This part collects 12-month pitch fault information from 30 2 MW-wind turbines in a real wind farm. The information includes the fault types, the number of faults and the working states of the wind turbine. Usually, the operation of wind turbines can be divided into 4 working states according to the level of the wind speed [34]. Figure 2 shows an example layout of the four working states in a wind turbine. The 1st working state is the start-up and grid-connected stage of the wind turbine. Demonstrated in this state, the wind speed is so small that the pitch angle keeps to the minimum and the generator speed rises steadily. The wind turbine cuts into the 2nd working state, as the wind speed increases, to capture the maximum wind energy. The 3rd working state is the constant power control stage where the pitch system needs to constantly adjust the pitch angle to ensure that the wind turbine works at the rated power. During the 4th working state, the wind speed is too high, and the wind turbine will cut out and shut down.

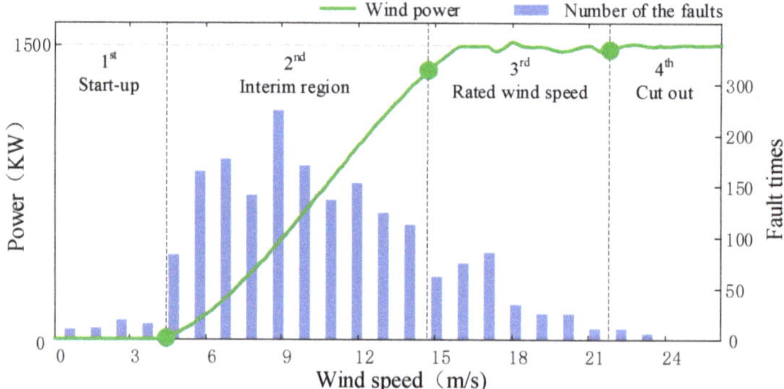

Figure 2. State partition and fault statistics for the 30 2 MW wind turbines.

Figure 2 also shows the average wind power curve of 30 2 MW-wind turbines and the fault statistics under different wind speeds. Taken from the statistical results shown in Figure 2, the distribution of pitch faults is directly related to the working states of the wind turbine. Once the working state is determined, in other words, the corresponding fault types also are determined., The random variation of wind speed, however, necessarily determines that the wind turbine must be switched frequently under different working states. It directly leads to the understanding that a certain fault easily might occur in one particular working state, while it might occur rarely in other ones, therefore, the SCADA data used to diagnose faults also need to be classified according to the working states.

The SCADA data is composed of multi-dimensional signal sources which are collected by multiple sensors. To accurately locate a fault, the most appropriate SCADA signals should be found first. The SCADA system can record more than 100 signals at the same time, and the highest recording frequency can reach 60 times per second. When all the signals are used for analysis and processing, the calculation is unimaginable. Additionally, faults cannot be accurately located, mainly due to the strong coupling between SCADA signals. When a fault occurs at a certain location, for example, multiple signals might alert at the same time. It also is found that there are very complex connections between SCADA signals and turbine components. Such connections will be reconnected with the change of working states. It is difficult to determine faults simply by analyzing the hardware structure of the wind turbines, therefore, it is important to find the correlation signals closely related to the faults, which is particularly helpful for the accurate identification of the pitch failures. A detailed explanation will be reported in the next section.

Taking the above considerations, Table 1 illustrates nine types of frequent faults (F1–F9) and one fault-free case (F10) which are the diagnostic targets of this paper according to the statistical analysis.

Table 1. Fault list.

Fault No.	Fault Name
F1	Pitch brake fuse
F2	Pitch control box connection
F3	Pitch drive error
F4	Pitch position error delay
F5	Pitch access time
F6	Pitch rationality
F7	Pitch stop time
F8	Pitch converter communication
F9	Synchronization of pitch position
F10	Fault-free

3. Selective Ensemble Monitoring Strategy Based on Small-World Neural Networks

A novel ensemble monitoring strategy is proposed to diagnose pitch faults by using multi-dimensional SCADA data. Such a strategy is a distributed diagnostic system, which takes four working states of the wind turbines as four parallel models. Each parallel model is a five-step selective ensemble model of SWNNs, in which the five steps are data partition, SWNN members' creation, SWNN training, ensemble members' selection, and ensemble output, respectively. It is noteworthy that, in the first step, the original SCADA data are divided into four sub-datasets according to the four working states of the wind turbine. To facilitate the description of the next steps, the architecture in the 2nd working state is selected as an example to display the proposed ensemble strategy, which is shown in Figure 3.

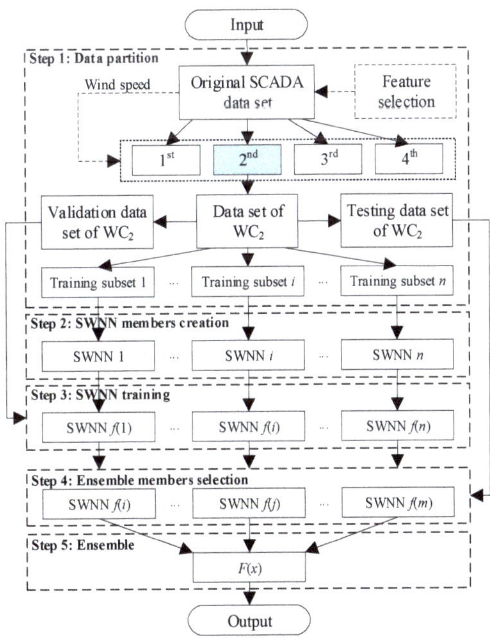

Figure 3. A five-step selective ensemble strategy for dynamic fault monitoring of pitch system in the 2nd working state.

3.1. Data Partition

Data processing is the decisive step in ensuring that the ensemble strategy can achieve excellent results. Section 3.1.1 shows the original SCADA data is first partitioned into four subsets based on the four working states, respectively. Section 3.1.2 explains how a Correlation Information Entropy (CIE) method is used to select correlation signals that are related to the faults from the multi-dimensional SCADA signals. Section 3.1.3 discusses 10 indicators which are designed to extract fault-causing features and normal features from the correlation signals.

3.1.1. Data Classification Based on the Dynamic Working States

To establish the distributed diagnostic system, the original SCADA signals are divided into four groups based on the four working states of wind turbines. Figure 4 gives the process of the data classification, where the wind speed is the decision-maker for state transitions. The labels of 1st, 2nd, 3rd and 4th represent the four working states respectively, and $C(v)$ is the division criterion which is calculated by Equation (1). The divided signals are used for further data processing and feature extraction, because it avoids confusion with other irrelevant data, especially at the beginning of data processing.

$$
C(v) = \begin{cases}
1, & 0 < v(t) < 3\text{m/s}, \\
2, & 3\text{m/s} \le v(t) < 12\text{m/s}, \\
3, & 12\text{m/s} \le v(t) < 25\text{m/s}, \\
4, & 25\text{m/s} \le v(t)
\end{cases}
\tag{1}
$$

where, $v(t)$ is the current wind speed, and $C(v)$ is the division criterion.

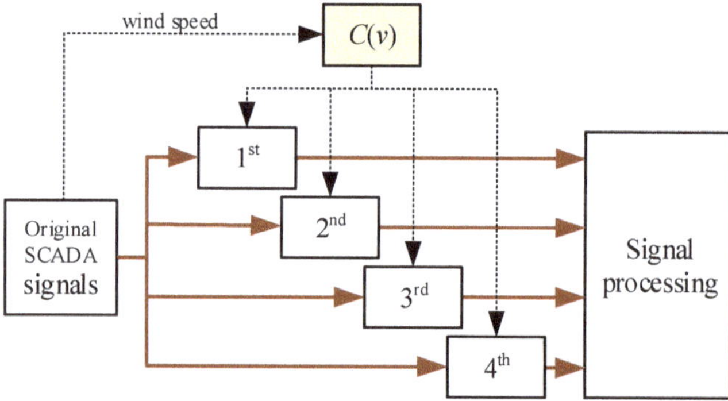

Figure 4. Data classification according to the working states.

3.1.2. Correlation Signals Selection under Dynamic Working States Based on CIE

Following the data partition, there are still many signals remaining and, as mentioned at the end of Section 2, not all of these signals are associated with the faults in a certain working state. It is necessary, therefore, to find the correlation signals related to the faults from the multi-dimensional SCADA signals. The Correlation Information Entropy (CIE) method is used to complete the above task.

CIE is an effective feature reduction approach. It can accurately measure the correlation between multiple signals on the basis of the high reliability with low calculations. Suppose that P is the SCADA output sequences of N signals within T time. Prior to computing, each signal should be centralized and normalized to ensure that all values are in the same order of magnitude. The centralized and normalized values are obtained by Equations (2) and (3), respectively.

$$
\hat{y}_n(t) = y_n(t) - \frac{1}{N} \sum_{n=1}^{N} y_n(t)
\tag{2}
$$

$$\overline{y}_n(t) \;=\; \frac{\hat{y}_n(t)}{\sqrt{\displaystyle\sum_{n=1}^{N} (\hat{y}_n(t))^2}} \tag{3}$$

P can be expressed as Equation (4):

$$P \;=\; \{y_n(t)\}_{1 \le n \le N, 1 \le t \le T}, P \in R^{T \times N} \tag{4}$$

where, P is the SCADA output sequences of the nth signal, $y_n(t)$ is the output value of the nth signal at time t ($t = 1, 2, 3, \ldots, T$). R is a matrix of real numbers.

Subsequently, the correlation matrix Q is generated by P. It contains the correlation information between N signals, which can be expanded as Equation (5):

$$Q \;=\; P^{\mathrm{T}}P \;=\; \begin{bmatrix} 1 & q_{12} & \cdots & q_{1N} \\ q_{21} & 1 & \cdots & q_{2N} \\ \vdots & \vdots & \ddots & \vdots \\ q_{N1} & q_{N2} & \cdots & 1 \end{bmatrix} \;=\; I + \widetilde{R}, R \in R^{N \times N} \tag{5}$$

where, P^{T} is the transposition matrix of P. q_{ij} ($q_{ij} \in [0, 1]$, $i \ne j$, $i = 1, 2, \ldots, n$, $j = 1, 2, \ldots, n$) donates the correlation degree of the ith signal to the jth signal. The 1 in the principal diagonal of Q represents the self-correlation coefficient of the signals. I is the autocorrelation matrix, and $\widetilde{R}$ is the co-correlation matrix that implies the overlap information of all signals.

The above correlation information in the Q is the correlation degree between any two signals. Next, calculate the contribution of one signal to all signals. The λ_i^R, λ_i^I and $\lambda_i^{\widetilde{R}}$ denote the eigenvalues of Q, I and $\widetilde{R}$, respectively. The *CIE* is defined as Equation (6), and its range is between $[0, 1]$. It is worth noting that the larger the correlation degree between signals, the smaller the corresponding *CIE*.

$$CIE \;=\; -\sum_{i=1}^{N} \frac{\lambda_i^R}{N} \log_N \frac{\lambda_i^R}{N} \tag{6}$$

To find the appropriate SCADA signals related to the pitch system, the following example lists 15 initial signals in Table 2 and uses CIE to calculate the correlation of all the signals. Additionally, the 15 initial signals are all captured from different working states and each one contains 2000 samples of normal data. Figure 5 shows the *CIE* results of the 15 signals for different working states:

Table 2. Signal list.

Signal No.	Signal Name
1	Wind speed
2	Active power
3	Reactive power
4	Power factor
5	Wind direction angle
6	Generator speed
7	Generator electrical power
8	Rotor speed
9, 10, 11	Pitch angle of the 3 blades
12	Generator stator temperature
13, 14, 15	Blade root moment of the 3 blades

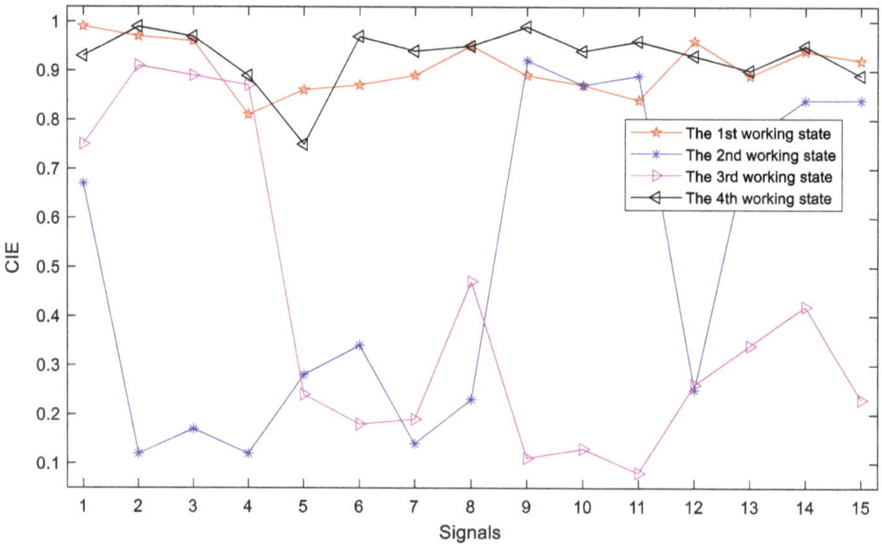

Figure 5. The Correlation Information Entropy (CIE) results of 15 initial signals.

Viewing Figure 5, select the signals with their *CIEs* below 0.3 as the appropriate signals in different working states. The 1st and 4th working states have no appropriate signals because the pitch system is not working and is not associated with the monitoring signals. Found in the 2nd and 3rd working states, 7 signals and 8 signals are selected as the appropriate signals respectively, where the signal numbers are 2, 3, 4, 5, 7, 8, 12 and 5, 6, 7, 9, 10, 11, 12, 15. Form these selected signals into two sets of $X_2 = \{2, 3, 4, 5, 7, 8, 12\}$ and $X_3 = \{5, 6, 7, 9, 10, 11, 12, 15\}$, in which X_2 and X_3 represent the 2nd and 3rd working states, respectively. Note that the 2nd and 3rd working states will continue to be studied in the following work, while the 1st and 4th working states are beyond the scope of this paper.

3.1.3. Discretized Fault Feature Extraction

Extracting fault features from the appropriate signals mainly is to find the changing characteristics from the time series. Specifically, it is the process of using discrete values to describe a limited sequence. A sliding window is used to intercept data from the appropriate signals. The abscissa of the window is a certain period of time, and the ordinate is the value of the signal. The intercepted data is called a "run". Each run should have a label, which is a normal label or a fault label. Moreover, 10 kinds of time-domain indicators (TDIs) are designed to calculate the features of the run. The 10 TDIs are independent but closely related, which are shown in Table 3. Actually, the runs with normal labels are the vast majority, and the runs with fault labels are the minority. This imbalanced distribution is undoubtedly counterproductive to further classification. A combination method combining over-sampling and under-sampling [33] is used in this case to expand the number of fault runs.

Table 3. 10 time-domain indicators.

TDI	Meaning	Function
$T_1 = \sqrt{\frac{1}{N}\sum\limits_{i=1}^{N} x_i^2}$	Root mean square value	The validity of data and size of the noise
$T_2 = \frac{1}{N}\sum\limits_{i=1}^{N}\left(x_i - \left(\frac{1}{N}\sum\limits_{i=1}^{N} x_i\right)\right)^2$	Variance	Dispersion of data
$T_3 = \frac{1}{N}\sum\limits_{i=1}^{N}\frac{(x_i-\bar{x}_i)^4}{\left((1/N)\sum_{i=1}^{N}(x_i)^2\right)^2},$ $\bar{x}_i = \frac{1}{N}\sum\limits_{i=1}^{N} x_i$	Kurtosis index	Sharpness or flatness of data
$T_4 = \frac{\max|x_i|}{T_1}$	Peak index	The ratio of peak to the root mean square value
$T_5 = \frac{\max|x_i|}{(1/N)\sum_{i=1}^{N}|x_i|}$	Impulse index	Degree of data mutation
$T_6 = \max\{x_i\} - \min\{x_i\}$	Peak to peak value	Mutation degree of the peak value
$T_7 = \left(\frac{1}{N}\sum\limits_{i=1}^{N}\left(\sqrt{|x_i|}\right)\right)^2$	Square root amplitude	Amplitude variation of data
$T_8 = \frac{1}{N}\sum\limits_{i=1}^{N}|x_i|$	Average amplitude	Amplitude size of data
$T_9 = \frac{T_1}{T_8}$	Waveform index	Waveform Amplitude of Data
$T_{10} = \frac{\max|x_i|}{T_8}$	Abundance index	Proportion of peak data in magnitude

According to the correlation signals selection in Section 3.1.2, the signals of $X_2 = \{2, 3, 4, 5, 7, 8, 12\}$ and $X_3 = \{5, 6, 7, 9, 10, 11, 12, 15\}$ are selected as the appropriate signals for future diagnosing of the pitch failures. Taking the 2nd working state as an example, the process of discretizing feature extraction for the seven signals in X_2 is described as follows:

(1) Define a new dataset X_t, including the $X_2 = \{2, 3, 4, 5, 7, 8, 12\}$ and the label set of $y^n(F_i)$.

$$X_t = \{x^n(t)\}, n = 2, 3, 4, 5, 7, 8, 12 \tag{7}$$

$$x^n(t) = [x^n(1), x^n(2), ..., x^n(t), y^n(F_i)] \tag{8}$$

where, t is the sampling size; n is the nth signal, which is shown in Table 2. $x^n(t)$ represents the tth value in the nth signal. $y^n(F_i)$ is the label information, F_i is the fault types which are shown in Table 1.

(2) Calculate the features of the signals based on the 10 TDIs separately, then combine the features to obtain a simplified discrete data matrix of $X_n(\text{II})$, which is defined as Equation (9).

(3) Repeating the above two steps, the feature matrix of the 3rd working state can be calculated simultaneously. The discrete data matrix $X_n(\text{III})$ is described in Equation (10).

$$X_n(\text{II}) = \begin{bmatrix} T_1^2 & T_2^2 & \cdots & T_{10}^2 & y^2(F_i) \\ T_1^3 & T_2^3 & \cdots & T_{10}^3 & y^3(F_i) \\ \vdots & \vdots & \ddots & \vdots & \vdots \\ T_1^{12} & T_2^{12} & \cdots & T_{10}^{12} & y^{12}(F_i) \end{bmatrix}_{7\times11} \tag{9}$$

$$X_n(\text{III}) = \begin{bmatrix} T_1^5 & T_2^5 & \cdots & T_{10}^5 & y^5(F_i) \\ T_1^6 & T_2^6 & \cdots & T_{10}^6 & y^6(F_i) \\ \vdots & \vdots & \ddots & \vdots & \vdots \\ T_1^{15} & T_2^{15} & \cdots & T_{10}^{15} & y^{15}(F_i) \end{bmatrix}_{8\times11} \tag{10}$$

Considering $X_n(\text{II})$ and $X_n(\text{III})$, n represents the nth signal in X_2 and X_3, respectively. The row vector represents the feature vector of the nth signal, including 10 TDI values and a label value. The column vector represents all correlation signals information for the 2nd and 3rd states. Specifically, there are 70 TDIs and 7 label values in the 2nd working state, 80 TDIs and 8 label values in the 3rd working state. These TDIs and label values will be used to train SWNNs.

3.2. SWNN Members Creation

Generally, if the ensemble members are accurate and diverse, the ensemble model will be more accurate than any of its individual members [28], however, for neural network ensemble members, one drawback is that the initial situations almost determine the effect of the network. Such situations include initial parameters, training data, topology, and the learning process. Fortunately, SWNNs have been optimized in these initial situations and are well suited to be the ensemble members.

The SWNN is a middle ground neural network between regularity and disorder networks [9,35]. The probability p ($0 < p < 1$) is used to probe the intermediate region. When $p = 0$ or $p = 1$, the SWNN are completely regular or completely random. While p increases from 0 to 1, the SWNN becomes increasingly disordered and all connections between neurons are rewired randomly. Additionally, once the number of input, output and hidden layer neurons of the network are determined, there will be a definite value of p to enable the whole network to achieve the highest clustering with the shortest characteristic path length. Quite the opposite, the SWNN also can randomly reconstruct diverse networks with different structures under the same value of probability p. Compared with the traditional neural networks, the SWNN easily can obtain various network structures by modifying p values rather than setting a large number of initial values or changing the number of neurons or layers.

The SWNN's structure, topology and the detailed training formulas are based on the existing study [9]. Figure 6 shows the example SWNN structure of the ensemble model in the 2nd working state, where the red thick lines are the rewiring edge, and the dashed lines are the rewired edges. The detailed parameters of the SWNN will be set in Section 4.

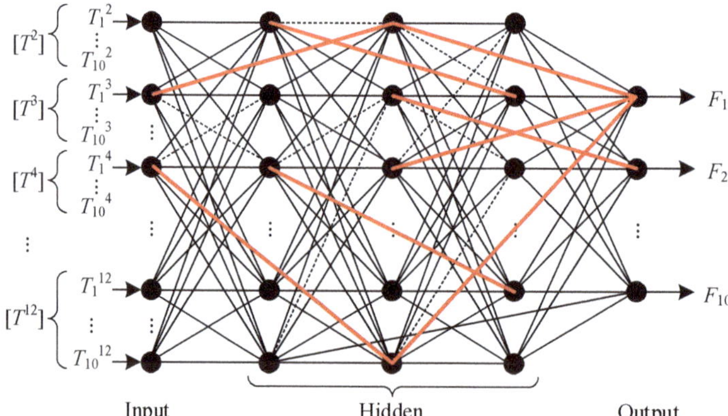

Figure 6. SWNN ensemble members in the 2nd working state.

When constructing the SWNN ensemble members for the 2nd working state, the number of input neurons is 70, which matches 70 TDIs in Equation (9). Three hidden layers are selected with 70 hidden neurons in each layer, and the activation function is a Logistic function. The number of output neurons is 10, representing 10 kinds of fault labels in Equation (9). Similarly, when constructing the SWNN ensemble members for the 3rd working state, the number of input hidden neurons are 80 to correspond the 80 TDIs in Equation (10). The training process of the SWNN will now be introduced.

3.3. SWNN Training

The SWNN ensemble members will be trained using different datasets. SWNN is a multi-layer forward neural network, which is trained by leaping forward-propagation and backward-propagation. Equation (11) shows the weight matrix of the SWNN, where the values on the diagonal line represent the weights of the regular network, while those not on the diagonal line represent the weights of random reconnection. The reconnection weights will determine the way of propagation, where Equation (11) gives the matrix space:

$$
W = \begin{bmatrix}
0 & 0 & \cdots & 0 & 0 \\
0 & W_{23} & \cdots & W_{2(w-1)} & W_{1(w)} \\
\vdots & \vdots & \ddots & \vdots & \vdots \\
0 & 0 & 0 & W_{(w-2)(w-1)} & W_{(w-1)w} \\
0 & 0 & 0 & 0 & W_{(w-1)w}
\end{bmatrix}
\tag{11}
$$

During the forward-propagation stage, suppose that P samples are given to the input layer of the SWNN, and the network outputs are obtained based on the weight vector W. The purpose is to minimize the error function E_{total} that is defined as:

$$
E_{\text{total}} = \frac{1}{2} \sum_{s=1}^{P} |Y_s - V_s|^2
\tag{12}
$$

where, Y_s is the actual output and V_s is a desired one.

During the back-propagation stage, the gradient descent method is used to obtain the optimal solutions. The direction and magnitude change Δw_{ij} can be computed as:

$$
\Delta w_{ij} = -\frac{\partial E_{total}}{\partial w_{ij}}
\tag{13}
$$

Each SWNN is trained by different datasets for training the ensemble members. Such datasets will be explained in the Experimental validation section. The above two stages are executed during each iteration of the back-propagation algorithm until E_{total} converges.

3.4. Selecting Appropriate Ensemble Members

Following training, each individual SWNN member has generated its own result, however, if there are a great number of individual members, a subset of representatives to improve ensemble efficiency needs to be selected. Existing research has proven that the selective ensemble technique can discern *many* members from *all* to achieve better classification accuracy [29]. An improved global correlation based on the Pearson Correlation Coefficient is proposed to select the appropriate SWNN members.

Suppose that there are n ensemble members ($f_1, f_2, \ldots, f_n$), and each member has p forecast values. Then the total error matrix E_{total} can be represented by Equation (14):

$$
E_{\text{total}} = \begin{bmatrix}
e_{11} & e_{12} & \cdots & e_{1n} \\
e_{21} & e_{22} & \cdots & e_{2n} \\
\vdots & \vdots & \ddots & \vdots \\
e_{p1} & e_{p2} & \cdots & e_{pn}
\end{bmatrix}_{p \times n}
\tag{14}
$$

where, $p = 10$ represents the fault types in Table 1. e_{pn} is the pth classification error of the nth ensemble member.

According to the E_{total}, the mean $\bar{e}_i$ and the covariance V_{ij} are described by Equations (15) and (16), respectively.

$$\bar{e}_i = \frac{1}{p}\sum_{k=1}^{p} e_{ki} \ (i = 1, 2, ..., n) \tag{15}$$

$$V_{ij} = \frac{1}{p}\sum_{k=1}^{p} (e_{ki} - \bar{e}_i)(e_{kj} - \bar{e}_j) \ (i, j = 1, 2, ..., n) \tag{16}$$

where, i and j ($i, j = 1, 2, \ldots, n$) represent the ith ensemble member f_i and the jth ensemble member f_j. Then, the correlation matrix R can be calculated by Equation (17):

$$R = (r_{ij}), r_{ij} = \frac{V_{ij}}{\sqrt{V_{ii}V_{jj}}} \ (i, j = 1, 2, ..., n) \tag{17}$$

where, r_{ij} is the correlation coefficient that describes the degree of correlation between f_i and f_j. V_{ii} and V_{jj} are the variances of the two members, which comes from the autocorrelation coefficient $r_{ii} = 1$ and $r_{jj} = 1$ ($i, j = 1, 2, \ldots, n$).

Further extended to calculate the global correlation, let ρ_{fi} denote the correlation between f_i and $(f_1, f_2, \ldots, f_{i-1}, f_{i+1}, \ldots, f_n)$. R is a symmetric matrix whose expansion is shown in Equation (18):

$$R = \begin{bmatrix} 1 & r_{12} & r_{13} & \cdots & r_{1j} \\ r_{12} & 1 & r_{23} & \cdots & r_{2j} \\ r_{13} & r_{23} & 1 & \ddots & \vdots \\ \vdots & \vdots & \ddots & \ddots & r_{ij} \\ r_{1j} & r_{2j} & \cdots & r_{ij} & 1 \end{bmatrix}_{n \times n} \tag{18}$$

Subsequently, the correlation matrix R is represented by a block matrix as shown in Equation (19):

$$R \rightarrow \begin{bmatrix} R_{-i} & r_i \\ r_i^T & 1 \end{bmatrix} \tag{19}$$

where, R_{-i} denotes the correlation matrix of lacking member f_i, and the transformation of R is shown in Figure 7:

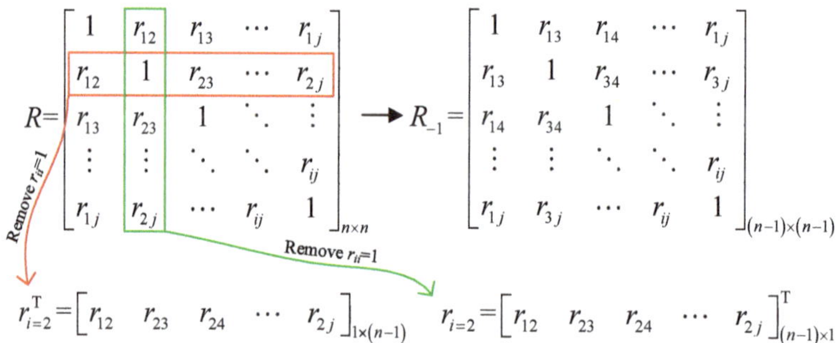

Figure 7. Transformation of correlation matrix R.

Then, the plural-correlation coefficient can be calculated by Equation (20):

$$\rho_i^2 = r_i^T R_{-i} r_i \ (i = 1, 2, ..., n) \tag{20}$$

Regarding a pre-specified threshold θ, if $\rho_i^2 < \theta$, the member f_i is removed from the member group, otherwise, the member f_i is retained. The procedure is shown in Figure 8. Additionally, the retained members can be re-selected by repeating the process until more satisfied members are obtained.

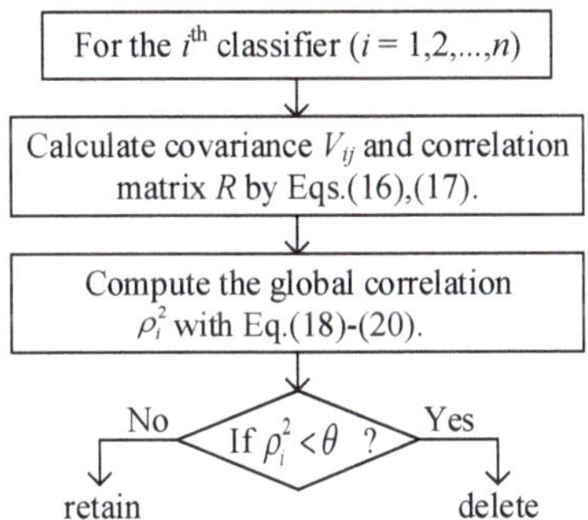

Figure 8. Procedure for selecting ensemble members.

3.5. Integrating the Multiple Members into an Ensemble Output

During the previous steps, several appropriate ensemble members of SWNN have been selected. Regarding the subsequent task, a final decision value is obtained by combining the results of the selected members based on the weighted integration method.

Suppose that there are m members retained, the outputs of the members could construct a column vector f_i according to the fault types in Table 1. f_i can be represented as Equation (21):

$$f_i = [f_i^1, f_i^2, ..., f_i^k]^{\mathrm{T}} \tag{21}$$

where, $i = 1, 2, \ldots, m$ stands for the ith ensemble member. f_i^k ($k = 1, 2, \ldots, p$) is the predictive probability of the kth output neurons in the ith member, whose ranges is in [0, 1]. All the outputs, therefore, can be constructed by a matrix f:

$$f = \begin{bmatrix} f_1^1 & f_2^1 & \cdots & f_m^1 \\ f_1^2 & f_2^2 & \cdots & f_m^2 \\ \vdots & \vdots & \ddots & \vdots \\ f_1^p & f_2^p & \cdots & f_m^p \end{bmatrix}_{p \times m} \tag{22}$$

Take out the row vectors from f and one-by-one and calculate the weight values of each row. Then, the calculated weight values are reconstructed into a row vector of $\begin{bmatrix} w_1^k & w_2^k & \cdots & w_m^k \end{bmatrix}$. Expand each w^k and construct a weight matrix w with k rows. The above processes are shown in Equations (23) and (24).

$$w_i^k = \frac{f_i^k}{\sum_{i=1}^m f_i^k} \Leftrightarrow \begin{bmatrix} w_1^k & w_2^k & \cdots & w_m^k \end{bmatrix} \tag{23}$$

$$w = \begin{bmatrix} w_1^1 & w_2^1 & \cdots & w_m^1 \\ w_1^2 & w_2^2 & \cdots & w_m^2 \\ \vdots & \vdots & \ddots & \vdots \\ w_1^k & w_2^k & \cdots & w_m^k \end{bmatrix}_{k \times m} \tag{24}$$

Combine the w and f to calculate the ensemble outputs of F with the results obtained by Equations (25) and (26).

$$F = \begin{bmatrix} f^1 & f^2 & \cdots & f^k & \cdots & f^p \end{bmatrix} \tag{25}$$

$$f^k = [f]^k \times [w^{\mathrm{T}}]^k \tag{26}$$

where, f^k ($k = 1, 2, \ldots , p$) is the integrated probability of the kth classification, whose ranges also are in $[0, 1]$. Set a threshold $\sigma = 0.5$. When $f^k \geq \sigma$, let $f^k = 1$. When $f^k < \sigma$, let $f^k = 0$.

To summarize, the multistage reliability-based SWNN ensemble model can be concluded in the following steps:

(1) Pretreat the original SCADA data and extract its features to construct n training datasets, TR_1, $\mathrm{TR}_2, \ldots , \mathrm{TR}_n$.
(2) Train n SWNNs ensemble members with n training datasets.
(3) Select m appropriate members based on a weighted integration method.
(4) Fuse multiple SWNN members' outputs into an aggregated value.

4. Experimental Validation

Actual data is used to train the ensemble model of SWNN and then applied to detect nine kinds of the abrupt and incipient faults. The training data is originated from the SCADA systems of 30 2MW-wind turbines in a wind farm for one year.

4.1. Case Preparation

Thirty SWNN members are established, respectively, for the 2nd and 3rd working states of the wind turbines, and their parameters are as shown in Table 4:

Table 4. Structure information of SWNN members.

Working State	Ensemble Members	Input Neurons	Hidden Neurons × Layers	Output Neurons	Probability p	Reconnection Number
2nd	30	70	70×3	10	0.08	1232
3rd	30	80	80×3	10	0.08	1600

Notably, for meeting the reliable performance, 26,000 runs of time-series datasets were captured to collect the training (TR), validation (VA) and testing (TE) data. Each run contains 3600 consecutive samples, whose sampling time is 1 second. The three types of data are distributed in accordance with the following principles:

(1) The TE data should not overlap with others, so 6000 runs were selected at first for the final test of the SWNN members.
(2) The remaining 20,000 runs were divided into 90% TR data and 10% VA data, using a bootstrap sampling method, and there were 12,640 runs and 7360 runs, respectively.
(3) The abnormal runs will account for one tenth of the TR, VA and TE data respectively, and the abnormal runs also need to be allocated averagely in the TR, VA, and TE data.
(4) Repeat sampling 30 times to get 30 groups' datasets for each working state, then the 30 groups' datasets are used to train the 30 SWNN members, respectively.

Table 5 shows an example data distribution in Group 1 for the 2nd and 3rd working states, in which the F1–F9 are the labels of the 9 fault types, and F10 is the label of the fault-free.

Table 5. Data distribution of time-series runs in Group 1.

Working State	Data	Runs										
		F1	**F2**	**F3**	**F4**	**F5**	**F6**	**F7**	**F8**	**F9**	**F10**	**Total**
2nd	TR	154	139	150	134	143	141	139	145	128	11,367	12,640
	VA	84	70	74	85	91	84	83	78	80	6631	7360
	TE	67	65	68	72	71	61	79	52	68	5397	6000
3rd	TR	140	142	133	145	151	128	135	142	140	11,384	12,640
	VA	80	83	84	71	74	93	86	80	79	6630	7360
	TE	63	70	71	68	61	65	72	57	79	5394	6000

4.2. Error Analysis

Subsequent to training the SWNN members, Figure 9 gives the calculation results of the Normalized Global Correlation (NGC) for a total 60 SWNN members, in which Figure 9a,b represent the 2nd and 3rd working state, respectively. Ranking the testing values of NGC, the first 8 SWNN members are selected to construct a selective ensemble model for each working state. Furthermore, the selective ensemble model is used to predict and classify 10 types of faults, and the fault prediction distribution and the classification rate are shown in Figure 10 and Table 6, respectively.

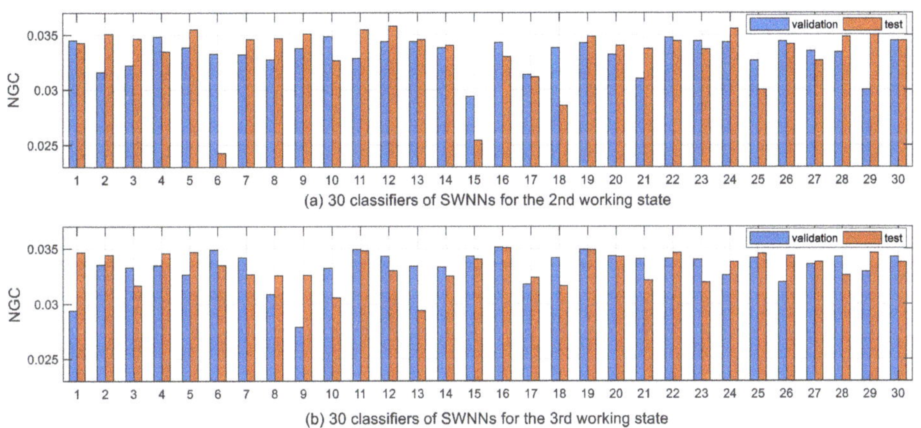

Figure 9. Normalized Global Correlation (NGC) of the ensemble members.

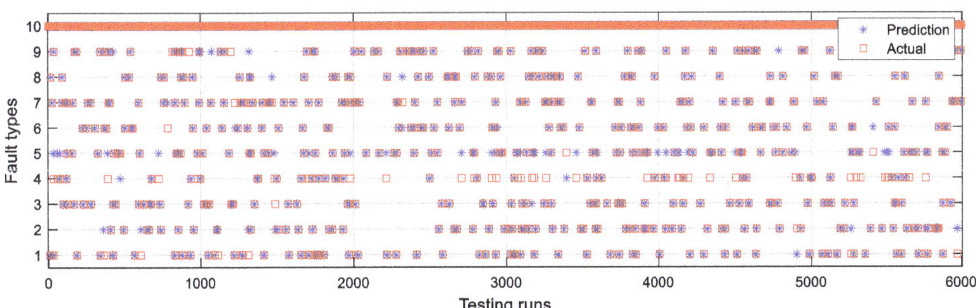

Figure 10. Fault prediction distribution of selective ensemble model.

Table 6. Correct classification rate of selective members.

Working State	Data Case	Ensemble Members' Number								Ensemble Rate
		1	2	3	4	5	6	7	8	
2nd	TR	98.45%	98.05%	97.9%	97.5%	97.5%	97.39%	97.34%	97.28%	98.20%
	TE	92.70%	90.78%	95.26%	93.51%	93.35%	92.88%	92.51%	92.41%	93.84%
3rd	TR	98.71%	98.58%	98.44%	98.44%	98.39%	98.34%	98.31%	98.26%	98.63%
	TE	94.47%	93.85%	93.22%	93.22%	92.87%	92.64%	93.12%	92.55%	94.11%

Figure 10 illustrates, the fifth fault has the lowest accuracy while among the other types of faults there are also many cases of misclassification and missed judgment. Table 6 shows, in the 2nd and 3rd working states, the training accuracy and testing accuracy of the 8 SWNN classifiers are over 97% and 92%, respectively. It can be illustrated that the SWNNs are close to each other in classification accuracy, which benefits from its special network structure, i.e., the homogeneous but structurally stochastic structure. Additionally, the SWNN ensemble model has a higher accuracy rate (93.8%) than that of the SWNN members. It proves that SWNN is very suitable as the ensemble members, and the proposed selective SWNN ensemble model can detect nine pitch faults effectively.

5. Comparison Validation

The proposed selective SWNN ensemble model is compared with three existing methods for online fault detection. The false alarm rate (FAR), the missed fault rate (MFR), and the mean fault diagnosis delay (MFD) are used as the evaluation indices to evaluate the performance of these models.

5.1. Comparison Approaches

Three comparison approaches are TRSWA-BP Neural Network (TRSWA-NN) [12], SWPSO-Support Vector Regression (SWPSO-SVR) [36] and SWPSO learning vector quantization (SWPSO-LVQ), which are presented briefly. They have shown good performance in wind power prediction and fault diagnosis of wind turbines, which are compared based on the former captured data.

TRSWA-NN: This neural network's learning process is based on an efficiency tabu, real-coded, small-world optimization algorithm (TRSWA), which combines EMD (empirical mode decomposition), PSR (phase space reconstruction), and EMD-based PSR to detect and isolate the faults of wind turbines.

SWPSO-SVR: This scheme uses the combination of support vector regression and small-world particle swarm optimization for fault detection and isolation in wind turbines.

SWPSO-LVQ: This scheme combines the LVQ network based on the small-world particle swarm optimization for detection and isolation, which has a good performance for all of the faults.

5.2. Evaluation Indices

The evaluation indices contain false alarm rate (FAR), missed fault rate (MFR) and mean fault diagnosis delay (MFD). The three indices are calculated as follows:

$$FAR = l_{i,10} \Big/ \sum_{j=1}^{10} l_{j,10} \tag{27}$$

$$MFR = \sum_{j=1,i=1}^{10} l_{j,i} \Big/ \sum_{j=1}^{10} l_{j,i} \tag{28}$$

$$MFD = t_{\text{fault occurrence}} - t_{\text{fault detection/isolation}} \tag{29}$$

where, $l_{j,i}$ is the number of samples from the ith class but classified to the jth class. MFD represents the delay time between the fault occurrence and fault detection/isolation.

5.3. Comparative Analysis

The same datasets above will be applied to train the three models and the proposed selective SWNN ensemble model. Monte-Carlo analysis [37] is used to calculate the indices and to test the robustness of the comparison approaches. Particularly, a rigorous test simulation based on 10,000 runs has been executed, during which realistic wind turbine uncertainties have been considered. Table 7 shows the performance of the selective SWNN ensemble model against other approaches.

The results from Table 7 show the overall efficacy of the proposed ensemble SWNN model. Particularly in the neural network-based approaches of the ensemble SWNN model, TRSWA-NN and SWPSO-LVQ, it seems to achieve interesting results with quite low FAR, MFR and MRD for all the fault cases. More specifically, when identifying the first 7 faults, the maximum FAR and MFR of the ensemble model are 0.037 and 0.031, respectively, and the maximum MFD is not more than 1.3 s; while, when identifying the 8th and 9th faults, the FAR and MFR are about 0.1 and 0.08, and the minimum MFD is 8.5 s. There are many differences between the two situations, meaning that the ensemble model has good accuracy and can maintain a fast corresponding speed for diagnosing the first 7 pitch faults, but it is not ideal for the latter 2 faults. The same situation occurs in the TRSWA-NN and SWPSO-LVQ models, which can easily monitor the first 7 faults. SWPSO-SVR has a very small MFD in diagnosing all kinds of the 9 faults, however. The SWPSO-SVR has no better FAR and MFR, which means it cannot guarantee high fault recognition accuracy when dealing with multi-dimensional and large amounts of SCADA data. Conversely, compared with TRSWA-NN and SWPSO-LVQ, the selective SWNN ensemble model also shows the optimal results with smaller FAR, MFR and MFD, especially when identifying and classifying the first seven pitch faults.

Table 7. Comparisons of the presented method and other approaches.

Fault Case	Index	Selective SWNN Ensemble Model	TRSWA-NN	SWPSO-SVR	SWPSO-LVQ
1	FAR	0.012	0.021	0.034	0.018
	MFR	0.009	0.019	0.048	0.013
	MFD(s)	0.08	0.12	0.02	0.11
2	FAR	0.018	0.043	0.021	0.031
	MFR	0.021	0.024	0.043	0.033
	MFD(s)	0.1	0.23	0.02	0.3
3	FAR	0.008	0.01	0.011	0.005
	MFR	0.005	0.008	0.02	0.009
	MFD(s)	0.03	0.06	0.01	0.08
4	FAR	0.037	0.042	0.113	0.068
	MFR	0.031	0.034	0.09	0.073
	MFD(s)	1.3	3.7	0.5	1.8
5	FAR	0.021	0.022	0.081	0.035
	MFR	0.017	0.026	0.09	0.01
	MFD(s)	0.13	0.11	0.02	0.15
6	FAR	0.011	0.015	0.065	0.045
	MFR	0.012	0.017	0.055	0.033
	MFD(s)	0.08	0.12	0.03	0.08
7	FAR	0.019	0.035	0.057	0.04
	MFR	0.022	0.018	0.055	0.017
	MFD(s)	0.1	2.3	0.01	0.18
8	FAR	0.08	0.062	0.153	0.091
	MFR	0.072	0.04	0.181	0.08
	MFD(s)	8.5	4.8	0.8	3.8
9	FAR	0.11	0.12	0.135	0.108
	MFR	0.067	0.094	0.19	0.124
	MFD(s)	11.8	8.3	1.5	6.9

6. Conclusions

Fault diagnosis of wind turbines, as a basic research project, plays a major role in today's electricity markets. The five-step ensemble strategy proposed is a novel technique that uses the small-world neural networks as the ensemble members to improve monitoring reliability and classification accuracy. Compared to the conventional methods, the proposed method can be summarized as follows:

(1) The proposed selective SWNN ensemble strategy provides a comprehensive monitoring mechanism for most fault types under different working states of the wind turbines.

(2) The proposed selective SWNN ensemble strategy decomposes the dynamic monitoring into four distributed ensemble models, and each model contains multiple SWNNs to selectively output multiple judgment results. These results can be coordinated to provide consistent fault classifications at a fixed prediction horizon. The classification performance is effective and accurate. Specifically, the proposed method under different working states shows an average training accuracy and testing accuracy of over 97% and 92%, respectively, when detecting the most pitch failures.

(3) The proposed selective SWNN ensemble strategy reveals stronger adaptability and sensitivity than that of the single SWNN, especially when classifying and identifying the abrupt and incipient faults, with a higher accuracy rate of over 93.8%. Additionally, it is able to use a short delay time while achieving a lower false alarm rate and lower missed fault rate than the conventional models.

Future works focus on: (1) Regarding data processing, the ratio of normal samples to abnormal samples needs to be studied. Such ratio will play a decisive role in the final classification results. (2) Concerning terms of ensemble member optimization, a deep small-world neural network will be proposed to diagnose the fault of wind turbines. It will be compared with other deep learning algorithms, such as the Convolutional Neural Network (CNN), the Long Short-Term Memory (LSTM) and the Recurrent Neural Network (RNN).

Author Contributions: S.W. planned and supervised the whole project; M.L. developed the optimization algorithm, designed the criterion and performed the simulation and experiments. M.L. and S.W. contributed to discussing the results and writing the manuscript.

Funding: This research was funded by National Natural Science Foundation of China, grant number 50776005, 51577008.

Conflicts of Interest: The authors declare no conflict of interest.

References

1. Lin, Y.G.; Tu, T.; Liu, H.W.; Li, W. Fault analysis of wind turbines in China. *Renew. Sustain. Energy Rev.* **2016**, *55*, 482–490. [CrossRef]

2. Nandi, T.N.; Herrig, A.; Brasseur, J.G. Non-steady wind turbine response to daytime atmospheric turbulence. *Phil. Trans. R. Soc. A* **2017**, *375*, 20160103. [CrossRef] [PubMed]

3. Churchfield, M.J.; Lee, S.; Michalakes, J.; Moriarty, P.J. A numerical study of the effects of atmospheric and wake turbulence on wind turbine dynamics. *J. Turbul.* **2012**, *13*, N14. [CrossRef]

4 Laks, J.; Pao, L.; Wright, A.; Kelley, N.; Jonkman, B. The use of preview wind measurements for blade pitch control. *Mechatronics* **2011**, *21*, 668–681. [CrossRef]

5. Kelley, N.; Hand, M.; Larwood, S.; McKenna, E. The NREL large-scale turbine inflow and response experiment- Preliminary results. In Proceedings of the 2002 ASME Wind Energy Symposium, Reno, NV, USA, 14–17 January 2002; pp. 412–426.

6. Jiang, Y.; Yin, S.; Kaynak, O. Data-driven monitoring and safety control of industrial cyber-physical systems: Basics and beyond. *IEEE Access* **2018**, *6*, 47374–47384. [CrossRef]

7. Rahimilarki, R.; Gao, Z.W.; Zhang, A.H.; Richard, J.B. Robust neural network fault estimation approach for nonlinear dynamic systems with applications to wind turbine systems. *IEEE Trans. Ind. Inf.* **2019**, *18*. [CrossRef]

8. Marugán, A.P.; Márquez, F.P.G.; Perez, J.M.P.; Ruiz-Hernández, D. A survey of artificial neural network in wind energy systems. *Appl. Energy* **2018**, *228*, 1822–1836. [CrossRef]

9. Wang, S.X.; Li, M.; Zhao, L.; Jin, C. Short-term wind power prediction based on improved small-world neural network. *Neural Comput. Appl.* **2018**, *29*, 1–13. [CrossRef]

10. Marugán, A.P.; Chacón, A.M.P.; Márquez, F.P.G. Reliability analysis of detecting false alarms that employ neural networks: A real case study on wind turbines. *Reliab. Eng. Syst. Saf.* **2019**, *191*, 106574. [CrossRef]

11. Pelletier, F.; Masson, C.; Tahan, A. Wind turbine power curve modelling using artificial neural network. *Renew. Energy* **2016**, *89*, 207–214. [CrossRef]

12. Wang, S.X.; Zhao, X.; Li, M.; Wang, H. TRSWA-BP neural network for dynamic wind power forecasting based on entropy evaluation. *Entropy* **2018**, *20*, 283. [CrossRef]

13. Cheng, F.Z.; Peng, Y.Y.; Qu, L.Y.; Qian, W. Current-based fault detection and identification for wind turbine drivetrain gearboxes. *IEEE Trans. Ind. Appl.* **2017**, *53*, 878–887. [CrossRef]

14. Stetco, A.; Dinmohammadi, F.; Zhao, X.Y.; Robu, V.; Flynn, D.; Barnes, M.; Keane, J.; Nenadic, G. Machine learning methods for wind turbine condition monitoring: A review. *Renew. Energy* **2019**, *133*, 620–635. [CrossRef]

15. Cheng, F.Z.; Wang, J.; Qu, L.Y.; Qian, W. Rotor current-based fault diagnosis for DFIG wind turbine drivetrain gearboxes using frequency analysis and a deep classifier. *IEEE Trans. Ind. Appl.* **2018**, *54*, 1062–1071. [CrossRef]

16. Zhao, H.S.; Liu, H.H.; Hu, W.J.; Yan, X.H. Anomaly detection and fault analysis of wind turbine components based on deep learning network. *Renew. Energy* **2018**, *127*, 825–834. [CrossRef]

17. Teng, W.; Cheng, H.; Ding, X.; Liu, Y.B.; Ma, Z.Y.; Mu, H.H. DNN-based approach for fault detection in a direct drive wind turbine. *IET Renew. Power Gener.* **2018**, *12*, 1164–1171. [CrossRef]

18. Grimshaw, S. An Introduction to the Bootstrap. *Technometrics* **2012**, *37*, 340–341. [CrossRef]

19. Breiman, L. Bagging predictors. *Mach. Learn.* **1996**, *24*, 123–140. [CrossRef]

20. Freund, Y. Boosting a weak algorithm by majority. *Inf. Comput.* **1995**, *121*, 256–285. [CrossRef]

21. Liu, Y.; Yao, X. Ensemble learning via negative correlation. *Neural Netw.* **1999**, *12*, 1399–1404. [CrossRef]

22. Qin, S.; Wang, K.; Ma, X.; Wang, W.; Li, M. Ensemble learning-based wind turbine fault prediction method with adaptive feature selection. In Proceedings of the International Conference of Pioneering Computer Scientists, Engineers and Educators, Changsha, China, 22–24 September 2017; pp. 572–582.

23. Sollich, P.; Krogh, A. Learning with ensembles: How over-fitting can be useful. In *Advances in Neural Information Processing Systems 8*; Touretzky, D.S., Mozer, M.C., Hasselmo, M.E., Eds.; MIT Press: Denver, CO, USA; Cambridge, MA, USA, 1996; pp. 190–196.

24. Schapire, R.E. The strength of weak learnability. *Mach. Learn.* **1990**, *5*, 197–227. [CrossRef]

25. Pashazadeh, V.; Salmasi, F.R.; Araabi, B.N. Data driven sensor and actuator fault detection and isolation in wind turbine using classifier fusion. *Renew. Energy* **2018**, *116*, 99–106. [CrossRef]

26. Dey, S.; Pisu, P.; Ayalew, B. A Comparative study of three fault diagnosis schemes for wind turbines. *IEEE Trans. Control Syst. Technol.* **2015**, *23*, 1853–1868. [CrossRef]

27. Hornik, K.; Stinchcombe, M.; White, H. Multilayer feedforward networks are universal approximators. *Neural Netw.* **1989**, *2*, 359–366. [CrossRef]

28. Hansen, L.K.; Salamon, P. Neural network ensembles. *IEEE Trans. Pattern Anal. Mach. Intell.* **1990**, *12*, 993–1001. [CrossRef]

29. Zhou, Z.H.; Wu, J.X.; Tang, W. Ensembling neural networks: Many could be better than all. *Artif. Intell.* **2002**, *137*, 239–263. [CrossRef]

30. Morelli, L.G.; Abramson, G.; Kuperman, M.N. Associative memory on a small-world neural network. *Eur. Phys. J. B-Condens. Matter Complex Syst.* **2004**, *38*, 495–500. [CrossRef]

31. Zhao, X.; Wang, S.X. Convergence analysis of the tabu-based real-coded small-world optimization algorithm. *Eng. Optim.* **2014**, *46*, 465–486. [CrossRef]

32. Suen, C.Y.; Lam, L. Multiple classifier combination methodologies for different output levels. *Lect. Notes Comput. Sci.* **2000**, *1857*, 52–66.

33. Wu, Z.; Lin, W.; Ji, Y. An integrated ensemble learning model for imbalanced fault diagnostics and prognostics. *IEEE Access* **2018**, *6*, 8394–8402. [CrossRef]

34. Wang, N. Advanced wind turbine control. *Adv. Wind Turbine Technol.* **2018**, *5*, 281–297.

35. Watts, D.J.; Strogatz, S.H. Collective dynamics of 'small-world' networks. *Nature* **1998**, *393*, 440–442. [CrossRef] [PubMed]
36. Wang, S.X.; Li, M.; Tian, T.T.; Zhang, J.H. Evaluation and prediction of operation status for pitch system based on SCADA data. In Proceedings of the 2017 Chinese Automation Congress (CAC), Jinan, China, 20–22 October 2017; pp. 3327–3332.
37. Simani, S.; Castaldi, P.; Farsoni, S. Data-driven fault diagnosis of a wind farm benchmark model. *Energies* **2017**, *10*, 866. [CrossRef]

Article

Hierarchical Fault-Tolerant Control using Model Predictive Control for Wind Turbine Pitch Actuator Faults

Donggil Kim [1] and Dongik Lee [2],*

[1] Department of Robot Engineering, Kyungil University, Gyeongsan 38428, Korea
[2] School of Electronics Engineering, Kyungpook National University, Daegu 41566, Korea
* Correspondence: dilee@ee.knu.ac.kr; Tel.: +82-53-940-8686

Received: 11 June 2019; Accepted: 7 August 2019; Published: 12 August 2019

Abstract: Wind energy is one of the fastest growing energy sources in the world. It is expected that by the end of 2022 the installed capacity will exceed 250 GW thanks to the supply of large scale wind turbines in Europe. However, there are still challenging problems with wind turbines. In particular, off-shore and large-scale wind turbines are required to tackle the issue of maintainability and availability because they are installed in harsh off-shore environments, which may also prevent engineers from accessing the site for immediate repair works. Fault-tolerant control techniques have been widely exploited to overcome this issue. This paper proposes a novel fault-tolerant control strategy for wind turbines. The proposed strategy has a hierarchical structure, consisting of a pitch controller and a wind turbine controller, with parameter estimations using the adaptive fading Kalman filter technique. The pitch controller compensates any fault with a pitching actuator, while the wind turbine controller computes the optimal reference command for pitching behavior so that the effect of the fault with a pitch actuator can be minimized. The performance of the proposed approach is demonstrated through a set of simulations with a wind turbine benchmark model.

Keywords: fault-tolerant control; Kalman filter; model predictive control; wind turbines

---

1. Introduction

The demand for sustainable and economical renewable energy has been increasing. Wind energy is considered as one of most important renewable energy sources thanks to the fact that wind is an infinite and free source of energy without harmful waste. According to the WindEurope report [1], 11.7 GW of wind power was newly installed in 2018 and there is a total 189 GW of installed wind energy capacity in Europe. This will exceed 250 GW by the end of 2022.

Today, wind turbines are responsible for a large part of the energy production. The high reliability of wind turbines is critical to reduce the cost of operation and maintenance [2,3]. However, wind turbines operate in harsh environments, which causes failure of subsystems, including actuators and sensors. Failure of subsystems or components may result in a change of the overall system structure and performance, either slowly or significantly. Therefore, an appropriate compensation of failure following an early diagnosis has significant impacts on the whole structural safety and the stable energy generation.

The pitch control system plays an important role for the operation of wind turbines. The main purposes of the pitch control include limiting the wind power capture in cases of high wind speed conditions, mitigating the operational load, stalling, and braking in control region 3 with effective wind speeds of 12.5–25 m/s, and stopping the wind energy generation in cases of low wind speed conditions in control region 2 with effective wind speeds of 3–12.5 m/s [4]. Therefore, any fault with the pitch control system may cause serious problems, such as asymmetric loads for blades, fluctuations in the speed of generator and power generation, and degradation of the stability of the wind turbine [5,6].

Recently, researchers have paid attention to fault tolerance control techniques for wind turbines. Fault-tolerant control can be used to improve the reliability of a wind turbine system in harsh operating environments and to guarantee the graceful degradation of performance, even in the event of faults with system components. Some important works include linear parameter variable control [7], adaptive gain sliding mode control [8], fuzzy-based adaptive control [9,10], and nonlinear geometric approach [11]. In previous work [7], the linear matrix inequality is used for active fault-tolerant control, while the bilinear matrix inequality is applied to passive fault-tolerant control. In another study [8], a sliding mode fault-tolerant controller based on gain adaptive control was designed along with the robust sliding mode observer for states, unknown outputs, and sensor faults. In previous work [9,10], fuzzy model reference adaptive control for failure of the torque actuator in a wind turbine was proposed. In another study [11], an active fault-tolerant controller using the nonlinear geometric approach was presented.

The model predictive control (MPC) technique systematically handles constraints and optimizes the controller to satisfy the control objectives [12]. Also, MPC has an implicit capability of fault-tolerance in terms of the prediction of models, constraints, and objective functions [13]. Many researchers have published MPC-based fault-tolerant control for wind turbines [14,15]. In a previous study [14], MPC worked as the nominal controller in normal conditions, while it worked as the pre-compensator in the presence of a fault. In another study [15], MPC was used to accommodate faults with a pitch actuator and a generator torque actuator using the Laguerre function for parameterization of control. Most of the existing research used an objective function that involves the effect of a fault. How to properly adjust the objective function of MPC in response to faults is an interesting research topic [16]. However, it is hard to find publications addressing the design of fault-tolerant MPC cooperating with a fault detection and diagnosis (FDD) module that provides the MPC controller with information regarding faults.

This paper presents a novel fault-tolerant control strategy for wind turbines with a hierarchical architecture in which MPC is cooperated with a FDD module using online estimations of fault parameters. Since MPC can effectively deal with the variation in parameters by using an internal model, the proposed strategy employs MPC as the baseline controller for both the pitch system and the wind turbine. The proposed method comprises two different control modes: (a) fault-tolerant control with the pitch system only; and (b) fault-tolerant control with the pitch system and the wind turbine. The performance of the proposed method is compared with proportional-integral (PI) control and sliding mode control through a set of simulations with a wind turbine benchmark model.

2. Statement of Wind Turbine Problem

In this work, we consider the wind turbine control system benchmark model, which was provided by kk-electronics for the IFAC Fault Detection, Supervision and Safety for Technical Processes (SAFEPROCESS) conference (Figure 1) [17]. The benchmark model represents a horizontal-axis wind turbine with a three-bladed rotor. The model was obtained through a standard approach given in the literature [18]. The wind turbine control system consists of seven subsystems: a wind model, aerodynamics, a drive train, a generator-converter unit, a pitch system, and a tower. In this paper, only necessary parts, including aerodynamics and a pitch system, are presented.

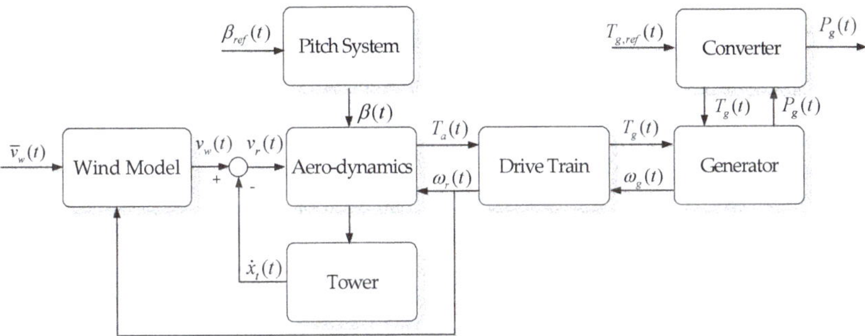

Figure 1. Wind turbine control system model [18].

2.1. Description of Wind Turbine Aerodynamics Model

The aerodynamic equations of wind turbines are given in previous work [4,19]. The power available from the wind passing through the entire swept rotor can be expressed as

$$P_w = \frac{1}{2}\dot{m}v_r^2 = \frac{1}{2}\rho\pi R^2 v_r^3 \tag{1}$$

where P_w is the power available from the wind (W), $\dot{m}$ is the mass flow rate of the wind, v_r is rotor effective wind speed (m/s), ρ is air density (kg/m^3), and R is the radius of rotor disc (m). Only a portion of power available P_w is converted to the rotor power, which is represented by the power coefficient, C_p, depending on the tip-speed ratio (λ) and the blade pitch angle (β). The power extracted from the available power P_w is given by.

$$P_a(t) = P_w(t)C_p(\lambda(t), \beta(t)) \tag{2}$$

The tip-speed ratio is defined as the ratio between the tip speed of blades and the rotor effective wind speed

$$\lambda = \frac{\omega_r(t)R}{v_r(t)} \tag{3}$$

where $w_r(t)$ is rotor speed (rad/s). The power coefficient C_p has a theoretical upper limit of $16/27 \approx 0.593$, known as the Betz theory. The aerodynamic torque (T_a) acting on the rotor is defined as

$$T_a(t) = \frac{P_a(t)}{\omega_r(t)} = \frac{1}{2\omega_r(t)}\rho\pi R^2 v_r^3 C_p(\lambda(t), \beta(t)) \tag{4}$$

In the benchmark model, the blade pitch actuation system has three identical hydraulic actuators. Each hydraulic pitch system is modeled by a second-order system

$$\frac{\beta(s)}{\beta_{ref}(s)} = \frac{\omega_n^2}{s^2 + 2\zeta\omega_n s + \omega_n^2} \tag{5}$$

where β_{ref}, ω_n, and ζ are the reference pitch angle, the natural frequency, and the damping factor of the pitch actuator, respectively.

The pitch control system plays an important role in generating the reference pitch angle to manage the power generation from the wind. At wind speeds below the rate value (<12.5 m/s), the reference pitch angle is set to zero. When wind speeds are above the rate value (>12.5 m/s), the pitch control system aims to keep the generated power around the rated value by adjusting the reference pitch angle. Therefore, this work focuses on the design of fault-tolerant control that can compensate for the effect of the fault with the pitch system.

2.2. Model of Pitch Actuator System Fault

In the benchmark model for a 4.8-MW wind turbine [17], each of the pitch systems is actuated with an identical pitch actuator and assumed to have the same dynamic behavior as described in Equation (5). To facilitate the subsequent controller design, the state space model of the pitch system is rewritten as follows:

$$
\begin{bmatrix} \dot{\beta} \\ \ddot{\beta} \end{bmatrix} = \begin{bmatrix} 0 & 1 \\ -\omega_n^2 & -2\varsigma\omega_n \end{bmatrix} \begin{bmatrix} \beta \\ \dot{\beta} \end{bmatrix} + \begin{bmatrix} \beta \\ \dot{\beta} \end{bmatrix} \beta_{ref}
$$
$$
y = \begin{bmatrix} 1 & 0 \end{bmatrix} \begin{bmatrix} \beta \\ \dot{\beta} \end{bmatrix}
$$

$$(6)$$

Each actuator has the physical constraints, the min/max magnitudes, and the rates, given as (−2 deg, 90 deg) and (−8 deg/s, +8 deg/s). Each pitch system may have faults due to hydraulic leakage, which will result in the change in dynamics; that is, the variations of ω_n and ς. This kind of fault is modeled as a convex combination of values at the nominal condition and the low-pressure fault, as follows [8]:

$$
\omega_n^2 = \omega_{n0}^2 + (\omega_{nf}^2 - \omega_{n0}^2)f
$$
$$
\varsigma\omega_n = \varsigma_0\omega_{n0} + (\varsigma_f\omega_{nf} - \varsigma_0\omega_{n0})f
$$

$$(7)$$

where ς_0 and ω_{n0} are the nominal values of ς and ω_n, and ς_f and ω_{nf} are their values at a low pressure fault; f represents the fault indicator. That is, $f = 0$ denotes the normal pressure $\varsigma_0 = 0.6$ rad/s and $\omega_{n0} = 11.11$ rad/s, and $f = 1$ denotes the low-pressure fault with $\varsigma_f = 0.9$ rad/s and $\omega_{nf} = 3.42$ rad/s [17].

3. Design of Hierarchical Fault-Tolerant Model Predictive Control

3.1. Control Principle for Wind Turbines

The goals of control for wind turbines are classified into two groups according to the wind speed rate value. For the above wind speed rate value (>12.5 m/s in the benchmark model), the controller aims to regulate the generate power (P_g) around the rated power and also to protect the generator from the limited rotation speed due to the excessive aerodynamic torque. On the contrary, for the below wind speed rate value (<12.5 m/s in the benchmark model), the wind turbine controller aims to drive the turbine rotor speed at the optimal rotating speed by regulating the pitch angle, $\beta_r = 0$. The regulation of P_g is achieved by regulating T_g (see Equation (4)). Therefore, the control inputs of the wind turbine system are $T_{g,ref}$ and β_{ref}. In this paper, two types of fault-tolerant MPC are designed: one for the generator system ($T_{g,ref}$) and another for the pitch system (β_{ref}).

3.2. Fault-Tolerant Model Predictive Control for Wind Turbines

The structure of the proposed fault-tolerant control strategy is illustrated in Figure 2. In this strategy, notice that fault accommodation occurs in two places: one is in the pitch system controller (local controller) and the other is in the wind turbine system controller (global controller). For this reason, the proposed method can be seen as "fault-tolerant control with a hierarchical structure". In practice, the fault-tolerant MPC for the pitch system compensates for the effects of faults in the pitch system as much as possible using online estimations of fault parameters. If the performance degradation persists, then the fault-tolerant MPC for the wind turbine system attempts to meet the overall performance requirements. This judgment is made by the FDD manager based on information of the wind turbine condition, including the fault index, the power generation, the generator speed, etc. The local FDD module provides specific information regarding faults, such as the fault index and estimations of fault parameters using the adaptive fading Kalman filter algorithm [20,21]. Online estimations of fault parameters and the decision regarding the fault-tolerant control form the FDD algorithm, which will be discussed in a separate paper. An extensive survey of FDD can be found in previous work [22].

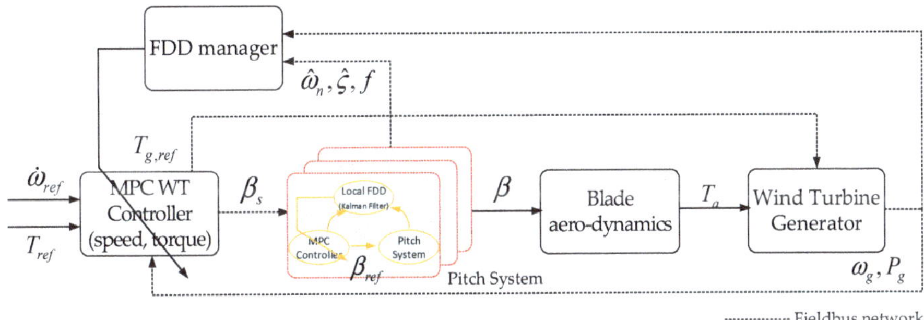

Figure 2. Hierarchical structure of the proposed fault-tolerant control.

3.2.1. Design of Fault-Tolerant MPC for the Pitch System

The purpose of the fault-tolerant MPC for the pitch system is to control the pitch system, even if any fault with the pitch actuator occurs, so that the discrepancy between the response of a nominal system and the response of a faulty system can be minimized. The fault-tolerant MPC for the pitch system can be formulated as follows if accurate estimates of the parameter are available:

$$
\min_{\Delta\beta_{ref}} J(k) = \sum_{i=i}^{Np} \|\hat{\beta}(k+i|k) - \beta_s(k+i|k)\|_Q^2 + \sum_{i=i}^{Nu} \|\Delta\beta_{ref}(k+i)\|_R^2
$$
$$
\text{subject to :}
$$
$$
\text{pitch system model in Equation (5)}
$$
$$
\beta_{min} < \beta_{ref} < \beta_{max}
$$
$$
\Delta\beta_{min} < \Delta\beta_{ref} < \Delta\beta_{max}
$$

$$(8)$$

where $\hat{\beta}$, β_s and $\Delta\beta_{ref}$ denote the estimated pitch angle, the pitch angle set-point given from the wind turbine system controller, and the variation in the pitch angle reference, respectively. N_p is the length of the prediction horizon, N_u is the length of the control horizon, and Q and R are weighting matrices When solving the optimization problem given in Equation (8), all of the system parameters are updated via the adaptive fading Kalman filter algorithm to reflect the effect of the fault with the pitch system at every sampling time.

3.2.2. Design of Fault-Tolerant MPC for the Wind Turbine System

The main purpose of the fault-tolerant MPC for the wind turbine system is to meet the operating requirements of the wind turbine and to generate the maximum electric power available at the current wind speed, which often varies. Furthermore, to achieve the capability of fault tolerance, the wind turbine system controller must be able to compensate for the effect of a fault that cannot be completely compensated by the fault-tolerant MPC for the pitch system. The wind turbine should be maintained at maximum power while not exceeding the safe electrical and mechanical loads. These requirements can be expressed in the optimization problem as follows:

$$
\min_{\Delta\beta_s} J(k) = \sum_{i=i}^{Np} \|y(k+i|k) - r(k+i|k)\|_Q^2 + \sum_{i=i}^{Nu} \|\Delta\beta_s(k+i)\|_R^2
$$
$$
\text{subject to :}
$$
$$
\text{wind turbine model in Section 2}
$$
$$
\beta_{min} < \beta_s < \beta_{max}
$$

$$(9)$$

where N_p and N_u are the lengths of the prediction horizon and the control horizon, respectively. Here, y is the output vector of the wind turbine system, r is the reference vector, which may change according to the wind speed; $\Delta \beta_s$ is the variation of the pitch angle set-point, and Q and R are weighting matrices. The output vector y and the reference vector r are defined as

$$y = \begin{bmatrix} \dot{\omega}_g & T_g \end{bmatrix}, r = \begin{bmatrix} 0 & \frac{P_{ref}}{\eta_g \omega_{ref}} \end{bmatrix} \tag{10}$$

where $\dot{\omega}_g$ is time rate of the change of the generator angular speed, T_g is the current generator torque, P_{ref} is the power generation reference (it is a constant value, which is 4.8 MW in this paper), η_g is a coefficient, and ω_g is the current generator angular speed at the given sampling time. Table 1 summarizes the control parameters for MPC.

Table 1. Model predictive control parameters for implementation.

Parameters	Notation	Value
Sampling Period	-	0.01
Prediction Horizon	Np	20
Control Horizon	Nu	2
Output Constraint Min/Max	β_{min}/β_{max}	−2/90
Output Constraint Rate Up/Down	$\Delta\beta_{min}/\Delta\beta_{max}$	8/−8
Input/Output Weight	R/Q	0.1/1

4. Simulation Results

4.1. Wind Turbine Benchmark Model

The effectiveness of the proposed method is verified through simulations with the wind turbine benchmark model in both the nominal case and the fault case with a pitch fault. The wind turbine benchmark model, which is developed in the MATLAB/Simulink® (R2014A, MathWorks, Natick, MA, USA) programming environment, was introduced in the design competition by IFAC SAFEPROCESS [17]. In the benchmark model, a basic PI controller for pitch control is implemented with proportional gain of 4 and integrator gain of 1 [6]. To evaluate the performance of the proposed approach, sliding mode control (SMC) for a wind turbine is also implemented [23]. Simulations were performed for three cases, which are: (a) no FTC for the wind turbine system, FTC for the pitch system; and (b) FTC for both the pitch and the wind turbine system. The parameters and variables used in the wind turbine benchmark model and fault-tolerant MPC are summarized as follows:

Symbol	Description
β	blade pitch angle for [°]
$\dot{\beta}$	change of pitch angle [°/s]
β_s	set-point of pitch angle for pitch controller [°]
β_{ref}	reference of pitch angle for pitch system [°]
$\hat{\beta}$	estimation of pitch angle [°]
$\hat{\omega}_n$	estimation of natural frequency pitch system [rad/s]
$\hat{\varsigma}$	estimation of damping factor pitch system [rad/s]
f	estimation of fault index for pitch system [rad/s]
P_{ref}	rated power generation (4.8×10^6) [W]
T_{ref}	rated generator torque [Nm]
ω_{ref}	rated generator angular speed (162) [rad/s]
P_g	current generated power [W]
$P_{g,e}$	error of generated power ($P_{ref}-P_g$) [w]
ω_g	current generator angular speed [rad/s]
$\omega_{g,e}$	error of generator angular speed ($\omega_{ref}-\omega_g$) [rad/s]

4.2. Simulations for Healthy/Fault-Free Condtion

Simulations are carried out to compare the performance of the reference PI controller and the proposed MPC for the healthy condition. The simulation is executed using the actual wind data that are provided with the benchmark model. Figure 3 is the actual wind profile of each blade (blade 1, blade 2, and blade 3) used throughout the simulations. For a more realistic model of aerodynamics, the wind profile includes the effects of tower shadow and wind shear. The simulation results are given in Figures 4 and 5.

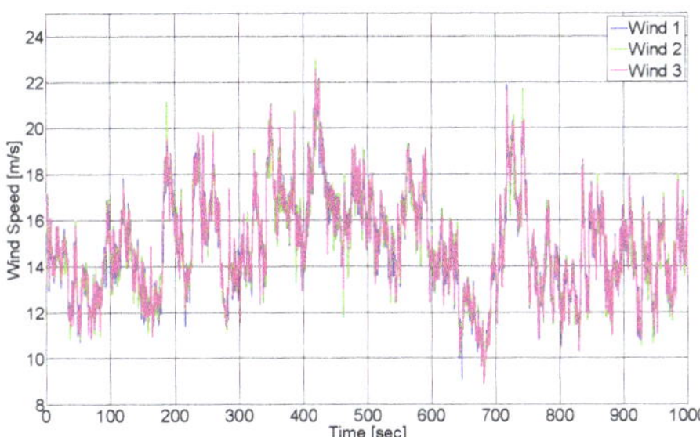

Figure 3. Actual wind profile of each blade, including the effects of tower shadow and wind shear.

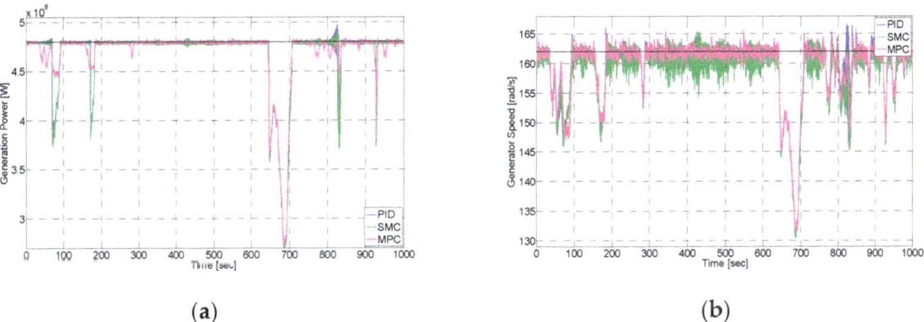

Figure 4. Simulation results with healthy conditions: (**a**) power generation; (**b**) generator speed.

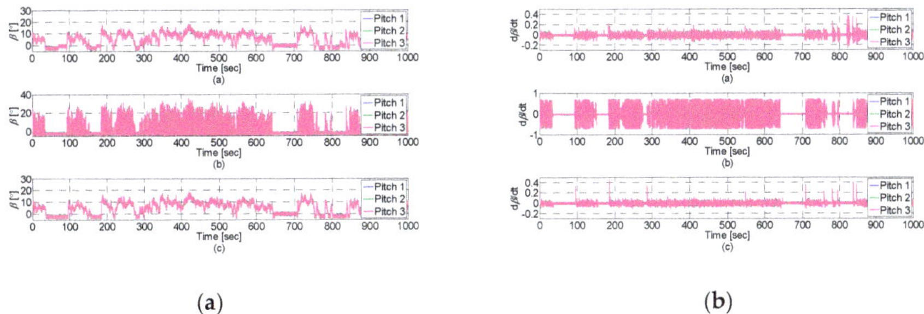

Figure 5. Simulation results with healthy conditions: (**a**) pitch angle; (**b**) pitch angular rate (top: PI; mid: SMC; bottom: MPC).

The responses with both controllers look similar. The power generation is well controlled near the rated power limit, which can be observed in Figure 4a. Table 2 summaries the simulation results. Three metrics are employed to compare the performance of the different control methods: (1) the sum of squared power generation error $(P_{g,e}) = P_{g,e}\,P_{g,e}$; (2) the sum of squared generator speed error $(\omega_{g,e})$;

and (3) the sum of squared second blade pitch rate ($\dot{\beta}_2$) to measure the energy consumption by actuators. As shown in Table 2, the MPC shows a better performance than PI and SMC. The MPC is superior to PI in all performance indices, the power generation, and the energy consumption. In particular, the level of energy consumption with MPC shows an outstanding result.

Table 2. Controller performance in healthy conditions (the numbers in parentheses denote normalized values with respect to the PI controller).

Controller	$\int (P_{g,e}(t))^2 dt$ [10^{16}]	$\int (\omega_{g,e}(t))^2 dt$ [10^6]	$\int (\dot{\beta}_2(t))^2 dt$ [10^2]
PI	1.0380 (1)	2.6868 (1)	1.2618 (1)
SMC	1.1479 (1.11)	3.0746 (1.14)	79.2850 (62.8)
MPC	0.97361 (0.938)	2.5886 (0.963)	0.5745 (0.455)

4.3. Simulations for Faulty Condtion

The simulation results with faults of the pitch system are presented in Figures 6 and 7. The fault occurs in the hydraulic pumps driving the pitch actuators for blades 2 and 3, and they occur abruptly at approximately $t = 200$ seconds. The same wind profile shown in Figure 3 is used. Note that PI and SMC controllers present the instability at $t = 300$ and 650 s, respectively. This behavior may be related to the relatively low wind speed at $t = 300$ s and the pitch actuation fault. Figure 6a shows the power generation curve. It can be observed that the power output with PI or SMC and MPC are quite different under the pitch actuation fault. In the results with SMC, it is obvious that the model uncertainties due to the pitch system fault are larger than the robust region of SMC. The power output with MPC tracks the rated value well, but there is a large variation in the power output curve for the PI and SMC controllers. These results are reflected in the generator speed curve, shown in Figure 6b. This is related to the sluggish response of the pitch system because of the fault and the decrease of the wind speed around $t = 300$ s. Finally, the proposed MPC outperforms the PI and SMC controllers with respect to the input actuation cost, as shown in Figure 7b. Although there is considerable activity in the pitch system in response to the fault, the PI and SMC controller do not work properly. Table 3 summarizes the performance metrics. As mentioned, PI and SMC show poor control performance. On the other hand, MPC shows a better performance than PI or SMC due to its intrinsic robustness.

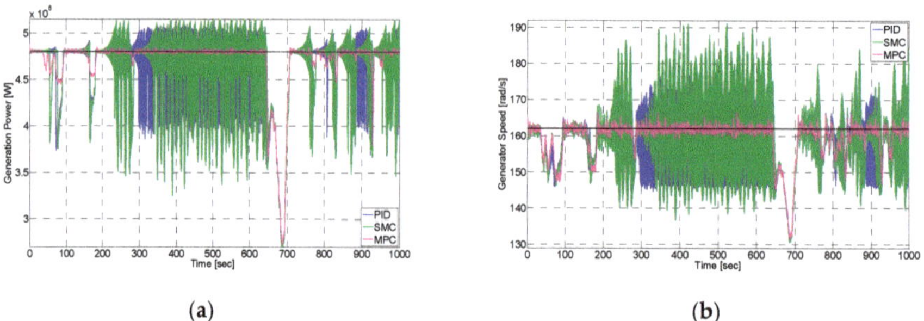

(a)   (b)

Figure 6. Simulation results with fault conditions: (**a**) power generation; (**b**) generator speed.

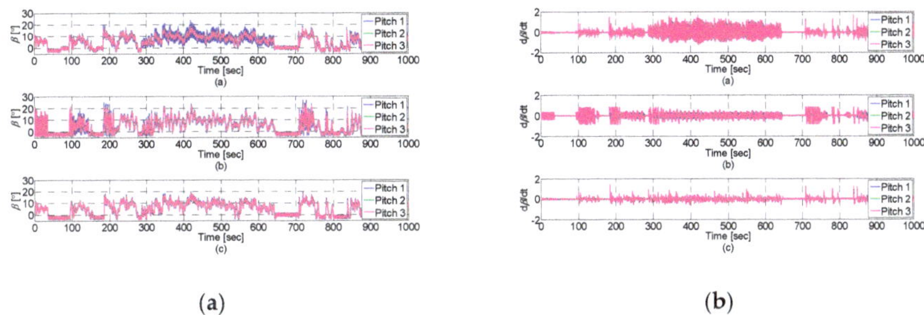

Figure 7. Simulation results with faulty conditions: (**a**) pitch angle; (**b**) pitch angular rate (top: PI; mid: SMC; bottom: MPC).

Table 3. Controller performance in faulty conditions (the numbers in parentheses denote normalized values with respect to the PI controller).

Controller	$\int (P_{g,e}(t))^2 dt$ $[10^{16}]$	$\int (\omega_{g,e}(t))^2 dt$ $[10^{6}]$	$\int \left(\dot{\beta}_2(t)\right)^2 dt$ $[10^{4}]$
PI	1.3809	5.7436	1.0195
	(1)	(1)	(1)
SMC	1.7812	8.4175	0.8687
	(1.29)	(1.47)	(0.852)
MPC	0.9736	2.6120	0.3444
	(0.705)	(0.455)	(0.338)

4.4. Simulations with Fault-Tolerant Control in Faulty Condtions

In this section, simulation results for fault-tolerant control in the pitch system fault are presented. The same fault scenario as described in the previous section is used to verify the effectiveness of the proposed fault-tolerant control strategy. In this section, two fault-tolerant strategies are evaluated, which are divided into FTC with the pitch system only and FTC with both the pitch system and the wind turbine system.

4.4.1. Fault-Tolerant Control with Pitch System Only

First, fault-tolerant control of the pitch system is considered. When the fault occurs, the change of parameter values due to the pitch system fault is estimated by using the adaptive fading Kalman filter. Then, MPC is re-synthesized based on the new set of pitch system parameters. The simulation results are given in Figures 8 and 9. To highlight the effectiveness, the simulation results with and without FTC are compared. Unfortunately, fault-tolerant control with the pitch system only does not work effectively. As shown in Figure 8, the power generation curves and the generator speed curves for the two controllers are similar. There is only slight improvements when FTC is applied. Moreover, Table 4 shows that the improvements are achieved at the expense of pitch actuation energy. The overuse of the faulty pitch system may cause other problems, such as the reduction in remaining useful lifetime (RUL). Therefore, it is concluded that fault-tolerant control with the pitch system only is not reasonable.

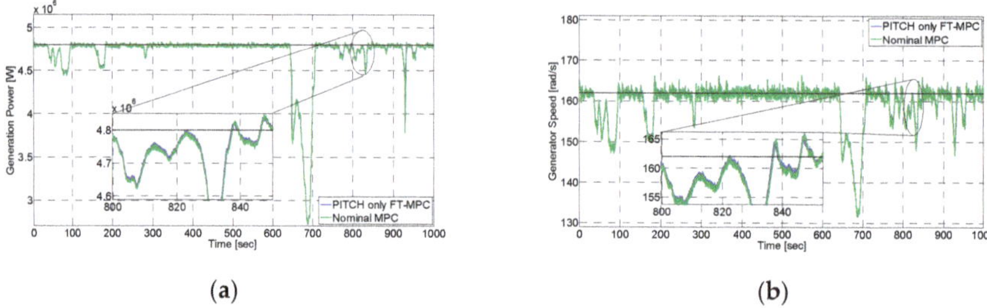

(a) **(b)**

Figure 8. Simulation results with fault-tolerant control with pitch system only: (**a**) power generation; (**b**) generator speed.

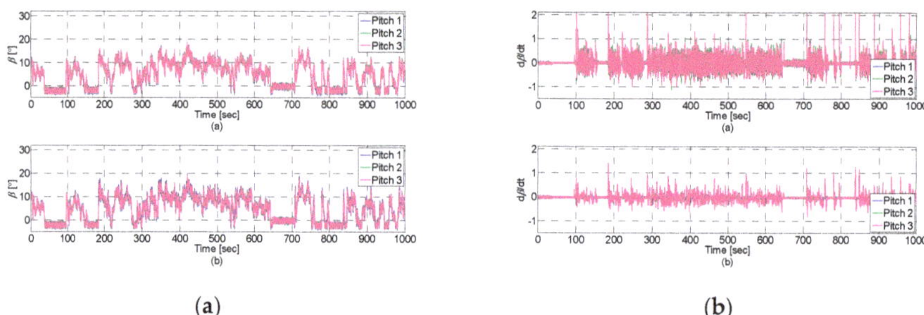

(a) **(b)**

Figure 9. Simulation results with fault-tolerant control with pitch system only: (**a**) pitch angle; (**b**) pitch angular rate (top: fault-tolerant with MPC (FT-MPC); bottom: nominal MPC).

Table 4. Controller performance with and without fault-tolerant control (the numbers in parentheses denote normalized values with respect to the MPC).

Controller	$\int (P_{g,e}(t))^2 dt\ [10^{15}]$	$\int (\omega_{g,e}(t))^2 dt\ [10^{6}]$	$\int (\dot{\beta}_2(t))^2 dt\ [10^{3}]$
Nominal MPC	9.7363 (1)	2.6120 (1)	3.4449 (1)
FTC with pitch system only	9.7085 (0.997)	2.5649 (0.982)	6.2457 (1.81)
FTC with pitch and wind turbine systems	9.6849 (0.995)	2.5407 (0.973)	1.6201 (0.470)

4.4.2. Fault-Tolerant Control with Pitch and Wind Turbine System

This section presents the simulation results of the proposed MPC with both the pitch system and the wind turbine system. The results are compared with and without reconfiguration. In Figures 10 and 11, the power generation curves and the generator speed curves for the two controllers are similar. However, notice the important difference in Figure 11. Figure 11a shows the individual blade pitch angle without and with fault-tolerant control, respectively. A significant difference between the two control methods is evident. In the nominal MPC method, the three pitch actuators are used equally to control the aerodynamic force, even though blades 2 and 3 have faults. This difference occurs because the nominal MPC controller does not have any information about the pitch system condition.

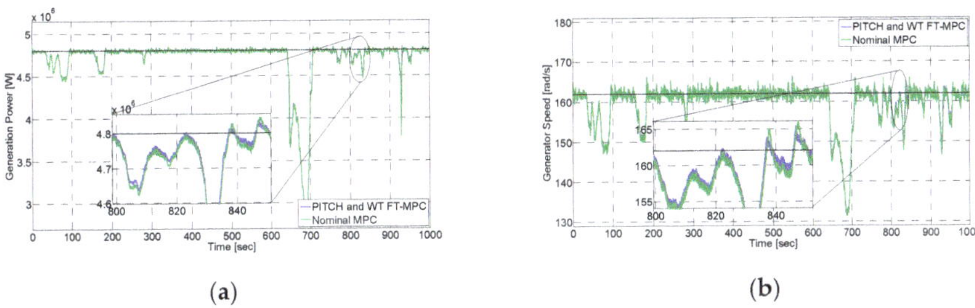

Figure 10. Simulation results with fault-tolerant control with pitch and wind turbine systems: (**a**) power generation; (**b**) generator speed.

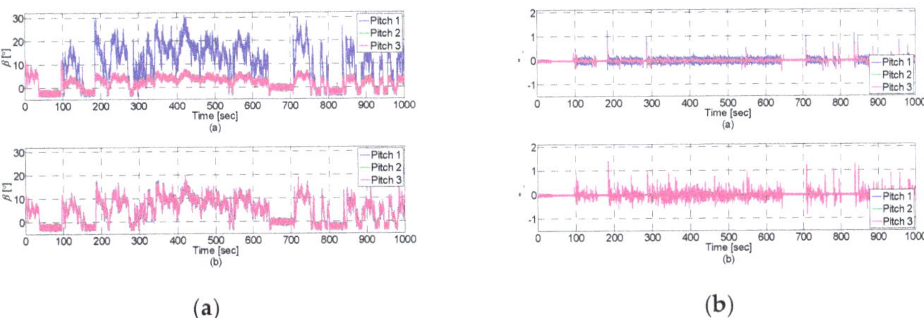

Figure 11. Simulation results with fault-tolerant control with pitch and wind turbine systems: (**a**) pitch angle; (**b**) pitch angular rate (top: FT-MPC; bottom: nominal MPC).

In contrast, FTC with both the pitch system and the wind turbine system increases the movement of healthy blade 1 to regulate the power generation, instead of blades 2 and 3, which are faulty. Therefore, this approach is more reasonable than FTC with the pitch system only. The results are summarized in Table 1. All performance indices show the benefit of FTC with the pitch system and the wind turbine system. The power generation and the generator speed regulation are slightly better than FTC with the pitch system only. Notably, the pitch actuation energy cost of blade 2 is reduced by more than 50% compared with the nominal MPC method. It is expected that saving the pitch actuation energy cost of the faulty pitch system will extend its RUL, which will lead to the lower maintenance costs.

4.5. Discussion

As shown in Tables 3 and 4, the proposed MPC shows better performance than existing methods in the case of a pitch system fault. Especially, the hierarchical (or two-stage) fault tolerant control strategy reduces the pitch actuation energy cost more than 50%. This will extend the remaining useful lifetime (RUL) of the pitch system, and thus reduce the maintenance costs. However, the large difference in pitch angles between the blades may cause a problem with asymmetric loads and fatigue. To tackle this problem, it is required to design a MPC optimization cost function covering the unbalanced load mitigation problem [24].

5. Conclusions

Wind turbines must satisfy a high degree of reliability to guarantee unrelenting power generation, while reducing the operational and maintenance costs. However, the harsh operating environments of wind turbines can cause failure of subsystems, including actuators and sensors. In this paper, a novel fault-tolerant control strategy for wind turbines has been proposed. The proposed strategy has a hierarchical structure, which consists of two fault tolerant controllers: one for the pitch system at the lower level, and another for the wind turbine system at the higher level. The proposed control strategy is based on the model predictive control (MPC) technique, thanks to its advantage in dealing with the model variations and the uncertainty caused by faults. A set of simulation results with a benchmark model demonstrated the performance of the proposed method in comparison with the existing PI and SMC controllers.

Author Contributions: Conceptualization, D.K.; methodology, D.K.; software, D.K.; writing—original draft preparation, D.K.; writing—review and editing, D.L.; supervision, D.L.; project administration, D.L.; funding acquisition, D.L.

Funding: This work was supported by the National Research Foundation of Korea (NRF) grant funded by the Korea government (MSIT) (No. NRF-2017R1C1B5076020 and NRF-2017R1A2B4003008).

Conflicts of Interest: The authors declare no conflict of interest.

References

1. WindEuripe. Available online: https://windeurope.org/ (accessed on 10 June 2019).
2. Kaldellis, J.K.; Zafirakis, D. The wind energy (r) evolution: A short review of a long history. *Renew. Energy* **2011**, *36*, 1887–1901. [CrossRef]
3. Njiri, J.G.; Söffker, D. State-of-the-art in wind turbine control: Trends and challenges. *Renew. Sustain. Energy Rev.* **2016**, *60*, 377–393. [CrossRef]
4. Burton, T.; Jenkins, N.; Sharpe, D.; Bossanyi, E. *Wind Energy Handbook*; John Wiley & Sons: Hoboken, NJ, USA, 2011.
5. Ribrant, J.; Bertling, L. Survey of failures in wind power systems with focus on Swedish wind power plants during 1997–2005. *IEEE Trans. Energy Convers.* **2007**, *22*, 1–8. [CrossRef]
6. Odgaard, P.F.; Stoustrup, J.; Kinnaert, M. Fault-tolerant control of wind turbines: A benchmark model. *IEEE Trans. Control Syst. Technol.* **2013**, *21*, 1168–1182. [CrossRef]
7. Sloth, C.; Esbensen, T.; Stoustrup, J. Robust and fault-tolerant linear parameter varying control of wind turbines. *Mechatronics* **2011**, *21*, 645–659. [CrossRef]
8. Lan, J.; Patton, R.; Zhu, X. Fault-tolerant wind turbine pitch control using adaptive sliding mode estimation. *Renew. Energy* **2018**, *116*, 219–231. [CrossRef]
9. Badihi, H.; Zhang, Y.; Hong, H. Wind turbine fault diagnosis and fault-tolerant torque load control against actuator faults. *IEEE Trans. Control Syst. Technol.* **2015**, *23*, 1351–1372. [CrossRef]
10. Badihi, H.; Zhang, Y.; Hong, H. Fuzzy Gain-Scheduled Active Fault-Tolerant Control of a Wind Turbine. *J. Frankl. Inst.* **2014**, *351*, 3677–3706. [CrossRef]
11. Simani, S.; Castaldi, P. Active Actuator Fault-Tolerant Control of a Wind Turbine Benchmark Model. *Int. J. Robust Nonlinear Control* **2014**, *24*, 1283–1303. [CrossRef]
12. Maciejowski, J. *Predictive Control with Constraints*; Prentice Hall: Upper Saddle River, NJ, USA, 2000.
13. Maciejowski, J. The Implicit Daisy-Chaining Property of Constrained Predictive Control. *Appl. Math. Comput. Sci.* **1998**, *8*, 101–117.
14. Yang, X.; Maciejowski, J. Fault-tolerant model predictive control of a wind turbine benchmark. In Proceedings of the 8th IFAC Symposium on Fault Detection, Supervision and Safety of Technical Processes (SAFEPROCESS), Mexico City, Mexico, 29–31 August 2012; pp. 337–342.
15. Benlahrache, M.A.; Othman, S.; Sheibat-Othman, N. Faults Tolerant Control of Wind Turbine Based On Laguerre Model Predictive Compensator. In Proceedings of the 2015 European Control Conference (ECC), Linz, Austria, 15–17 July 2015; pp. 3653–3658.
16. Shi, T.; Xiang, X.; Wang, L.; Zhang, Y.; Sun, D. Stochastic Model Predictive Fault Tolerant Control Based on Conditional Value at Risk for Wind Energy Conversion System. *Energies* **2018**, *11*, 193.
17. Odgaard, P.; Stoustrup, J.; Kinnaert, M. Fault tolerant control of wind turbines—A benchmark model. *IFAC Proc. Vol.* **2009**, *42*, 155–160. [CrossRef]
18. Esbensen, T.; Sloth, C. Fault Diagnosis and Fault-tolerant Control of Wind Turbines. Master's Thesis, Faculty of Engineering, Aalborg University, Aalborg, Danmark, 2009.
19. Bianchi, F.D.; Battista, H.; Mantz, R.J. *Wind Turbine Control Systems: Principles, Modelling and Gain Scheduling Design*; Springer: London, UK, 2007.

20. Kim, J.; Kiss, B.; Lee, D. An adaptive unscented Kalman filtering approach using selective scaling. In Proceedings of the IEEE International Conference on Systems, Man and Cybernetics (SMC), Budapest, Hungary, 9–12 October 2016; pp. 784–789.

21. Kim, K.; Lee, J.; Park, C. Adaptive Two-Stage Extended Kalman Filter for Fault-Tolerant INS-GPS Loosely Coupled System. *IEEE Trans. Aerosp. Electron. Syst.* **2009**, *45*, 125–137.

22. Gao, Z.; Cecati, C.; Ding, S. A Survey of Fault Diagnosis and Fault-Tolerant Techniques-Part I: Fault Diagnosis with Model-Based and Signal-Based Approaches. *IEEE Trans. Ind. Electron.* **2015**, *62*, 3757–3767. [CrossRef]

23. Lee, S.; Joo, Y.; Back, J.; Seo, J.; Choy, I. Sliding mode controller for torque and pitch control of PMSG wind power systems. *J. Power Electron.* **2011**, *11*, 342–349. [CrossRef]

24. Liu, Y.; Patton, R.; Lan, J. Fault-tolerant Individual Pitch Control using Adaptive Sliding Mode Observer. In Proceedings of the 10th IFAC Symposium on Fault Detection, Supervision and Safety of Technical Processes (SAFEPROCESS), Warsaw, Poland, 29–31 August 2018; pp. 1127–1132.

Article

Use of Markov Decision Processes in the Evaluation of Corrective Maintenance Scheduling Policies for Offshore Wind Farms

Helene Seyr * and Michael Muskulus

Department of Civil and Environmental Engineering, Norwegian University of Science and Technology NTNU, NO-7491 Trondheim, Norway
* Correspondence: helene.seyr@gmail.com; Tel.: +47-400-867-61

Received: 18 June 2019; Accepted: 2 August 2019; Published: 3 August 2019

Abstract: Optimization of the maintenance policies for offshore wind parks is an important step in lowering the costs of energy production from wind. The yield from wind energy production is expected to fall, which will increase the need to be cost efficient. In this article, the Markov decision process is presented and how it can be applied to evaluate different policies for corrective maintenance planning. In the case study, we show an alternative to the current state-of-the-art policy for corrective maintenance that will achieve a cost-reduction when energy production prices drop below the current levels. The presented method can be extended and applied to evaluate additional policies, with some examples provided.

Keywords: maintenance planning; maintenance strategy; maintenance; corrective maintenance; repair; offshore wind energy; maintenance scheduling; optimization; modeling

---

1. Introduction

Offshore wind energy is an established form of energy generation in Europe and is globally gaining interest in countries all around the world, especially in East Asia. However, in most places, electrical energy produced by offshore wind farms (OWFs) is still more expensive than other electricity generation methods. Many improvements have already been achieved for different factors influencing the cost of electricity generated from wind. The size of the wind turbine support structures has been optimized, by e.g., minimizing the use of expensive materials. Turbine efficiency has been improved by e.g., optimizing the shape and materials of turbine blades. The overall production of a wind farm can be improved by studying wake effects and optimizing the control of individual turbines. Within the offshore wind research community, the optimization of operation and maintenance has recently been gaining interest from researchers all around the world. One reason for this is the high share of operation and maintenance cost in the overall energy costs up to a third of the price of electricity produced is due to operations and maintenance [1]. Reducing these operation and maintenance costs will improve the total cost of energy production and help achieve cost competitiveness with other generation methods, such as onshore wind or solar energy. Different groups have developed simulators that model the operation of OWFs, as reviewed in [2]. With these simulators, the researchers are able to investigate different maintenance scheduling policies by comparing the simulation results of different policies. Existing models depend almost exclusively on Monte Carlo simulations, i.e., running a large number of simulations with the same inputs, in order to investigate uncertainties and variations in different inputs and variables (wave heights, wind speeds or failure occurrence) used. Additionally, different policies can only be evaluated manually, by implementing each strategy individually in the model. Some exceptions to the dependency on Monte Carlo simulations are some newer approaches, using stochastic models [3] or genetic algorithms [4]. The influence of uncertainties on the optimal scheduling of corrective maintenance has been investigated for the repair time [5], and the weather forecast [3,6].

In this paper, we present a method that can be applied to compare different maintenance scheduling policies for an OWF at a given location (with known weather) for a specific failure type with a known repair time. In contrast to most of the existing tools and models, no simulations are required and expected values for different performance indicators [7], like downtime or production losses can be compared for the different policies. With the presented method, including uncertainties is straightforward—it can be included directly into the model as opposed to running Monte Carlo simulations with different parameters, as has been done by most of the existing models. Uncertainty in the sea state is included in the presented case study. Section 2 explains the details of the method used and gives the details about the mathematical structure. Implementation of the method is explained in Section 3. Section 4 presents a case study applying the presented method. Discussions and an outlook on alternative policies that could be evaluated with this framework are given in Section 5.

2. Methodology

2.1. Markov Decision Process

The method we present in this paper is based on a Markov decision process (MDP). A MDP is a stochastic control process that can be seen as an extension of a Markov chain, adding actions and rewards [8]. The MDP can be described as a 5-tuple: $(\mathcal{S}, \mathcal{A}, \mathcal{P}, \mathcal{R}, \gamma)$, where $\mathcal{S}$ is a set of states, $\mathcal{A}$ a set of actions, $\mathcal{P}$ the transition probabilities between states, given actions, $\mathcal{R}$ a real-valued reward (or penalty) function that calculates the reward (or penalty) of any given state and γ a discount factor. As the name suggests, this process assumes the Markov property, therefore the effects of an action taken in a state only depend on that state and not the prior history of the process. An example of a Markov decision process is presented in Figure 1. In the present framework, the set of states $\mathcal{S}$ includes an finite number of states—in the example in Figure 1, six states are shown. (Infinite sets of states are possible in the framework of Markov decision processes. For more information about the mathematical concept, please refer to the literature, e.g., [8].) Each state can be described by one or multiple properties. These can be e.g., a location (distance from some fixed point), reward given in the respective state, or, in the case of offshore wind farm maintenance, the status of the turbine, a sea state observation, or the time needed to complete a repair. Each state differs from all other states in at least one characteristic, so no duplicates exist. The actions in the MDP can be either deterministic or stochastic. Deterministic actions lead to a (fixed) new state that the process will continue in after the current state. A stochastic action specifies a probability distribution over the next states. The transition probabilities between states depend on the action undertaken in that state and specify the new state, subject to that action. Therefore, for each state and possible action in that state, there is at least one positive transition probability to another state. For each state and action, the transition probabilities sum to one. A deterministic action is a special case of a stochastic action, with exactly one positive transition probability equal to one. The example in Figure 1 includes two stochastic actions and the associated transition probabilities. The reward function is a real valued function, assigning a value to each state and action combination. When a negative value is assigned by the reward function, it is often called a penalty function instead. In the example in Figure 1, each of the six states has one of two reward values, namely 1 and 0.

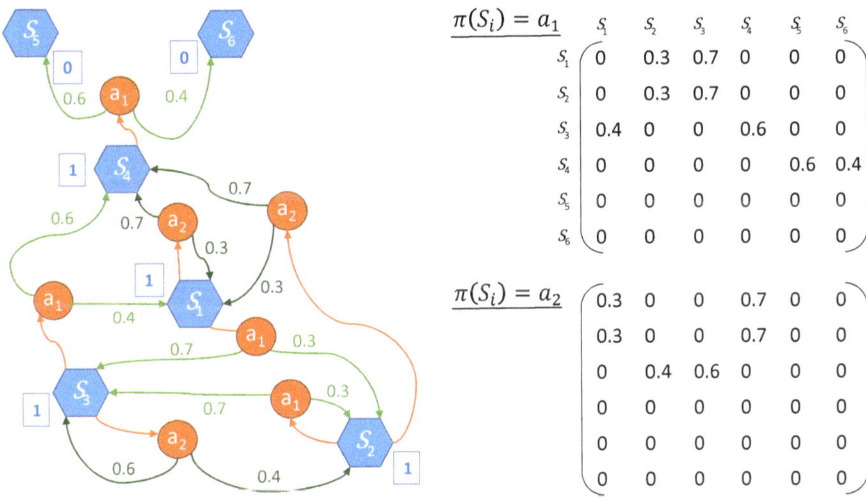

Figure 1. An example of a Markov decision process (**left**). The blue hexagons represent the states, with their rewards indicated in blue boxes next to the state. Orange circles indicate the two actions that can be taken in each state. Subject to the action, the transition probabilities are indicated with green arrows and the value displayed next to the arrow. Transition probabilities following action a_1 are shown in light green, while transition probabilities following action a_2 are shown in dark green. The policies "choose action a_1 in all states" (upper) and "choose action a_2 in all states" (lower) are presented by their transition matrices (**right**).

In addition to the Markov decision process that describe how the system works, our setup contains a set of policies Π. A policy $\pi \in \Pi$ is a mapping from $\mathcal{S}$ to $\mathcal{A}$, and can be understood as a decision makers rule for choosing one of the possible actions $a \in \mathcal{A}$ in each state. In order to follow a policy, one must (a) determine the current state s, (b) determine the action to be executed in that state $a = \pi(s)$, (c) determine the new state s' and continue, alternating (b) and (c). The goal of using a MDP is of course to find an optimal (or at least better than existing) maintenance strategy. In the framework of the MDP, this is done by finding an optimal policy π. In order to evaluate a policy (and ultimately finding the optimal policy), it is necessary to determine expectation of the total reward gained by following it (in order to optimize it). Intuitively, one could try to sum all rewards obtained in the MDP when following the policy, but this can quickly become overwhelming. (Typically, summing all rewards will yield an infinite sum, namely for all MDPs with either infinite state space or for MPDs with infinite horizon. For more information about these cases, refer to e.g., [8].) The solution is to use an objective function to map the sequence of rewards to (single, real) utility values. Options to obtain an objective function are (1) setting a finite horizon, (2) using discounting to favour earlier rewards over later rewards and (3) averaging the reward rate in the limit.

Instead of optimizing the policy, in some cases, it might be desirable to compare different policies with each other. When combining an MDP with a fixed policy that chooses exactly one action for each state, the result is a Markov chain. This is because all of the actions are defined by the policy and one is left with the transition probabilities between states. One example of a resulting Markov chain is visualized in Figure 2. In this Markov chain, the value of each state S_i can be calculated based on the reward $\mathcal{R}(S_i)$ of that state and based on the values of the states that can be reached. It is calculated as

$$V(S_i) = \mathcal{R}(S_i) + \sum_j P_{ij} V(S_j), \tag{1}$$

where P_{ij} is the transition probability between state S_i and state S_j from $\mathcal{P}$. The equations in (1) are known as Bellman equations, named after Richard Bellman. We can solve the linear equation system (LES) defined by the transition probabilities and reward function to find the values $V(S_i)$ for each state. When comparing two policies, one can look up the value of a specific state one is interested in, usually a 'starting' point. In the case of OWF maintenance, this could e.g., be a state in which a failure occurs and the value could then be representative of the time it takes for this failure to be corrected, with a penalty incurred for each step taken without resolving the failure. A case study comparing different policies is presented in Section 4.

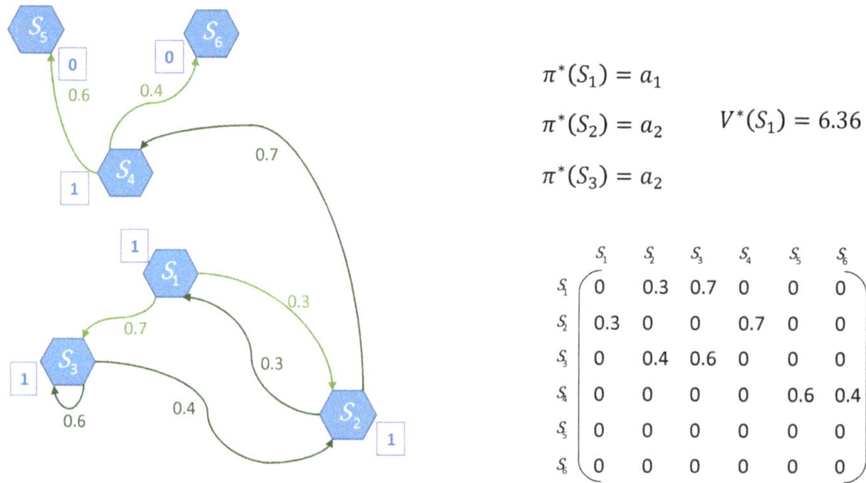

Figure 2. An example of a Markov chain (MC), as the result of selecting one policy in the setup of Figure 1 (**left**). The policy displayed on the right-hand side is the optimal policy for this MDP, when starting in S_1. The transition probabilities for the MC are shown in the matrix (**right**).

In the example shown in Figure 1, a possible policy would be to always choose action a_1. The corresponding Markov chain is presented on the right-hand side of the figure in the form of its transition probabilities. The rewards (presented in the figure in blue next to the states) are $\mathcal{R}(S_1) = \mathcal{R}(S_2) = \mathcal{R}(S_3) = \mathcal{R}(S_4) = 1$, and $\mathcal{R}(S_5) = \mathcal{R}(S_6) = 0$. In order to calculate the value of each of the states, we solve the equation system defined by the Bellman Equation (1):

$$
\begin{pmatrix} V(S_1) \\ V(S_2) \\ V(S_3) \\ V(S_4) \\ V(S_5) \\ V(S_6) \end{pmatrix}
=
\begin{pmatrix} 1 \\ 1 \\ 1 \\ 1 \\ 0 \\ 0 \end{pmatrix}
+
\begin{pmatrix}
0 & 0.3 & 0.7 & 0 & 0 & 0 \\
0 & 0.3 & 0.7 & 0 & 0 & 0 \\
0.4 & 0 & 0 & 0.6 & 0 & 0 \\
0 & 0 & 0 & 0 & 0.6 & 0.4 \\
0 & 0 & 0 & 0 & 0 & 0 \\
0 & 0 & 0 & 0 & 0 & 0
\end{pmatrix}
\begin{pmatrix} V(S_1) \\ V(S_2) \\ V(S_3) \\ V(S_4) \\ V(S_5) \\ V(S_6) \end{pmatrix},
$$

$$
\begin{pmatrix}
-1 & 0.3 & 0.7 & 0 & 0 & 0 \\
0 & -0.7 & 0.7 & 0 & 0 & 0 \\
0.4 & 0 & -1 & 0.6 & 0 & 0 \\
0 & 0 & 0 & -1 & 0.6 & 0.4 \\
0 & 0 & 0 & 0 & 0 & 0 \\
0 & 0 & 0 & 0 & 0 & 0
\end{pmatrix}
\begin{pmatrix} V(S_1) \\ V(S_2) \\ V(S_3) \\ V(S_4) \\ V(S_5) \\ V(S_6) \end{pmatrix}
=
\begin{pmatrix} -1 \\ -1 \\ -1 \\ -1 \\ 0 \\ 0 \end{pmatrix}.
$$

The values of the states are then

$$
\begin{pmatrix} V(S_1) \\ V(S_2) \\ V(S_3) \\ V(S_4) \\ V(S_5) \\ V(S_6) \end{pmatrix} = \begin{pmatrix} 5.05 \\ 5.05 \\ 3.62 \\ 1 \\ 0 \\ 0 \end{pmatrix}.
$$

If one is interested in comparing the value of a specific state under two different policies, the calculation is repeated for that policy and the values compared. It is also possible to find the *optimal* policy without comparing the values based on the resulting Markov chains. In order to find the optimal policy, we define the optimal value function V^* by the recursive set of equations

$$
V^*(S_i) = \mathcal{R}(S_i) + max_{a \in \mathcal{A}} \left[\sum_j P(S_j|S_i, a) V^*(S_j) \right], \tag{2}
$$

so the optimal value of a state S_i is the reward in the state, plus the maximum over all actions we could take in the state. This is a the generalized form of the Bellman Equation (1) for policies. In the example shown in Figure 1, the possible actions are a_1 and a_2. The idea behind this maximum is that in every state we aim to choose the action that maximizes the value of the future. The optimal value function V^* can be found by e.g., value iteration. When V^* is known, the optimal policy π^* can be found by picking the action that maximizes the expected optimal value:

$$
\pi^*(S_i) = argmax_{a \in \mathcal{A}} \left[\sum_j P(S_j|S_i, a) V^*(S_j) \right]. \tag{3}
$$

In the example shown in Figure 1, the optimal policy is to conduct action a_1 in states S_1 and S_4, and a_2 in states S_2 and S_3. Figure 2 shows the Markov chain corresponding to the optimal policy.

2.2. Probabilistic Weather Input

In the modeling of offshore wind farm maintenance, the uncertainty in the local weather, and more specifically the wave height, is often included. In the given framework, the local weather can be included into the model by adding a sea state (wave height) property in the definition of the states. Some of the existing maintenance approaches [2] use Markov chains to model the weather. Similar to these approaches, transition probabilities between different wave height bins can be used in the MDP. Given some source of historical weather data, the wave height data are sorted into so-called "bins"-categories summarizing wave heights in a given interval. The size of these bins should be adjusted based on the application, for offshore wind farm maintenance 0.4 m is a useful interval step [5]. Then, the probabilities to transition from one bin to all other bins are calculated based on the number of occurrences of transitions between these bins in the given data source. In order to be able to investigate seasonality, one can calculate separate matrices for e.g., each month of the year.

2.3. Repair Time Modelling

Another factor that influences the decision-making in offshore wind farm maintenance is the time it takes to bring a wind turbine component that has failed back to an operational state. Throughout this article, we will use the term "repair time". This repair time is the cumulative amount of time spent during maintenance actions, and we assume a fixed repair time, without uncertainty. The repair time should not be confused with the time between a failure and its resolution, which we will refer to as "downtime". The repair time can be included into the MDP as a parameter to the states. During maintenance, the MDP will move from states with a high remaining repair time, to states with a lower remaining repair time, until a state with no remaining repair time is reached, where the process will stop.

2.4. Calculation of Production Loss

We want to use the MDP to evaluate different policies for offshore wind farm maintenance. One aspect to compare is the production loss of a turbine or wind farm under a given policy. The production loss can only be estimated, as explained in [7], as we cannot measure the absence of production. Therefore, a method to estimate the production loss is needed. Given information about the wind speed from e.g., measurements and a knowledge about how the power production dependents on the wind speed, it is straightforward to find an estimate of the production losses. One could include the wind speed as a parameter into the states of the MDP as was done with the wave height. As we do not need to use the wind speed as a decision criterion for the policies, we use a matrix with conditional probabilities of wind speed values given a wave height, similar to how it has been presented in [5]. Given some source of weather data, the wind speeds are first sorted into bins—a bin size of 1 m/s is sufficient for production loss calculations. For each state in the MDP with a given wave height parameter, the conditional probability for each wind bin is populated in a matrix that can later be used to look up these values. The expected production loss for each state in the MDP can be calculated based on these relative probabilities and a power curve for the turbine type of interest. A power curve can be either obtained directly from the manufacturer or a linearized power curve can be used, based on the turbine model. If one does not have information about an actual turbine model, reference turbines like [9] or [10] can be used. In order to obtain the (expected) production loss, the production values for each discrete wind speed bin, as obtained from the power curve, are weighted (multiplied) with the (conditional) probability from the matrix. The sum of these weighted production values is then the production loss for the state. For a state S_i, given n discrete wind speed steps with conditional probabilities $P(u_k|S_i)$ for $k = 1 \ldots n$, and a power curve $p(\cdot)$ the expectation of the production loss $L(S_i)$ is

$$E(L_{S_i}) = \sum_{k=1}^{n} p(u_k)P(u_k|S_i). \tag{4}$$

3. Implementation

In order to use the presented method to evaluate different maintenance policies, it is necessary to implement it in a programming language. The resulting program can then be used to evaluate well-known policies and compare them to alternative options. In our analysis, the implementation was conducted in Python 3.

In order to define the MDP, we define the states, actions, policies and the reward function. The set of states S can be generated, by defining the composition of a state and then generating a list of possible states. A state could e.g., be a tuple of several parameters $S_i = (p_1(i), p_2(i), p_3(i))$, where each of the parameters can take different values (e.g., $p_1(i) \in \{0, 0.4, 0.8, 1.2, \ldots, 10.0, 10.4\}$ a wave height, $p_2(i) \in \{5, 4, 3, 2, 1, 0\}$ the number of remaining repair hours, and $p_3(i) \in \{\text{'at shore', 'offshore'}\}$ the vessel location). The different actions $a \in \mathcal{A}$, will have different outcomes depending on the state. Possible actions for an implementation for the offshore wind maintenance planning could be "go out to the wind farm", "repair the turbine", and "return to shore". As described above, each policy $\pi \in \Pi$ is a set of rules, defining which action should be taken in which state. It can be implemented as a set of conditional expressions ensuring that only transitions between states which correspond to the actions defined by the policy are possible. When investigating multiple policies with similar rules, a high level policy can be implemented first, and the characteristic parameter changed for each individual policy in the evaluation. The reward function is a function assigning a real value to each state. It is also possible for the reward value to be dependent on the action taken to reach the state. Implementation of this reward function highly depends on the structure of the states; in most cases, it will be a function depending on one or more parameters of the state.

In order to calculate the value of a policy, the first step is to define the equation system resulting from plugging the policy into the MDP, thereby forming a Markov chain. The LES has been observed to follow some rules and the matrix defining it can be produced following these steps:

1. Find the number of states n, and find a mapping of the states, assigning each of them a natural number, effectively applying an order to the states.
2. Create the $(n \times n)$ matrix containing the transition probabilities for the investigated policy.
3. Calculate the entries of the reward vector, where the i-th entry corresponds to $\mathcal{R}(S_i)$.
4. The matrix $\mathcal{P}$ and vector $\mathcal{R}$ define an equation system

$$(\mathcal{P} - I_n)V = -\mathcal{R},$$

   which can be solved using a linear algebra routine in e.g., Matlab or Python. Depending on the structure of the matrix and vector, different algorithms might be used to achieve fast computation.
5. In order to investigate different properties of a policy, the same matrix is used in a LES combined with different reward functions for each property.

4. Case Study

4.1. MDP Definition

In this case study, the states $S \in \mathcal{S}$ of the MDP are tuples of the form $S =$ (location, wave height, repair time left, steps waited), where 'location' can take on either of the values 'port' or 'turbine'. The significant wave height ('wave height') takes values in steps of 0.4 m between 0 m and 10.4 m. The 'repair time' starts off with an initial value, specific to the turbine component that is investigated. The values for repair time are taken from [11], the most recent source for offshore wind turbine failure and repair data. Different components and types of repairs have been investigated in this case study, each with a distinct mean time to repair and worker requirement. For the example of the major blade repair, with 21 h mean time to repair, the values for the 'repair time' range from 0 to 21 h in steps of 1 h. For other components and repair times, the values have a different range. The steps are, however, set to 1 h, for all repair types and components investigated. This results in a different number of states for different types of repair. The 'steps waited' also take steps of 1, starting at 0 and ranging up to 3 depending on the maintenance policy. A summary of the parameters for the states is shown in Table 1. The set of actions $\mathcal{A} = \{$stay, wait, reset wait time, go out, repair, return$\}$, where the actions 'wait' and 'reset wait time' are only used in some of the policies. How the actions are used in the different policies is detailed below in Section 4.4, a summary of the possible actions is provided in Table 2. The transition probabilities between states $\mathcal{P}$ depend on the transition probabilities of the significant wave height values. These probabilities are calculated based on the weather data from FINO 1 [12]. More details on how the probabilities are calculated are given in Section 4.2. The reward function $\mathcal{R}$ is used to evaluate different aspects of the maintenance policies. To evaluate the influence of the policy-change on the expected downtime of the turbine, a penalty is used for the steps it takes to end up in a repaired state. To calculate the expected production losses, the reward function $\mathcal{R}$ represents a penalty of the production losses. These are calculated based on the correlation of wind speeds and wave height and a linearized power curve for the NREL 5 MW turbine [9]. The details of this calculation are presented below in Section 4.3. Discounting is not used in this case study and hence the discount factor set to $\gamma = 1$. To evaluate a maintenance policy, we investigate the value of the initial states. These states are those in which the failure occurred and hence the repair has not started. As we assume cumulative repairability (i.e., when a repair has to be interrupted, progress is kept and the repair can be continued at a later stage), these are all states with the initial repair time values. Since the failure can occur at any wave height, multiple states with this repair value exist. These are weighted with their probability of occurrence and the values summed before reporting.

Table 1. The different parameters of the states and their possible values.

Parameter	Possible Values	Comment
Location	$\{port, turbine\}$	
Wave height [m]	$\{0, 0.4, 0.8, \ldots, 10, 10.4\}$	
Repair time [h]	$\{0, 1, 2, \ldots, max - 1, max\}$	maximum (*max*) depends on component
Steps waited	$\{0, 1, 2, 3\}$	depends on strategy

Table 2. The different actions and how they influence the parameters of the next state.

Action	Parameters for Next State
stay	
wait	steps waited +1
reset wait time	steps waited = 0
go out	location = turbine
repair	repair time −1
return	location = port

4.2. Weather Input

In this case study, the weather data used to calculate the transition probabilities between wave heights (and subsequently states) comes from the FINO 1 measurement campaign. The data from the FINO 1 measurements have some missing observations. Additionally, the wind speeds are provided in 10 min aggregated means while wave height measurements are provided for 30 min intervals, which is not convenient for the calculation of production losses (Section 4.3). The transition probabilities have therefore been calculated based on the interpolated time series also used in [13,14]. In order to calculate the transition probabilities, the significant wave height is categorized in steps of 0.4 m first. This means that all wave height observations between 0 m and 0.4 m will be collected in one so-called bin. The same is done for the values between 0.4 m and 0.8 m, and so on. We have chosen to calculate separate matrices with the transition probabilities for each month, by sorting the data beforehand. This has the advantage that we can investigate and observe how the season affects the optimal policy.

4.3. Calculation of Production Loss

As described above in Section 2.4, the expected production loss for each state is calculated based on probabilities of the wind speed given the sea state, and a power curve. The probabilities are based on data from the FINO 1 measurement campaign. The same 1 h-interpolated FINO 1 data [13] that was used to calculate the wave height transition probabilities was used. For each observation point, a wave height value and a wind speed value are known. The wave height and wind speed are then categorized. For the wind speed, the step size is 1 m/s, so each observation for wind speeds between 0 m/s and 1 m/s will be collected together. The same is done for wind speeds between 1 m/s and 2 m/s and so on. The wave heights are categorized as described in Section 4.2. Then, the conditional probabilities of these wind speeds subject to the wave height at the same point in time are gathered. For the calculation of the production loss, information about the power curve is needed in addition to the weather. In our case study, a linearized power curve is used for the NREL 5 MW turbine [9], as was also done in [5]. The linearized power curve is based on the cut-in and cut-off wind speed as well as the wind speed where the rated power (5 MW) is reached. When solving the MDP, the production loss values are used to calculate the reward of each state. The weighted values of the initial states are then summed and reported, as described above in Section 4.1. The loss of production is calculated in terms of electric power (kWh). If one is interested to compare this directly to the cost of maintenance, the energy needs to be valued in terms of money. This can be done by either using a (variable) electricity market price or a (fixed) feed-in-tariff.

4.4. Policies

This section presents the different maintenance policies that are investigated and compared in the case study. As described in Section 2, a single policy assigns an action $a \in \mathcal{A}$ to each state $S \in \mathcal{S}$. A summary of all policies, with different parameters is shown in Table 3. For each policy, the possible actions under this policy are listed.

Table 3. Names of the different policies investigated in this article, as well as the maximum number of steps that can be waited under this policy and the possible actions.

Policiy Name	Max (Steps Waited)	Possible Actions
go-right-away	0	stay, go out, repair, return
wait-1-step	1	stay, wait, reset wait time, go out, repair, return
wait-2-steps	2	stay, wait, reset wait time, go out, repair, return
wait-3-steps	3	stay, wait, reset wait time, go out, repair, return
0.8 m-limit	0	stay, go out, repair, return
1.2 m-limit	0	stay, go out, repair, return
2.0 m-limit	0	stay, go out, repair, return
2.4 m-limit	0	stay, go out, repair, return
2.8 m-limit	0	stay, go out, repair, return

4.4.1. Go-Right-Away

In order to be able to conduct maintenance, a vessel has to be at the turbine and the wave height needs to be below a defined threshold of 1.6 m. This is a value, based on the often presented wave height limit of 1.5 m for vessel access [15], modified to fit the wave height resolution of the case study. In this strategy, as soon as the wave height is below the threshold of 1.6 m, the vessel is sent to the wind turbine. We assume a travel time of one step (1 h) in this case study, which might be short compared to some wind farms. However, since we are using weather data from FINO 1, which is next to the Alpha Ventus wind farm in the North Sea, we are already assuming a wind farm relatively close to shore which will have a shorter travel time. Once the vessel reaches the turbine, repair is conducted if the wave height is still below the threshold. As soon as the wave height crosses the threshold, the repair is interrupted and the vessel returns to port. We assume that the repair is cumulative, i.e., when the repair is interrupted, it can be continued at a later stage without any loss of progress. The return to port takes one step (1 h) again. As soon as the wave height crosses below the limit again, another access is made until the turbine is repaired. We do not take into account any restrictions to the working time of the maintenance crew or vessel crew, so it is possible to have one access and conduct the full repair without ever returning to port. This is a simplification that could be justified, if the boat has living quarters and enough personnel on board to rotate in shifts. Figure 3 shows a decision diagram for this policy. In every state of the Markov decision process, the diagram can be used to find the action that the policy prescribes for that state. In Figure 4, a minimal MDP is shown for this policy. Here, two steps of repair are required and two wave heights are considered, namely below and above the limit. The probability to stay below the limit is denoted as $P(-,-)$, the probability to change wave height from below the limit to above the limit is denoted as $P(-,+)$ and so on. Assuming the state is ('port', 'above limit', '2'), the first check is whether the repair time is greater than zero, which it is. The next check is whether the vessel is at the turbine, which it is not. Thus, the next inquiry is whether the wave height is below the limit, which it is not. The action is then 'stay'. The state will be the same in the next step with a probability of $P(+,+)$ and will change to ('port', 'below limit', '2') with a probability of $P(+,-)$. In this state, the action will be 'go out'.

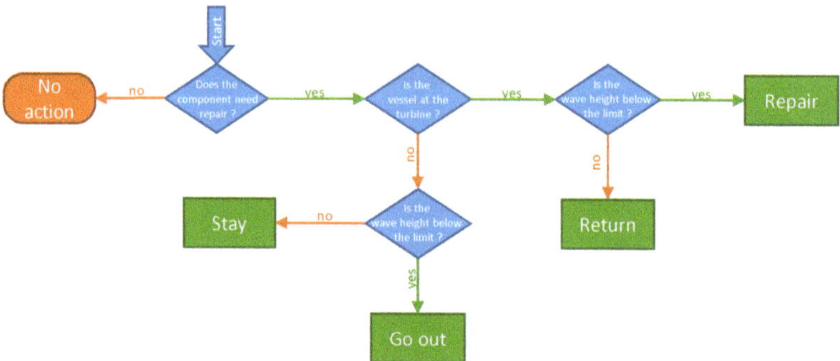

Figure 3. The decision tree for the original (go-right-away) policy. This assessment is conducted for each state and influences the transition probabilities in the MDP, by choosing an action for each state. An example of how the policy is applied can be seen in Figure 4.

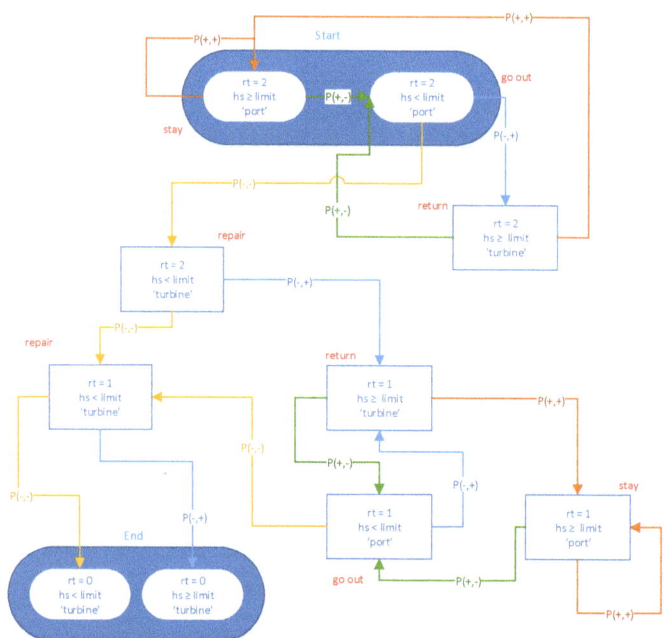

Figure 4. A minimal example of the process for the original (go-right-away) policy. Here, only two steps of repair are shown. The process starts in states with repair time (rt) equal to 2 h (rt = 2), which is then reduced to 1 h (rt = 1) and finally 0 h (rt = 0). The wave height (hs) is categorized as being below (<) or above (≥) the threshold (limit), and transition probabilities are adjusted to accommodate this simplification. P(+,−) is the probability to get from a wave height above threshold (hs ≥ limit) to a wave height below threshold (hs < limit). With 'start', we mark the states in which a maintenance decision maker would start the decision of when to repair, i.e., the point in time when the failure occurs/is reported. The decision taken in each state is marked in red, next to the respective state. How the decision is made, based on the state and maintenance policy can be understood from Figure 3.

4.4.2. Wait-n-Steps

An alternative to accessing the wind farm as soon as the wave height is below the threshold is to wait a certain number of steps in good weather, before going out with the vessel to conduct maintenance. The intuition behind this policy is that, if the sea has been calm for several time-steps, it is more likely to stay calm (i.e., below the wave height limit) due to persistence. Waiting a certain amount of time in good weather assures that the observation below the limit was not just an outlier and one can avoid interrupting the maintenance operations. In the investigated policies, the number of waiting steps is fixed and independent of the observed wave height in the state. In our case study, we investigated wait-times of one step, two steps and three steps. Each step represents 1 h. The other aspects of the strategy remain as before. Again, the repair is assumed to be cumulative, so, if the repair is interrupted, progress is kept and it can be continued and completed at a later stage. The maintenance is aborted and the vessel returns to shore as soon as the wave height is above the threshold. The time it takes to access the turbine and return to port respectively is one step (1 h). The decision diagram for this policy is shown in Figure 5.

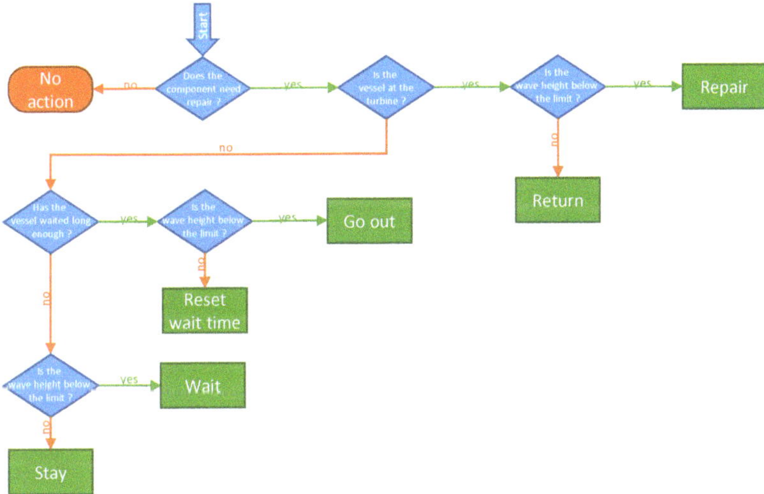

Figure 5. The decision tree for the wait-n-steps policy. First, the decision maker checks, whether a repair is necessary (repair time > 0). Depending on the location (at turbine), a wait-time check is conducted. This depends on the number of wait steps specified by the policy (1 h, 2 h, 3 h). Finally, the weather is checked and the correct action chosen for this state. This assessment is conducted for each state and influences the transition probabilities in the MDP, by choosing an action for each state.

4.4.3. Different-Limits

The third type of policy that is being investigated in this article has a second wave height threshold. One limit (new) is used for the decision of going out to the wind turbine and the other (original) threshold of 1.6 m is used for the decision to start and continue the repair. It is also used for triggering a possible return of the vessel to the harbour. We investigate both lower (stricter) and higher (laxer) wave height limits for access (new limits), specifically we investigate the limits 0.8 m, 1.2 m, 2 m, 2.4 m, and 2.8 m. The repair is again assumed to take a fixed amount of time and can be completed by accumulating enough maintenance (repair) actions. Again, as soon as the wave height is above the (original) wave height threshold, the repair is aborted and the vessel returns to shore. The decision tree for this policy is identical to the one of the go-right-away policy shown in Figure 3, only that the weather check uses a different threshold in a port and at the turbine. The go-right-away policy is a special case of this strategy, where both limits are 1.6 m.

4.5. Repair Data Input

For the repair time values, data from [11] are used. They present the mean time to repair [h], number of workers needed and mean annual failure rates for 19 wind turbine components. For each component, three types of failures are distinguished, namely 'major replacement', 'major repair' and 'minor repair'. Each of these have their own values, leading to a total of 57 different combinations of component and repair type, with specific repair time and worker requirements. We have investigated some selected turbine components and failure types, namely major gearbox replacement, major blade repair, and minor electrical repair. The repair time value is used to generate the possible states for the MDP, whereas the worker requirement is used for cost calculations. The values that have been used are summarized in Table 4.

Table 4. Turbine components that are investigated in the case study and their repair parameters.

Component	Cumulative Repair Time [h]	Average Number of Workers Needed
Gearbox, major replacement	231	17.2
Blade, major repair	21	3.3
Electrical, minor repair	5	2.2

4.6. Cost Data Input

The costs for vessel and workers are dependent on the number of accesses, the total operation time (travel time and working time combined), the vessel charter costs, the vessel hourly costs, the number of workers needed for the repair, and the worker hourly wages. In the case study, these costs are all set according to values from the literature, provided in Table 5. In order to value the production losses in terms of money, the market price for electricity or feed-in-tariffs can be used. Since the price of electricity varies a lot, both between seasons, time of the day and countries, not a single "correct" electricity price can be used to analyze the production losses. In order to show the variation in electricity prices and their influence on the optimal maintenance policy, we include an analysis of the corrective maintenance cost in the case study. In order to gain some insight into the electricity prices in Europe, we used [16,17].

Table 5. Input used to calculate the maintenance cost.

Hourly vessel	287.5 €	calculated from daily cost from [13]
Hourly worker	55.3 €	calculated from annual worker salary
Mobilisation vessel	1000 €	arbitrary (similar to number from [18])

4.7. Results

In this section, some aspects under which the different maintenance policies have been compared are presented. Some of these aspects are similar to the key performance indicators presented in [7].

4.7.1. Repair Actions

The number of repair actions with each policy can be used as a control in order to detect possible mistakes in the implementation of the strategy. All policies include cumulative repair, no degeneration and the work is not continued after the repair is completed. Therefore, the expected repair time calculated and returned by each maintenance policy should be equal to the repair time needed to bring the investigated component back to a state as-good-as-new. Due to memory and rounding errors, this differs insignificantly between policies, in the magnitude of 10^{-10} h in our study.

4.7.2. Downtime

The expected downtime of a maintenance action or repair can be used to evaluate a maintenance policy. The downtime of a turbine is defined as the time the turbine is in a non-operational state caused by either a fault, or by a maintenance action. With an increase in downtime, the time-based availability of the turbine is reduced, often leading to lost production and a lower energy-based availability [7]. The downtime will incur production losses and the decision maker is therefore most likely interested to reduce it. In order to calculate the expected downtime for a policy, the reward function of the MDP is modified such that every step the process takes (i.e., every transition from one state to the next) gets a penalty of 1, representing the time that is lost in this step. When the MDP is then solved, thus the value of each state is calculated, and the average of values of the starting states weighted by the probability of occurrence gives the expected downtime until the cause of downtime (in this case a failure) is resolved. The starting states are those states with a repair time equal to the expected repair time and can be understood as the time of occurrence of the failure. The turbine downtime is, unsurprisingly, higher for the more restrictive policies. For the 'wait-n-steps'-policies, downtime is always higher than for the original 'go-right-away' strategy. For the 'different-limits' policies, those with a less restrictive limit are observed to have a slightly lower downtime than the original strategy. Due to the threshold for access being less restrictive, the vessel is more often at the turbine location. It can be avoided to "waste" one time step of calm weather for the access. This increases the likelihood of the vessel being already at the turbine location when the weather is calm enough to conduct a repair and therefore a faster resolution of the failure. For policies with a stricter limit than 1.6 m, the downtime increases, depending on the repair time and month, to up to three times the downtime of the original policy. Figure 6 shows the downtime for each policy and each month for the major gearbox replacement with a repair time of 231 h.

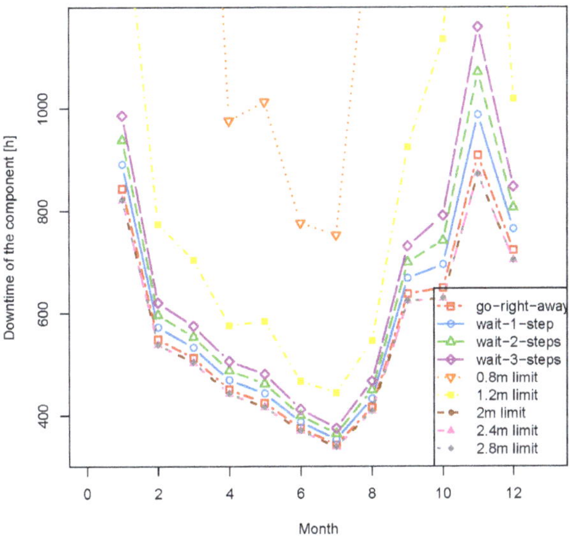

Figure 6. Downtime of the wind turbine due to a major gearbox replacement for different policies. Policies with a less restrictive wave height threshold for the vessel access have a lower downtime than more restrictive policies.

4.7.3. Production Losses

The second aspect that is used to evaluate a maintenance policy is the production lost due to the downtime of the turbine. As explained in Section 2.4, we calculate the production loss based on the wave height in each state. In the MDP, the expected production loss for each policy can be calculated, by using the lost production as 'reward' in the process. Then, the value of the starting states represents the production loss that can be expected by using the evaluated policy. Results for the production loss are shown in Figure 7, for a minor repair of the electrical system. It can be observed that the policies with a laxer wave hold threshold for vessel access have a slightly lower production loss than the original policy. The more restrictive policies on the other hand lead to an increase in lost production, up to more than three times the values of the original policy. For the calculation of the losses in terms of monetary value, different electricity prices have been used in this case study, based on data from Eurostat [16] for various countries. These results are shown combined with other maintenance costs below in Section 4.7.5.

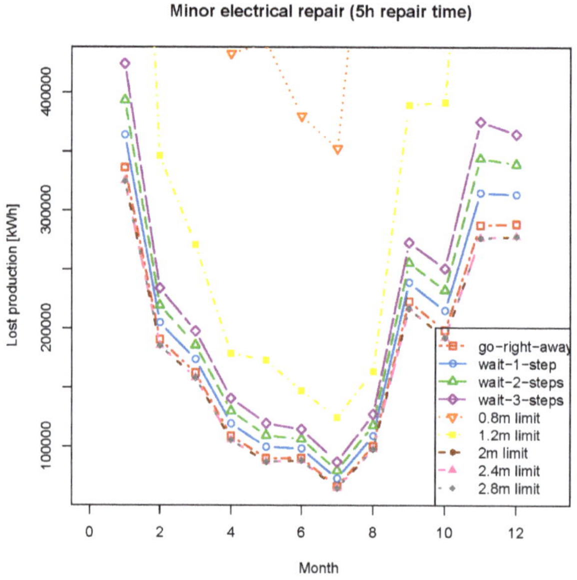

Figure 7. Lost production in kWh for a minor repair of the electrical system, with a repair time of 5 h. The policies with a higher (less restrictive) wave height threshold for vessel access show slightly lower losses in production than the original 'go-right-away'-policy.

4.7.4. Number of Vessel Accesses and Returns

Another aspect that can be used to compare different maintenance policies is the number of vessel accesses. This number is of interest, since usually each vessel mobilization induces a fixed cost for the maintenance provider or wind farm operator. Hence, the decision maker is interested in keeping the total number of vessel mobilizations low, while still trying to conduct a repair as fast as possible. The number of vessel accesses for each policy can be monitored, again by modifying the reward function. The reward is set to 1 for each state in which the selected action is 'go out'. Each time a vessel is sent from the port to the turbine, the reward will increase by one and after the process has finished, the expected number of vessel accesses can be calculated in the same way as the number of repair actions or downtime. As the MDP is stopped as soon as the repair is complete, the number of vessel returns will always be one less than the number of accesses, and can be calculated by following the same logic as for the vessel accesses, switching the action from 'go-out' to 'return'. The number of accesses needed before a completed repair implies vessel and worker costs. Figure 8 shows that the policies with a wave height threshold of 1.6 m ('go-right-away' and 'wait-n-steps') perform very similar in terms of number of vessel mobilizations. The policies with a more restrictive wave height threshold (0.8 m and 1.2 m) show fewer vessel mobilizations. The policies with a higher threshold for waves (2 m, 2.4 m, 2.8 m) show very high numbers of vessel mobilizations, up to 10-times the values of the 'go-right-away'-policy. This is likely caused by the wave height limit for repairs, which remains at 1.6 m also for those policies. When the vessel goes out in harsher weather than is allowed during repairs, and this weather persists for longer than the travel time, the vessel has to return to port right away and no repair can be conducted.

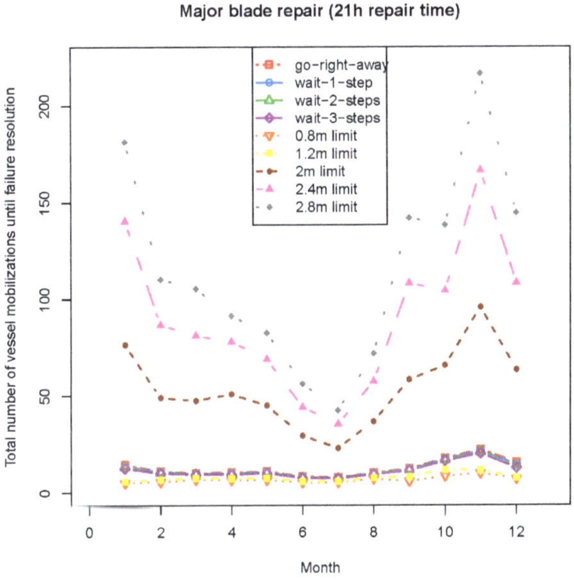

Major blade repair (21h repair time)

Figure 8. Total number of vessel accesses until the major blade repair is completed—for different policies.

4.7.5. Total Cost of Maintenance

The results for the total cost calculation are naturally the most complex, as they combine the cost calculations with the production losses. In the given framework with cumulative repair and no penalty for an unsuccessful repair attempt, one expects that the 'go-right-away' strategy will be the cheapest option, as this strategy leads to the fastest resolution of the failure. Our case study confirms this under the current electricity prices and assumed worker and vessel costs. Figure 9 shows the example of a major blade repair, and the total cost of maintenance for different policies. Should, however, the electricity price drop, and reach levels below 2.4 Euro-cent, the 'wait-1' strategy surpasses the original (go-right-away) strategy, as the avoidance of unnecessary vessel mobilizations will outbalance the losses due to turbine downtime. This can be seen from Table 6, for the major gearbox replacement for the month of June. As the production losses highly depend on the repair time and weather, no universal "cut-off" point between policies exists, but has to be investigated on an individual basis. Should the current trends of dropping yield for the electricity producer continue, we expect to see novel policies surpassing the cost-performance of the current state-of-the-art policy.

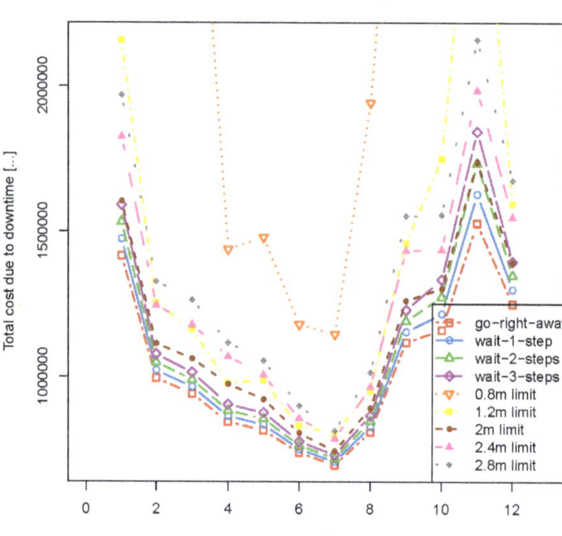

Figure 9. Results for the total costs for the major repair of a turbine blade. We assume an electricity price of 30.84 Euro-cent (Germany second half 2017 from [16]. The original strategy is the cheapest option independent of the season.

Table 6. Electricity price at which a break-even is reached between two policies in €. A value is calculated for each month of the year, for the major gearbox replacement, with a repair time of 231 h. This comparison is not complete and solely meant as an example to show that novel policies indeed become cheaper than the original policy for low enough electricity prices.

Month	simple = wait1	simple = wait2	simple = wait3	wait1 = wait2	wait1 = wait3	wait2 = wait3
1	0.0117	0.0114	0.0110	0.0110	0.0106	0.0103
2	0.0163	0.0153	0.0144	0.0143	0.0134	0.0125
3	0.0118	0.0108	0.0099	0.0098	0.0090	0.0082
4	0.0211	0.0192	0.0175	0.0173	0.0158	0.0142
5	0.0193	0.0178	0.0164	0.0163	0.0150	0.0137
6	0.0247	0.0228	0.0212	0.0210	0.0194	0.0178
7	0.0180	0.0168	0.0157	0.0156	0.0146	0.0136
8	0.0175	0.0158	0.0143	0.0141	0.0127	0.0113
9	0.0113	0.0104	0.0096	0.0095	0.0088	0.0081
10	0.0095	0.0086	0.0078	0.0077	0.0070	0.0063
11	0.0075	0.0070	0.0066	0.0066	0.0061	0.0057
12	0.0223	0.0209	0.0196	0.0195	0.0182	0.0169

5. Discussion

The most important takeaway from this paper should be the methodology that has been presented. The Markov decision process is a powerful tool and yet so versatile that it can be modified to fit a multitude of use cases. Uncertainties in different parameters can be included, by adding a parameter to the state, representing e.g., the probability of a successful repair, or the occurrence of a new failure.

The results presented in the case study Section 4.1 show that the Markov decision process is a valid approach to assess different maintenance policies for offshore wind farms. It has shown that, depending on the circumstances, the current state-of-the-art maintenance policy is indeed optimal. We have further shown that, with an electricity price below 3 Euro-cent, the 'wait-1-step'-policy becomes better than the original strategy in the given framework. This is assuming a crew transfer vessel with the presented values for maintenance costs can be used for the given repair and weather probabilities based on FINO 1 [12].

According to Fraunhofer ISE [17], the wind specific electricity prices in Germany are currently between 8 and 14 Euro-cent, while consumer end prices for electricity were at 31.23 Euro-cent in Germany in the second half of 2018 according to Eurostat [16]. This shows that only a small fraction of the end-consumer price is paid to the wind farm operator, roughly between 26–46% of the consumer end price goes to the energy producer in Germany. Fraunhofer ISE [17] predict the prices in Germany to further drop to around 5 to 11 Euro-cent by 2030.

When applying these percentages to other European countries, like Lithuania with an electricity price of 0.1097 Euro-cent in the second half of 2018 [16], a yield of 3–5 Euro-cent becomes realistic. This is without the prediction of a drop in the share of the percentage of the electricity price that goes to the producer. Factoring that into the previous calculation, a yield between 1.8–0.3 Euro-cent for Lithuania in 2030 can be predicted. Therefore, wind farm operators might soon be interested to look beyond the state-of-the-art strategy and investigate other policies.

Another aspect to consider is the limitation of this study concerning different vessel types. A gearbox replacement usually requires a lifting vessel with a crane, which generally have higher mobilization and hourly hire rates than the ones investigated in the current framework. In reality, the maintenance policy of waiting for a persistently calm sea might therefore already be economically viable in some cases for the current electricity prices.

A similar argument can be observed for wind farms that are far offshore, with longer travel times. The example presented here was based on FINO 1 [12] data, a measurement mast close to the Alpha Ventus wind farm very close to the coast. With an increasing travel time, the cost of a failed maintenance attempt (an unnecessary vessel mobilization) increases and it is expected that another policy than 'go-right-away' will be economically better and possibly already for current electricity prices.

The Markov decision process can be used to study and compare many different maintenance policies that have not been discussed here. It is also very straightforward to use the same process for wind farms with a longer travel time, other site-specific weather conditions, turbine types or cost numbers. Some examples for investigations in the future include:

- Maintenance policies taking into account work-time restrictions or shift lengths.
- Policies including multiple vessels.
- Policies including different vessel types.
- Investigations of multiple failures or turbines.
- A framework that takes into account incomplete repair actions or loss of repair progress in case of interrupted maintenance.
- Wind farms further from shore, with a longer travel time and harsher weather.

6. Materials and Methods

Wind and wave data from the FINO 1 project are provided by the Bundesministerium für Wirtschaft und Energie (BMWi), Federal Ministry for Economic Affairs and Energy and the Projektträger Jülich, project executing organization (PTJ). They can be downloaded from http://fino.bsh.de/ by users from Europe, for research purposes.

The implementation of the method, as used for the case study, is freely available from https://github.com/helenese/MDP, licensed under Creative Commons Attribution—NonCommercial 4.0 International (CC BY-NC 4.0).

7. Conclusions

In this article the Markov decision process (MDP) has been presented as a useful method for offshore wind farm maintenance modeling. The method can be adapted to fit many use cases and uncertainties can be included without relying on Monte Carlo simulations. The case study has validated the use of this concept and further indicates that under a hypothetically, lower electricity price alternative policies for the scheduling of repair will become more efficient than the current state of the art.

Author Contributions: H.S. and M.M. contributed to the research idea, H.S. conducted the analysis, M.M. supervised the analysis, H.S. wrote the paper, and M.M. reviewed the paper.

Funding: Part of the work leading to this publication was financed by the AWESOME project (awesome-h2020.eu), which has received funding from the European Union's Horizon 2020 research and innovation programme under the Marie Skłodowska-Curie Grant No. 642108. Most of the work of the first author was completed without funding after the funding period was over.

Acknowledgments: The authors would like to acknowledge the two anonymous reviewers, whose comments and suggestions helped improve the quality of this paper.

Conflicts of Interest: The authors declare no conflict of interest. The funders had no role in the design of the study; in the collection, analyses, or interpretation of data; in the writing of the manuscript, or in the decision to publish the results.

References

1. Feng, Y.; Tavner, P.; Long, H. Early experience with UK round 1 offshore wind farms. *Energy* **2010**, *163*, 167–181. [CrossRef]
2. Seyr, H.; Muskulus, M. Decision Support Models for Operations and Maintenance for Offshore Wind Farms: A Review. *Appl. Sci.* **2019**, *9*, 278. [CrossRef]
3. Gintautas, T.; Sørensen, J.D. Improved Methodology of Weather Window Prediction for Offshore Operations Based on Probabilities of Operation Failure. *J. Mar. Sci. Eng.* **2017**, *5*, 20. [CrossRef]
4. Stock-Williams, C.; Swamy, S.K. Automated daily maintenance planning for offshore wind farms. *Renew. Energy* **2018**. [CrossRef]
5. Seyr, H.; Muskulus, M. Value of information of repair times for offshore wind farm maintenance planning. *J. Physics: Conf. Ser.* **2016**, *753*, 092009. [CrossRef]
6. Seyr, H.; Muskulus, M. How Does Accuracy of Weather Forecasts Influence the Maintenance Cost in Offshore Wind Farms? In Proceedings of the 27th International Ocean and Polar Engineering Conference, International Society of Offshore and Polar Engineers, San Francisco, CA, USA, 25–30 June 2017 .
7. Gonzalez, E.; Nanos, E.M.; Seyr, H.; Valldecabres, L.; Yürüşen, N.Y.; Smolka, U.; Muskulus, M.; Melero, J.J. Key Performance Indicators for Wind Farm Operation and Maintenance. *Energy Procedia* **2017**, *137*, 559–570. [CrossRef]
8. Puterman, M.L. *Markov Decision Processes: Discrete Stochastic Dynamic Programming*; John Wiley & Sons: Hoboken, NJ, USA, 2014.
9. Jonkman, J.; Butterfield, S.; Musial, W.; Scott, G. *Definition of a 5-MW Reference Wind Turbine for Offshore System Development*; National Renewable Energy Laboratory: Golden, CO, USA, 2009. [CrossRef]
10. Bak, C.; Zahle, F.; Bitsche, R.; Kim, T.; Yde, A.; Henriksen, L.C.; Hansen, M.H.; Blasques, J.P.A.A.; Gaunaa, M.; Natarajan, A. The DTU 10-MW reference wind turbine. In *Danish Wind Power Research*; Sound/Visual Production, Trinity: Fredericia, Denmark, 2013.
11. Carroll, J.; McDonald, A.; McMillan, D. Failure rate, repair time and unscheduled O&M cost analysis of offshore wind turbines. *Wind. Energy* **2015**, *19*, 1107–1119.
12. Bundesamt für Seeschifffahrt und Hydrographie (BSH). FINO Datenbank. Available online: http://fino.bsh.de (accessed on 23 January 2018)
13. Dinwoodie, I.; Endrerud, O.E.V.; Hofmann, M.; Martin, R.; Sperstad, I.B. Reference Cases for Verification of Operation and Maintenance Simulation Models for Offshore Wind Farms. *Wind. Eng.* **2015**. [CrossRef]
14. Dornhelm, E.; Seyr, H.; Muskulus, M. Vindby—A Serious Offshore Wind Farm Design Game. *Energies* **2019**, *12*, 1499. [CrossRef]
15. van Bussel, G.; Bierbooms, W. The DOWEC Offshore Reference Windfarm: Analysis of Transportation for Operation and Maintenance. *Wind. Eng.* **2003**, *27*, 381–391. [CrossRef]
16. Eurostat, the Statistical Office of the European Union. Energy Statistics-Electricity Prices for Domestic and Industrial Consumers, Price Components. Available online: https://ec.europa.eu/eurostat/web/energy/data/database (accessed on 4 June 2019).

17. Kost, C.; Shammugam, S.; Jülch, V.; Nguyen, H.T.; Schlegl, T. *Stromgestehungskosten Erneuerbare Energien*; Fraunhofer-Institut für Solare Energiesysteme ISE: Freiburg, Germany, 2014. (In German)
18. Martin, R.; Lazakis, I.; Barbouchi, S.; Johanning, L. Sensitivity analysis of offshore wind farm operation and maintenance cost and availability. *Renew. Energy* **2016**, *85*, 1226–1236. [CrossRef]

Review

A Survey of Condition Monitoring and Fault Diagnosis toward Integrated O&M for Wind Turbines

Pinjia Zhang * and Delong Lu

Department of Electrical Engineering, Tsinghua University, Beijing 100084, China
* Correspondence: pinjia.zhang@ieee.org; Tel.: +86-10-6278-8629

Received: 15 June 2019; Accepted: 16 July 2019; Published: 20 July 2019

Abstract: Wind power, as a renewable energy for coping with global climate change challenge, has achieved rapid development in recent years. The breakdown of wind turbines (WTs) not only leads to high repair expenses but also may threaten the stability of the whole power grid. How to reduce the operation and the maintenance (O&M) cost of wind farms is an obstacle to its further promotion and application. To provide reliable condition monitoring and fault diagnosis (CMFD) for WTs, this paper presents a comprehensive survey of the existing CMFD methods in the following three aspects: energy flow, information flow, and integrated O&M system. Energy flow mainly analyzes the characteristics of each component from the angle of energy conversion of WTs. Information flow is the carrier of fault and control information of WT. At the end of this paper, an integrated WT O&M system based on electrical signals is proposed.

Keywords: condition monitoring; fault diagnosis; survey; wind turbine (WT); electrical signal

1. Introduction

1.1. Background

The global desire for clean energy is driving the rapid development of wind power, and wind turbine (WT) manufacturing techniques have made tremendous progress in recent decades. As shown in Figure 1, the reports show that the installed capacity of wind power in the United States and P.R. China accounted for more than half of the total worldwide in 2017. Furthermore, the USA wind power market surged in 2017 with 9598 MW of new capacity added, bringing the cumulative total to 89,077 MW as shown in Figure 2 [1]. Meanwhile, China surpassed the United States in cumulative installations in 2017, and the latest data show that China led both offshore and onshore wind energy installation in 2017 [2]. The rapid increase of installed capacity of WTs has brought hope for solving the problem of fossil energy shortage, especially the utilization of offshore wind energy.

Despite the rapid development of WTs, the operation and the maintenance (O&M) techniques of WTs are lagging behind, and little research has been done. Compared with traditional steam turbines and hydraulic turbines, WTs are usually located in remote and harsher environments with high humidity, high salt-fog, large temperature fluctuation and even snow-covered conditions. Long-term and large-scale load fluctuations also bring great uncertainty to prognostic and health management of WTs. The WT is complex in structure, and its nacelle is located above a tower hundreds of meters high and difficult to access. The O&M data from the first-line wind farms show that the high failure rates of WTs seriously squeezes the economic benefits of the wind farm. The condition monitoring and fault diagnosis (CMFD) technique of a wind farm is the core issue of timely discovery of incipient faults and arranging scheduled maintenance to reduce the O&M cost of the wind farm.

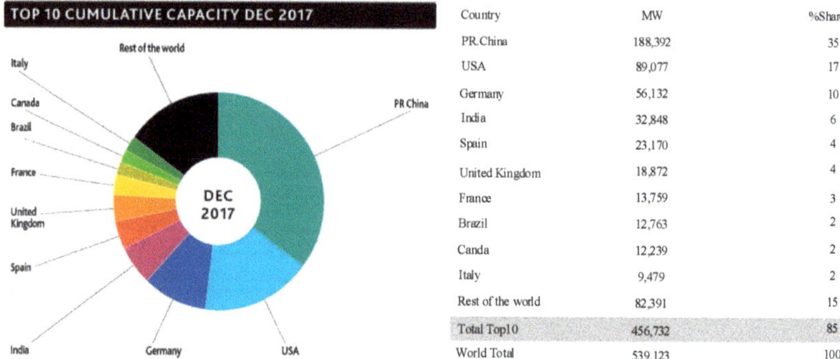

Country	MW	%Share
PR.China	188,392	35
USA	89,077	17
Germany	56,132	10
India	32,848	6
Spain	23,170	4
United Kingdom	18,872	4
France	13,759	3
Brazil	12,763	2
Canda	12,239	2
Italy	9,479	2
Rest of the world	82,391	15
Total Top10	456,732	85
World Total	539,123	100

Figure 1. Annual and cumulative growth in USA wind power capacity [1].

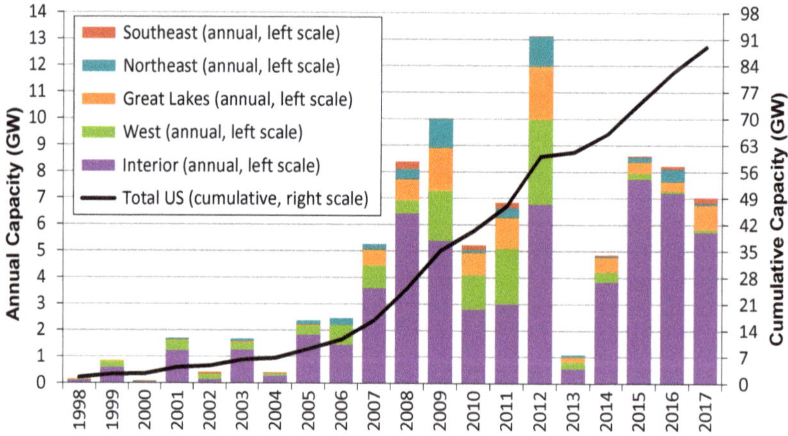

Figure 2. Annual and cumulative growth in USA wind power capacity [2].

1.2. Overview of the Survey

This paper introduces the related techniques of the WT CMFD from a comprehensive perspective. Energy flow, information flow, and an integrated WT O&M system based on electrical signal are discussed in detail in this paper. In the design and the production stage of WTs, the fault modes and the fault characteristic parameters of each component of WTs need to be considered beforehand. Firstly, the various components of the WT are divided into wind energy subset, mechanical energy subset, and electrical energy subset from the perspective of energy flow. As the highlight of the paper, the failure modes and the characteristic parameters of each component are considered comprehensively. Secondly, the various signals of the WT are divided into vibration, torque, electrical, temperature, and acoustic emission (AE) from the perspective of information flow. The intrinsic link between the fault information carriers is discussed and summarized in detail. Thirdly, since the electrical signal is non-intrusive in nature, it is not necessary to add redundant sensors to convert other signals into electrical signals during fault diagnosis. The electrical signal is the optimal fault information carrier in the O&M information flow of the wind farm. A wind farm O&M method based on the electrical signals is proposed.

Compared with the existing literature survey, the contribution of this paper is to focus on the intrinsic link between the various components of the energy flow and the various fault information carriers of the information flow. The manifestations of each faulty component in each fault information carrier is discussed and summarized in detail. Considering that the purpose of all sensors is to convert other kinds of signals into electrical signals, the O&M methods based on electrical signals can directly skip the sensor link, thus the wind farm O&M based on electrical signals with the integration of control and fault diagnosis is emphasized and separated into a single section.

2. WT Component

WTs are a complex and highly coupled power conversion system from wind energy to mechanical energy to electrical energy. In order to obtain maximum wind energy and ensure safe and stable operation of WTs, many sensors and auxiliary equipment are installed in the nacelle of WTs, which increase the complexity of WT. Figures 3 and 4 show the typical structure of WTs. The difference between two structures is that doubly fed induction generator (DFIG) WTs have drivetrain gearboxes, and the generator is directly connected with the blades in a permanent magnet synchronous generator (PMSG). The components of WTs can be divided into three subsets: wind energy subset, mechanical energy subset, and electrical energy subset according to energy flow. The blades are rotating devices that are exposed to the outside and directly capture wind energy. The tower is the supporting device for the entire WT nacelle and blades. The harsh environment will cause cracks from the point to the surface of the tower and may even cause the entire tower to fall down. The pitch system and the yaw system are auxiliary devices for adjusting the pitch angle and the nacelle direction in order to obtain maximum wind energy. These devices are all a subset of the wind energy in the energy flow. Both the shaft and the gearbox are mechanical drivetrain in the nacelle, which belong to the subset of the mechanical energy in the energy flow. Since the vibration sensor is easily attached to the surface of the mechanical drivetrain, the vibration signal is generally preferred in the industry for fault diagnosis. However, unfortunately, the narrow nacelle of the WT creates obstacles to the installation of sensors. A generator and its auxiliary power electronic device transform mechanical energy into uniform and standard electric energy belonging to the electric energy subset of energy flow. Since there is a steel shell on the outside of the motor, it is not advisable to go deep into the motor for monitoring, which would probably cause damage to the electromagnetic reaction. In addition, the more widely used a power electronic device is, the more fragile it will be, which poses a challenge for the fault diagnosis of the electric energy subset.

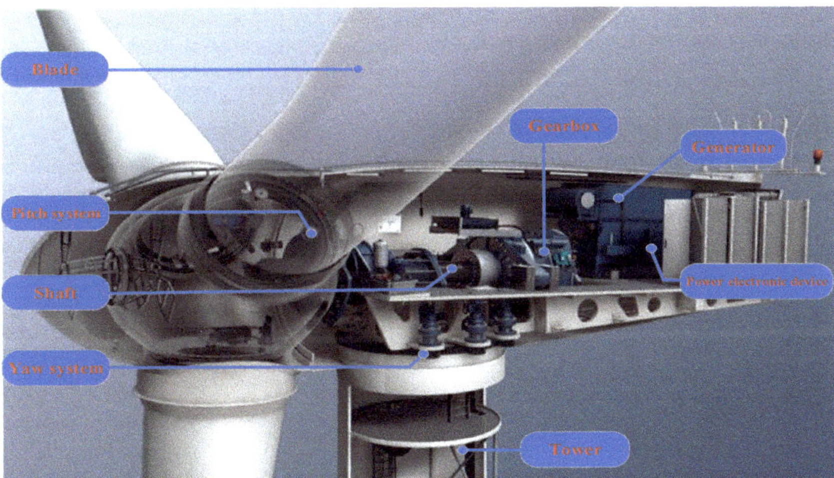

Figure 3. Doubly fed induction generator (DFIG).

Figure 4. Permanent magnet synchronous generator (PMSG).

The CMFD is an effective way to realize intelligent O&M and reduce the cost of wind power. Condition monitoring is a process of monitoring the operation parameters of electromechanical systems. By monitoring the change of parameters and analyzing the physical relationship between parameters and faults, the diagnosis and the prediction of faults can be realized in a timely manner. The wind farm supervisory control and data acquisition (SCADA) system is a remote O&M cloud platform realizing the functions of monitoring, controlling, recording, and statistics of the operation status of each important unit and accessory systems [3,4]. Vibration, electric, AE, temperature, and oil are all possible information carriers of fault parameters. The existing condition monitoring techniques are mainly divided into two parts: offline condition monitoring and online monitoring. Offline condition monitoring requires the WT to be separated from the working condition in order to extract fault information. The advantages of online condition monitoring over offline condition monitoring are obvious—online condition monitoring is performed, whereas the WTs are in service. This reduces the loss of energy production costs in the process of WT condition monitoring. At the same time, it should be emphasized that online monitoring is to extract fault parameters during the operation of WTs. How to extract small fault from massive data under normal conditions is still a great challenge. The harsh environment of the wind farm combined with variable speed and variable load conditions poses a severe challenge to improving the signal-to-noise ratio (SNR) of online monitoring and to achieving decoupling between fault and normal conditions.

As shown in Figure 5, data surveyed from three wind farms of Europe show the average annual failure of each WT component and the average time of each failure. From this, a conclusion was drawn that the failure rates of blades, generators, electrical system, electrical control, and gearboxes are higher. When WT fault occurs, yaw system, blades, drivetrain, and gear box failures often lead to a long downtime of WT. The components with high failure rate and long downtime caused by each failure are the core issues of CMFD for WTs.

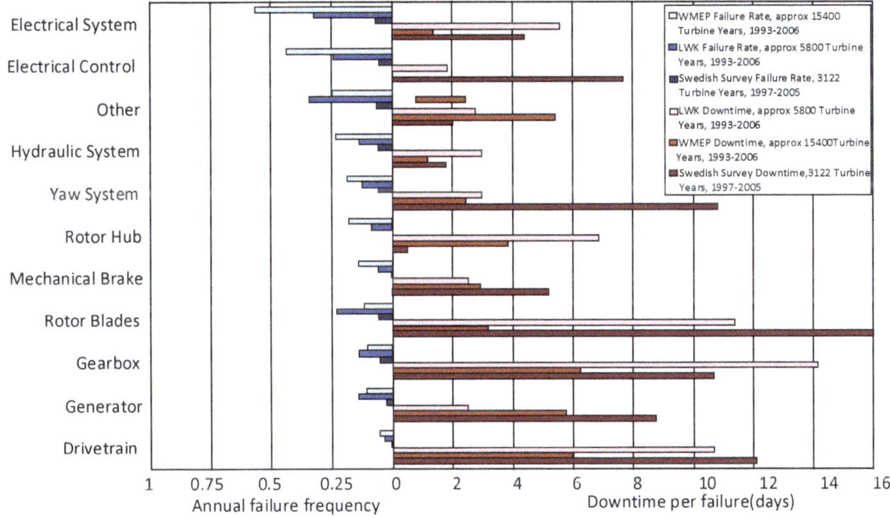

Figure 5. Annual failure rate and downtime per failure for surveyed wind farms in Europe [5].

3. WT Energy Flow

3.1. Wind Energy Subset

Components used to capture wind energy or serve wind energy in WTs belong to the wind energy subset in the energy flow. Blades, pitch systems, yaw systems, and towers are indispensable components. Components belonging to the wind energy subset are exposed to the wild, resulting in inevitably higher failure rates. Moreover, due to the change of wind speed and the inertia of the rotation of its own blades, the working condition is unstable. These characteristics of the wind energy subset components bring new challenges to CMFD of WTs, which are discussed below.

3.1.1. Blade

The blade is a device for WTs to capture wind energy. It is the only mechanical component that is widely oscillating and exposed to the outside for all components of the WT. Exposed to the natural environment, the blade is prone to various faults due to the influence of aerodynamic and adverse environmental factors [6,7]. The imbalance of blades threatens the stability of the whole WT and can even lead to the collapse of the wind tower. In recent years, as the focus of wind energy development has shifted from land to sea while bringing more clean energy, it has also led to the size of blades becoming larger and larger [8]. With high wind speed and high salinity and humidity, the risk of failure of larger blades increases exponentially. The maintenance cost of offshore WTs is high, and the blades damaged seriously need to be repaired by transporting ships back to factories. The CMFD of offshore wind farms have a broader application prospect.

The imbalanced fault of the blade is found by using the frequency spectrum of the vibration signal [9–12]. The imbalanced faults of WT blades can also be monitored by analyzing the frequency spectrum of the shaft torque [13]. Although the generator of a WT is far from the blade in the drivetrain chain, it can also analyze the health of the blade by analyzing the power spectrum [14]. More directly, the researchers used fiber bragg grating (FBG) sensors to monitor the probable unseen crack faults in the blades [15,16].

Although researchers all over the world have proposed various methods to deal with WT blade failures, the challenges are still great. For example, the measurement of shaft torque needs a torque sensor, and the measurement power needs to simultaneously sample the three-phase voltage and the current of the stator, thus the calculation ability and the storage space are big challenges. The blade fault monitoring is greatly affected by the surrounding environment and working conditions. The data collected by the strain sensor are non-periodic, and it is difficult to form a unified fault diagnosis standard.

3.1.2. Pitch System and Yaw System

In order to obtain the maximum wind energy and realize the real-time tracking of wind direction, WTs are equipped with a yaw system and a pitch system. An example of energy conversion for WTs is shown in Figure 6.

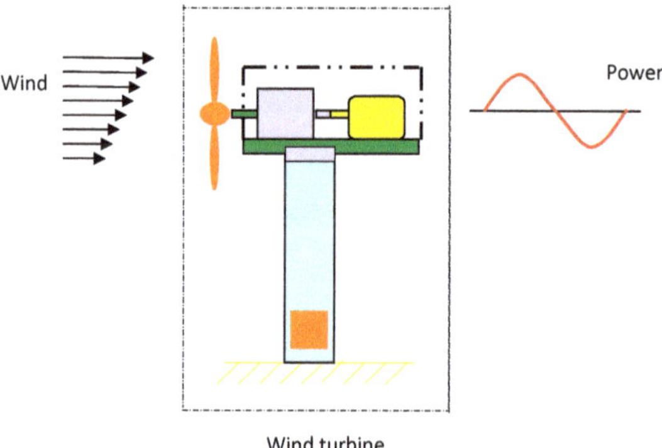

Figure 6. Energy conversion of WTs

The pitch system is a device to control the rotation of the blade around the hub, while the yaw system is a device to control the rotation of the nacelle around the tower [5]. The pitch system and the yaw system structures are similar to the main system of WTs, which consists of a motor, a deceleration gearbox, and a bearing. Therefore, most of the CMFD methods applied to the main drive system are applicable to the pitch system. A data-driven fault diagnosis technique is introduced to solve the pitch system fault of WTs [17]. Aiming at the special working conditions of variable speed and variable load in the pitch system, researchers proposed a gear fault diagnosis technique based on an order tracking algorithm [18]. The research on CMFD of pitch bearing with high failure rates has not been reported yet. Similar to the failure causes of the pitch system, installation misalignment, encoder failure, and lack of lubricant all could lead to yaw system failure. By using power curve copula modeling, the fault diagnosis of the pitch and the yaw system is realized [19]. There has been little research on CMFD of the yaw system until now. With the increase of WT capacity, the weight of nacelle increases, thus the research on CMFD of the yaw system deserves more attention. The problem faced by the pitch system and the yaw system is that the actual working condition is variable speed and variable load. The frequency of the fault signature in the spectrogram causes aliasing due to this special condition. The traditional steady-state analysis method is difficult to apply to this special working condition. The small space in the WT nacelle and the hub limits the installation of the sensor, thus the sensorless fault diagnosis techniques of the pitch system and the yaw system is the direction of the effort.

3.1.3. Tower

The tower is the support component of the whole nacelle. Long-term exposure to the outside world and the fluctuation of wind speed will usually affect the stability of the tower, which also threatens the stability of the upper nacelle and the blades. The vibration signals collected in the SCADA system are used to monitor the health of WT towers. Based on the data transmitted from the strain sensor directly installed on the tower, the cloud platform for wind farm O&M can monitor the subtle changes of the tower in time [20,21]. Researchers introduced the method of measuring strain and bending deformation of a WT tower based on FBG to evaluate the load on the tower in real time [22]. This method not only realizes fault diagnosis but also estimates the tower condition of WT, which greatly reduces the threshold of tower design and wind farm construction cost. Early tower failures may even lead to tower collapse if they are not detected in time when there is a strong wind. Because the collapse of towers is usually sudden, timely detection of early micro-faults is the key to CMFD of WT towers.

3.2. Mechanical Energy Subset

The mechanical energy subset includes two components, the shaft and the gearbox, which together create the tie between the wind energy and the electrical energy that can be used by human beings. Meshing and friction of mechanical components are collectively the process of transmitting energy, but once the mechanical components are subjected to more than the load they can bear, cracks and fracture will occur.

3.2.1. Shaft

Shaft is the connecting device between the hub, the gearbox, and the generator, which plays the function of energy flow transmission. The shaft constantly collides and rubs against various components during the journey of sharing energy. The change of wind speed causes the rotation speed of the shaft to follow the change, which leads to a large fluctuation of force in the transmission process and greatly increases failure rates. There are few studies on CMFD of WT shafts. By using the characteristic parameters corresponding to the faults in vibration and AE, the fault condition of the shaft is detected [23]. When the shaft is imbalanced, the load torque of the generator will fluctuate periodically. Experiments proved that load torque of the motor could be used to monitor the health of the shaft [24]. However, it should be noted that the load of the motor is greatly affected by the fluctuation of wind speed, and the diagnostic results are greatly affected by the working conditions. Similarly, acoustic signals are used to monitor the cracked shaft [25]. Generator current signals are used to monitor the misalignment of the shaft [26]. The difficulty of CMFD regarding shafts lies in the fact that shafts are usually used in conjunction with gearboxes or bearings, and it is difficult to determine the specific source of faults in actual working conditions.

3.2.2. Gearbox

A gearbox is a drivetrain device to realize speed matching between blade and generator. Due to the gears being relatively fragile and the fact that subtle faults are difficult to detect, the gearbox is considered to be the component with the highest failure rates, which often puts doubly fed induction WTs at a disadvantage when competing with permanent magnet direct drive WTs. Installation dislocation, fatigue wear, and lack of lubricant can affect the working conditions of the gearbox. Compared with the fault of the generator body, the gear fault has its own characteristics [27,28]. Because it is a drivetrain component between mechanical components and electric components, it is possible to monitor the condition based on vibration, electrical signals, AE, and temperature signal [6,29].

The influence of the torque fluctuation caused by gear failure on the stator current of the generator is analyzed. After fast Fourier transform, the fault gear can be converted into a fault sideband in the current spectrum. The fault sideband interval is f_{cc}.

$$f_{cc} = \left\{ f | f = k f_s \pm \sum_{i=1}^{I} p_i f_{sh,i} \pm \sum_{j=1}^{J} q_j f_{m,j}; p_i, q_j = 0, 1, 2, \ldots \right\} \tag{1}$$

where f_s is the fundamental frequency, k is a positive integer representing the fundamental and the possible harmonics of the current, $f_{sh,i}$ is the rotating frequency of the ith shaft in the gearbox, $f_{m,j}$ is the jth gear meshing frequency, and I and J are the numbers of the shafts and the gear pairs in the gearbox, respectively. When a gear failure occurs in the drivetrain chain, the corresponding side frequency band in the spectrum will be significantly strengthened. Similar to the characteristic frequency in the current, the characteristic frequency f_{gb} in the vibration signal can be expressed as:

$$f_{gb} = \left\{ f | f = \sum_{i=1}^{I} l_i f_{sh,i} \pm \sum_{j=1}^{J} m_j f_{m,j}; l_i, m_j = 0, 1, 2, \ldots \right\} \tag{2}$$

By comparing the characteristic frequencies of the current signal with those of the vibration signal, it can be found that the current signal is affected by the fundamental wave, which leads to the low SNR of the fault characteristic frequency. However, at the same time, it should be noted that the vibration signal needs to be equipped with sensors, and the sensor itself has the risk of failure. In the case of gearbox failure, noise will be generated, and friction will also lead to temperature rise. AE signal and temperature signal are also worth considering [30,31].

3.3. Electrical Energy Subset

The capture of wind energy and the transmission of mechanical energy are all for the production of electrical energy. The generator and its auxiliary power electronic devices are indispensable for the electrical energy subset. The CMFD for the electrical energy subset components determines whether a qualified electrical energy can be produced.

3.3.1. Generator

A typical WT generator is shown in Figure 7. Several surveys of fault types of induction motors, conducted by IEEE-IAS [32–34], EPRI [35], and Allianz [36], are compared in Table 1. Generator components can be divided into four parts: bearing, stator, rotor, and others. Because the existing WT is rarely a squirrel cage structure, the rotor fault is not considered here. The motor bearing fault and the stator inter-turn short circuit fault are selected as two typical motor faults for detailed analysis.

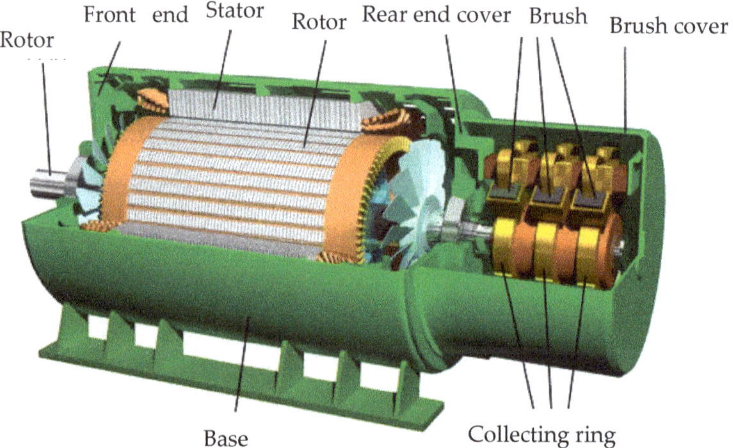

Figure 7. Typical WT generator structure.

Table 1. Percentage of failure by component in induction motor.

Major Components	IEEE-IAS % of Failures	EPRI % of Failures	Allianz % of Failures
Bearing related	44	41	13
Stator related	26	36	66
Rotor related	8	9	13
Others	22	14	8

Generator bearings are composed of an inner ring, an outer ring, a ball, and a cage. The characteristic frequencies of each part are explained in detail elsewhere [37].

$$f_o = 0.5 \cdot N \cdot f_r \cdot (1 - \frac{D_b \cdot \cos\theta}{D_p}) \tag{3}$$

$$f_i = 0.5 \cdot N \cdot f_r \cdot (1 + \frac{D_b \cdot \cos\theta}{D_p}) \tag{4}$$

$$f_b = 0.5 \cdot N \cdot \frac{D_p}{D_b} \cdot f_r \cdot (1 - (\frac{D_b \cdot \cos\theta}{D_p})^2) \tag{5}$$

$$f_i = 0.5 \cdot f_r \cdot (1 - \frac{D_b \cdot \cos\theta}{D_p}) \tag{6}$$

where f_o, f_i, f_b, and f_c are the fault characteristic frequencies of the outer race, the inner race, the ball, and the cage faults, respectively, f_r is the rotational frequency of the bearing, N is the number of balls, D_b and D_p are the ball diameter and the ball pitch diameter, respectively, and θ is the ball contact angle with the races.

The CMFD of WT bearings is mainly based on vibration signals [38–40]. A bearing fault diagnosis technique based on electrical signals was developed recently. The stator current signal of the generator is used to monitor the WT bearing [41]. The difficulty of bearing fault diagnosis is that the bearing is composed of many parts that interact with each other. Bearing faults are often not single but are superimposed on each other, thus it is difficult to analyze the characteristic frequencies. It is also necessary to consider how to improve the SNR when using electrical signals.

Generator stator is the static part of the generator. The stator consists of three parts: stator core, stator winding, and machine base. The main function of the stator is to produce a rotating magnetic field for the pushing rotor. The stator inter-turn short circuit fault is one of the most representative stator faults. Because of the internal fault of the generator, the CMFD based on electrical signals becomes the first choice. The stator current waveform of the generator is extracted to judge whether the stator is healthy or not [42–44]. For example, consider that a researcher proposes the negative sequence component method to extract the characteristic parameters of stator inter-turn short circuit fault and carry out the relevant theoretical derivation [42]; another group proposes to monitor the stator winding faults of doubly fed induction WTs by using the rotor side current spectrum signal and finds the characteristic frequency, which is not affected by load conditions [43]. Stator reactive power is used to solve this problem [44]. The fault characteristic frequencies f_ω in electrical signals can be expressed by the following formula:

$$f_\omega = \left\{ f | f = \frac{\left[k \pm \frac{n(1-s)}{p} \right]}{f_s}; k = 1,3; n = 1,2,\ldots,(2P-1) \right\} \tag{7}$$

where p is the number of pole pairs, f_s is the fundamental frequency, and s is the slip. The stator inter-turn short-circuit fault can cause temperature rise and noise, and some literature has carried out research based on this phenomenon [45]. The difficulty of stator faults lies in how to find incipient faults and arrange scheduled maintenance.

3.3.2. Power Electronic Device

WTs are often located in a harsh environment with high temperature heating, oil and water pollution, and dust, which not only affects the performance of the power electronic device but also easily leads to device failure. Unlike the simple and reliable mechanical components discussed above, power electronic devices are more precise and fragile, and slight changes in the environment will affect the working condition [46]. Figure 8 shows the structure and the fault proportion of the power electronic device.

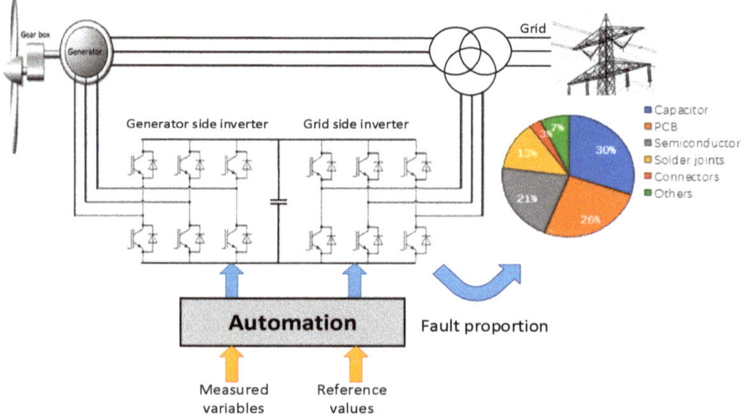

Figure 8. WT power electronic device and fault proportion.

Power electronics devices are used to connect the power generated by DFIG to the grid [47,48]. From Figure 8, a conclusion was drawn that capacitors, printed circuit boards (PCBs), and power semiconductors [e.g., insulated gate bipolar transistor (IGBT) modules] are the three main reliability-critical components. In power electronic devices, CMFD based on vibration and AE signals is obviously not applicable. Electrical signals and temperature changes due to faults have become better choices [49–51]. The types of capacitor faults can be divided into numerical deviation, excessive leakage, short circuit, and even explosion. The types of PCB faults can be divided into fixed faults, bridge faults, delay faults, crosstalk faults, etc. [52]. WTs, especially permanent magnet direct drive WTs, which need full power IGBT energy change modules, call for a high reliable quality of power electronic devices. High power IGBT module failure risks are higher. The failure mechanism of an IGBT module of the converter mainly includes aluminum bond line shedding, weld layer fatigue, bond line root fracture, and aluminum oxide reconstruction [53].

In view of the severe reality of low reliability and high maintenance costs of WT power electronic devices, how to evaluate its power electronic module and how to evaluate the residual life of its power electronic module are the keys to condition-based O&M [54]. The mechanical vibration of WTs and the random fluctuation of wind speed will affect the reliability of the power electronic devices. For power electronic devices, researchers lack a deep understanding of their pre-failure aging mechanism, thus the condition monitoring technique reflecting their health level has been stagnant. At present, many engineers and designers often give priority to the reliability improvement methods, such as the crimping packaging technique, fault tolerance, and built-in redundancy, while ignoring condition monitoring as a standby option. The main principle of condition monitoring based on electrical signals is that the end characteristics of IGBT are closely related to its failure degree. A paper published by researchers shows that the three electrical parameters of gate valve voltage, transconductance, and on-voltage drop can be used as characteristic parameters of IGBT condition monitoring [55]. With the development of the power electronic power system and the deep integration of power electronics and power grids, the monitoring of power electronic devices such as converters has become increasingly prominent.

4. WT Information Flow

During the operation of WTs, there is not only the conversion of energy forms but also the transmission of information flow for O&M. The information caused by mechanical failure will be transmitted to the torque signal and the electrical signal along with the flow of energy, which is called information flow. At the same time as mechanical vibration, nonlinear changes in the parameters (such as AE and temperature) are caused, and fault diagnosis based on these signals is also a possible choice.

4.1. Strain

Strain sensor is a kind of sensor based on measuring the strain produced by the force and the deformation of the object. FBG is the most commonly used sensor. It is a kind of sensor that can transform the change of strain on mechanical components into an electrical signal. When a crack occurs in the blade or the tower, the strain sensor detects the stress change at a certain point and converts it into an electrical signal. FBG strain sensor is used to diagnose the fault of WT blades [15,16]. Relevant work describes the usage of strain sensors to detect the health of wind towers [21]. In order to ensure the safety of the tower in a wind field, the monitoring method based on the strain sensor can monitor the status of the tower in real time without excessive design margins, thus saving on construction costs [22]. For those components that are not covered by CMFD based on electrical and vibration signals, such as the blade and the tower faults, the CMFD based on strain signal becomes a better choice. However, the CMFD based on strain signal is affected by the environment and the working conditions. Other key problems are that it is difficult to establish a unified standard and there is a lack of theoretical support.

4.2. Vibration

Vibration signal is the most widely used and mature CMFD signal in the industry. After decades of development, CMFD techniques based on vibration signals have formed their own standards (ISO10816). The acceleration sensor is the most widely used vibration sensor, which converts the received acceleration signal into an electrical signal and transmits it to the data acquisition system. Sensitivity, frequency range, and test environment are all factors to be considered in selecting sensors. A scientist illustrated the gearbox fault of WTs in detail and introduced the relevant algorithms for gearbox fault in the WT industry [29]. The test of a WT drivetrain system based on the National Renewable Energy Laboratory (NREL) WT test platform was carried out completely [29]. The algorithms used include frequency domain and cepstrum analysis, time synchronization average narrowband and residual methods, and methods based on analysis and spectral kurtosis. Similarly, the health of gearboxes was evaluated based on the RMS and the peak values of vibration signals [56,57]. Researchers used a decision tree for feature selection, the optimal priority tree algorithm and the function tree for algorithm and feature classification, and subsequently proposed a better fault diagnosis algorithm for WT blades [58]. By analyzing the difference between vibration signal and other signals, a result was proposed and verified by experiment using empirical mode decomposition and wavelet transform [59]. Finally, the related work to improve the SNR was carried out. By using binary empirical mode decomposition, the SCADA system extracted the fault feature of the vibration signal to realize the fault analysis of the WT shaft [60]. Although the theory, the algorithm, and the experience of CMFD based on vibration signal are relatively mature, their shortcomings are obvious. The acquisition of vibration signals requires the installation of redundant sensors, which increases the cost and occupies the already narrow space. At the same time, the vibration signal is relatively isolated and can only be measured at a single point. It is difficult to form a comprehensive, systematic, integrated, and interactive O&M system.

4.3. Torque

When a mechanical fault occurs in a component of a WT, a periodic fluctuation of fault torque may occur on the shaft. The periodic torque fluctuation on the shaft provides the possibility of CMFD based on the torque signal. Torque signal was used to detect the blade imbalance fault of WT. Using the shaft torque observer, the healthy condition of WT was judged [61]. Analyzing the causal chain between vibration signal, torque signal, and electrical signal, researchers put forward the practical significance of using torque to monitor the fault of WTs. The limitations and the shortcomings of the torque signal have some similarities with the vibration signal. Similarly, special sensors need to be installed, which increases the cost of O&M and occupies narrow space. The bigger problem of the torque signal is affected by the normal working conditions. The proportion of fault torque is not high, thus it is difficult to detect the incipient fault. At the same time, if the voltage and the current signals are used to calculate the torque signal, the accumulative error between the signals will greatly hurt the signal accuracy. The accuracy of the torque sensor is worse than that of the vibration signal, which is the reason why the practical application of the torque signal is rare.

4.4. Temperature, Oil, and AE

When mechanical or electrical faults occur, kinetic energy and energy can be converted into heat energy, resulting in abnormal temperature rise. Therefore, monitoring temperature signal to achieve CMFD of WTs has become a choice. The non-linear relationship between temperature and faults is analyzed based on data and realizes the prediction of early faults [62,63]. The usage of the intelligent system of predictive maintenance allows the health of the WT gearbox and generator to be monitored through temperature signal [64,65]. By detecting the change of temperature signal, the misalignment of the shaft can be detected [66]. Researchers achieved non-intrusive thermal monitoring by signal injection without needing redundant thermocouples, which has a certain reference value for thermal protection of WTs [67]. The biggest challenge facing the temperature signal is that it is greatly influenced by the surrounding environment. Environmental temperature, humidity, and other conditions are obstacles to its application. How to eliminate the influence of the environment usually needs to be combined with other collected signals. How to clarify the relationship between fault signal and temperature rise and eliminate the environment is a hot research topic to be addressed in the future. Oil is used in WTs for lubrication, hydraulic pressure, etc. The monitoring of these components cannot make use of electrical signals, and there is no periodic vibration in the fault, thus it is unrealistic to use vibration signals. The monitoring of oil-related parameters has become a possible choice. Monitoring the leakage of hydraulic oil of the pitch system is a method to know its health [68]. The quality of oil is a tool to realize online monitoring of the WT gearbox [69]. Similarly, gear wear can also be known by detecting the quality of the oil [70]. By monitoring the particulate contaminated areas of lubricants, researchers were able to predict the remaining useful life of WTs [71]. The aspect that restricts monitoring based on oil signal is that oil is not widely used, and there is no uniform standard for the quality of oil. As a by-product of vibration signals, AE is also a useful method for fault prediction and diagnosis. A method for predicting the residual life of the gearbox and the shaft by combining AE with vibration signal was proposed [23]. AE can also be used to monitor the sound when the gearbox fails individually [31]. The advantage of AE is that it does not need to be close to the fault components, and it is a non-invasive methods of fault detection. At the same time, the shortcomings of AE are obvious, and the diagnosis results are greatly affected by surrounding noise. Moreover, it is difficult to determine the fault source when fault information occurs in AE. The characteristic parameters of each fault component based on AE also need to be further studied. Temperature, oil, and AE signal based CMFD of WTs are all by-products of vibration faults, thus they are classified into one category.

5. WT Algorithm Flow

The algorithm applied to CMFD of WTs is a means to improve the SNR and find fault information as soon as possible. When the monitored components are in a working state, how to eliminate the influence of operating conditions on fault information extraction and realize online monitoring requires the intervention of an algorithm. After more than a hundred years of development, researchers have proposed a variety of algorithms. Finding fault diagnosis algorithms for wind turbines is discussed in this section.

5.1. FFT and Wavelet Transform

In 1965, J. W. Cooley and J. W. Tukey proposed a fast algorithm for computing discrete Fourier transform [72]. Fast Fourier transform (FFT) greatly simplifies the calculation of discrete Fourier transform and promotes its practicality. FFT is used to analyze vibration signals to judge the main bearing faults of WTs [73]. Rotor fault is detected by FFT through the current waveform of the generator [74]. By using FFT to analyze the blade angle collected by the encoder, researchers realized the diagnosis of the blade crack fault [75]. FFT is the basis of modern signal processing. It extends the time domain signal to the frequency domain signal and analyzes the fault characteristics from another aspect. Wavelet transform is a new transform analysis method compared with FFT. It inherits and develops the idea of short-time Fourier transform localization and overcomes the shortcomings of window size not changing with frequency. It can provide a "time–frequency" window that changes with frequency. It is an ideal tool for signal time–frequency analysis and processing. Its main features are that it can fully highlight the characteristics of some aspects of the problem by transformation, can localize the time (space) frequency analysis, and can gradually refine the signal by scaling translation operation. Finally, it achieves time subdivision at high frequency and frequency subdivision at low frequency and can automatically adapt to the requirements of time–frequency signal analysis, thus focusing on arbitrary details of the signal. It solves the difficult problem of Fourier transform and has become a major breakthrough in scientific methods since Fourier transform. Wavelet transform is used to monitor generator stator winding faults and rotor imbalance faults [76]. Researchers proposed the concept of the wavelet energy transfer equation to analyze vibration signals for monitoring bearing faults [77]. The problem of wavelet transform is that the selection of wavelet bases is very difficult, and different wavelet bases have great influence on the results. At the same time, time–frequency analysis based on wavelet transform is not suitable for variable speed and variable load operation.

5.2. Order Tracking (OT)

For conventional vibration, frequency is usually used to describe the number of times of vibration in a second, which is called vibration frequency. The order represents the number of times an event occurs for each rotation of a rotating component (360 degrees). We define order tracking O as:

$$O = \frac{60 \times f}{n} \tag{8}$$

where f is the fault information carriers collected, and n is the instantaneous speed of the rotating component. Because the instantaneous speed n is divided in the definition, the effect of speed variation is eliminated. Order tracking (OT) transforms the time domain signal into the angle domain signal and successfully realizes the fault diagnosis under variable speed and variable load conditions. The relevant schematic diagram is shown in Figure 9 below. OT needs to use the time domain signal and the instantaneous speed signal. Current signal is used to estimate instantaneous speed for the OT algorithm [39]. Vibration signal is divided by instantaneous speed to get the order signal, which successfully solves bearing fault diagnosis. The highlight of literature [78–80] is that, for the first time, the concept of order tracking was introduced into the current signal, and the SNR was successfully improved by converting the current to the rotor side. The unsteady state fault diagnosis based on OT is also the focus of future research.

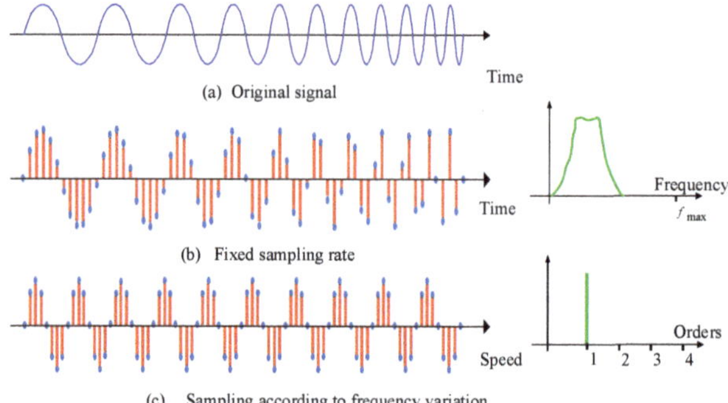

Figure 9. OT equal angular interval sampling.

5.3. Artificial Intelligence (AI)

As a new hotspot, artificial intelligence (AI) has its own application in the field of CMFD of WTs. The ultimate help of big data and AI is to make the construction of a wind farm more perfect and replace possible extensive management modules with big data and AI methods. Therefore, the intelligent management of the whole life cycle digital module of a wind farm can be realized, and a wise forward operation and a wise rear operation can be realized. The health evaluation system of WTs based on an artificial intelligence algorithm was constructed [81]. Broader research of CMFD for a machine based on artificial intelligence was systematically introduced [82]. The intelligent O&M of WTs based on current signal was developed and applied to real wind farms [83]. For the fault diagnosis of WT bearing, support vector machine (SVM) can be used to predict its remaining useful life [84]. Similarly, the k-means algorithm can also be used to predict the remaining useful life of WTs [85]. Additionally, the k-nearest neighbors (kNN) algorithm has application in gearbox fault of WTs [86]. AI algorithms are just emerging, and many theories need to be studied urgently. At the same time, we need to pay attention to whether or not the research of the algorithm is flawed, and the mechanism and the reasons behind the algorithm are still worthy of our great attention.

6. Wind Farm Integrated O&M System Based on Electrical Signal

6.1. Theory Analysis

Above all, a comprehensive discussion about the techniques of CMFD of WTs based on multiple signals was carried out. It is noteworthy that they all convert fault information into electrical signals and store them in the data acquisition system. If the electrical signals can be collected directly for CMFD, additional sensors can be removed to realize CMFD of WTs without sensors. Another advantage of CMFD based on electrical signals is obvious, which is that it can realize the integration of the control system and the fault diagnosis system, reducing cost and occupy less space. Therefore, the CMFD system of WTs based on electrical signals is the goal researchers need to strive for. Especially in recent years, the rise of flexible WTs provides a broad space for integrated O&M of WTs based on electrical signals. The stator current signal of the motor is used to estimate the instantaneous mechanical speed under variable speed and variable load conditions [39]. A method of fault diagnosis of bearing based on the EEMD method using the stator current of the motor was introduced [41]. The fault of stator winding could be diagnosed by the stator current waveform signal [42]. Researchers used the rotor current signal of the motor to diagnose the inter-turn fault, eliminate the influence of the fundamental wave of the stator current, and significantly improve the SNR [43]. By reactive power monitoring, the inter-turn short circuit faults could be found [44]. A method of fault diagnosis of bearing based on current-demodulated signals was introduced [87]. Researchers collected the voltage and the current signals of generators and improved the SNR of gearbox faults by using the new algorithm [88]. As for the fault of the generator body, fault monitoring is usually carried out by using its own electrical signals. It is noteworthy that the existing monitoring of power electronic devices is mainly based on electrical signals [46,47,52].

The CMFD technique of WTs based on electrical signals is the most promising and worthy of further study. The advantages of CMFD based on electrical signals are not necessary for redundant sensors to achieve the coordination of the fault diagnosis system and the control system and multi-dimensional input to realize the mutual monitoring of sensors themselves. For the fault of the motor body, it can be monitored well because of its strong relationship with the electric signal. The results of CMFD based on electrical signals are not very good for blades and towers, which are weakly connected with electrical signals. In addition to eliminating the influence of fundamental wave by using the rotor current signal, other methods to improve the SNR by eliminating fundamental wave and clutter wave influence are also the focus of future research work. From the point of view of energy flow and information flow, potential causal chains of CMFD of WTs were revealed. Aiming at an electric signal that can realize the integration of the control system and the fault diagnosis, this paper makes a detailed analysis and elaborates upon it. The corresponding WT flow chart is shown in Figure 10. Taking energy flow, information flow and algorithm flow as the main lines, the O&M of wind farms are divided into three parts.

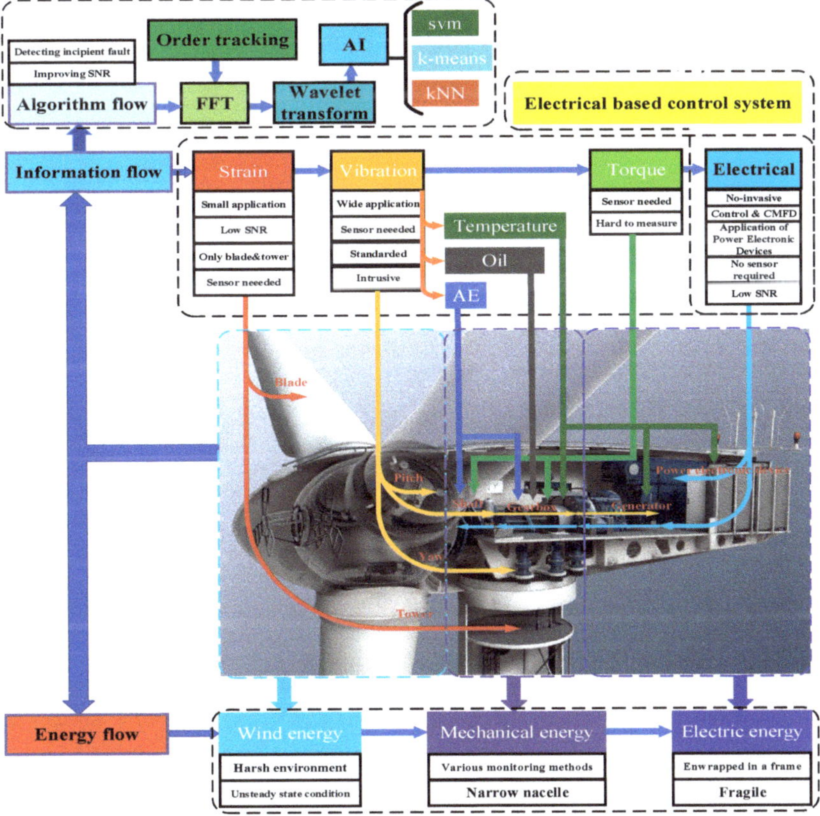

Figure 10. WT causal chains and monitoring components.

6.2. Experimental Setup

The theory of integrated O&M of WTs based on electrical signals was put forward, and many works in this field were carried out. Electrical signal was the only signal we needed to collect. Vibration signal was used as a comparison signal. Rogowski coils are used to collect current signals, and voltage probes are used to collect voltage signals. Relevant work proved the validity of the proposed theory as shown in Figure 11 [88]. Because the WT nacelle is tens of meters or even hundreds of meters high, coupled with the harsh environment of the wind farm, a special data acquisition (DAQ) system for such conditions has been developed. If necessary, the DAQ system can also be combined with the control system to realize on-line CMFD of WTs. Based on this, a complete integrated O&M system of wind farms based on electrical signals has been built.

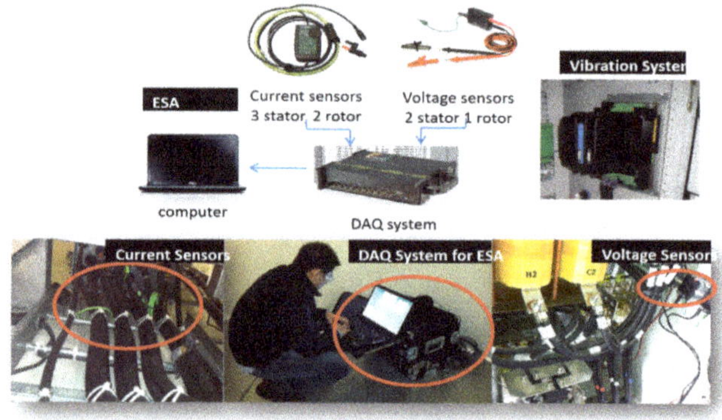

Figure 11. Typical electrical CMFD system for the WT gearbox.

In addition, we provide the safety chain of WT fault diagnosis in actual wind farms. The flow chart is shown in Figure 12.

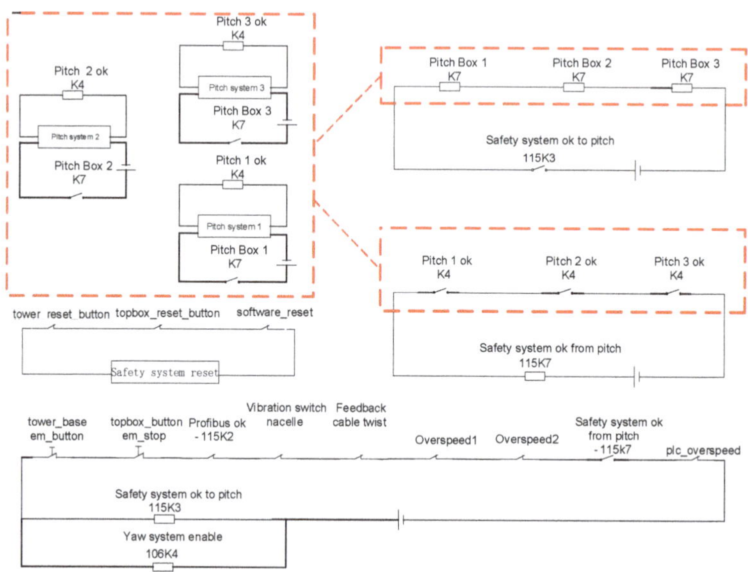

Figure 12. WT causal chains and monitoring components.

From the figure above, we can see that the pitch system affects the safety chain of the main control system through the contact of the K4 relay in each pitch cabinet, and the safety chain of the main control system influences the pitch system through the coil of the K7 relay in each pitch cabinet. The safety chain of the pitch and the safety chain of the main control are independent and interactive. When a node in the safety chain of the main control system is disconnected, the relay-115K3 coil of the safety chain to the pitcher loses power, and its contacts are disconnected. The coil of the K7 relay in each pitch cabinet loses power, and the pitch system enters the mode of emergency shutdown. When the pitch system fails (such as loss of OK signal of pitch converter, action of 90 degree limit switch, etc.), the pitch system cuts off the power supply of the K4 relay, and the contact of the K4 relay is disconnected, which makes the relay-115K7 coil of the self-pitching safety chain lose power; consequently, its contact is disconnected, and the whole safety chain of the main control system is disconnected. At the same time, the safety chain to the pitch relay-115K3 coil loses power, thus its contact is disconnected, the coil of the K7 relay in each pitch cabinet loses power, its contact is disconnected, and the control system of the blade without fault in the pitch system enters the mode of emergency shutdown. This design makes the safety chain link, which can maximize the protection of the WT.

In the actual connection, the nodes in our security chain are not really connected in series but are connected by the relationship between "and" in the security chain module. Each input is logically high level 1. After several signals are connected, its output must be high level 1, but as long as one input signal becomes low level 0, its output must be low level 0. Logical output is actually controlled by the output module of the safety chain, which controls the -115K3 and the -106K4 relays, respectively. The input is realized by the actual switch contacts and Boolean variables in the program. The actual switching state of the switch contacts is collected by the input module of the safety chain module. The Boolean variables in the program are controlled by the program. The corresponding logic structure of the WT safety system is shown in Figure 13.

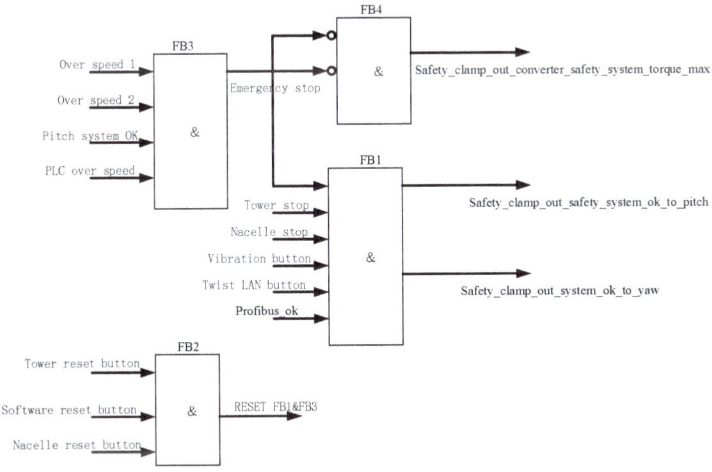

Figure 13. Logic structure of the WT safety system.

7. Conclusions and Discussion

In view of the predicament of high O&M costs of WTs, this paper comprehensively investigates the existing CMFD techniques of WTs. Especially in recent years, with the rapid development of offshore WTs, the cost of damage and replacement of important components of offshore WTs is even higher than the construction of new WTs. This paper explains the CMFD of WTs from three aspects: energy flow, information flow, and integration of the WT control system and fault diagnosis based on electrical signals. In view of the internal relationship of the WT structure, the concepts of energy flow, information flow and algorithm flow are proposed in this paper. Energy flow is a form of energy in WTs, which is divided into a subset for capturing wind energy, a subset for transferring mechanical energy, and a subset for generating electrical energy. Energy flow is the purpose and the fundamental significance of WTs. In order to realize the energy flow and the safe operation of WTs, the control and monitoring system of WTs based on information flow is indispensable. In order to improve SNR of information flow, algorithm flow is essential. A sensor is an indispensable component for acquiring information of WTs. Providing special sensors for CMFD systems is not only expensive but also takes up limited space. Therefore, a sensorless and non-intrusive WT CMFD system based on electrical signals is the direction of our efforts.

Through the above discussion, the CMFD of WTs is shown from three aspects of energy flow, information flow and algorithm flow. A CMFD system for WTs O&M based on electrical signals is proposed. The corresponding conclusions are as follows:

1. Wind farms are generally located in remote areas with harsh environments. WTs have complex structures, and nacelle is difficult to access. The high failure rates of WTs call for more advanced techniques for CMFD of WTs.

2. The purpose of WTs is to realize the conversion of wind energy to mechanical energy and then to electrical energy by energy flow. In order to better control and monitor the WT, the acquisition of information flow based on sensors is indispensable. In order to improve SNR of information flow, algorithm flow is essential.

3. The CMFD based on electrical signals can use the signals acquired by the control system to realize sensorless CMFD of WTs and integration of control system and fault diagnosis, which is the direction of future efforts.

Author Contributions: Conceptualization, P.Z.; methodology, P.Z., D.L.; validation, D.L.; formal analysis, D.L.; investigation, P.Z., D.L.; resources, P.Z.; data curation, D.L.; writing—original draft preparation, D.L.; writing—review and editing, P.Z.

Funding: This work is funded by Beijing Municipal Natural Science Foundation (L161002) and National Natural Science Foundation of China (51822705,51777112,61703227).

Conflicts of Interest: The authors declare no conflict of interest.

References

1. Sawyer, S.; Liming, Q.; Fried, L. Global Wind Report—Annual Market Update 2017. Available online: https://www.researchgate.net/publication/324966225_GLOBAL_WIND_REPORT_-_Annual_Market_Update_2017 (accessed on 4 May 2018).

2. Ryan, W.; Mark, B. Wind Technologies Market Report. 2017. Available online: https://www.energy.gov/sites/prod/files/2018/08/f54/2017_wind_technologies_market_report_8.15.18.v2.pdf (accessed on 20 August 2018).

3. Schlechtingen, M.; Santos, I.F.; Achiche, S. Wind turbine condition monitoring based on SCADA data using normal behavior models. Part 1: System description. *Appl. Soft Comput.* **2013**, *13*, 259–270. [CrossRef]

4. Schlechtingen, M.; Santos, I.F. Wind turbine condition monitoring based on SCADA data using normal behavior models. Part 2: Application examples. *Appl. Soft Comput.* **2014**, *14*, 447–460. [CrossRef]

5. Matthews, P.C.; Chen, B.; Tavner, P.J. Automated on-line fault prognosis for wind turbine pitch systems using supervisory control and data acquisition. *IET Renew. Power Gener.* **2015**, *9*, 503–513.

6. Qiao, W.; Lu, D. A Survey on Wind Turbine Condition Monitoring and Fault Diagnosis—Part I: Components and Subsystems. *IEEE Trans. Ind. Electron.* **2015**, *62*, 6536–6545. [CrossRef]

7. Qiao, W.; Lu, D. A Survey on Wind Turbine Condition Monitoring and Fault Diagnosis—Part II: Signals and Signal Processing Methods. *IEEE Trans. Ind. Electron.* **2015**, *62*, 6546–6557. [CrossRef]

8. Zhao, M.; Jiang, D.; Li, S. Research on fault mechanism of icing of wind turbine blades. In Proceedings of the World Non-grid-connected Wind Power & Energy Conference, Nanjing, China, 24–26 September 2009.

9. Ramlau, R.; Niebsch, J. Imbalance Estimation without Test Masses for Wind Turbines. *J. Sol. Energy Eng.* **2009**, *131*, 011010. [CrossRef]

10. Kusnick, J.; Adams, D.E.; Griffith, D.T. Wind turbine rotor imbalance detection using nacelle and blade measurements. *Wind Energy* **2015**, *18*, 267–276. [CrossRef]
11. Niebsch, J.; Ramlau, R.; Nguyen, T.T. Mass and Aerodynamic Imbalance Estimates of Wind Turbines. *Energies* **2010**, *3*, 696–710. [CrossRef]
12. Gardels, D.J.; Qiao, W.; Gong, X. Simulation studies on imbalance faults of wind turbines. In Proceedings of the Power & Energy Society General Meeting, Providence, RI, USA, 25–29 July 2010.
13. Caselitz, P.; Giebhardt, J. Rotor Condition Monitoring for Improved Operational Safety of Offshore Wind Energy Converters. *J. Sol. Energy Eng.* **2005**, *127*, 445–447. [CrossRef]
14. Gong, X.; Qiao, W. Simulation investigation of wind turbine imbalance faults. In Proceedings of the International Conference on Power System Technology, Hangzhou, China, 24–28 October 2010.
15. Rubert, T.; Perry, M.; Fusiek, G. Field demonstration of real-time wind turbine foundation strain monitoring. *Sensors* **2018**, *18*, 97. [CrossRef] [PubMed]
16. Arsenault, T.J.; Achuthan, A.; Marzocca, P. Development of a FBG based distributed strain sensor system for wind turbine structural health monitoring. *Smart Mater. Struct.* **2013**, *22*, 075027. [CrossRef]
17. Kusiak, A.; Verma, A. A data-driven approach for monitoring blade pitch faults in wind turbines. *IEEE Trans. Sustain. Energy* **2011**, *2*, 87–96. [CrossRef]
18. Lu, D.; Zhang, P. MCSA-based Fault Diagnosis Technology for Motor Drivetrains. In Proceedings of the 2018 IEEE International Power Electronics and Application Conference and Exposition (PEAC), Shenzhen, China, 4–7 November 2018.
19. Gill, S.; Stephen, B.; Galloway, S. Wind Turbine Condition Assessment through Power Curve Copula Modeling. *IEEE Trans. Sustain. Energy* **2012**, *3*, 94–101. [CrossRef]
20. Guo, P.; Xu, M.; Bai, N. Wind turbine tower vibration modeling and monitoring driven by SCADA data. *Zhongguo Dianji Gongcheng Xuebao/Proc. Chin. Soc. Electr. Eng.* **2013**, *33*, 128–135.
21. Benedetti, M.; Fontanari, V.; Zonta, D. Structural health monitoring of wind towers: Remote damage detection using strain sensors. *Smart Mater. Struct.* **2011**, *20*, 055009. [CrossRef]
22. Bang, H.J.; Kim, H.I.; Lee, K.S. Measurement of strain and bending deflection of a wind turbine tower using arrayed FBG sensors. *Int. J. Precis. Eng. Manuf.* **2012**, *13*, 2121–2126. [CrossRef]
23. Soua, S.; Van Lieshout, P.; Perera, A. Determination of the combined vibrational and acoustic emission signature of a wind turbine gearbox and generator shaft in service as a pre-requisite for effective condition monitoring. *Renew. Energy* **2013**, *51*, 175–181. [CrossRef]
24. Wilkinson, M.R.; Spinato, F.; Tavner, P.J. Condition Monitoring of Generators & Other Subassemblies in Wind Turbine Drive Trains. In Proceedings of the IEEE International Symposium on Diagnostics for Electric Machines, Cracow, Poland, 6–8 September 2007.
25. Eftekharnejad, B.; Addali, A.; Mba, D. Shaft crack diagnostics in a gearbox. *Appl. Acoust.* **2012**, *73*, 723–733. [CrossRef]
26. Abusaad, S.; Benghozzi, A.; Smith, A. The Detection of Shaft Misalignments Using Motor Current Signals from a Sensorless Variable Speed Drive. *Mech. Mach. Sci.* **2014**, *23*, 173–182.
27. Nie, M.; Wang, L. Review of Condition Monitoring and Fault Diagnosis Technologies for Wind Turbine Gearbox. *Procedia CIRP* **2013**, *11*, 287–290. [CrossRef]
28. Feng, Y.; Qiu, Y.; Crabtree, C.J. Monitoring wind turbine gearboxes. *Wind Energy* **2013**, *16*, 728–740. [CrossRef]
29. Sheng, S. Wind Turbine Gearbox Condition Monitoring Round Robin Study—Vibration Analysis. *Off. Sci. Tech. Inf. Tech. Rep.* **2012**, *68*, 856–860.
30. Zhong, X.Y.; Zeng, L.C.; Zhao, C.H. Research of Condition Monitoring and Fault Diagnosis Techniques for Wind Turbine Gearbox. *Appl. Mech. Mater.* **2012**, *197*, 206–210. [CrossRef]
31. Qin, H.W.; Cao, F.C.; Fan, Q.Y. Use of AE Testing Data for Condition Monitoring in Wind Turbine Gearbox. *Adv. Mater. Res.* **2014**, *1070*, 1893–1897. [CrossRef]
32. Bell, R.N.; Mcwilliams, D.W.; O'Donnell, P. Report of Large Motor Reliability Survey of Industrial and Commercial Installations, Part I. *IEEE Trans. Ind. Appl.* **1985**, *4*, 853–864.
33. Bell, R.N.; Mcwilliams, D.W.; O'Donnell, P. Report of Large Motor Reliability Survey of Industrial and Commercial Installations, Part II. *IEEE Trans. Ind. Appl.* **1985**, *4*, 865–872.
34. Bell, R.N.; Mcwilliams, D.W.; O'Donnell, P. Report of Large Motor Reliability Survey of Industrial and Commercial Installations, Part III. *IEEE Trans. Ind. Appl.* **1987**, *23*, 153–158.

35. Albrecht, P.F.; Appiarius, J.C.; Mccoy, R.M. Assessment of the Reliability of Motors in Utility Applications —Updated. *IEEE Trans. Energy Convers.* **1986**, *1*, 39–46. [CrossRef]

36. Zhang, P.; Du, Y.; Habetler, T.G. A Survey of Condition Monitoring and Protection Methods for Medium-Voltage Induction Motors. *IEEE Trans. Ind. Appl.* **2011**, *47*, 34–46. [CrossRef]

37. Zarei, J.; Poshtan, J. Bearing fault detection using wavelet packet transform of induction motor stator current. *Tribol. Int.* **2007**, *40*, 763–769. [CrossRef]

38. Chen, J.; Pan, J.; Li, Z. Generator bearing fault diagnosis for wind turbine via empirical wavelet transform using measured vibration signals. *Renew. Energy* **2016**, *89*, 80–92. [CrossRef]

39. Wang, J.; Peng, Y.; Qiao, W. Current-Aided Order Tracking of Vibration Signals for Bearing Fault Diagnosis of Direct-Drive Wind Turbines. *IEEE Trans. Ind. Electron.* **2016**, *63*, 6336–6346. [CrossRef]

40. Yang, B.; Liu, R.; Chen, X. Fault Diagnosis for Wind Turbine Generator Bearing via Sparse Representation and Shift-invariant K-SVD. *IEEE Trans. Ind. Inform.* **2017**, *13*, 1321–1331. [CrossRef]

41. Amirat, Y.; Choqueuse, V.; Benbouzid, M. EEMD-based wind turbine bearing failure detection using the generator stator current homopolar component. *Mech. Syst. Signal Process.* **2013**, *41*, 667–678. [CrossRef]

42. Wang, L.; Zhao, Y.; Jia, W. Fault diagnosis based on current signature analysis for stator winding of Doubly Fed Induction Generator in wind turbine. In Proceedings of the International Symposium on Electrical Insulating Materials, Niigata, Japan, 1–5 June 2014.

43. Shah, D.; Nandi, S.; Neti, P. Stator-Interturn-Fault Detection of Doubly Fed Induction Generators Using Rotor-Current and Search-Coil-Voltage Signature Analysis. *IEEE Trans. Ind. Appl.* **2008**, *45*, 1831–1842. [CrossRef]

44. Abadi, M.B.; Cruz, S.M.A.; Gonçalves, A.P. Inter-turn fault detection in doubly-fed induction generators for wind turbine applications using the stator reactive power analysis. In Proceedings of the Renewable Power Generation Conference (RPG 2014), Naples, Italy, 24–25 September 2014.

45. Zaher, A.; Mcarthur, S.D.J.; Infield, D.G. Online wind turbine fault detection through automated SCADA data analysis. *Wind Energy* **2009**, *12*, 574–593. [CrossRef]

46. Ma, K.; Yang, Y.; Wang, H.; Blaabjerg, F. Design for reliability of power electronics in renewable energy systems. In *Use, Operation and Maintenance of Renewable Energy Systems*; Springer: Cham, Switzerland, 2014; pp. 295–338.

47. Lu, B.; Sharma, S.K. A Literature Review of IGBT Fault Diagnostic and Protection Methods for Power Inverters. *IEEE Trans. Ind. Appl.* **2009**, *45*, 1770–1777.

48. Avenas, Y.; Dupont, L.; Baker, N. Condition Monitoring: A Decade of Proposed Techniques. *IEEE Ind. Electron. Mag.* **2015**, *9*, 22–36. [CrossRef]

49. Yang, S.; Xiang, D.; Bryant., A. Condition Monitoring for Device Reliability in Power Electronic Converters: A Review. *IEEE Trans. Power Electron.* **2010**, *25*, 2734–2752. [CrossRef]

50. Xue, S.; Quan, Z.; Jian, L. Reliability evaluation for the DC-link capacitor considering mission profiles in wind power converter. In Proceedings of the IEEE International Conference on High Voltage Engineering & Application, Chengdu, China, 19–22 September 2016.

51. Boettcher, M.; Reese, J.; Fuchs, F.W. Reliability comparison of fault-tolerant 3L-NPC based converter topologies for application in wind turbine systems. In Proceedings of the IECON 2013—39th Annual Conference of the IEEE Industrial Electronics Society, Vienna, Austria, 10–13 November 2013.

52. Ghimire, P.; de Vega, A.R.; Beczkowski, S. Improving power converter reliability: Online monitoring of high-power IGBT modules. *IEEE Ind. Electron. Mag.* **2014**, *8*, 40–50. [CrossRef]

53. Patil, N.; Das, D.; Goebel, K. Identification of failure precursor parameters for Insulated Gate Bipolar Transistors (IGBTs). In Proceedings of the International Conference on Prognostics & Health Management, Denver, CO, USA, 6–9 October 2008.

54. Yang, L.; Agyakwa, P.A.; Johnson, C.M. Physics-of-Failure Lifetime Prediction Models for Wire Bond Interconnects in Power Electronic Modules. *IEEE Trans. Device Mater. Reliab.* **2013**, *13*, 9–17. [CrossRef]

55. Xiang, D.; Ran, L.; Tavner, P. Condition monitoring power module solder fatigue using inverter harmonic identification. *IEEE Trans. Power Electron.* **2011**, *27*, 235–247. [CrossRef]

56. Siegel, D.; Zhao, W.; Lapira, E. A comparative study on vibration-based condition monitoring algorithms for wind turbine drive trains. *Wind Energy* **2014**, *17*, 695–714. [CrossRef]

57. Igba, J.; Alemzadeh, K.; Durugbo, C. Analysing RMS and peak values of vibration signals for condition monitoring of wind turbine gearboxes. *Renew. Energy* **2016**, *91*, 90–106. [CrossRef]

58. Joshuva, A.; Sugumaran, V. A data driven approach for condition monitoring of wind turbine blade using vibration signals through best-first tree algorithm and functional trees algorithm: A comparative study. *ISA Trans.* **2017**, *67*, 160–172. [CrossRef] [PubMed]

59. Abouhnik, A.A. *An Investigation into Vibration Based Techniques for Wind Turbine Blades Condition Monitoring*; Manchester Metropolitan University: Manchester, UK, 2014.

60. Yang, W.; Court, R.; Tavner, P.J. Bivariate empirical mode decomposition and its contribution to wind turbine condition monitoring. *J. Sound Vib.* **2011**, *330*, 3766–3782. [CrossRef]

61. Perišić, N.; Kirkegaard, P.H.; Pedersen, B.J. Cost-effective shaft torque observer for condition monitoring of wind turbines. *Wind Energy* **2015**, *18*, 1–19. [CrossRef]

62. Gong, X. *Online Nonintrusive Condition Monitoring and Fault Detection for Wind Turbines*; University of Nebraska Lincoln: Lincoln, Nebraska, 2012.

63. Guo, P.; Infield, D.; Yang, X. Wind Turbine Generator Condition-Monitoring Using Temperature Trend Analysis. *IEEE Trans. Sustain. Energy* **2011**, *3*, 124–133. [CrossRef]

64. Garcia, M.C.; Sanz-Bobi, M.A.; Pico, J.D. SIMAP: Intelligent System for Predictive Maintenance: Application to the health condition monitoring of a wind turbine gearbox. *Comput. Ind.* **2006**, *57*, 552–568. [CrossRef]

65. Abdusamad, K.B.; Gao, D.W.; Muljadi, E. A condition monitoring system for wind turbine generator temperature by applying multiple linear regression model. In Proceedings of the North American Power Symposium, Manhattan, KS, USA, 22–24 September 2013.

66. Tonks, O.; Wang, Q. The detection of wind turbine shaft misalignment using temperature monitoring. *CIRP J. Manuf. Sci. Technol.* **2017**, *17*, 71–79. [CrossRef]

67. Cheng, S.; Du, Y.; Restrepo, J.A. A Nonintrusive Thermal Monitoring Method for Induction Motors Fed by Closed-Loop Inverter Drives. *IEEE Trans. Power Electron.* **2012**, *27*, 4122–4131. [CrossRef]

68. Wu, X.; Li, Y.; Li, F. Adaptive Estimation-Based Leakage Detection for a Wind Turbine Hydraulic Pitching System. *IEEE-ASME Trans. Mechatron.* **2015**, *17*, 907–914. [CrossRef]

69. Coronado, D.; Kupferschmidt, C. Assessment and Validation of Oil Sensor Systems for On-line Oil Condition Monitoring of Wind Turbine Gearboxes. *Procedia Technol.* **2014**, *15*, 748–755. [CrossRef]

70. Sheng, S. Monitoring of Wind Turbine Gearbox Condition through Oil and Wear Debris Analysis: A Full-Scale Testing Perspective. *Tribol. Trans.* **2015**, *59*, 149–162. [CrossRef]

71. Zhu, J.; Yoon, J.M.; He, D. Online particle-contaminated lubrication oil condition monitoring and remaining useful life prediction for wind turbines. *Wind Energy* **2015**, *18*, 1131–1149. [CrossRef]

72. Cooley, J.W. An Algorithm for the machine calculation of complex Fourier series. *Math. Comput.* **1965**, *19*, 297–301. [CrossRef]

73. Bodla, M.K.; Malik, S.M.; Rasheed, M.T. Logistic regression and feature extraction based fault diagnosis of main bearing of wind turbines. In Proceedings of the 2016 IEEE 11th Conference on Industrial Electronics and Applications (ICIEA), Hefei, China, 5–7 June 2016.

74. Ibrahim, R.; Watson, S.J. Advanced Algorithms for Wind Turbine Condition Monitoring and Fault Diagnosis. In Proceedings of the Windeurope Summit, Hamburg, Germany, 27–29 September 2016.

75. Lee, J.K.; Park, J.Y.; Oh, K.Y. Transformation algorithm of wind turbine blade moment signals for blade condition monitoring. *Renew. Energy* **2015**, *79*, 209–218. [CrossRef]

76. Yang, W.; Tavner, P.J.; Wilkinson, M. Condition monitoring and fault diagnosis of a wind turbine with a synchronous generator using wavelet transforms. In Proceedings of the 2008 4th IET Conference on Power Electronics, Machines and Drives, York, UK, 2–4 April 2008.

77. Zhang, L.; Lang, Z.Q. Wavelet Energy Transmissibility Function and its Application to Wind Turbine Bearing Condition Monitoring. *IEEE Trans. Sustain. Energy* **2018**, *9*, 1833–1843. [CrossRef]

78. Sapenabano, A.; Pinedasanchez, M.; PuchePanadero, R. Harmonic Order Tracking Analysis: A Novel Method for Fault Diagnosis in Induction Machines. *IEEE Trans. Energy Convers.* **2015**, *30*, 833–841. [CrossRef]

79. Sapena-Bano, A.; Riera-Guasp, M.; Puche-Panadero, R. Harmonic Order Tracking Analysis: A Speed-Sensorless Method for Condition Monitoring of Wound Rotor Induction Generators. *IEEE Trans. Ind. Appl.* **2016**, *52*, 4719–4729. [CrossRef]

80. Sapena-Bano, A.; Burriel-Valencia, J.; Pineda-Sanchez, M. The Harmonic Order Tracking Analysis Method for the Fault Diagnosis in Induction Motors under Time-Varying Conditions. *IEEE Trans. Energy Convers.* **2017**, *32*, 244–256. [CrossRef]

81. Morshedizadeh, M. *Condition Monitoring of Wind Turbines Using Intelligent Machine Learning Techniques*; University of Hawai'i at Manoa: Honolulu, Hawaii, 2017.

82. Ali, Y.H. *Artificial Intelligence Application in Machine Condition Monitoring and Fault Diagnosis Artificial Intelligence-Emerging Trends and Applications*; IntechOpen: Mosul, Iraq, 2018.

83. Pattison, D.; Segovia Garcia, M.; Xie, W. Intelligent integrated maintenance for wind power generation. *Wind Energy* **2016**, *19*, 547–562. [CrossRef]

84. Cheng, F.Z.; Qu, L.Y.; Qiao, W.; Hao, L. Enhanced Particle Filtering for Bearing Remaining Useful Life Prediction of Wind Turbine Drivetrain Gearboxes. *IEEE Trans. Ind. Electron.* **2019**, *66*, 4738–4748. [CrossRef]

85. Durbhaka, G.K.; Selvaraj, B. Predictive maintenance for wind turbine diagnostics using vibration signal analysis based on collaborative recommendation approach. In Proceedings of the 2016 International Conference on Advances in Computing, Communications and Informatics (ICACCI), Jaipur, India, 21–24 September 2016.

86. Lin, D.F.; Chen, P.H.; Williams, M. Measurement and Analysis of Current Signals for Gearbox Fault Recognition of Wind Turbine. *Meas. Sci. Rev.* **2013**, *13*, 89–93. [CrossRef]

87. Gong, X.; Qiao, W. Bearing fault diagnosis for direct-drive wind turbines via current-demodulated signals. *IEEE Trans. Ind. Electron.* **2013**, *60*, 3419–3428. [CrossRef]

88. Zhang, P.; Neti, P. Detection of Gearbox Bearing Defects Using Electrical Signature Analysis for Doubly Fed Wind Generators. *IEEE Trans. Ind. Appl.* **2014**, *51*, 2195–2200. [CrossRef]

 *energies*

Article

An Imbalance Fault Detection Algorithm for Variable-Speed Wind Turbines: A Deep Learning Approach

Jianjun Chen [1], Weihao Hu [1], Di Cao [1], Bin Zhang [1], Qi Huang [1], Zhe Chen [2] and Frede Blaabjerg [2],*

[1] School of Mechanical and Electrical Engineering, University of Electronic Science and Technology of China, Chengdu 611731, China
[2] Department of Energy Technology, Aalborg University, DK-9220 Aalborg, Denmark
* Correspondence: fbl@et.aau.dk

Received: 18 June 2019; Accepted: 17 July 2019; Published: 18 July 2019

Abstract: Wind power penetration has increased rapidly in recent years. In winter, the wind turbine blade imbalance fault caused by ice accretion increase the maintenance costs of wind farms. It is necessary to detect the fault before blade breakage occurs. Preliminary analysis of time series simulation data shows that it is difficult to detect the imbalance faults by traditional mathematical methods, as there is little difference between normal and fault conditions. A deep learning method for wind turbine blade imbalance fault detection and classification is proposed in this paper. A long short-term memory (LSTM) neural network model is built to extract the characteristics of the fault signal. The attention mechanism is built into the LSTM to increase its performance. The simulation results show that the proposed approach can detect the imbalance fault with an accuracy of over 98%, which proves the effectiveness of the proposed approach on wind turbine blade imbalance fault detection.

Keywords: imbalance fault detection; LSTM; attention mechanism; blades with ice

---

1. Introduction

As a clean and renewable energy, wind power has developed rapidly in recent years [1]. With the increasing penetration of wind power, the problems of high maintenance costs of wind turbines and high failure rate have been highlighted [2,3]. Forty percent of the maintenance cost of a wind farm is related to wind turbine component failure [4,5]. Wind turbines are generally installed on the mountain or along the coastline, thus it is difficult to obtain the daily operating state of wind turbines. Wind turbine failure mainly includes the mechanical failure of the gearbox, various bearing and rotor [6], breakage of blades [7], an abnormal working state of generator and power electronics [8], etc. When the wind turbines fail, the fault of which will arise power oscillation in the power system. These problems lead to high maintenance costs and damage to the power grid. Therefore, it is necessary to diagnose the potential danger of wind turbines to avoid more serious accidents before the wind turbine has a devastating failure [9,10].

Traditional fault diagnosis methods are sensor-based monitoring. Installing a large number of sensors in different parts of the wind turbine increases the investment cost. Therefore, it is necessary to apply a more effective data-driven method in fault diagnosis to reduce the investment costs [11]. Various methods have been proposed in literature to solve this problem: Fault detection based on the improved temporal constraint network method [12], the history-driven differential evolution approach [13], cointegration residuals analysis [14], generator current signals [15], and machine learning method [16], etc. The machine learning-based approach has especially been applied in many fields in recent years. Deep learning (DL) is one of the most important parts of this hot topic. Google's AlphaGo [17] and AlphaGo Zero [18] use the deep neural network to train themselves, and have had great breakthroughs in recent years. Because of the great learning ability of DL, it can also be applied in a power system.

Since 2006, DL has appeared as a new research field in machine learning research field [19]. DL can be used to extract the features of a large number of data [20–22]. Due to the high calculation costs and the features being difficult to be extracted, it is not applicable to obtain the features by using traditional mathematical methods [21]. Various DL-based approaches have been proposed in literature for wind turbine fault detection due to the strong feature extraction ability of DL [23]. In general, the fault detection of the DL-based approach consists of two steps: First, one extracts the fault features by neural network, and second, one realizes the classification based on the extracted features [24]. Reference [25] applied the sparse auto-encoder in fault detection of the wind power system transmission line, which realized the wind farm transmission line faults identification, with an accuracy of 99%. A neural network-based approach for gearbox bearings fault detection was proposed in [26]. Study [27] successfully applied an auto-encoder-based method in wind turbine gearbox fault diagnosis. Convolutional neural network is used in fault detection of the wind turbine gearbox [28]. A deep auto-encoder-based method for wind turbine blade breakage diagnose is proposed in study [29], and the accuracies of the detection results reach 100%. All these prove that the application of DL in a power system is feasible. The imbalance fault caused by icing on wind turbine blades is difficult to detect. It is necessary to find a feasible method to detect the fault.

This paper proposes a new method for the wind turbine blade fault detection by combining long short-term memory (LSTM) with the attention mechanism. The contributions of this paper are as follows:

(1) This paper proposes a data-driven method to solve the imbalance fault detection of wind turbine blades which considers the imbalance faults caused by the ice accretion.
(2) A novel method based on LSTM and attention mechanism is proposed to solve the problem of wind turbine blade imbalance fault diagnosis, and it overcomes the problem that traditional methods have in extracting fault features.

The rest of this paper is structured as follows. Section 2 presents the working condition and the imbalance fault of wind turbine. The deep learning framework and the fault diagnosis method are shown in Section 3. Section 4 presents the case study of this research. Finally, the conclusion and summary of this paper is shown in Section 5.

2. Wind Turbines Imbalance Fault

The imbalance fault of wind turbine blades accounts for the majority of the wind turbine failures [4]. Under ideal conditions, the quality of the three wind turbine blades is equal. However, the mass of the wind turbine blades is imbalanced in real-world scenarios due to various factors. For example:

(a) Due to the technical problems in the production process, there are some mass errors among the blades.
(b) Wind turbines are usually installed at heights of tens or even hundreds of meters and the locations are usually at the peaks of mountains or offshore. Wind turbine blades will be corroded by exposure to harsh environments for a long time, which causes the imbalance of wind turbine blades.
(c) In addition, in dusty or extreme cold weather conditions, wind turbine blades will be covered with dust or ice. When the dust or ice accumulates to a certain level, the imbalance fault of wind turbines will occur.

This paper assumes that the imbalance faults of a wind turbine are caused by the wind turbine blades, which are covered with ice. Wind turbines operate at variable wind speed conditions and the wind speed curve is shown in the following Figure 1.

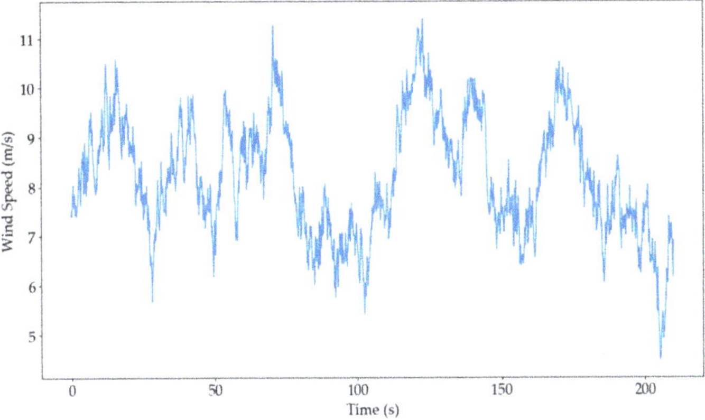

Figure 1. The variable wind speed of test data.

In this research, in order to obtain the data of the wind turbine under different conditions, a 2MW wind turbine with a doubly fed induction generator (DFIG) was built by G. H. Bladed simulation software to verify the proposed method [30,31]. Figure 1 shows that the wind turbines were operating under the variable wind speed which ranges from 4 to 11 m/s. Under this condition, the output power curves of the wind turbine in normal state and fault state are shown in Figure 2.

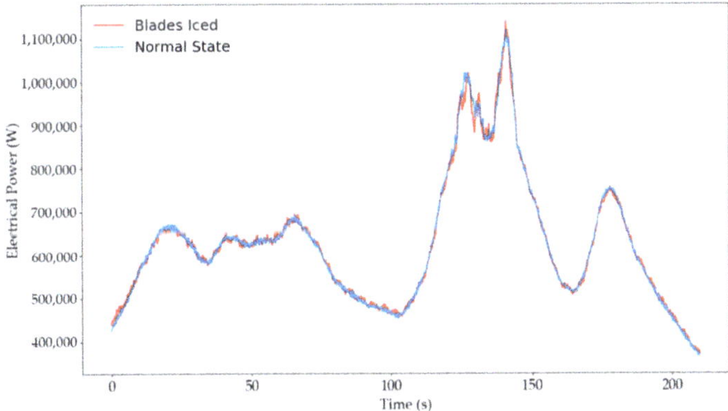

Figure 2. The output power curves of the wind turbine: Blue line represents the output electrical power of normal state under the variable wind speed. Red line presents the output electrical power of blade in iced fault state under the variable wind speed.

It shows that the trends of the two curves are almost the same and it is difficult for traditional fault analysis methods to distinguish the difference between normal state and iced state. Compared with the traditional mathematical methods, this study adopts a neural network, which has proved to be effective in extracting features and detecting the imbalance faults of wind turbine. At the same time, the method of neural network could also reduce the calculation costs.

3. Deep Learning Framework and Training Process

A DL framework is shown in this section. Traditional DL mainly includes the following basic network frameworks: fully connected neural network (FNN), convolutional neural network (CNN) and recurrent neural network (RNN) [20]. RNN has advantages in processing time series data, LSTM is an improved version of RNN and is good at extracting long-term dependency features. LSTM is used to extract features in the proposed approach of this paper. Compared with ordinary neural networks, LSTM network can solve the vanishing gradient problem [32]. In addition, LSTM is outstanding in feature extraction from temporal dependencies data [33]. When processing the time series data, LSTM has high efficiency in the field of machine learning [34,35]. But with the length of data increasing, LSTM has difficulty in feature extracting [33]. In order to enhance the learning ability of LSTM, this study adds the attention mechanism after LSTM. The attention mechanism helps LSTM learn the temporal dependencies data [36].

The DL framework proposed in this paper contains two parts: LSTM and attention mechanism. The details are described in the following subsections respectively.

3.1. Recurrent Neural Network (RNN)

RNN is a neural network with a special structure. Compared with FNN and CNN, RNN can be regarded as a network with memory. It stores the features of the previous moment which is used as the next moment's input. Thus RNN can better obtain the characteristics of time series data than CNN and FNN.

A simple RNN include 3 layers: Input layer, hidden layer and output layer. The standard RNN structure is shown in Figure 3:

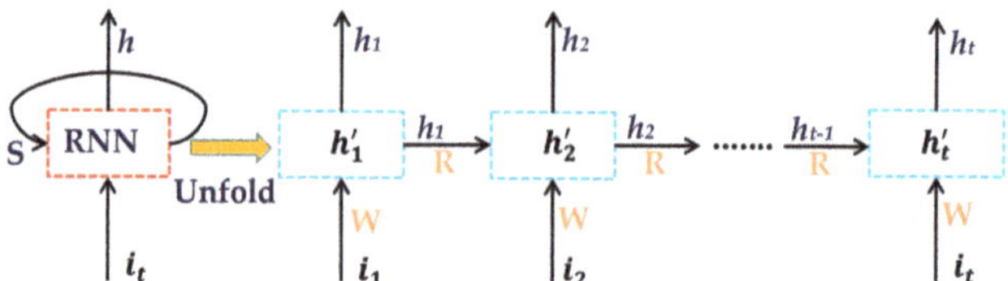

Figure 3. A simple unfold recurrent neural network (RNN) structure.

Where i_t represents the input vector at the t-th moment. In hidden layers, every cell A has an activation function. At each time step of the model, the RNN cell outputs an eigenvalue, which will be sent to the next cell. The specific function of RNN is shown in Equations (1) and (2):

$$h'_t = W \times i_t + R \times h_{t-1} + b,\tag{1}$$

$$h_t = sigmoid(h'_t),\tag{2}$$

where h'_t represents the hidden state of the neural network at time t; W, R and b represent the weight matrices and the bias vector, respectively. h_t is the output of the t-th RNN cell and *sigmoid* is the activation function. Because of this structure, RNN has memory function that can memorize the features of the time series data.

The learning ability of RNN decreases with the increase of dimension and amount of data, however, LSTM can solve this disadvantage of RNN.

3.2. The Overall Framework

In Section 2, preliminary analysis shows that the features of the fault data are not obvious enough. In order to detect the fault effectively, the combination of LSTM and attention mechanism to realize wind turbines imbalance fault detection and classification is a feasible option.

In order to make the neural network more sensitive to the wind turbine imbalance faults, this paper also considers other parameters of wind turbines, not including power and current, which are shown in Equation (3), where v is the hub wind speed magnitude, ω is the rotor speed, p is the electrical power, i is the turbine current and t_m is the generator torque. The v, ω, p, i and t_m are all column vectors. In order to obtain the torque information during wind turbine blade rotation and better reflect the operation characteristics of the wind turbine, the sampling time interval of wind turbine data in this research is 0.08 s.

$$X_t = [v\ \omega\ p\ i\ t_m],\tag{3}$$

The overall structure of the imbalance fault detection is shown in Figure 4. It shows that the attention mechanism is added after the output of LSTM cells, and the softmax function completes the fault detection.

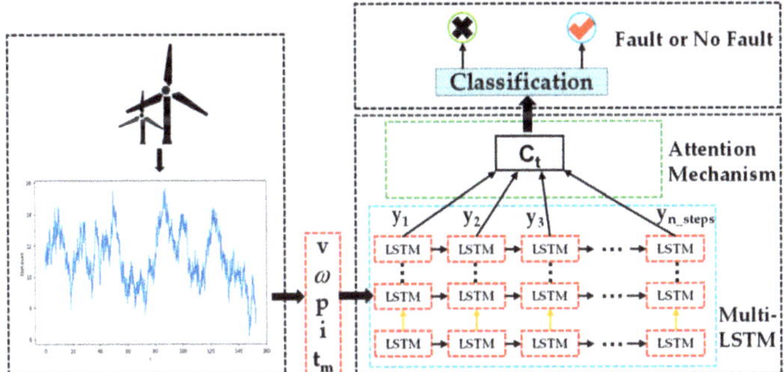

Figure 4. The overall structure of the imbalance fault detection.

3.2.1. LSTM

Since standard RNN just thinks about neighboring states, if the state is too far from the current RNN, the data may be forgotten which could lead the neural network loss learning ability. However, the LSTM doesn't have that problem. Compared with RNN, LSTM has added three special gates in its cell, the forget gate, input gate and output gate. The inner structure of LSTM is shown in Figure 5. The most important part to the LSTM network is the cell state L_t [37]. The LSTM can control the three gates to decide whether the outside data should be written in the cell or not.

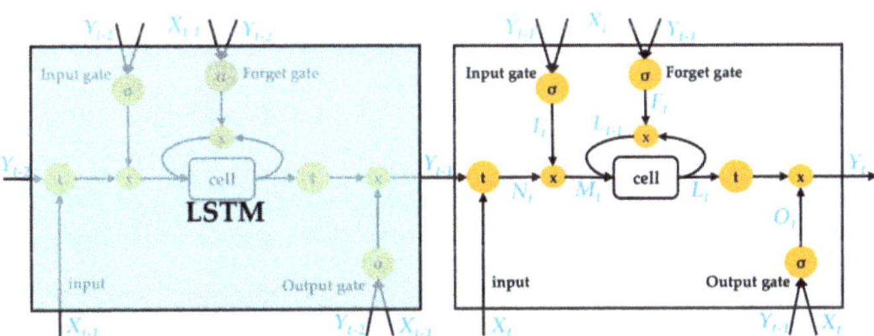

Figure 5. The inner structure of long short-term memory (LSTM), where a circle with a σ represents an activation function and a circle with a x represents multiply function. I_t, F_t, O_t are the output information of input, forget and output gates; and these three control value are all connected with the input X_t and the output of the previous moment Y_{t-1}.

The functions of these three gates are shown in Equations (4)–(6), respectively,

$$I_t = \sigma(W_i \times X_t + Z_i \times Y_{t-1} + b_i),\tag{4}$$

$$F_t = \sigma\big(W_f \times X_t + Z_f \times Y_{t-1} + b_f\big),\tag{5}$$

$$O_t = \sigma(W_o \times X_t + Z_o \times Y_{t-1} + b_o),\tag{6}$$

where the activation function σ is *sigmoid* function, and W_i, W_f, W_o, Z_i, Z_f and Z_o are the weight of each gate respectively, the shapes of which are all matrices, and b_i, b_f, b_o are the biases vector of these three gates. The input data N_t is show as Equation (7),

$$N_t = th(W_t \times X_t + Z_t \times Y_{t-1} + b_t), \tag{7}$$

where th is the activation function, *tanh*, W_t and Z_t are the weight matrices, and b_t is the input biases vector. After obtaining the three gates state and the input information, the intermediate variable M_t can be described as below:

$$M_t = I_t \times N_t, \tag{8}$$

where $\times$ denotes matrix multiplication. The state value of input gate, I_t, is range from 0 to 1, it determines proportionately how much input to pass to the next step. Therefore, the LSTM cell information L_t and the output state Y_t can be formulated as Equations (9) and (10). Like I_t, the state value of forget gate and output gate, F_t and O_t, both range from 0 to 1.

$$L_t = M_t + F_t \times L_{t-1}, \tag{9}$$

$$Y_t = th(L_t) \times O_t, \tag{10}$$

After getting the output information of the LSTM, they all will be sent into the attention mechanism for further processing. Attention mechanism multiplies different time series data by a weight coefficient then obtains the final dynamic characteristics.

In this research, the updating of training parameters is based on gradient descent method and the specific algorithms are shown as below:

$$W_{new} = W_{old} - lr \cdot \frac{\partial E}{\partial W_{old}}, \tag{11}$$

$$Z_{new} = Z_{old} - lr \cdot \frac{\partial E}{\partial Z_{old}}, \tag{12}$$

$$b_{new} = b_{old} - lr \cdot \frac{\partial E}{\partial b_{old}}, \tag{13}$$

where W_{new} and Z_{new} represent the new weights W_i, W_f, W_o, W_t or Z_i, Z_f, Z_o, Z_t after updating of the neural network. Similarly, b_{new} represents the new bias of the network. W_{old}, Z_{old} and b_{old} are the weights and bias of the previous training. lr is the learning rate of the neural network and E is the loss function value. In this research, the loss function of the model is sparse softmax cross entropy with logits [38]. This loss function is a combination of softmax and cross entropy functions. Comparing with softmax cross entropy with logits and cross entropy, the calculation speed of the selected function is faster.

3.2.2. Attention Mechanism

When dealing with long input sequence, only the output of the LSTM neural network, y_t, is used as the information representation of the entire input sequence, that means all information of the input sequence is compressed into a fixed length vector. As the length of the input sequence continues to increase, the ability of the overall model to process information will be limited and weakened. In order to solve this problem, this research has introduced the attention mechanism in the decoding phase. Attention mechanism can be considered as a simple three-layer neural network, which includes input layer, hidden layer and output layer. The input in this paper is the last layer's output of a multi-layer LSTM, which is a vector and the length of this vector is equal to the time steps of LSTM.

Attention mechanism has great advantages on time series learning, and the core goal of attention mechanism is turning the fixed output Y_t into a dynamic context vector C_t. Its characteristic equation can be broken down into the following three steps:

1. The first step is calculating the parameter at i-th time, $u^{i,t}$, which is described as Equation (14):

$$u^{i,t} = V^T \times tanh(W_a \times Y_t + b_a), i, t = 1, 2, \cdots, n\_steps,\qquad(14)$$

where $u^{i,t}$ is a model which scores how well the input of i-th moment and the output of t-th moment match, V^T, W_a, b_a. are the pending training parameters, and $tanh$ is the activation function.

2. The second step is normalizing the data obtained at step one, then getting the weight score $\alpha_{i,t}$ of each state, which is shown as Equation (15),

$$\alpha_{i,t} = \frac{e^{u_{i,t}}}{\sum_{k=1}^{n\_steps} e^{u_{i,k}}}, i, t = 1, 2, \cdots, n\_steps,\qquad(15)$$

where $\alpha_{i,t}$ is a weight coefficient, which is the normalized probability distribution of $u^{i,t}$ at each time step based on Equation (14).

3. Obtaining the dynamic characteristics vector C_t by multiplying the output of LSTM by the probability, which is shown in Equation (16),

$$C_t = \sum_{t=1}^{n\_steps} \alpha_{i,t} \cdot Y_t, i, t = 1, 2, \cdots, n\_steps,\qquad(16)$$

After getting the dynamic context vector C_t, the process of decoding is almost the same as the traditional sequence classification based on LSTM.

3.3. The Training Process

The training process of the algorithm is shown as in Table 1. All parameters have been described in the previous subsection.

Table 1. The training process of the algorithm of the neural network.

Algorithm	Training Process of the Network
Input:	**The parameters of the wind turbineX_t.**
Output:	**The kind of imbalance faults and the accuracies of network.**
1	Randomly initialize the weights W and biases b of the network model.
2	**for** i in max-iterations:
3	Obtain the accuracy and loss value of training network.
4	Error back propagation (E), update the weights and biases based on gradient descent method: $W_{new} = W_{old} - lr \cdot \frac{\partial E}{\partial W_{old}}$, $Z_{new} = Z_{old} - lr \cdot \frac{\partial E}{\partial Z_{old}}$, $b_{new} = b_{old} - lr \cdot \frac{\partial E}{\partial b_{old}}$
5	**for** i%200 = 0:
6	Test and obtain the kind of faults, accuracies and error value of network.
7	**end**
8	**end**

More details regarding the working principle of the proposed method can be described as follows: Firstly, the raw data of wind turbine under normal and imbalance fault operation state are generated by simulation software. The shape of raw data is a two dimensional matrix: $[v\ \omega\ p\ i\ t_m]$, which has been described in Equation (3). But the shape of input data of LSTM must be a three-dimensional array, so that the first task of the model need to do is reshape the raw data into a three dimensional array of shape [**batch size, time step, n-inputs**], where batch size and time step are the training parameters of the neural network and can be adjusted, and n-inputs represents the number of different kinds of wind turbine operation data, which in this paper is five. After the raw data has been reshaped, one then mixes fault data with normal data as the dataset of the model. The dataset will be randomly divided into a training set and testing set. Finally, after learning the features of these dataset, the model can classify the fault signals and normal signals by sparse softmax cross entropy with logits function.

4. Case Study

This paper uses the G. H. Bladed software to simulate the wind power generator with different kinds of imbalance fault, then collects the main information by this software to do the following data processing. This study randomly chooses the 80% of each dataset as a training set, and the remaining 20% of the dataset is divided into 10% for the validation set and 10% for the testing set.

Hardware environment and software platform: The training of network is completed on a PC with Intel(R) Core i9-7900X @ 3.30GHz CPU, 64G DDR4 RAM and Nvidia GeForce RTX 2080 Ti (11GB VRAM). And the software platforms are WINDOWS-10 (Professional) operating system and Pycharm 3.6 (64 bit). This paper uses the GPU version of TensorFlow to build the LSTM neural network and accelerate the hardware.

Data pre-processing: Firstly, add different labels to the different imbalance faults data obtained from G. H. Bladed. Then divide the data into appropriate time-step length as a batch.

4.1. Experimental Results

Figure 6 shows that the imbalance fault occurs at the 10,000th sampling points and disappears at the 20,000th points. Figure 7 shows that when the imbalanced fault is detected, the model gives a pulse signal with a value of 1. When the fault disappears, the value of the pulse drops to 0. Because of signal transmission and data calculation, there will be a short time (1.8 s) delay which is shown in Figure 7.

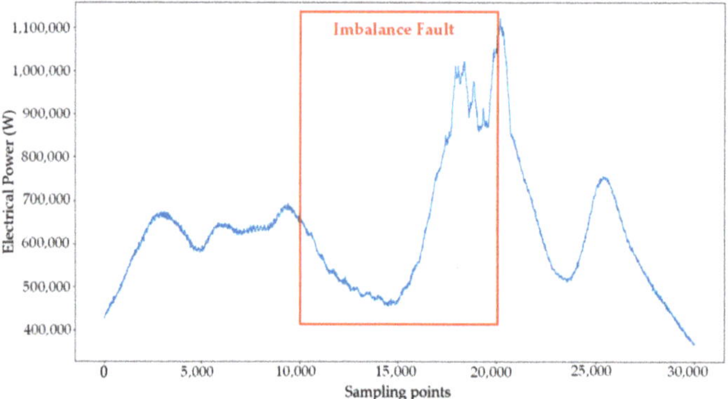

Figure 6. Imbalance fault occurs from the 10,000th to 20,000th sampling points.

In order to prove the feasibility of the proposed method, this paper provides the detection results under different imbalanced fault conditions. The number of iced wind turbine blades ranges from one to three and the mass of ice is also variable. The detection results of network under one wind turbine blade iced condition are shown in Figure 8, and the parameter of imbalance fault is obtained every 200 iterations. The fault detection accuracy of the neural network is more than 99%. The result shows that the proposed DL-based approach is effective in detecting the wind turbine fault.

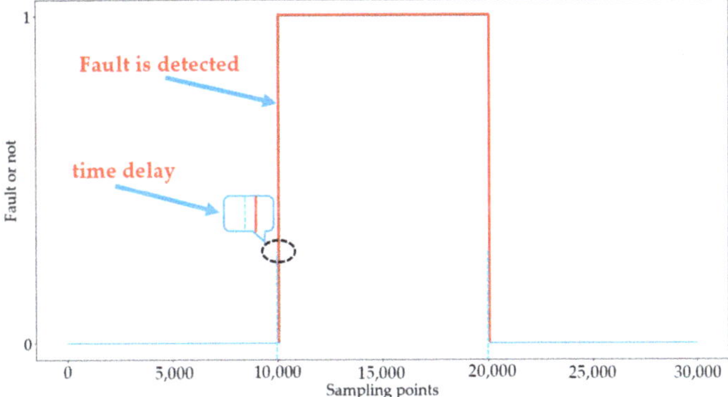

Figure 7. The fault is detected by the proposed model.

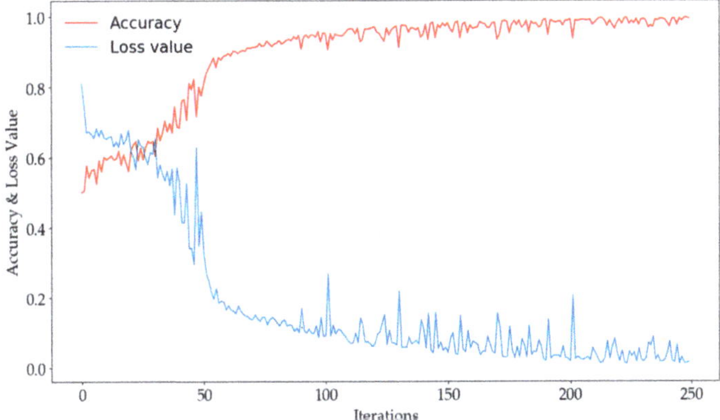

Figure 8. The accuracy and loss value of the neural network.

The accuracies of a neural network with 256 attention size and the accuracy of LSTM without attention mechanism are shown in Figure 9. It can be observed that LSTM combined with attention mechanism can increase the convergence rate. In the early stages, the accuracy of the neural network with attention mechanism hardly changes but the accuracy of the network without attention mechanism slowly rises. As the attention mechanism can hold more features of the time series data, when the network finds the best gradient descent direction, the accuracy of the neural network with attention mechanism rises rapidly. Finally, the accuracy of the network model with attention mechanism is higher than the LSTM without attention mechanism. It proves that the performance of the neural network can be improved by adding attention mechanism.

The accuracies of neural network with different attention size are listed in Table 2. With the increase in attention size, the accuracies of neural network increase. The results in this research show that the best attention size of LSTM combines with attention mechanism is 256, the accuracy of which reaches 99.8%.

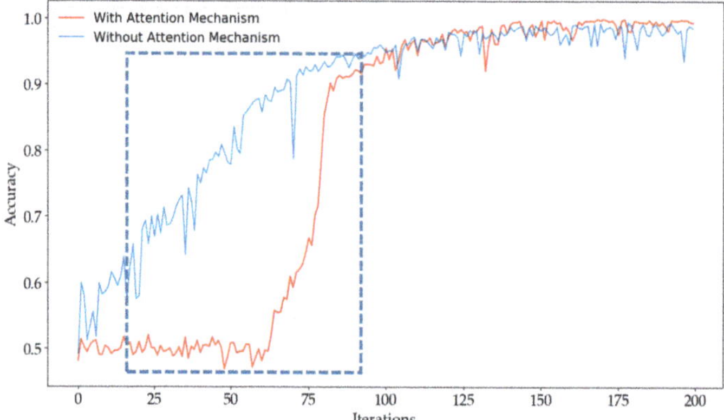

Figure 9. The accuracy curves: The red curve is the accuracy of model with 256 attention size, and the blue curve is the accuracy of LSTM without attention mechanism.

Table 2. The accuracies of models with different attention size.

Attention Size	Iced Number	Accuracy
50	One blade	98.8%
	Two blades	99.2%
	Three blades	99.0%
128	One blade	98.7%
	Two blades	99.0%
	Three blades	98.3%
256	One blade	99.6%
	Two blades	99.8%
	Three blades	99.3%

The accuracies of the neural network with different time-step are shown in Figure 10. It can be observed that in the early stage of the learning process, the accuracy of model with one time-step rises rapidly; but in the end, the accuracy of model with only one time-step is much lower than others with a larger time-step. The reason for this phenomenon is that the datasets are temporal dependencies and only one time-step leads the neural networks can't obtain the temporal correlation characteristics commendably.

The accuracies of models with different time-step length are listed in Table 3. With the increase of time-step, the accuracy of network also increases.

The accuracies of models with different batch size are listed in Table 4. It shows that the highest accuracy of neural networks with batch size of 48 is not more than 88%. Because the batch size of the dataset will determine the direction of gradient descent, a too small batch of dataset will make the direction of gradient descent uncertain, which decreases the learning ability of the neural network. When the batch size of the model increases, the accuracy improves significantly.

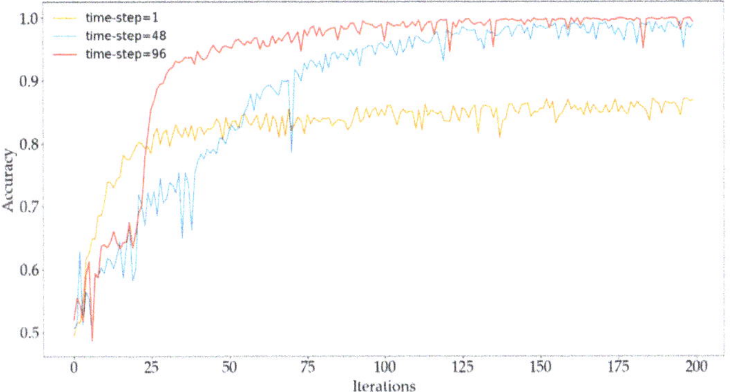

Figure 10. The accuracies of neural network with different time-step under two blades with ice accretion condition.

Table 3. The accuracies of models with different time-step.

Time-Step	Iced Number	With Attention Mechanism	Without Attention Mechanism
1	One blade	87.5%	83.4%
	Two blades	83.5%	85.5%
	Three blades	86.9%	85.6%
48	One blade	97.2%	93.4%
	Two blades	99.0%	95.6%
	Three blades	99.2%	94.9%
96	One blade	99.6%	98.1%
	Two blades	99.8%	98.3%
	Three blades	99.8%	98.6%

Table 4. The accuracies of models with different batch size.

Batch Size	Iced Number	With Attention Mechanism	Without Attention Mechanism
48	One blade	87.5%	81.2%
	Two blades	85.4%	77.1%%
	Three blades	83.3%	79.2%
2048	One blade	97.8%	94.1%
	Two blades	98.6%	94.3%
	Three blades	99.1%	97.3%
4096	One blade	98.1%	97.9%
	Two blades	99.8%	98.2%
	Three blades	100%	98.8%

It can be observed from Tables 3 and 4 that time-step and batch-size are important parameters for neural network: When their values are too large, the memory is heavily occupied and the training time of neural networks increase significantly. The best time-step and batch size of the model in this paper are 96 and 4096 respectively.

When the mass of ice accretion of the wind turbine blades increases, the features of imbalance fault of wind turbine blades are becoming more and more obvious. Compared with 15 kg, the accuracy curves of model with 15 kg and 30 kg ice accretion of each blade are shown in Figure 11. It shows that the accuracy of 30 kg ice accretion of each blades reaches 100%.

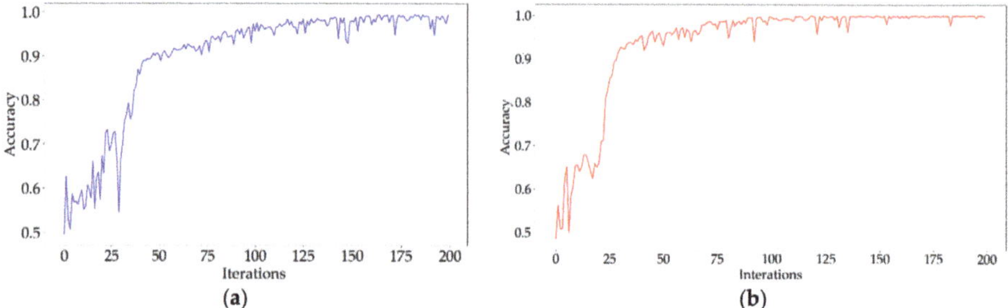

Figure 11. The accuracies of models under different mass of ice accretion condition: (**a**) 15 kg ice on each blade, (**b**) 30 kg ice on each blade.

4.2. Methods Comparison

In order to prove the validity of the method proposed in this paper, this simulation compares the proposed method with standard RNN network. Take the icing on the surface of two blades of a wind turbine as an example, the results of a standard RNN compared with the LSTM with attention mechanism (LSTMAM) are shown in Figure 12.

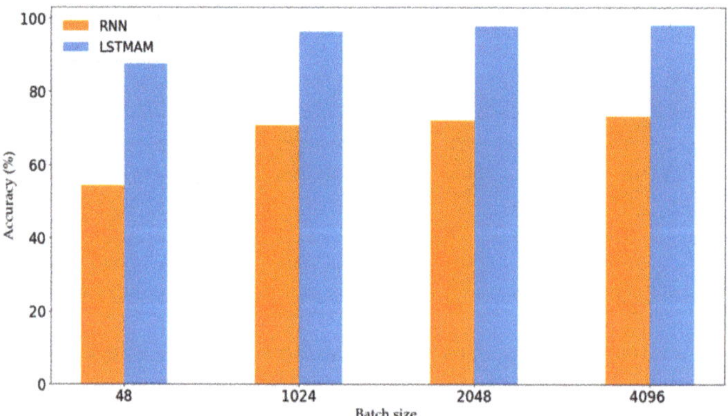

Figure 12. The accuracies of recurrent neural network (RNN) and LSTM with attention mechanism (LSTMAM) with different Batch size.

It is obvious that in Figure 12, no matter how the batch size increases, the accuracies of RNN are no more than 74%; but the lowest accuracy of the proposed method is 87.5%. This paper also compares the proposed method with other methods, such as support vector machines (SVM) and Gaussian processes classification (GPC). Take the icing on the surface of two blades of a wind turbine as an example, the results are shown in Table 5. It shows that the accuracies of SVM and GPC are much lower than LSTMAM. Because traditional SVM and GPC are applicable to a small-scale dataset, but when the dimension and complexity of data increase, it is difficult to classify the faults by these methods. The results show that the proposed method outperforms various benchmark methods.

Table 5. The accuracies of different methods.

Approach	Accuracy
RNN	71.3%
SVM	65.0%
GPC	48.3%
LSTMAM	99.8%

A high sampling frequency is required in this research. When the imbalance fault occurs, the variation will occur on the low speed shaft torque and the rotating frequency of shaft is called 1 P [39]. Meanwhile, there is the fluctuation on aerodynamic torque on hub and effect on rotor speed caused by tower shadow. The spectra of the shaft torque or the output electric power of wind turbine with three blades will have fluctuation at 3 P frequency, which is three times the shaft rotating frequency. It is necessary to judge the frequency of 1 P and 3 P to detect whether the wind turbine has imbalance fault. The rotor speed of the wind turbine shown in this research is from 9 to 18 r/min, which corresponds to the 1 P and 3 P oscillation frequency from 0.15 to 0.3 and 0.45 to 0.9 Hz respectively. And the sampling frequency in this research is 12.5 Hz. According to Nyquist Sampling Theory [40], if the sampling frequency is too low, it is difficult to observe 1 P and 3 P frequency, which leads to inaccurate or the inability to detect the fault by the proposed method.

The noise of raw data can influence the learning of neural networks, which makes the model misjudge the signal. There are some artificial intelligence methods which can deal with the noise problem and with relatively mature technology, such as the auto-encoder [41], variational auto-encoder, stacked denoising auto-encoder [42], etc. These methods can effectively improve the robustness of the model to the noise.

5. Conclusions

This paper proposes an DL-based method which combines LSTM and an attention mechanism for wind turbine imbalance fault detection and classification. Compared with the standard LSTM, combining the LSTM and an attention mechanism can improve the learning ability and the convergence rate. This paper not only analyzes the voltage and current signals, but also considers other factors, such as wind speed and the torque of the hub in the dataset. Furthermore, compared with standard RNN, SVM and Gaussian Processes classification methods, the proposed method has a better performance in imbalance fault detection. The simulation results show that the proposed method is feasible in wind turbine blade imbalance detection and the highest accuracy of the proposed method is 100%.

Author Contributions: Conceptualization, W.H., D.C., J.C. and Q.H.; Methodology, W.H., D.C. and J.C.; Software, D.C. and J.C.; Validation, D.C. and J.C.; Formal Analysis, W.H., D.C., J.C. and B.Z.; Investigation, D.C. and J.C.; Data Curation, W.H and J.C.; Writing-Original Draft Preparation, D.C. and J.C.; Writing-Review & Editing, W.H., F.B., B.Z. and D.C.; Visualization, D.C. and J.C.; Supervision, W.H., Z.C. and F.B.

Funding: This research was funded by the National Natural Science Foundation of China, grant number 51707029.

Acknowledgments: The authors gratefully acknowledge the National Natural Science Foundation of China and appreciate the insightful comments and suggestions from the reviewers and the editor.

References

1. Salvador, S.; Costoya, X.; Sanz-Larruga, F.; Gimeno, L. Development of Offshore Wind Power: Contrasting Optimal Wind Sites with Legal Restrictions in Galicia, Spain. *Energies* **2018**, *11*, 731. [CrossRef]
2. Zhao, M.; Jiang, D.; Li, S. Research on fault mechanism of icing of wind turbine blades. In Proceedings of the 2nd World Non-Grid-Connected Wind Power and Energy Conference, Nanjing, China, 11–12 September 2009.
3. Wang, N.; Li, J.; Hu, W.; Zhang, B.; Huang, Q.; Chen, Z. Optimal reactive power dispatch of a full-scale converter based wind farm considering loss minimization. *Renew. Energy* **2019**, *139*, 292–301. [CrossRef]
4. Entezami, M.; Hillmansen, S.; Weston, P.; Papaelias, M.P. Fault detection and diagnosis within a wind turbine mechanical braking system using condition monitoring. *Renew. Energy* **2012**, *47*, 175–182. [CrossRef]
5. Li, J.; Wang, N.; Zhou, D.; Hu, W.; Huang, Q.; Chen, Z.; Blaabjerg, F. Optimal Reactive Power Dispatch of Permanent Magnet Synchronous Generator-Based Wind Farm Considering Levelised Production Cost Minimisation. *Renew. Energy* **2019**, *145*, 1–12. [CrossRef]
6. Tabatabaeipour, S.M.; Odgaard, P.F.; Bak, T.; Stoustrup, J. Fault Detection of Wind Turbines with Uncertain Parameters: A Set-Membership Approach. *Energies* **2012**, *11*, 2424–2448. [CrossRef]

7. Hur, S.; Recalde-Camacho, L.; Leithead, W.E. Detection and compensation of anomalous conditions in a wind turbine. *Energy* **2017**, *124*, 74–86. [CrossRef]

8. Yang, W.; Liu, C.; Jiang, D. An unsupervised spatiotemporal graphical modeling approach for wind turbine condition monitoring. *Renew. Energy* **2018**, *127*, 230–241. [CrossRef]

9. Hameed, Z.; Hong, Y.S.; Cho, Y.M.; Ahn, S.H.; Song, C.K. Condition monitoring and fault detection of wind turbines and related algorithms: A review. *Renew. Sustain. Energy Rev.* **2009**, *13*, 1–39. [CrossRef]

10. Ozgener, O.; Ozgener, L. Exergy and reliability analysis of wind turbine systems: A case study. *Renew. Sustain. Energy Rev.* **2007**, *11*, 1811–1826. [CrossRef]

11. Deng, F.; Chen, Z.; Khan, M.R.; Zhu, R. Fault Detection and Localization Method for Modular Multilevel Converters. *IEEE Trans. Power Electron.* **2015**, *30*, 2721–2732. [CrossRef]

12. Cui, Y.; Shi, J.; Wang, Z. Power System Fault Reasoning and Diagnosis Based on the Improved Temporal Constraint Network. *IEEE Trans. Power Deliv.* **2016**, *31*, 946–954. [CrossRef]

13. Zhao, J.; Xu, Y.; Luo, F.; Dong, Z.Y.; Peng, Y. Power system fault diagnosis based on history driven differential evolution and stochastic time domain simulation. *Inf. Sci.* **2014**, *275*, 13–29. [CrossRef]

14. Dao, P.B.; Staszewski, W.J.; Barszcz, T.; Uhl, T. Condition monitoring and fault detection in wind turbines based on cointegration analysis of SCADA data. *Renew. Energy* **2018**, *116*, 107–122. [CrossRef]

15. Gong, X.; Qiao, W. Imbalance Fault Detection of Direct-Drive Wind Turbines Using Generator Current Signals. *IEEE Trans. Energy Convers.* **2012**, *27*, 468–476. [CrossRef]

16. Marugán, A.P.; García Márquez, F.P.; Perez, J.M.P.; Ruiz-Hernández, D. A survey of artificial neural network in wind energy systems. *Appl. Energy* **2018**, *228*, 1822–1836. [CrossRef]

17. Silver, D.; Huang, A.; Maddison, C.J.; Guez, A.; Sifre, L.; Van Den Driessche, G.; Schrittwieser, J.; Antonoglou, L.; Panneershelvam, V.; Lanctot, M.; et al. Mastering the game of Go with deep neural networks and tree search. *Nature* **2016**, *529*, 484–489. [CrossRef]

18. Silver, D.; Schrittwieser, J.; Simonyan, K.; Antonoglou, I.; Huang, A.; Guez, A.; Hubert, T.; Baker, L.; Lai, M.; Bolton, A.; et al. Mastering the game of Go without human knowledge. *Nature* **2017**, *550*, 354–359. [CrossRef] [PubMed]

19. Schmidhuber, J. Deep learning in neural networks: An overview. *Neural Netw.* **2015**, *61*, 85–117. [CrossRef]

20. LeCun, Y.; Bengio, Y.; Hinton, G. Deep learning. *Nature* **2015**, *521*, 436–444. [CrossRef]

21. Wang, H.; Li, G.; Wang, G.; Peng, J.; Jiang, H.; Liu, Y. Deep learning based ensemble approach for probabilistic wind power forecasting. *Appl. Energy* **2017**, *188*, 56–70. [CrossRef]

22. Cremer, J.L.; Konstantelos, I.; Tindemans, S.H.; Strbac, G. Data-Driven Power System Operation: Exploring the Balance Between Cost and Risk. *IEEE Trans. Power Syst.* **2019**, *34*, 791–801. [CrossRef]

23. Helbing, G.; Ritter, M. Deep Learning for fault detection in wind turbines. *Renew. Sustain. Energy Rev.* **2018**, *98*, 189–198. [CrossRef]

24. Lei, J.; Liu, C.; Jiang, D. Fault diagnosis of wind turbine based on Long Short-term memory networks. *Renew. Energy* **2019**, *133*, 422–432. [CrossRef]

25. Chen, K.; Hu, J.; He, J. Detection and Classification of Transmission Line Faults Based on Unsupervised Feature Learning and Convolutional Sparse Autoencoder. *IEEE Trans. Smart Grid* **2018**, *9*, 1748–1758.

26. Bangalore, P.; Tjernberg, L.B. An Artificial Neural Network Approach for Early Fault Detection of Gearbox Bearings. *IEEE Trans. Smart Grid* **2015**, *6*, 980–987. [CrossRef]

27. Jiang, G.; He, H.; Xie, P.; Tang, Y. Stacked Multilevel-Denoising Autoencoders: A New Representation Learning Approach for Wind Turbine Gearbox Fault Diagnosis. *IEEE Trans. Instrum. Meas.* **2017**, *66*, 2391–2402. [CrossRef]

28. Jiang, G.; He, H.; Yan, J.; Xie, P. Multiscale Convolutional Neural Networks for Fault Diagnosis of Wind Turbine Gearbox. *IEEE Trans. Ind. Electr.* **2019**, *66*, 3196–3207. [CrossRef]

29. Wang, L.; Zhang, Z.; Xu, J.; Liu, R. Wind Turbine Blade Breakage Monitoring With Deep Autoencoders. *IEEE Trans. Smart Grid* **2018**, *9*, 2824–2833. [CrossRef]

30. Hassan, G. Bladed Theory Manual. Co: Garrad Hassan & Partners Ltd, 2011. Available online: https://max.book118.com/html/2013/1030/4863004.shtm (accessed on 12 March 2019).

31. Burton, T.; Sharpe, D.; Jenkins, N.; Bossanyi, E. *Wind Energy Handbook*; Wiley: Hoboken, NJ, USA, 2001. [CrossRef]

32. Hochreiter, S.; Schmidhuber, J. Long Short-Term Memory. *Neural Comput.* **1997**, *9*, 1735–1780. [CrossRef]

33. Karim, F.; Majumdar, S.; Darabi, H.; Chen, S. LSTM Fully Convolutional Networks for Time Series Classification. *IEEE Access* **2018**, *6*, 1662–1669. [CrossRef]
34. Graves, A.; Mohamed, A.; Hinton, G. Speech recognition with deep recurrent neural networks. In Proceedings of the IEEE International Conference on Acoustics, Speech and Signal Processing, Vancouver, BC, Canada, 26–27 May 2013.
35. Zheng, C.; Wang, S.; Liu, Y.; Liu, C.; Xie, W.; Fang, C.; Liu, S. A Novel Equivalent Model of Active Distribution Networks Based on LSTM. *IEEE Trans. Neural Netw. Learn. Syst.* **2019**. [CrossRef]
36. Gao, L.; Guo, Z.; Zhang, H.; Xu, X.; Shen, H.T. Video Captioning With Attention-Based LSTM and Semantic Consistency. *IEEE Trans. Multimed.* **2017**, *19*, 2045–2055. [CrossRef]
37. Wan, Z.; Li, H.; He, H.; Prokhorov, D. Model-Free Real-Time EV Charging Scheduling Based on Deep Reinforcement Learning. *IEEE Trans. Smart Grid* **2018**. [CrossRef]
38. Ahmed, N. Data-Free/Data-Sparse Softmax Parameter Estimation With Structured Class Geometries. *IEEE Signal Process. Lett.* **2018**, *25*, 1408–1412. [CrossRef]
39. Arany, L.; Bhattacharya, S.; Macdonald, J.H.G.; Hogan, S.J. Closed form solution of Eigen frequency of monopile supported offshore wind turbines in deeper waters incorporating stiffness of substructure and SSI. *Soil Dyn. Earthq. Eng.* **2016**, *83*, 18–32. [CrossRef]
40. Mishali, M.; Eldar, Y.C. From Theory to Practice: Sub-Nyquist Sampling of Sparse Wideband Analog Signals. *IEEE J. Sel. Top. Signal Process.* **2010**, *4*, 375–391. [CrossRef]
41. Deng, Y.; Bao, F.; Kong, Y.; Ren, Z.; Dai, Q. Deep Direct Reinforcement Learning for Financial Signal Representation and Trading. *IEEE Trans. Neural Netw. Learn. Syst.* **2017**, *28*, 653–664. [CrossRef]
42. Vincent, P.; Larochelle, H.; Lajoie, I.; Bengio, Y.; Manzagol, P.A. Stacked Denoising Autoencoders: Learning Useful Representations in a Deep Network with a Local Denoising Criterion. *J. Mach. Learn. Res.* **2010**, *11*, 3371–3408.

 *energies*

Article

Framework for Managing Maintenance of Wind Farms Based on a Clustering Approach and Dynamic Opportunistic Maintenance

Juan Izquierdo [1,2,*], Adolfo Crespo Márquez [2], Jone Uribetxebarria [1] and Asier Erguido [1,2]

[1] Ikerlan Technology Research Centre, Operations and Maintenance Technologies Area,
    20500 Gipuzkoa, Spain; juribetxebarria@ikerlan.es (J.U.); aerguido@ikerlan.es (A.E.)
[2] Industrial Organization and Business Management I, School of Engineering, University of Seville,
    Camino de los Descubrimientos s/n, 41092 Seville, Spain; adolfo@us.es
* Correspondence: jizquierdo@ikerlan.es

Received: 15 April 2019; Accepted: 20 May 2019; Published: 28 May 2019

Abstract: The growth in the wind energy sector is demanding projects in which profitability must be ensured. To fulfil such aim, the levelized cost of energy should be reduced, and this can be done by enhancing the Operational Expenditure through excellence in Operations & Maintenance. There is a considerable amount of work in the literature that deals with several aspects regarding the maintenance of wind farms. Among the related works, several focus on describing the reliability of wind turbines and many set the spotlight on defining the optimal maintenance strategy. It is in this context where the presented work intends to contribute. In the paper a technical framework is proposed that considers the data and information requisites, integrated in a novel approach a clustering-based reliability model with a dynamic opportunistic maintenance policy. The technical framework is validated through a case study in which simulation mechanisms allow the implementation of a multi-objective optimization of the maintenance strategy for the lifecycle of a wind farm. The proposed approach is presented under a comprehensive perspective which enables the discovery an optimal trade-off among competing objectives in the Operations & Maintenance of wind energy projects.

Keywords: maintenance management; wind turbines; clustering; reliability; dynamic opportunistic maintenance; simulation

---

1. Background and Introduction

The attention drawn by renewable energy has increased considerably over recent years. This increasing importance has nurtured an important growth, which is especially prominent in the wind energy sector [1]. Wind energy is one of the main sources of power generation in Europe with a vast majority of installed capacity in the form of onshore Wind Farms (WFs) [2]. For the profitability of wind energy projects to be ensured, it is essential to reduce the levelized cost of energy (LCoE) to its minimum [2,3]. Aiming at increasing the energy yield of WFs, the LCoE should be reduced, and this can be done by directly addressing the Capital Expenditure (CAPEX) and the Operational Expenditure (OPEX) of the projects [2].

The scope of the research here presented abides in OPEX reduction; more specifically, this paper is intended to reduce the Operations and Maintenance (O&M) costs. How to reduce the O&M costs has become an ongoing challenge for WFs; the figures associated with these costs are notorious [4,5] and may rise by up to 32% and 12–30% for offshore and onshore WFs, respectively [6,7]. However, aiming at lowering the LCoE by reducing O&M costs is a two-fold challenge since it also entails minimizing the lost energy production [8] for the entire lifecycle, which oscillates around 20 years perspective [9]. According to the International Renewable Agency, the costs associated with O&M

account for up to one quarter of the proportion of the LCoE, with 80% of this expenses directly attributed to maintenance [10–12].

In view of the maintenance role in the LCoE and thus in the profitability of WFs, it is important to consider models for optimization of O&M plans and decisions [2]. The evolution of maintenance models and methodologies have kept pace with the constant technological evolution of Wind Turbines (WTs) [3]. According to [3], the goal of all the approaches and methodologies is determining the most adequate maintenance plan, the management of the resources, and the aspects related to Reliability, Availability, and Maintainability (RAM) of the WTs.

Within such context, the WF operators are bound to develop new techniques and decision-support tools for optimal maintenance strategies, if they strive to maximize the profitability of the investment [8]. Accordingly, maintenance management discipline acquires a highly significant position since it provides a comprehensive perspective for the management of WTs, allowing for optimal maintenance strategies which reduce maintenance costs while maximizing availability [3,5].

The decision-making process in the asset management field has been divided into strategic (long-term), tactic (medium-term), and operational (short-term) to achieve excellence in maintenance [13–15]. It is the purpose of the research in this paper to address the different aspects of the maintenance decision-making process. This is done by answering the research question of whether it is possible to achieve maintenance excellence for the lifecycle of WFs by a maintenance strategy which considers the different behaviors of the WTs and integrated business-related objectives.

Aiming at providing an answer for the aforementioned research question, a technical framework for managing maintenance is proposed in the paper. The framework is a comprehensive proposal that considers different aspects regarding the maintenance of a WF. Within the possibilities offered by the current trend for big data, certain key aspects to create a failure database are integrated in the framework, which enables a clustering approach based on the failure behaviors of the different failure modes of each WT. This approach supports an opportunistic maintenance policy with dynamic thresholds that regard not only to the reliability of the assets but also business considerations. Besides, the framework also incorporates the strategic view through a Life-Cycle Cost (LCC) perspective integrated by means of a multi-objective optimization and supported by simulation techniques that provide valuable information to find an attractive trade-off between cost and performance.

1.1. Related Works

In the context of maintenance, several studies have proven the utility of reliability approaches to optimize it [5,16,17]. If seeking to take forward a reliability study, the information needed as an input to the decision should be at hand, in the right format, and on time [18]. The idea of a common database in the wind energy sector is not novel. In [19] the objectives of a RAM database can be seen and in [10] different sources of WT data are analyzed. The challenges faced when building such database with quality data have been addressed by the integration of data coming from different sources [4,10,19,20]. Nonetheless, to translate the data into information and exploit its inherent value, it is essential a correct assessment of the failure process and therefore the selection of the proper time-to-failure model, which will enable optimization of the maintenance plan [21].

The reliability study of a WF and the creation of the failure database involves combining data from similar assets, in this case WTs. This combination of data is known as data-pooling, and in [21], as well as in [22], several conditions to be met by the equipment subject to data-pooling are stated. Nonetheless, this may lead to combining data from assets with unlike behaviors. The heterogeneity in the failure behavior among WTs may be caused by the presence of different models of WTs, which entail different technical solutions [23]. Moreover, this multiplicity in the failure frequencies may also be the result of the different operational conditions of the assets [16,17,24,25].

To deal with the diversity in failure behaviors when managing a considerable number of assets, similarity-based approaches have been proposed in recent research works [26–28]. The work in [27] where a spectral clustering approach is proposed to later address the maintenance optimization problem to all the assets belonging to the same cluster is especially interesting. Another interesting work is the research in [29], which provides a very similar approach using a different clustering algorithm. In the wind energy sector, and with the purpose of enhancing maintenance management, approaches based on clustering concepts and algorithms can also be found in the recent literature. A fuzzy clustering approach based on Mahalanobis distance is proposed in [30] for warnings and failure detection and it is applied to a real WF. Another interesting proposal is the cluster analysis combined with Frequent Pattern Mining presented in [31] for WT fault detection. In addition, it has also been proposed along with Artificial Neural Networks in [32] for developing optimal maintenance strategies.

The selected approach highly conditions the adequacy of the maintenance strategy [33]. In the case of WTs, it is important to select one that takes into account the multi-component nature of the turbines [34,35]. The WTs are composed of several subsystems with dependencies among them that can be classified as (i) economic, where the simultaneous performance of maintenance activities implies different economic consequences that implement them individually [36]; (ii) structural, where maintenance actions on one system may imply maintenance activities on others [37]; and (iii) stochastic, where the failure hazard of two different systems are not independent [38]. In this context, opportunistic maintenance policies are the most suitable, and therefore they have been widely researched [33].

Opportunistic maintenance policy makes the most of short-term situations to perform the maintenance of non-failed systems when a failure has already happened in another one, grounding its decision on a threshold regarding a system's age, reliability or health condition [5]. Opportunistic maintenance has proven its utility in the sector through several works in the recent literature [39], and it has been highly related to multi-criteria approaches [40]. The consideration of opportunistic maintenance under a multi-criteria perspective allows the handling of some main conflicting objectives such as maintenance cost, availability, or manager preferences [41]. This advantage is especially beneficial in the wind energy sector where it is important to bear in mind the maximization of revenue, power, and reliability, and the minimization of Operations and Maintenance costs [42]. Some of the latest research in the wind energy sector is in this area. The application of opportunistic maintenance in [43] takes advantage of low wind-speed periods to perform corrective actions. The research in [44] provides an opportunistic maintenance policy based on remaining useful life estimation according to condition-monitoring data. The opportunistic maintenance is also proposed with different types of maintenance actions, e.g., the works in [34,45–47]. In addition, the authors of [48,49] integrate the opportunistic maintenance policy with multi-objective optimization.

1.2. Overview

Given the importance of the reviewed works in the literature of the Wind Energy sector, it is the scope of this paper to provide a managing framework that supports the maintenance management of WF operators. The proposed framework is a cross-functional value proposition that starts by considering the requisites of the failure database to later define a maintenance policy. The opportunistic maintenance policy will be supported by a clustering approach that allows the addressing of different behaviors of the WTs within the same WF. Besides, the framework is considered under a lifecycle perspective integrated through simulation techniques and multi-objective optimization algorithms.

The proposed framework is presented in the following Section 2. An initial and brief introduction is first provided along with a representation according to IDEF0 methodology, then every function is explained in detail in each corresponding subsection. The principles of the creation of a failure data base for the study is explained in Section 2.1, the reliability considerations of WTs are presented in Section 2.2, and finally, the opportunistic maintenance policy is explained in Section 2.3. In Section 3 the case study is presented, Section 3.1 describes the development of the case study and the initial assumptions and then Section 3.2 consists of the results obtained in the implementation of the framework. Finally, Section 4 summarizes the main conclusions obtained through the research process, its output represented as the framework, and the application to a real case study in the wind energy sector.

2. Framework and Methods

The proposed technical framework, Figure 1, is a comprehensive integration of different technical solutions and methods to ease maintenance management. The framework proposes an initial data treatment, considering different information sources, to create a RAM database that will enable modeling of the reliability of WTs. To undertake the reliability modeling, a clustering algorithm is proposed along with the Kullback–Leibler Divergence measure, which approach addresses the difficulty derived from the heterogeneity in the failure behaviors of the WTs. The reliability models, which describe the failure behaviors within the WF, as well as data from the RAM database and information regarding the cost structure, will serve as input for the methods proposed for defining an optimal maintenance strategy. These methods regard multi-objective optimization algorithms and simulation software under both a lifecycle perspective and governed by the principles of opportunistic maintenance.

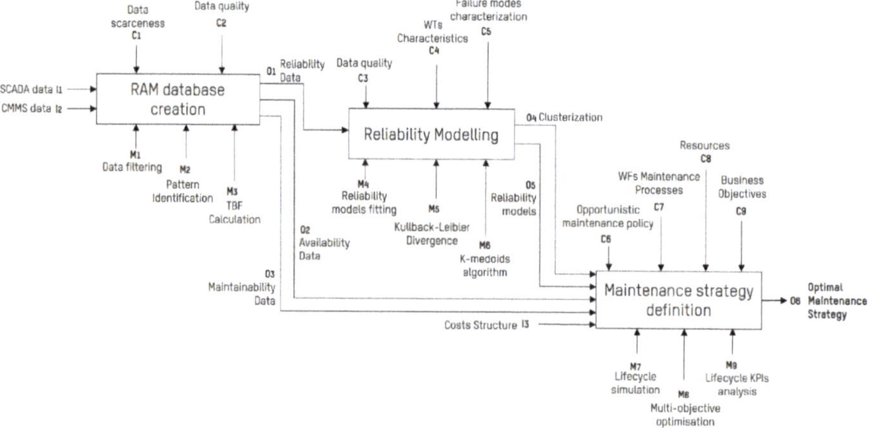

Figure 1. Technical framework.

The technical framework is represented in Figure 1 following the IDEF0 methodology [50], it consists of three functions: RAM database creation, Reliability modeling, and Maintenance strategy definition. The functions are associated with interfaces represented by arrows, which have been referred to as INCOMs [51]: Inputs (I) of the function box enter from the left, they are transformed into Outputs (O) leaving the box by means of the Mechanisms (M), which enter though the bottom and refer to tools, algorithms, or resources that enable the function; and the Controls (C) enter the box from the top and constrain the function.

In particular, Inputs and Outputs of the herein-depicted framework pertain to data and decisions, whereas Mechanisms correspond to specific algorithms, models, and means that allow creation of the RAM database, to model the reliability of WTs and to select the optimal maintenance strategy. The function application is constrained by the Controls, which in this research represents business aspects, data requisites, or maintenance process specificities. By defining the INCOMs in such manner, the technical framework flow can be conceived as a flow in which the output of a function constitutes the input of another until an optimal maintenance strategy is reached. In the successive subsections, each one of the functions with their corresponding INCOMs is thoroughly explained, besides Table 1 gathers all the mathematical symbols utilized in the subsections.

Table 1. Nomenclature.

Reliability Modeling

β	Shape parameter of Weibull distribution	d_i	Diagonal entries of Degree matrix
α	Scale parameter of Weibull distribution	$\overline{\overline{I}}$	Identity matrix
γ	Euler–Mascheroni constant	L_{sym}	Graph Laplacian matrix
Γ	Gamma function	$\lambda_1, ..., \lambda_C$	Graph Laplacian matrix eigenvalues
$\overline{\overline{W}}$	Similarity matrix	$\bar{u}_1, ..., \bar{u}_C$	Graph Laplacian matrix eigenvectors
w_{ij}	Entries of the similarity matrix	$\overline{\overline{U}}$	Matrix of eigenvectors
$d_{KL}(\mu_i \| \mu_j)$	Kullback–Leibler Divergence among two distributions	u_{ic}	Entries of $\overline{\overline{U}}$ matrix
d_{KL}^{sym}	Symmetric Kullback–Leibler Divergence	$\overline{\overline{T}}$	Transformed $\overline{\overline{U}}$ matrix
$\overline{\overline{D}}$	Degree matrix	t_{ic}	Entries of $\overline{\overline{T}}$ matrix

Maintenance Strategy Definition

k_a	Time value for money constant	GP_t	Generated power in t
q^c	Restoration factor of CM	NT	Number of MTs
q^{pr}	Restoration factor of PM	c^{disp}	Cost of dispatching a team
c^c	Corrective cost	θ_t	Binary variable for corrective dispatch of MT
c^{pr}	Preventive cost	γ_t	Binary variable for preventive dispatch of MT
c^{na}	Opportunity cost for not-produced energy	c^{team}	Cost of MT
c^p	Penalty cost	DRT	Dynamic reliability threshold
z_{hikt}	Binary variable for CM	SRT	Static reliability threshold
y_{hikjt}	Binary variable for PM	T^{wt}	Available working time of MTs
m^c	Maintainability of corrective actions	t	Time period (days)
m^{pr}	Maintainability of preventive actions		

2.1. RAM Database Creation

Information is an essential pillar in every decision-making process and it is highly conditioned by the characteristics of the data it comes from. In the decision-making process, to select the optimal maintenance strategy herein presented, the information utility will be highly conditioned by the quantity and quality of the available data (Constraints C1 and C2). As stated by [21], many data are recorded for maintenance management purposes rather than reliability; hence, the information content may be misleading without the proper scrutiny and cleaning. In the proposed technical framework, building up the RAM database from the data coming from the SCADA (Supervisory Control And Data Acquisition) and the CMMS (Computerized Maintenance Management Software) systems is considered.

- SCADA Data (Input I1): this data is intended to provide feedback of a high-level overview of the performance of the WTs. The data coming from the SCADA system can be categorized into three types of information recorded in time intervals: operational data, availability data, and alarms data. The operational data usually pertains to different variables which characterize the operation of the WTs, e.g., power output, wind speed, temperature of components, or environmental conditions. Availability is a measure of the total time the WT is operational and ready to produce power, independently of external factors such as the weather, the grid state, or maintenance activities. In addition, the data coming from the alarms is sensor information which indicates the state of the WTs and have an associated severity level and usually are responsible for triggering corrective maintenance actions.

- CMMS Data (Input I2): generally the operators of WFs keep track of the maintenance actions performed in the WTs in a variety of forms. This record of maintenance activities is usually known as *Work Orders* or *Maintenance Logs*. The work orders may be handwritten in predefined forms, but presently it is more common to find the data as digital work order inputs which have detailed information regarding the materials and resources consumed.

To construct a proper RAM database from this data, it is essential to invest a considerable amount of effort in filtering the data, bearing in mind the objective of the data. Therefore, a proper determination of failure modes underpins the data-quality enhancement (Mechanism M1). As the data comes from two different sources, it is essential to match the information contained in both. With that aim, the identification of the patterns that link SCADA information with the maintenance work orders of the CMMS (Mechanism M2) is proposed. Once the data is properly filtered according to the failure mode definition and the coherence among the two sources of data is ensure, it is possible to calculate the Time Between Failures (TBF) for the failure modes of the WTs (Mechanism M3).

2.2. Reliability Modeling

Considering that the database contains the failure information (output from previous functional box O1) with certain quality (Constraint C3), it is possible to address the modeling of the reliability following a spectral cluster approach, which is proposed to address the differences in the failure behaviors of the same failure mode in each WT (Constraints C4 and C5). The methodological process followed in this functional unit is represented in Figure 2.

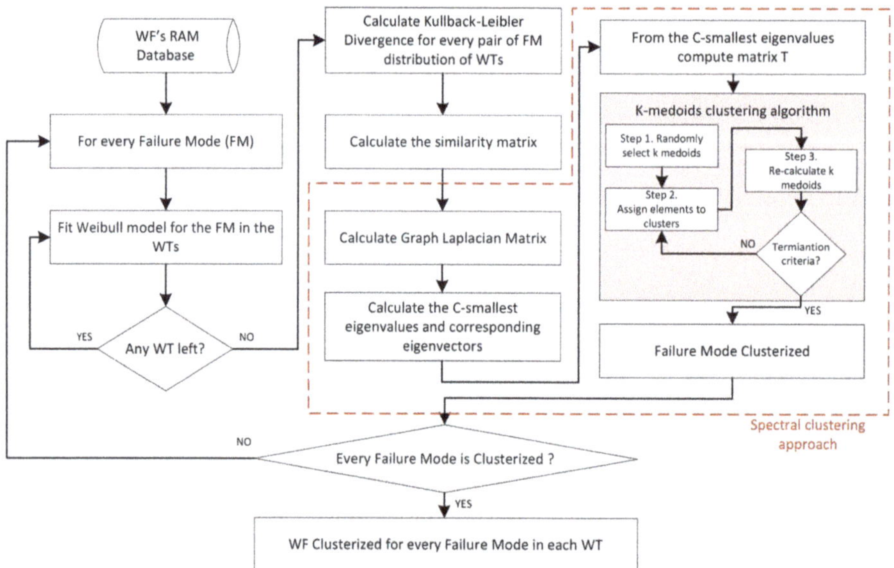

Figure 2. Methodological process of the Reliability Modeling functional box.

For every failure mode, the process considers fitting a Weibull distribution to the TBF data for each WT (Mechanism M4). With the Weibull distributions of each WT and failure mode, it is the purpose of the algorithms herein presented to cluster them following a similarity-based approach inspired by the work in [26,27].

It is possible to measure how similar two probability distributions are according to the Kullback–Leibler Divergence (Mechanism M5), which can be calculated by Equation (1) in the case of two Weibull distributions being $\gamma \approx 0.5772$ the Euler–Mascheroni constant and $\Gamma - \int_{-\infty}^{+\infty} t^{z-1} e^{-t} dt \ z \geq 0$ the gamma function. Nonetheless the Kullback–Leibler Divergence is a non-symmetric measure which can be converted to symmetric by Equation (2) and therefore the similarity among two Weibull distributions can be described in a generic and symmetric form by Equation (3). Expressing similarities in such way allows the definition of a similarity matrix $\overline{\overline{W}}$ in which each element is w_{ij}.

$$d_{KL}(\mu_i\|\mu_j) = \log\left(\frac{\beta_i\alpha_j^{\beta_j}}{\beta_j\alpha_i^{\beta_i}}\right) - (\beta_i - \beta_j)\left(\log(\alpha_i) - \frac{\gamma}{\beta_i}\right) + \left(\frac{\alpha_i}{\alpha_j}\right)^{\beta_j}\Gamma\left(\frac{\beta_j}{\beta_i} + 1\right) - 1 \tag{1}$$

$$d_{KL}^{sym} = \frac{1}{2}(d_{KL}(\mu_i\|\mu_j) + d_{KL}(\mu_j\|\mu_i)) \tag{2}$$

$$w_{ij} = \frac{1}{1 + d_{KL}^{sym}} \tag{3}$$

From the similarity matrix $\overline{\overline{W}}$ of size $[N, N]$ it is possible to construct the graph $G = (V, E)$ where each of the nodes v_i represents the baseline Weibull distribution of certain failure mode of each turbine, and each edge e_{ij} is the similarity between two distributions based on the Kullback–Leibler Divergence. Having defined such a graph, the problem is finding the partition of the graph such that the weights of the edges are small and large for intercluster and intracluster connections, respectively. To fulfil such an aim, the spectral clustering approach is proposed, therefore it is necessary to compute the normalized Graph Laplacian matrix. From $\overline{\overline{W}}$ the degree matrix $\overline{\overline{D}}$ is calculated, which is a diagonal matrix whose diagonal entries are defined according to Equation (4), and then the normalized Graph Laplacian matrix is calculated as described in Equation (5), where $\overline{\overline{L}} = \overline{\overline{D}} - \overline{\overline{W}}$ and $\overline{\overline{I}}$ is the identity matrix of size $[N, N]$.

$$d_i = \sum_{j=1}^{N} w_{ij}, \quad i = 1, ..., N \tag{4}$$

$$L_{sym} = \overline{\overline{D}}^{-1/2}\overline{\overline{L}}\,\overline{\overline{D}}^{-1/2} = \overline{\overline{I}} - \overline{\overline{D}}^{-1/2}\overline{\overline{W}}\,\overline{\overline{D}}^{-1/2} \tag{5}$$

To extract the information of the graph, the C smallest eigenvalues $\lambda_1, ..., \lambda_C$ are selected along with their corresponding eigenvectors $\bar{u}_1, ..., u_C$, with C the desired number of clusters. The relevant information is considered by transforming matrix $\overline{\overline{W}}$ into a reduced matrix $\overline{\overline{U}}$ of size $[N, C]$. The columns of $\overline{\overline{U}}$ are the C eigenvectors $\bar{u}_1, ..., \bar{u}_C$ which contain the information regarding the similarities among the i-th baseline distribution and others. It has been proven that it is possible to enhance cluster properties of the data by normalizing the rows of the matrix $\overline{\overline{U}}$ and forming matrix $\overline{\overline{T}}$ [52], where every element is computed following Equation (6).

$$t_{ic} = \frac{u_{ic}}{\left(\sum_{c=1}^{C} u_{ic}^2\right)^{0.5}}, \quad i = 1, ..., N, \ c = 1, ..., C \tag{6}$$

Once the matrix $\overline{\overline{T}}$ is obtained, a k-medoids algorithm (Mechanism M6) is proposed as an unsupervised clustering to partition the dataset into C clusters, which has proven to perform better for large datasets and to be more stable against possible outliers [53].

By repeating these calculations for every failure mode, it is possible to address the definition of the maintenance strategy of a WF. The effectiveness of the maintenance strategy is expected to increase since the WF is now characterized by different clusters (Output O4) of every failure mode, with their corresponding reliability model (Output O5) which directly tackles the issue of heterogeneity in the failure frequencies.

2.3. Maintenance Strategy Definition

The clustering and their corresponding reliability models will support the definition and optimization of the maintenance strategy. In line with the insights provided in the literature review, opportunistic maintenance management is especially suitable for the wind energy sector, since it takes advantage of short-term information and dependencies among the WTs to make optimal maintenance decisions. Accordingly, the framework proposes to implement a reliability-based dynamic opportunistic maintenance policy (C8) summarized in Figure 3, which as well as considering the economic dependencies of WTs, enables taking advantage of more favorable weather conditions to enhance WF production outcome. To fulfil this aim, based on the concept of dynamic reliability thresholds—acting as decision variables of the maintenance model—maintenance activities are fostered at low wind-speed periods, and hindered at high wind-speed periods, thus limiting the production losses caused by maintenance downtimes.

While the interested reader may address the dynamic opportunistic maintenance policy in [5], the building blocks of the multi-level maintenance optimization model developed for this research are summarized as follows (the definition of the intermediate variables of the model is in Table 2):

- **Problem definition.** The wind farm (WF) to be considered in the case study involves the maintenance of H wind turbines (WTs) and their systems ($i = 1,2,...,N$), connected in series. Each of the systems might fail in k different failure modes (FMs), which are classified according to their consequences and require different corrective maintenance (CM) activities ($k = 1,2,...,K$). Likewise, WF managers may decide to maintain the WTs' system before failure occurrence, where per-FM different preventive maintenance (PM) levels can be performed ($j = 1,2,...,J$). This classification depends on the restoration factor (q). If the PM activity restores the system to a state in which its operating and reliability behavior is as good as new, i.e., replacement activities, it is considered to be perfect maintenance ($j = J$). On the contrary, when PM activity partially restores the condition of the system to an operational condition worse than the new one but better than just before the performance of maintenance, it is considered to be imperfect maintenance (see [54] for further information), with $j = 1$ the most imperfect repair.
- **Objective functions.** The main objectives pursued by the maintenance strategy are defined. They should be aligned with the objectives of the business (C9) and should consider the whole lifecycle of the assets. In the case of WFs, special emphasis is placed on optimizing the OPEX (considering time value of money by means of k_a) and production losses entailed by the maintenance strategy (See Equations (12) and (13)). The core parts of these indicators are:

  - Corrective and preventive cost (I3), subjected to the restoration effect (q) of the maintenance activity carried out, as in [34,49]. They consider the materials and tools needed to perform them (c_{ik}^c, c_{ik}^{pr}), as well as the opportunity cost that they entail, represented by the amount of power that could not be produced because of performing such maintenance activities (c^{na}). Likewise, in the case of CM, failures usually avoid distribution of the committed energy, entailing a penalty cost (c^p).

$$z_{hikt} \cdot \left[c_{ik}^c \cdot \left(q_{ik}^c \right)^2 + m_{ik}^c \cdot GP_t \cdot \left(c^{na} + c^p \right) \right] \tag{7}$$

$$y_{hikjt} \cdot \left[c_{ikj}^{pr} \cdot \left(q_{ikj}^{pr} \right)^2 + m_{ikj}^{pr} \cdot GP_t \cdot c^{na} \right] \tag{8}$$

  - Maintenance resources cost (I3). They consider the number of maintenance teams hired (NT) and the cost entailed by dispatching them to the WFs $\left(c^{disp} \right)$ either preventively (γ_t) or correctively (θ_t).

$$(\gamma_t + \theta_t) \cdot c^{disp} \tag{9}$$

$$NT \cdot c^{team} \tag{10}$$

- Production losses. They consider the maintainability (O3) of PM and CM activities (m_{ik}^c, m_{ik}^{pr}) as well as the power that would have been generated (GP_t).

$$GP_t \cdot \left(m_{ik}^c \cdot z_{hikt} + m_{ikj}^{pr} \cdot y_{hikjt} \right) \tag{11}$$

- **Maintenance strategy.** An imperfect maintenance strategy where decisions are triggered by dynamic reliability thresholds is defined (see Equation (14)):

  1. DRT_{ikt}, if the reliability of any of the FMs (O5) is below this threshold, a maintenance team is compulsorily dispatched to the WFs to perform PM.
  2. SRT_{ikjt}, once a maintenance team is dispatched to the WF, either preventively or correctively, and according to the reliability of the specific FM (O5), it determines whether PM level j should be performed during period t for preventing FM k of system i.

- **Capacity constrains (C8).** Maintenance teams' availability and their working time (T^{wt}) is defined (see Equation (15)).
- **Maintenance process constraints (C7).** Only one maintenance task per WT and time period is allowed (see Equation (16)).

Once the building block of the model has been defined, the mathematical formulation of the model is expressed in the following equations regarding the objective functions (*OF*), the constraints of the model, and the intermediate binary variables (in Table 2).

$$OF_{Opex} = min \left[\sum_t (\gamma_t + \theta_t) \cdot c^{disp} + \sum_h \sum_i \sum_k \sum_t z_{hikt} \left[c_{ik}^c \left(q_{ik}^c \right)^2 + m_{ik}^c \cdot GP_t \left(c^{na} + c^p \right) \right] + \right.$$
$$\left. \sum_h \sum_i \sum_k \sum_j \sum_t y_{hikjt} \left[c_{ikj}^{pr} \left(q_{ikj}^{pr} \right)^2 + m_{ikj}^{pr} \cdot GP_t \cdot c^{na} \right] + \sum_t NT \cdot c^{team} \right] \cdot (1 + k_a)^{-t} \tag{12}$$

$$OF_{LP} = min \sum_t GP_t \cdot \left(\sum_h \sum_i \sum_k m_{ik}^c \cdot z_{hikt} + \sum_h \sum_i \sum_k \sum_j m_{ikj}^{pr} \cdot y_{hikjt} \right) \tag{13}$$

S.T.

$$0 \leq DRT_{ikt} \leq SRT_{ik1t} \leq \ldots \leq SRT_{ikjt} \leq$$
$$\leq \ldots \leq SRT_{ikJt} \leq 1 \ i\epsilon I, k\epsilon K, j\epsilon J; t\epsilon T \tag{14}$$

$$\sum_i \sum_k \sum_j m_{ikj}^{pr} \cdot y_{ikjt} + \sum_i \sum_k m_{ik}^c \cdot z_{ikt} \leq NT \cdot T^{wt} \ \forall t\epsilon T \tag{15}$$

$$\sum_j y_{hikjt} + z_{hikt} \leq 1 \ h\epsilon H, i\epsilon I, k\epsilon K, t\epsilon T \tag{16}$$

$$z_{hikt}, y_{hikjt} \epsilon \{0, 1\} \ h\epsilon H, i\epsilon I, k\epsilon K, t\epsilon T, \forall j = 1, 2$$

Table 2. Intermediate binary variables used in the model.

$z_{hikt} = \begin{cases} 1 & \text{if } CM\,k\,is\,performed\,in\,system\,i \\ & of\,WT\,h\,in\,period\,t \\ 0 & otherwise \end{cases}$	$y_{hikjt} = \begin{cases} 1 & \text{if } PM\,j\,is\,performed\,in\,FM \\ & k\,of\,system\,i\,of\,WT\,h\,in\,period\,t \\ 0 & otherwise \end{cases}$
$\theta_t = \begin{cases} 1 & \text{if } a\,MT\,is\,correctively\,dispatched\,to\,WF \\ & in\,period\,t \\ 0 & otherwise \end{cases}$	$\gamma_t = \begin{cases} 1 & \text{if } a\,MT\,is\,preventively\,dispatched \\ & to\,WF\,in\,period\,t \\ 0 & otherwise \end{cases}$

Due to the numerous stochastic processes that must be considered within maintenance management models, such as repair processes or climate conditions, it is difficult to solve them analytically. Therefore, in line with previous research [46,49,55], the analytically derived model has been implemented in a simulation model (M7) that enables both the maintenance processes characterization and its optimization. The specific simulation and decision processes, as well as the restrictions considered, may be addressed by the reader in the flowchart of Figure 3.

As may be noticed, maintenance decisions are triggered according to the dynamic reliability thresholds (conditioned by weather conditions), which define the maintenance strategy (C6). Whenever a maintenance activity is performed, their reliability is updated, as are the values of the OF considered (C9). Likewise, specific maintenance processes (C7) and available resources (C8) are considered to analyze whether maintenance activities may be triggered. Such decision processes are repeated for each time period, until the end of the WT lifecycle, where the final OF are achieved.

Since more than one objective is pursued by the modeled maintenance problem, i.e., minimizing OPEXs and production losses, multi-objective optimization algorithms (M8) must be implemented to solve them. To this respect, the multi-objective meta-heuristic NSGA II [56] has been implemented, which offers high-quality non-dominated solutions and diversity on the Pareto Front [57].

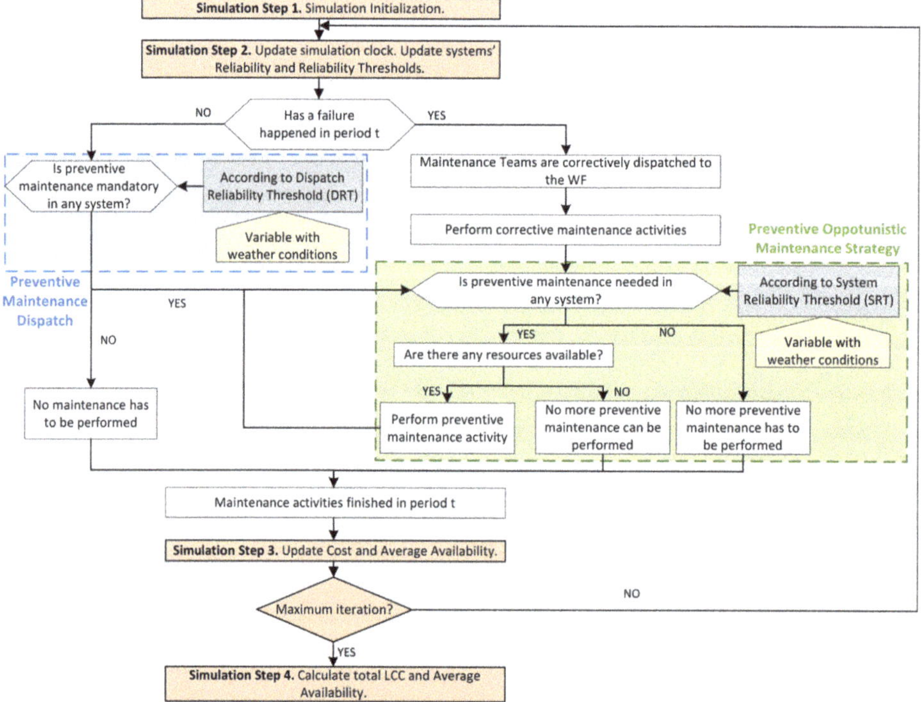

Figure 3. Dynamic Opportunistic Maintenance simulation flowchart.

In this context, the optimal maintenance strategies will be found by a joint use of the simulation model developed and the NSGA II optimization algorithm. While the former allows evaluation of the outcome of the selected maintenance strategy according to the OF, the latter will guide what maintenance strategies should be selected following the logic underlying behind the algorithm. Once the optimization process is finished, the aforementioned Pareto Front will be obtained with its corresponding Key Performance Indicators (KPI). From the Pareto Front and according to the lifecycle KPIs (M9), decision-makers will be able to represent their decision preferences through the maintenance strategy they choose (O6).

3. Case Study

To test the suitability of the proposed research presented in the paper, a case study based on real data is presented. The case study is especially focused on the reliability modeling and the maintenance strategy definition so the readers can see how to apply the proposed algorithms to manage the maintenance of the fleet. The proposed approach is compared in the case study against a static opportunistic maintenance policy which is in itself an advanced maintenance policy. By the application of the clustering approach and the dynamic opportunistic maintenance, it can be seen that the good results rendered by the static opportunistic maintenance can be further improved.

3.1. Description

The application of the technical framework is considered within a case study of an onshore WF consisting of 100 WTs ($H = 100$) whose behavior has been simulated for 20 years based on real field data. The real data correspond to over 300 WTs of 1.67 megawatt (MW) operating in the north of Spain for a time span of 12 years. According to the importance of wind speed for the energy-based availability and the dynamic opportunistic maintenance model, the simulation has been fed with wind data from the location assumed.

RAM data has been provided by a wind energy OEM and pertains to eight FMs. The FMs correspond to minor and major failures ($K = 2$) of four components ($N = 4$) which are critical from the maintenance perspective, either in terms of availability or cost: the gearbox, the blades, the yaw system, and the pitch system. Likewise, for each failure mode, both perfect and imperfect maintenance levels are considered ($J = 2$), where perfect maintenance has a restoration factor of $q_{ik2}^{pr} = 1$ and imperfect maintenance of $q_{ik1}^{pr} = 0.75$, according to the maintenance routine adopted.

In terms of the cost structure required to analyze the maintenance management from a lifecycle perspective, the main costs considered are as follows. The maintenance team costs, consisting of 2 workers each, are assumed to be 800 € /day, the opportunity cost 105 € /MWh, and the penalization cost 35 € /MWh. Likewise, the reader may address the specific material costs considered in [58]. Cost of PM is assumed to be lower than CM to avoid obvious results (30 % lower). Finally, an interest rate of 5% has been determined for the LCC analysis.

3.2. Results

From the simulated database, the behavior of each one of the FMs of every WT has been characterized by fitting a two-parameter Weibull distribution. Following the proposed methodology, the differences among each of the WTs for every failure mode have been assessed by means of the Kullback–Leibler Divergence. This enables the expression of the similarities among the turbines for each failure mode by the weight matrix, which can be represented as an undirected graph. Then, for each failure mode graph, the partition that maximizes the similarities among elements in the same cluster and minimizes the similarities outside the cluster has been found by the k-medoids algorithm. The partition provides the number of clusters for every failure mode, hence each cluster is formed of WTs with like behaviors for each failure mode. For every cluster, a joint reliability Weibull model can be fitted for the failure data that correspond to the WTs in the cluster; the model with which they will be managed will be developed later.

The partitions of the FMs have also been found by the k-means algorithm to compare the performance of both clustering algorithms. The results yielded by the k-means algorithm are very similar to the results yielded by the k-medoids. In fact, 11 out of 31 clusters remain the same. In the differing clusters, the practical deviations are very slight; the usage of one algorithm over the other entails differences of 0.49% and 2.77% on average in the obtained scale and shape parameters, respectively. Therefore, the remainder of the case study, consisting of the simulation-based optimization, has been carried out with the results yielded by the k-medoids algorithm, since it is more stable against outliers and performs better for large datasets according to previously cited literature.

In Figure 4, as an example, the undirected graph of the minor gearbox failure of every WT is represented. In it can also be seen the reliability functions corresponding to every cluster. In the depicted example, six clusters can be appreciated and the nodes in each cluster correspond to the WTs identified by a number. It can be seen that the edges connecting the nodes have different color intensity, which corresponds to the weight that quantifies the similarity among the behaviors of the same failure mode in different turbines, i.e., the higher the intensity, the more alike they are. By this representation, it can also be seen from a general perspective the similarities among the clusters. The reliability functions corresponding to each one of the clusters can also be seen in Figure 4, and it can be seen how the closer two are, the higher the intensity of the connections of the nodes from the clusters in the graph.

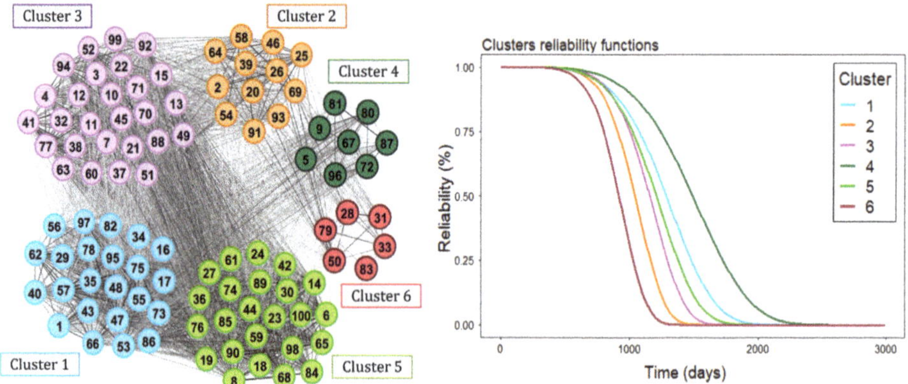

Figure 4. Clusterization example. Minor failure mode of the gearbox.

The results obtained for the clusterization of each one of the FMs have been summarized in Table 3. In the table, the components and the FMs of the number of clusters can be seen, as well as the number of WTs belonging to every cluster. The definition of minor and major FMs has been made according to the repairing details:

-   *Minor repairing.* It is provoked by a system failure that implies a maintenance team is dispatched for repairing. Costs of materials and tools are not very high, as minor components are going to be repaired. Thus, there is a reparation impact on system lifecycle but it is neither too high in terms of time, nor too costly. Repair time varies between 4 and 24 h.
-   *Major repairing.* Replacement of important components or complete systems is done. Therefore, repairs have a great impact on system lifetime, being able to return the system state to an as-good-as-new state. Costs of materials and tools are high, and repair time is above 24 h.

From Table 3, and also the graph of Figure 4, it is possible to conclude that the proposed approach is stable when identifying clusters with a low number of nodes in it. This is an important implication, since it is possible to identify small numbers of turbines behaving differently, and sometimes this may imply a cause which can be addressed or benchmarked.

Once the behaviors of the FMs in the WF have been assessed, it is possible to define the maintenance strategy according to the reliability of each one of the clusters of every failure mode. In the framework, the proposed maintenance strategy is an opportunistic maintenance policy defined by a simulation-based optimization. The optimization is multi-objective so the two-fold aforementioned problem of reducing maintenance costs is addressed. The simulation-based optimization provides several non-dominated solutions, which entails a trade-off among costs and production loss. With the solutions rendered by the optimization, it is possible to construct a Pareto Front in which the maintenance strategy can be selected according to the trade-off desired by the customers.

Table 3. Clusterization results.

Component	Failure Mode	Number of Clusters	Number of WTs in Each Cluster
Gearbox	Minor	6	23-12-27-8-24-6
	Major	3	52-27-21
Blades	Minor	4	45-33-18-4
	Major	2	39-61
Yaw	Minor	4	34-33-17-16
	Major	3	30-47-23
Pitch	Minor	5	24-27-28-7-14
	Major	3	29-20-51

In Figure 5, a comparison of two Pareto optimals is presented. One is obtained by addressing the failures through the clusterization and then defining maintenance strategy through dynamic opportunistic policy; the other one is obtained through a static opportunistic policy with a generic reliability Weibull model for each failure mode of the turbines. In Figure 5a, it can be seen that the proposed approach provides solutions which are unreachable for the static opportunistic policy. Furthermore, all the solutions provided by the clusterization with the dynamic opportunistic maintenance strategy are better in terms of costs and production loss. As an example to provide further insights, arbitrary customer requisites have been defined, which correspond to 122,000,000€ as a maximum OPEX and 120,000 MW/h lost for the 20 years of the lifecycle of the WF. In Figure 5b it can be seen that these requisites defined a feasible area of options and their intersection sets where is the Minimum Viable Offer (MVO).

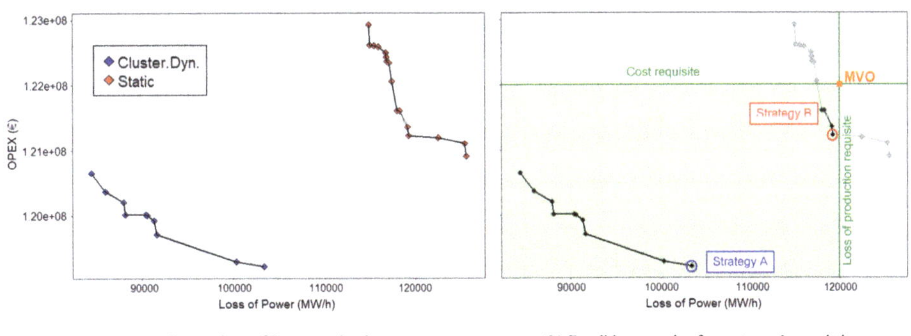

(a) Comparison of Pareto optimals

(b) Feasible strategies for customer's requisties

Figure 5. Pareto Front and Pareto Front with customer requisites.

According to the strategies represented in Figure 5, two of them have been selected (one from each policy) to compare their lifecycle performance. The selected strategies are the ones that fulfil the production loss requisite at a lower cost, and their performance in terms of OPEX and production loss is compared in Figure 6. It can be seen that the cost savings among the two strategies are over 3 million €, despite not being very significant. These savings come from the opportunity cost of producing energy at profitable wind-speed periods and from more accurate reliability estimates, which avoid some corrective actions. This result is reasonable and was expected since the dynamic opportunistic policy compared with the static does not reduce the amount of maintenance needed, but it performs it in more suitable moments.

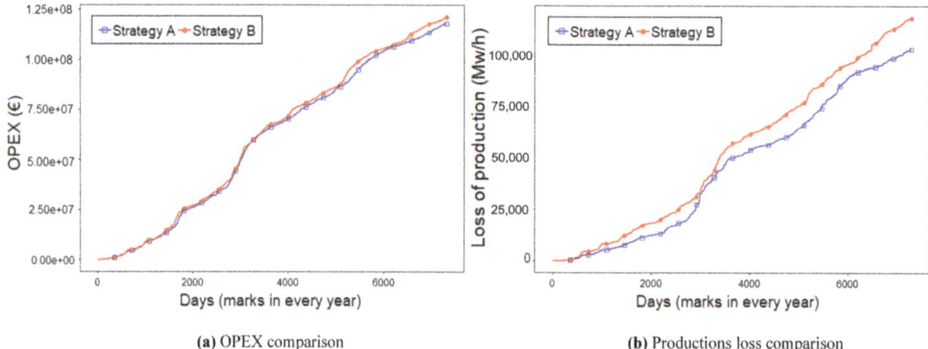

(a) OPEX comparison   (b) Productions loss comparison

Figure 6. Performance comparison of costs (**a**) and production loss (**b**) of strategies A and B.

Accordingly, the production loss difference is more significant, reaching an improvement of 14.5% by the end of the lifecycle. This improvement in the production loss is due to the performance of maintenance activities in the most favorable wind conditions. Besides, the better reliability estimates enhance the calculation of the failure probability, enabling the avoidance of unexpected CM, which takes more time to perform. Therefore, it can be seen that the proposed approach improves the performance of the static opportunistic maintenance with a general reliability model, not only in terms of cost, but in terms of production loss as well.

4. Concluding Remarks

In the present paper, a technical framework for managing the maintenance of WFs is proposed, which integrates three modules that depart from data considerations, providing reliability analysis tools to define the maintenance strategy. The framework considers the creation of a RAM database to provide a solid basis of information which is transformed into knowledge by reliability modeling. The second module aims at modeling the reliability of the WF; to do so, a clustering approach is proposed. By means of the clustering approach, the different behaviors of the WTs, which are caused by either different technical solutions or different working conditions, are properly addressed. This accurate description of the WF failures enables the definition of a more effective maintenance strategy, which is defined by a dynamic opportunistic policy optimization finding a trade-off among costs and production loss.

The suitability of the comprehensive framework has been validated through a case study based on real field data. The proposed approach has been tested against a static opportunistic maintenance which estimates the reliability of the WTs with a generic reliability model. It has been shown that by answering the research question, the integration of state-of-the-art techniques in the proposed framework provides a step forward in the achievement of maintenance excellence. The proposed approach outperforms the alternative in terms of cost and production loss. The improvements are due to two reasons: (i) the reliability estimates are more accurate because the behavior of the WTs is better characterized; and (ii) the maintenance activities are performed at the most convenient moments, fostering the maximization of energy production.

The research here presented is a step forward in the maintenance management of WFs due to its practical nature. However, to enhance its applicability and implementation in the wind energy industry, further efforts should focus on the integration of the strategy here presented with condition-based maintenance. Moreover, it would be interesting to also explore models that directly address the influence of changing operational conditions on the reliability of WTs. This holistic perspective entails potential benefits in the field of O&M of WFs, and it is worth researchers' and industry practitioners' attention.

274

Author Contributions: Conceptualization, J.I., A.C.M. and J.U.; Formal analysis, J.I. and A.E.; Investigation, J.I. and A.E.; Methodology, J.I.; Project administration, J.I.; Software, J.I.; Supervision, A.C.M. and J.U.; Validation, A.C.M. and J.U.; Writing—original draft, J.I.; Writing—review & editing, A.C.M., J.U. and A.E.

Funding: This research work was performed within the context of SustainOwner EmaitekPlus 2018–2019 Program of the Basque Government.

Conflicts of Interest: The authors declare no conflict of interest.

Abbreviations

The following abbreviations are used in this manuscript:

WF	Wind Farm
LCoE	Levelized Cost of Energy
CAPEX	Capital Expenditure
OPEX	Operational Expenditure
O&M	Operations & Maintenance
WT	Wind Turbine
RAM	Reliability, Availability, and Maintainability
LCC	Life-cycle Cost
INCOM	Inputs, Mechanisms, Controls and Outputs
SCADA	Supervisory Control And Data Acquisition
CMMS	Computerized Maintenance Management System
TBF	Time Between Failures
FM	Failure Mode
CM	Corrective Maintenance
PM	Preventive Maintenance
MT	Maintenance Team
NSGA	Non-dominated Sorted Genetic Algorithm
KPI	Key Performance Indicator
MW	Megawatt
MVO	Minimum Viable Offer

References

1. Ackermann, T. *Wind Power in Power Systems*; Wiley-Blackwell: Hoboken, NJ, USA 2005; [CrossRef]
2. Asgarpour, M.; Sørensen, J.D. Bayesian Based Diagnostic Model for Condition Based Maintenance of Offshore Wind Farms. *Energies* **2018**, *11*, 300. [CrossRef]
3. Merizalde, Y.; Hernández-Callejo, L.; Duque-Perez, O.; Alonso-Gómez, V. Maintenance Models Applied to Wind Turbines. A Comprehensive Overview. *Energies* **2019**, *12*, 225. [CrossRef]
4. Zhao, Y.; Li, D.; Dong, A.; Kang, D.; Lv, Q.; Shang, L. Fault Prediction and Diagnosis of Wind Turbine Generators Using SCADA Data. *Energies* **2017**, *10*, 1210. [CrossRef]
5. Erguido, A.; Márquez, A.C.; Castellano, E.; Fernández, J.G. A dynamic opportunistic maintenance model to maximize energy-based availability while reducing the life cycle cost of wind farms. *Renew. Energy* **2017**, *114*, 843–856. [CrossRef]
6. Kaldellis, J.; Kapsali, M. Shifting towards offshore wind energy—Recent activity and future development. *Energy Policy* **2013**, *53*, 136–148. [CrossRef]
7. Byon, E. Wind turbine operations and maintenance: a tractable approximation of dynamic decision making. *IIE Trans.* **2013**, *45*, 1188–1201. [CrossRef]
8. Ozturk, S.; Fthenakis, V.; Faulstich, S. Assessing the Factors Impacting on the Reliability of Wind Turbines via Survival Analysis—A Case Study. *Energies* **2018**, *11*, 3034. [CrossRef]
9. Nielsen, J.S.; Sørensen, J.D. Bayesian Estimation of Remaining Useful Life for Wind Turbine Blades. *Energies* **2017**, *10*, 664. [CrossRef]
10. Leahy, K.; Gallagher, C.; O'Donovan, P.; O'Sullivan, D.T.J. Issues with Data Quality for Wind Turbine Condition Monitoring and Reliability Analyses. *Energies* **2019**, *12*, 201. [CrossRef]

11. IRENA, I.R.E.A. *Renewable Power Generation Costs in 2012: An Overview*; International Renewable Energy Agency (IRENA): Abu Dhabi, UAE, 2012.

12. PowerTech eV, V. *Levelised Cost of Electricity (LCOE 2015)*; Verfugbar unter; VGB PowerTech e.V.: Essen, Germany, 2015.

13. Jardine, A.K.; Tsang, A.H. *Maintenance, Replacement, and Reliability: Theory And Applications*, 1st ed.; CRC Press: Boca Raton, FL, USA, 2005; doi:10.1201/9781420044614.

14. Bertling, L.; Wennerhag, P. *Wind Turbine Operation and Maintenance*; Technical Report; Elforsk: Stockholm, Sweden, 2012.

15. Shafiee, M. Maintenance logistics organization for offshore wind energy: Current progress and future perspectives. *Renew. Energy* **2015**, *77*, 182–193. [CrossRef]

16. Izquierdo, J.; Márquez, A.C.; Uribetxebarria, J. Dynamic artificial neural network-based reliability considering operational context of assets. *Reliabil. Eng. Syst. Saf.* **2019**, *188*, 483–493. [CrossRef]

17. Izquierdo, J.; Crespo, A.; Uribetxebarria, J.; Erguido, A. Assessing the impact of operational context variables on rolling stock reliability. A real case study. In *Safety and Reliability—Safe Societies in a Changing World*; CRC Press, Ed.; Taylor & Francis Group: London, UK; Trondheim, Norway, 2017; Proceedings of ESREL 2018, pp. 571–578.

18. Rausand, M.; Höyland, A. *System Reliability Theory: Models, Statistical Methods, and Applications*; John Wiley & Sons Hoboken, NJ, USA, 2004; Volume 396.

19. Hameed, Z.; Vatn, J.; Heggset, J. Challenges in the reliability and maintainability data collection for offshore wind turbines. *Renew. Energy* **2011**, *36*, 2154–2165. [CrossRef]

20. Nguyen, T.H.; Prinz, A.; Friisø, T.; Nossum, R.; Tyapin, I. A framework for data integration of offshore wind farms. *Renew. Energy* **2013**, *60*, 150–161. [CrossRef]

21. Louit, D.; Pascual, R.; Jardine, A. A practical procedure for the selection of time-to-failure models based on the assessment of trends in maintenance data. *Reliabil. Eng. Syst. Saf.* **2009**, *94*, 1618–1628. [CrossRef]

22. Stamatelatos, M.; Veseley, W.; Dugan, J.; Fragola, J.; Minarik, J.; Rialsback, J. *Fault Tree Handbook with Aerospace Applications*; NASA Office of Safety and Mission Assurance: Washington, DC, USA, 2002.

23. Li, H.; Yang, C.; Zhao, B.; Wang, H.; Chen, Z. Aggregated models and transient performances of a mixed wind farm with different wind turbine generator systems. *Electric Power Syst. Res.* **2012**, *92*, 1–10. [CrossRef]

24. Cao, L.; Qian, Z.; Zareipour, H.; Wood, D.; Mollasalehi, E.; Tian, S.; Pei, Y. Prediction of Remaining Useful Life of Wind Turbine Bearings under Non-Stationary Operating Conditions. *Energies* **2018**, *11*, 3318. [CrossRef]

25. Zimroz, R.; Bartelmus, W.; Barszcz, T.; Urbanek, J. Diagnostics of bearings in presence of strong operating conditions non-stationarity—A procedure of load-dependent features processing with application to wind turbine bearings. *Mech. Syst. Signal Process.* **2014**, *46*, 16–27. [CrossRef]

26. Cannarile, F.; Compare, M.; Di Maio, F.; Zio, E. Handling reliability big data: A similarity-based approach for clustering a large fleet of assets. In *Safety and Reliability of Complex Engineered Systems-Proceedings of the 25th European Safety and Reliability Conference*; ESREL: Zurich,Switzerland, 2015; Volume 2015, pp. 891–896.

27. Cannarile, F.; Compare, M.; Di Maio, F.; Zio, E. A clustering approach for mining reliability big data for asset management. *Proc. Inst. Mech. Eng. Part O J. Risk Reliabil.* **2018**, *232*, 140–150. [CrossRef]

28. Baraldi, P.; Di Maio, F.; Rigamonti, M.; Zio, E.; Redouane, S. Unsupervised clustering of vibration signals for identifying anomalous conditions in a nuclear turbine. *J. Intell. Fuzzy Syst.* **2015**, *28*, 1723–1731. [CrossRef]

29. Jiang, B.; Pei, J.; Tao, Y.; Lin, X. Clustering uncertain data based on probability distribution similarity. *IEEE Trans. Knowl. Data Eng.* **2013**, *25*, 751–763. [CrossRef]

30. Ruiz de la Hermosa, R. Wind farm monitoring using Mahalanobis distance and fuzzy clustering. *Renew. Energy* **2018**, *123*, 526–540. [CrossRef]

31. Elijorde, F.; Kim, S.; Lee, J. A Wind Turbine Fault Detection Approach Based on Cluster Analysis and Frequent Pattern Mining. *KSII Trans. Internet Inf. Syst.* **2014**, *8*, 664–677. [CrossRef]

32. Hameed, Z.; Wang, K. Development of optimal maintenance strategies for offshore wind turbine by using artificial neural network. *Wind Eng.* **2012**, *36*, 353–364. [CrossRef]

33. Wang, H.; Pham, H. Optimal Preparedness Maintenance of Multi-unit Systems with Imperfect Maintenance and Economic Dependence. In *Springer Series in Reliability Engineering*; Springer Science + Business Media, Berlin, Germany: 2006; pp. 135–150. [CrossRef]

34. Ding, F.; Tian, Z. Opportunistic maintenance optimization for wind turbine systems considering imperfect maintenance actions. *Int. J. Reliabil. Q. Saf. Eng.* **2011**, *18*, 463–481. [CrossRef]

35. Nicolai, R.; Dekker, R. A review of multi-component maintenance models. In Proceedings of the European Safety and Reliability Conference, Stavanger, Norway, 25–27 June 2007; pp. 289–296.
36. Dekker, R.; Wildeman, R.E.; van der Duyn Schouten, F.A. A review of multi-component maintenance models with economic dependence. *Math. Methods Oper. Res.* **1997**, *45*, 411–435. [CrossRef]
37. Sasieni, M.W. A Markov Chain Process in Industrial Replacement. *OR* **1956**, *7*, 148. [CrossRef]
38. Nakagawa, T.; Murthy, D. Optimal replacement policies for a two-unit system with failure interactions. *RAIRO-Oper. Res.* **1993**, *27*, 427–438. [CrossRef]
39. De Almeida, A.T.; Ferreira, R.J.P.; Cavalcante, C.A.V. A review of the use of multicriteria and multi-objective models in maintenance and reliability. *IMA J. Manag. Math.* **2015**, *26*, 249–271. [CrossRef]
40. Cavalcante, C.A.; Lopes, R.S. Multi-criteria model to support the definition of opportunistic maintenance policy: A study in a cogeneration system. *Energy* **2015**, *80*, 32–40. [CrossRef]
41. Cavalcante, C.; Lopes, R. Opportunistic Maintenance Policy for a System with Hidden Failures: A Multicriteria Approach Applied to an Emergency Diesel Generator. *Math. Probl. Eng.* **2014**, *2014*, 157282. [CrossRef]
42. Iqbal, M.; Azam, M.; Naeem, M.; Khwaja, A.; Anpalagan, A. Optimization classification, algorithms and tools for renewable energy: A review. *Renew. Sustain. Energy Rev.* **2014**, *39*, 640–654. [CrossRef]
43. Besnard, F.; Patrikssont, M.; Strombergt, A.B.; Wojciechowskit, A.; Bertling, L. An optimization framework for opportunistic maintenance of offshore wind power system. In Proceedings of the 2009 IEEE Bucharest PowerTech, Bucharest, Romania, 28 June–2 July 2009; [CrossRef]
44. Tian, Z.; Jin, T.; Wu, B.; Ding, F. Condition based maintenance optimization for wind power generation systems under continuous monitoring. *Renew. Energy* **2011**, *36*, 1502–1509. [CrossRef]
45. Ding, F.; Tian, Z. Opportunistic maintenance for wind farms considering multi-level imperfect maintenance thresholds. *Renew. Energy* **2012**, *45*, 175–182. [CrossRef]
46. Sarker, B.R.; Faiz, T.I. Minimizing maintenance cost for offshore wind turbines following multi-level opportunistic preventive strategy. *Renew. Energy* **2016**, *85*, 104–113. [CrossRef]
47. Zhu, W.; Fouladirad, M.; Bérenguer, C. A multi-level maintenance policy for a multi-component and multifailure mode system with two independent failure modes. *Reliab. Eng. Syst. Saf.* **2016**, *153*, 50–63. [CrossRef]
48. Atashgar, K.; Abdollahzadeh, H. Reliability optimization of wind farms considering redundancy and opportunistic maintenance strategy. *Energy Convers. Manag.* **2016**, *112*, 445–458. [CrossRef]
49. Abdollahzadeh, H.; Atashgar, K.; Abbasi, M. Multi-objective opportunistic maintenance optimization of a wind farm considering limited number of maintenance groups. *Renew. Energy* **2016**, *88*, 247–261. [CrossRef]
50. IEEE, C.S. *IEEE Standard for Functional Modeling Language—Syntax and Semantics for IDEF0*; IEEE Std 1320.1-1998; IEEE Computer Society: Washington DC, USA, 1998; p. i, [CrossRef]
51. Kusiak, A.; Larson, T.N.; Wang, J.R. Reengineering of design and manufacturing processes. *Comput. Ind. Eng.* **1994**, *26*, 521–536. [CrossRef]
52. Von Luxburg, U. A tutorial on spectral clustering. *Stat. Comput.* **2007**, *17*, 395–416. [CrossRef]
53. Velmurugan, T.; Santhanam, T. Computational complexity between K-means and K-medoids clustering algorithms for normal and uniform distributions of data points. *J. Comput. Sci.* **2010**, *6*, 363. [CrossRef]
54. Pham, H.; Wang, H. Imperfect maintenance. *Eur. J. Oper. Res.* **1996**, *94*, 425–438. [CrossRef]
55. Zille, V.; Bérenguer, C.; Grall, A.; Despujols, A. Modelling multicomponent systems to quantify reliability centred maintenance strategies. *Proc. Inst. Mech. Eng. Part O J. Risk Reliabil.* **2011**, *225*, 141–160. [CrossRef]
56. Deb, K.; Pratap, A.; Agarwal, S.; Meyarivan, T. A fast and elitist multiobjective genetic algorithm: NSGA-II. *IEEE Trans. Evol. Comput.* **2002**, *6*, 182–197. [CrossRef]
57. Salazar, D.; Rocco, C.M.; Galván, B.J. Optimization of constrained multiple-objective reliability problems using evolutionary algorithms. *Reliabil. Eng. Syst. Saf.* **2006**, *91*, 1057–1070. [CrossRef]
58. Martin-Tretton, M.; Reha, M.; Drunsic, M.; Keim, M. Data collection for current us wind energy projects: Component costs, financing, operations, and maintenance. *Contract* **2012**, *303*, 275–300. [CrossRef]

is presented in Section 2. In Section 3, the mathematical model for the described problem is presented. In Section 4, a case study is presented to demonstrate the use of the maintenance decision-making model. Some concluding remarks and the possible future work suggestions are given in Section 5.

2. Problem Description

Each component failure of a wind turbine have different maintenance/repair severities, i.e., the effort needed from the maintenance personnel, the cost associated with the maintenance work and the time needed to perform the repair vary for each component failure. It is reported in [11] that the grouping of turbine component failures with similar maintenance severity is done to develop failure classifications and the reported methodology will be followed in our study. The offshore turbine component failures may be classified into a finite set of failure classifications and each failure classification have a maintenance rank and a probability of occurrence. The "maintenance rank" of a failure classification is defined as "the natural number assigned to each failure classification based on the severity of maintenance involved in solving component failures, with 1 assigned to the failure classification of lowest maintenance severity and N assigned to the failure classification of highest maintenance severity". As each failure classification is assigned a maintenance rank, the total number of ranks is same as the total number of failure classifications. The "probability of occurrence of a failure classification" is defined as "the sum of all the individual failure probabilities of turbine components under a specific failure classification".

Irrespective of the type of maintenance, certain resources are required to perform the intended maintenance task. Resources needed to complete a maintenance activity are an access vessel, maintenance personnel and spare parts. The right combination of maintenance personnel, access vessel and spare part to address the offshore turbine failure is termed as "resource combination". In the case of an offshore wind turbine, different resource combinations are required to solve component failures under different failure classifications. For example, to solve the failure of a gearbox under a given failure classification, more maintenance personnel, expensive vessel and spare gearbox parts (assembled or individual spare parts) are required, whereas to solve the failure of a brake shoe falls under another failure classification, and less number of maintenance personnel, inexpensive vessels and brake shoe spare parts are required. Hence, two failure classifications could potentially result in two resource combinations. The failure of both the brake shoe and gearbox could also be addressed using one resource combination.

This provides us an intuitive understanding that there may exist two types of resource combinations to address the offshore turbine failure. We assume that the first type are, resource combinations that are dedicated to address component failures under only one specific failure classification and are referred as "A-type Resource Combinations" or simply "A-type RC's" throughout the paper. A-type RC is defined as "the combination of maintenance personnel, spare parts and vessels which can identify and solve component failures under single failure classification". A-type RC's cannot solve the failures occurred in turbine components under other failure classifications. We assume that the second type are, the resource combinations that are capable of solving turbine component failures under multiple failure classifications within a specified maintenance rank and are referred as "B-type Resource Combinations" or simply "B-type RC's" throughout the paper. The B-type RC for the nth ranked failure classification is defined as "the combination of maintenance personnel, spare parts and vessels which can solve component failures under the rank "1 to n" failure classifications". From the definition, it is understood that, if a B-type RC is sent to address the nth ranked failure classification it cannot solve component failures under rank "n + 1 to N" failure classifications.

Though today's turbines are usually equipped with condition monitoring (CM) systems, we consider the scenario that such condition monitoring systems are unable to indicate the exact failure classification upon a turbine failure. That is, no information on the kind of needed repair/failure classification and spare parts requirements are obtained from the CM systems. Such scenarios arise when natural events, including but not limited to storms, icing, and waves occur and these natural events account for 60% of the offshore turbine failures [12]. The occurrence of these natural events is unpredictable and leads to failure of both the turbine components and the CM systems, respectively. The human-influenced events are generally reliability related issues of the CM systems. It is reported in [13] that the reliability of the CM system is not 100% and the CM systems sometimes fail to produce an alarm when the turbine component requires immediate attention for maintenance. The event of the CM systems not producing an alarm leads to the component failure and apparently turbine failure. During this CM system unreliability event, the information failed turbine component is not obtained from the CM systems. Hence, these random natural and human influenced events (of failure) leads to situation where the O&M team will have no direct information from the CM systems to make resource related maintenance decisions. In this paper, we focus on this scenario of corrective maintenance where the information on failed turbine component and its failure classification is not known.

A wind farm may have many turbines in operation, which may fail anytime in the future. If any wind turbine at an offshore wind farm failed suddenly and, no information on the failed turbine component and its failure classification could be obtained from the CM systems, the O&M team do not know the exact resource combination to address the failed turbine. In this situation, the O&M team is unsure about which type of vessel to use, how many maintenance personnel to send, whether to take spare parts or not and which spare parts to take. This creates uncertainty in making decision on the resource combination for maintenance execution. The hypothesized problem situation is "a corrective maintenance trip to an offshore wind turbine with unknown turbine failure information". The aim of our study is "to find the cost-effective resource combination for the hypothesized problem situation". In this problem, the failure classification is not known at the time of maintenance initiation and all the resource combinations that are available in the onshore port turn out to be decision choices for the O&M team. The resource combination to be selected by the O&M team might solve the unknown failure in one trip or might not solve the unknown failure in one trip and necessitate an additional trip to solve the identified failure known from the first trip. Therefore, the O&M team is put into a situation to select only one resource combination among all the available resource combinations considering the two possible results of their decision. In order to make a decision, the cost associated with each decision choice must be evaluated taking into account the probability of occurrences of different failure classifications. Then, the resource combination with least cost could be selected as the cost-effective resource combination to address the unknown turbine failure. The objectives are to propose a simple and useful mathematical model to aid decision-making and to demonstrate the use of the proposed model through a case study.

3. Mathematical Model

In this section, the mathematical model for the described problem is proposed. If the offshore wind turbine have a finite number of failure classifications and each classification has a probability of occurrence, then:

$$\sum_{i=1}^{N} P_i = 1 \tag{1}$$

where P_i denotes the probability of occurrence of the ith failure classification. The probabilities of occurrences of all the failure classifications are assumed known.

To address the component failures under respective failure classifications of offshore wind turbine, two different types of resource combinations are described earlier in Section 2. In our model, both the types of resource combinations are considered as decision choices. Therefore, the selection of one resource combination among the available resource combinations (both A-type and B-type) is the only decision for the described problem. The decision is represented as a finite set of binary variables in our model:

$$S_{ij} = \begin{cases} 1, & \textit{use type } j \textit{ RC for failure classification } i \\ 0, & \textit{don't use type } j \textit{ RC for failure classification } i \end{cases} \tag{2}$$

$$\text{Constraint: } \sum_{i=1}^{N} \sum_{j=1}^{2} S_{ij} = 1 \tag{3}$$

where S_{ij} denotes the type j RC for the ith ranked failure classification. The above constraint ensures that only one S_{ij} is selected among the available N number of $S'_{ij}s$, to solve the unknown failure. All the type j RC's that are dedicated to address their respective i^{th} ranked failure classifications are assumed known.

The uncertainty in turbine failure information brings in two possible situations namely trip success and trip failure. The "trip success" is defined as the situation where the unknown turbine failure is solved in a single maintenance trip using either an A-type RC or a B-type RC. The "trip failure" is defined as the situation where the unknown turbine failure cannot be solved in a single maintenance trip and necessitates an additional trip to solve the identified known failure using an appropriate A-type RC. Both the probability of trip success and trip failure depends on the decision and the probability of occurrences of the failure classifications. The trip success and failure situations along with their probabilities are considered in the model.

When an A-type RC which is dedicated for the ith failure classification, is sent to address the unknown failure, the trip is successful when the failure classification is i and the trip is a failure when the failure classification is not i. For A-type RC, the probability of the maintenance trip to be a success is P_i and the probability of the maintenance trip to be a failure is $1-P_i$. If the failure classification is not i, we are able to identify that the failure is k and a single next trip with an A-type RC for k will solve the failure. When a B-type RC that is dedicated for the nth failure classification is sent to address the unknown failure, the trip is successful when the failure classification is $1, 2, 3, \ldots, n$ and, trip is a failure when the failure classification is k $(k > n)$. For B-type RC, the probability of the maintenance trip to be a success is $P_1 + P_2 + P_3 + \ldots + P_n$ and the probability of the maintenance trip to be a failure is $P_{n+1} + P_{n+2} + P_{n+3} + \ldots + P_N$. A single next trip with an A-type RC for k will solve the failure.

The objective is to find the expected total maintenance cost of the decision, to figure out the cost-effective decision and solve the unknown turbine failure. The total maintenance cost in our model includes the maintenance personnel cost, access vessel cost, special maintenance vessel cost (jack-up, crane, etc.), spare parts cost and, production losses due to downtime. The maintenance personnel and vessels are in use from the point of time they get ready to execute maintenance to the point of time they get back to shore after the maintenance activity. In addition, the turbine is unavailable until the maintenance crew get the turbine back to operation. Therefore, the mathematical model formulation involves various deterministic time elements of maintenance namely lead-time, logistic time, waiting time, travel time, failure identification time and repair time.

The time to get the vessel ready for maintenance is the lead-time and, the time to get the spare parts is the logistics time. It is assumed that all the resources (the vessels, the personnel and the spare parts) are always available in the onshore port for maintenance execution. This assumption eliminates the lead-time of vessels and the logistic time of spare parts in our model. The total delay in maintenance execution due to weather and sea-state conditions is the waiting time and is the sum of "the delay before travel starts" and "the delay at the turbine" [14]. It is dependent on weather and does not depend on the decision. Hence, the waiting time is a constant in our model. The time to identify the failure occurred at the turbine and figure out the component that requires maintenance is the failure identification time. The failure identification time does not depend on the decision and is a constant in our model. The time taken to travel back and forth the turbine using vessels is called the "travel time" and is the sum of the "travel time to the turbine" and "travel time from the turbine". The travel time is dependent on the decision, as the vessel speed may differ for different resource combinations. To calculate the travel time, the average distance of the turbines from the shore is considered in our model. The wind speed and wave height variations in the sea may affect the travel speed, which in turn affects the travel time. To simplify our analysis and exclude the hydrodynamics of the sea, the travel time is assumed to be independent of the wave height and wind speed in this paper.

The time it takes to perform the actual maintenance work is the repair time. In the case of trip success, the repair activity is completed successfully and the turbine failure is solved in one trip. In our model, the trip success situation includes the repair time. In the case of trip failure, the component failure is only identified and is not repaired in the first trip. The certain amount of time spent to identify the failure in the first trip (waiting time, failure identification time and travel time) along with the fixed cost for an additional trip to solve the known failure using an A-type RC is considered for trip failure. The fixed cost/purchase cost of spare parts are not considered in our model, instead the cargo handling costs of spare parts is considered as the spare parts cost in our model. The spare parts cost is the total tonnage of spare parts in a resource combination times the cargo handling cost per tonnage. To simplify our analysis, the weight of the spare parts is considered the only cargo weight in our model. Other weights such as the weight of the maintenance tools, technicians are not considered. The mathematical model for the described problem is given in Equation (4) as:

$$Z = \sum_{i=1}^{N} \sum_{j=1}^{2} S_{ij} \times g_{ij} \times D + \sum_{i=1}^{N} \sum_{j=1}^{2} S_{ij} \times H_{ij} + \sum_{i=1}^{N} \sum_{j=1}^{2} S_{ij} \times t_{ij} \times C_{ij} + \sum_{i=1}^{N} \sum_{j=1}^{2} S_{ij} \times \alpha_{ij} \times r_{ij} \times C_{ij} + \sum_{i=1}^{N} \sum_{j=1}^{2} S_{ij} \times \beta_{ij} \times A \quad (4)$$

$$C_{ij} = V_{ij} + (n_{ij} \times M) + R \quad (5)$$

$$\alpha_{ij} = P_i \text{ for } j = 1 \quad (6)$$

$$\alpha_{ij} = \sum_{k=1}^{i} P_k \text{ for } j = 2 \quad (7)$$

$$\beta_{ij} = 1 - P_i \text{ for } j = 1 \quad (8)$$

$$\beta_{ij} = \sum_{k=i+1}^{N} P_k \text{ for } j = 2 \quad (9)$$

Z	Expected total maintenance cost for S_{ij}
g_{ij}	Weight of spares for S_{ij} in tons
D	Cost per tonnage of spares
H_{ij}	Cost of special vessel for S_{ij}
t_{ij}	Travel time for S_{ij} in hours
C_{ij}	Cost of vessel, maintenance personnel, and revenue loss per hour for S_{ij}
V_{ij}	Vessel cost per hour for S_{ij}
n_{ij}	Number of maintenance personnel for S_{ij}
M	Maintenance personnel cost per hour
R	Revenue loss per hour
r_{ij}	Repair time for S_{ij} in hours
α_{ij}	Probability of trip success for S_{ij}
B_{ij}	Probability of trip failure for S_{ij}
P_i	Probability that the failure is of classification i
A	Fixed additional trip cost of sending an A-type RC to solve known failure, which includes vessel cost, personnel cost, spare parts cost, and revenue loss due to downtime

The above mathematical model describes the expected total maintenance cost of sending S_{ij} to address the unknown failure. The first two terms in the model, is the sum of the spare parts cost and fixed special vessel cost of S_{ij}. The third term in the model is the total cost including vessel cost, personnel cost and revenue loss incurred because of the travel to and from the turbine using S_{ij}. The fourth term in the model is the trip success using S_{ij}. The trip success considers the total cost including the vessel cost, personnel cost ad revenue loss incurred because of the repair activity at the turbine using S_{ij} and, the probability that the turbine failure could be solved by S_{ij}. The fifth term in the model is the trip failure using S_{ij}. The trip failure considers the total cost including the vessel cost, personnel cost ad revenue loss to solve the known failure using an appropriate A-type RC and, the probability that the turbine failure could not be solved by S_{ij}. The waiting time and failure identification time are constants in our proposed model and both the time elements does not affect the decision and the results. Therefore, the waiting time and failure identification time are not included in the model. In the Equations (6)–(9), $j = 1$ represents the A-type RC and $j = 2$ represents the B-type RC.

With appropriate inputs, the proposed model is capable of calculating the expected cost of each decision choice. Utilizing the enumeration method, the expected total cost of all the resource combinations are evaluated and, the resource combination with minimum expected cost is selected as the cost effective option to address the unknown turbine failure. The mathematical model formulated above includes both types of resource combinations described earlier in Section 2, as decision choices and this allows the decision makers to consider all the available resource combinations for decision-making. In addition, the simplicity of the model ensures that it takes less time and less technical effort to solve the model. Hence, all the OWF stakeholders could use the model anytime. Given the failure classifications, their probabilities and resource combinations (decision choices) and, using the proposed model, the O&M team at any OWF would be able to figure out the cost-effective resource combination to address the unknown turbine failure.

4. Case Study

The objective of the case study is to demonstrate the use of the proposed model for offshore wind turbine maintenance. To simplify our analysis, a wind farm model with identical turbines is selected for our case study.

4.1. Wind Farm Models

The OWEZ wind farm model reported in [11] is selected for the study. The OWEZ wind farm has 36 identical VESTAS 3 MW wind turbines with a total capacity of 108 MW. The wind farm is in the North Sea at 10–18 km distance from the harbor and the turbines are installed to a maximum depth of 20 m. Four failure classifications for corrective maintenance reported in [11] for a 3 MW wind turbine is applicable for the selected OWEZ wind farm model and is given in Table 1.

Table 1. Failure classifications for a 3 MW offshore wind turbine [11].

Maintenance Rank	Failure Classification	Definition
1	Imperfect maintenance	An imperfect maintenance operation where there is no requirement for spare parts.
2	Minimal replacement	A minimal replacement of small sized sub-components with a maximum weight of 1 tonne.
3	Perfect replacement I	A perfect replacement of medium weight sub-components with a maximum weight of 50 tonnes.
4	Perfect replacement II	A perfect replacement of medium or large sized sub-components, with weight 50 tonnes to 100 tonnes.

In accordance with the vessel characteristics reported in [15] and the weight of spares under each failure classification reported in [11], the A-type RC's and B-type RC's for corrective maintenance is given in Table 2. From Table 2, it could be observed that S_{11} and S_{12} have identical resource elements, which means both A-type and B-type RC's are identical for imperfect maintenance in this study.

The probabilities of different failure classifications reported in [11] is applied to the OWEZ wind farm model. The reported probabilities are considered as the base case model in the study. It can be observed that majority of the corrective maintenance for the base case model is imperfect maintenance.

Using the model in Section 3, the model inputs in Sections 4.1 and 4.2, and using the explicit enumeration method the expected total maintenance cost is calculated for all the available resource combinations (decision choices) for the different wind farm models of Table 3 and the results are shown in matrix form.

The cost matrix for the base case model (Z_0) is:

$$Z_0 = \begin{bmatrix} 91951 & 91951 \\ 515401 & 922110 \\ 621730 & 1072754 \\ 637456 & 1087414 \end{bmatrix}$$

The cost matrix for the wind farm model 1 (Z_1) is:

$$Z_1 = \begin{bmatrix} 513354 & 513354 \\ 922366 & 922110 \\ 621935 & 1072960 \\ 637250 & 1087414 \end{bmatrix}$$

The cost matrix for the wind farm model 2 (Z_2) is,

$$Z_2 = \begin{bmatrix} 513931 & 513931 \\ 515045 & 512740 \\ 1039273 & 1072388 \\ 637821 & 1087414 \end{bmatrix}$$

The cost matrix for the wind farm model 3 (Z_3) is:

$$Z_3 = \begin{bmatrix} 513931 & 513931 \\ 515045 & 512740 \\ 622300 & 653925 \\ 1054794 & 1087414 \end{bmatrix}$$

The minimum value of the cost matrix Z_n for the wind farm model n represents the optimal solution, that is, the corresponding resource combination is identified to be the cost-effective resource combination.

To prove the effectiveness of the proposed model, it is appropriate to compare the results of the proposed model with the traditional practice of solving the described problem. When no information on the failed turbine is obtained from the CM systems, generally the offshore O&M team send technicians to inspect the failed turbine in a small Crew Transfer Vessel, identify the failure classification and then send the required resource combination to solve the turbine failure. In order to compare the results of the proposed model with the general practice, the cost of the general practice is assumed as the sum of the inspection activity cost using S_{11} and the fixed cost of corrective maintenance trip for offshore wind turbine. All the inputs presented in Sections 4.1 and 4.2 are used to calculate this cost of traditional practice and is found to be $514,353. The estimated cost of traditional practice is used to compare the results of the proposed model and to find the cost savings, if any.

The cost-effective resource combination for each wind farm model considered in this study with the total expected maintenance cost and, the cost savings in comparison with the traditional practice are given in Table 6.

Table 6. Cost-effective resource combination for different wind farm models given in Table 3.

Wind Farm Model	Cost-Effective Resource Combination	Expected Total Maintenance Cost (in \$'s)	Cost Savings (in Comparison with Traditional Practice)
Base case	S_{11}	91591	82.12%
Model 1	S_{11}	513354	0.19%
Model 2	S_{22}	512740	0.31%
Model 3	S_{22}	512740	0.31%

The optimal resource combination can be directly selected from Table 6. From the results, it could be observed that, S_{11} (which is same as S_{12} in this study) is the cost-effective option to address the corrective maintenance for turbines that are in operation for less than 10 years (base case model and wind farm model 1). In addition, S_{22} is the cost-effective option to address the corrective maintenance for turbines that are in operation for more than 10 years (wind farm model 2 and 3). Comparing the results of the proposed model with the traditional practice, the proposed model produces very high cost savings of 82.12% for the base case model and a considerable cost savings for the other three different wind farm models. It has to be noted that the proposed model is for one corrective maintenance trip and when there are multiple corrective maintenance problem instances with no information from CM systems, the cost savings will be more for the wind farm models 1, 2 and 3.

The results that are generated from the model are not only dependent on the probability of failure classifications (given in Table 3) but also on the cost estimates (given in Tables 4 and 5). The value of the "fixed cost for corrective maintenance trip for an offshore wind turbine" in Table 5 is assumed to be the same for all types of corrective maintenance because of insufficient data and this affects both the estimated cost of the general practice and also the results generated from the models. This assumption on the fixed cost for corrective maintenance is a key reason that the base case has a huge amount of savings in comparison with the other three wind farm models. More accurate fixed costs for different types of corrective maintenance will result in better estimates for the general practice and, more accurate results for the wind farm models 1, 2 and 3. Accurate cost data in maintenance decision-making and sensitivity analysis of the proposed model to the cost estimates (in Tables 4 and 5) will be studied in our future work.

The case study provides a better understanding of the use of the proposed model to address a corrective maintenance situation when there is no information on turbine failure type. Three different wind farm models are considered in addition to the base case and the powerfulness of the model for different OWFs is demonstrated. The case study also gives us an understanding that when the number of failure classifications for an OWT/OWF increase, then the complexity in finding the cost-effective resource combination also increases.

5. Summary and Conclusions

In this paper, a short-term resource decision problem for corrective maintenance at offshore wind turbine is identified and described. A simple mathematical model is proposed to solve the decision problem. The model is proposed in such a way that the expected cost of the decision is mainly dependent on the probabilities of occurrences of failure classifications. The maintenance team at all offshore wind farm will have their own failure classifications, resource combinations and access to accurate failure data and, this model will assist the maintenance team in making resource decisions to address the corrective maintenance problem stated in this paper. Possible future work includes the lead-time and logistic time in the decision model and consider the uncertainty in weather and sea-state conditions and the hydrodynamics of the sea in the model.

Author Contributions: Conceptualization, S.N. and M.J.Z.; Problem formulation and model, S.N., M.J.Z. and Y.D.; Validation, S.N and M.J.Z.; Writing—original draft preparation, S.N.; Writing—review and editing, S.N., M.J.Z. and Y.D.; Supervision, M.J.Z. and Y.D.; Funding acquisition, M.J.Z. and Y.D.

Funding: This research is supported by Future Energy Systems under Canada First Research Excellent Fund (FES-T11-P01), the Natural Sciences and Engineering Research Council of Canada (Grant #RGPIN-2015-04897), and the National Natural Science Foundation of China (Grant #51577167).

Conflicts of Interest: The authors declare no conflict of interest.

References

1. Esteban, M.; Diez, J.; Lopez, J.; Negro, V. Review: Why offshore wind energy? *Renew. Energy* **2011**, *36*, 444–450. [CrossRef]

2. Tabassum, A.; Premalatha, M.; Abbasi, T.; Abbasi, S. Wind energy: Increasing deployment, rising environmental concerns. *Renew. Sustain. Energy Rev.* **2014**, *31*, 270–288. [CrossRef]

3. Leung, D.; Yang, Y. Wind energy development and its environmental impact: A review. *Renew. Sustain. Energy Rev.* **2012**, *16*, 1031–1039. [CrossRef]

4. Musial, W.; Beiter, P.; Schwabe, P.; Tian, T.; Stehly, T.; Spitsen, P. 2016 Offshore Wind Technologies Market Report. Technical Report; US Department of Energy, Office of Energy Efficiency & Renewable Energy, 2016. Available online: https://www.energy.gov/sites/prod/files/2017/08/f35/2016%20Offshore%20Wind%20Technologies%20Market%20Report.pdf (accessed on 22 June 2018).

5. Karyotakis, A. On the Optimisation of Operation and Maintenance Strategies for Offshore Wind Farms. Ph.D. Thesis, University College London, London, UK, February 2011.

6. Shafiee, M.; Sorensen, J.D. Maintenance Optimization and Inspection Planning of Wind Energy Assets: Models, Methods and Strategies. *Reliab. Eng. Syst. Saf.* **2017**, in press. [CrossRef]

7. Nachimuthu, S.; Zuo, M.J.; Ding, Y. Modelling factors affecting operation and maintenance costs of offshore wind farms. In Proceedings of the 2018 IISE Annual Conference, Orlando, FL, USA, 19–22 May 2018.

8. Besnard, F. On maintenance Optimization for Offshore Wind Farms. Ph.D. Thesis, Chalmers University of Technology, Gothenburg, Sweden, February 2013.

9. Besnard, F.; Patriksson, M.; Stromberg, A.B.; Wojciechowski, A.; Bertling, L. An Optimization Framework for Opportunistic Maintenance of Offshore Wind Power System. In Proceedings of the 2009 IEEE Bucharest Power Tech Conference, Bucharest, Romania, 28 June–2 July 2009.

10. Ravindranath, B. Modelling Short Term Decision Support Tool for O&M of Offshore Wind Farms. Master's Thesis, University of Lisbon, Lisbon, Portugal, December 2016.

11. Dewan, A. Logistics and Service Optimization for O&M of Offshore Wind Farms—Model Development and Output Analysis. Master's Thesis, Delft University of Technology, Delft, The Netherlands, March 2014.

12. Santos, F.; Teixeira, A.P.; Soares, C.G. Modelling and simulation of the operation and maintenance of offshore wind turbines. *Proc. Inst. Mech. Eng. Part O J. Risk Reliab.* **2015**, *229*, 385–393. [CrossRef]

13. May, A.; McMillan, D. Condition based maintenance for offshore wind turbines: The effects of false alarms from Condition Monitoring systems. In Proceedings of the European Safety and Reliability Conference, Amsterdam, The Netherlands, 29 September–2 October 2013.

14. Dowell, J.; Zitrou, A.; Walls, L.; Bedford, T.; Infield, D. Analysis of wind and wave data to assess maintenance access to offshore wind farms. In Proceedings of the European Safety and Reliability Conference, Amsterdam, The Netherlands, 29 September–2 October 2013.

15. Carroll, J.; McDonald, A.; McMillan, D. Failure rate, repair time and unscheduled O&M cost analysis of offshore wind turbines. *Wind Energy* **2016**, *19*, 1107–1119.

16. Dewan, A.; Asgarpour, M. Reference O&M Concepts for Near and Far Offshore Wind Farms. Technical Report. Energy Research Centre of the Netherlands (ECN), 2016. Available online: https://www.ecn.nl/publications/PdfFetch.aspx?nr=ECN-E--16-055 (accessed on 18 July 2018).

17. Maples, B.; Saur, G.; Hand, M.; Pietermen, R.V.; Obdam, T. Installation, Operation, and Maintenance Strategies to Reduce the Cost of Offshore Wind Energy. Technical Report; National Renewable Energy Laboratory and Energy Research Centre of the Netherlands, 2013. Available online: https://www.nrel.gov/docs/fy13osti/57403.pdf (accessed on 25 July 2018).

18. Seedah, D.; Harrison, R.; Boske, L.; Kruse, J. Container Terminal and Cargo-Handling Cost Analysis Toolkit. Technical Report. Texas A&M Transportation Institute, 2013. Available online: https://library.ctr.utexas.edu/ctr-publications/0-6690-ctr-p2.pdf (accessed on 23 July 2018).
19. Stehly, T.; Heimiller, D.; Scott, G. 2016 Cost of Wind Energy Review. Technical Report; National Renewable Energy Laboratory, US Department of Energy, Office of Energy Efficiency & Renewable Energy, 2017. Available online: https://www.nrel.gov/docs/fy18osti/70363.pdf (accessed on 4 July 2018).

in that framework for extracting feature descriptors with high discriminative power. The suggestion model is trained on drone inspection images of different wind turbines. We also employed advanced augmentation methods (as described in details in the Materials and Methods Section) to generalize the learned model. The more generalized model helps the system perform better on challenging test images during inference. Inference is the process of applying the trained model to an input image to receive the detected or, in our case, the suggested object in return.

Figure 1 illustrates the flowchart of the proposed method. To begin with, damages on wind turbine blades that are imaged using drone inspections are annotated in terms of bounding boxes by field experts. Annotated images are also augmented with the proposed advanced augmentation schemes (such as the pyramid, patching, and regular augmentations, as described in details in the Materials and Methods Section) to increase the number of training samples. The faster R-CNN deep learning object detection framework is applied to train from these annotated and augmented annotated images. Within the faster R-CNN framework, the backbone CNN in this case is the deep one, called the Inception-ResNet-V2 architecture. CNN converts images into high-level spatial features called the feature map. The region proposal network tries to estimate where the objects could be located, and ROI pooling is used to extract relevant features from the feature map for that particular region and based on that classifier, making the decision of whether an object of that particular class is present or not. After the training, the deep learning framework produces a detection model that can be applied for new inspection image analysis.

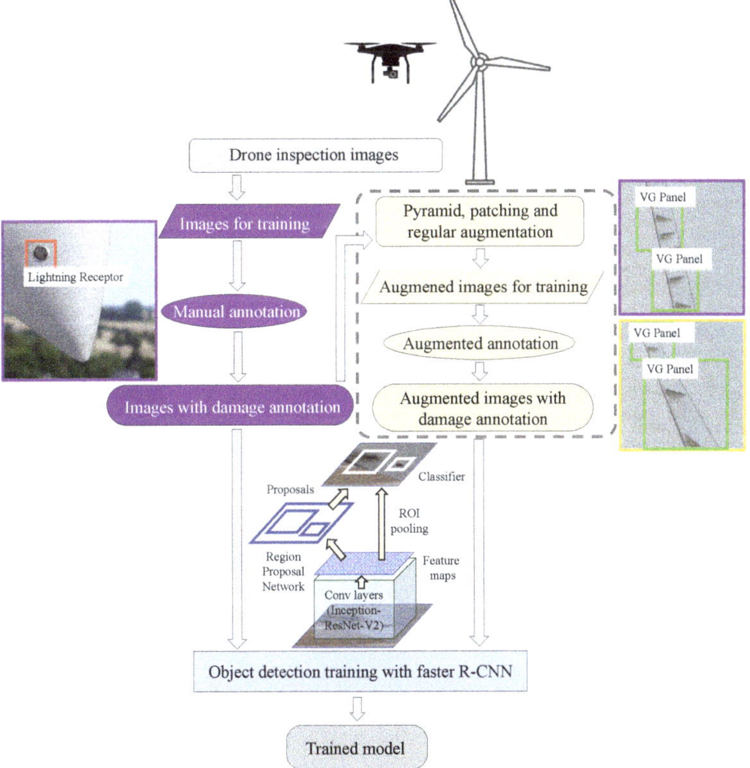

Figure 1. Flowchart of the proposed automated damage suggestion system. VG, Vortex Generator.

2. Materials and Methods

2.1. Manual Annotation

For this work, four classes were defined: Leading Edge erosion (LE erosion), Vortex Generator panel (VG panel), VG panel with missing teeth, and lightning receptor. Examples of each of these classes are illustrated in Figure 2. Figure 2a,g,h illustrates the examples of leading edge erosion annotated by experts using bounding boxes. Figure 2b,d,e illustrates the lightning receptors. Figure 2c shows the example of a VG panel with missing teeth. Figure 2c,f show the examples of well-functioning VG panels. These classes served as the passive indicators of the health condition of the wind turbine blades. The reason behind choosing these classes for experimentation was that all these types of damages produce specific visual traits recognizable by humans and would be valuable if a machine could learn to do the same.

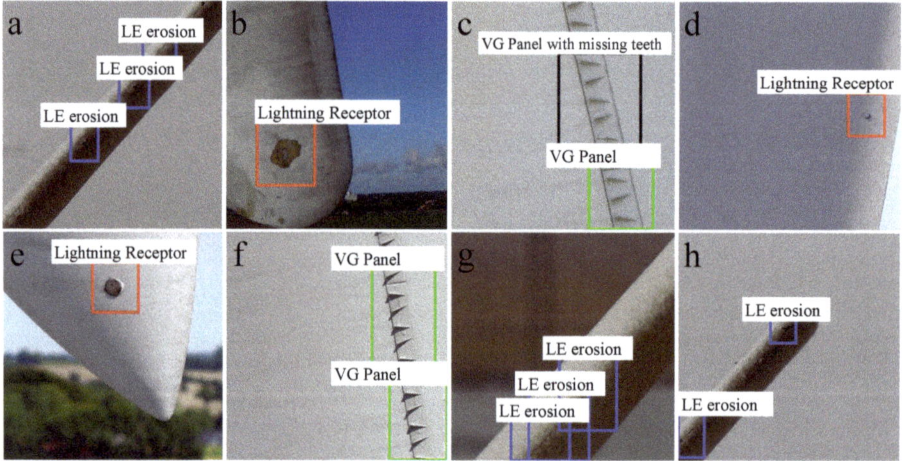

Figure 2. Examples of manually-annotated damages related to wind turbine blades. LE, Leading Edge.

A VG panel and a lightning receptor are not specific damage types, but rather external components on the wind turbine blade that often have to be visually checked during inspections. For inspection purpose, it is of value to detect the location of the lighting receptors and then identify if they are damaged/burned or not. In such cases, the machine learning task could be designed to identify damaged and non-damaged lightning receptors automatically. In this work, we simplified the task into first detecting the lightning receptor and, afterward, if needed, classifying them into damaged or non-damaged ones.

Table 1 summarizes the number of annotations for each class, which are annotated by experts and considered as ground truths. These annotated images were taken from the dataset titled "EasyInspect dataset" owned by EasyInspect ApS company, comprising drone inspection of different types of wind turbine blades located in Denmark. From the pool of available images, 60% were used for training and 40% for testing. As there was a limited number of examples of some damage types such as damaged lightning receptors, 40% of the samples were kept for reliable testing. To evaluate the performance of the developed system it is important to have a substantial number of unseen samples to examine. Annotations in the training set comprised of the annotations from full resolution images that were randomly selected. Annotations in the testing set comprised of the annotations from full-resolution images and also from cropped images containing objects of interest. This was done to make the testing part challenging by varying the scale of an object compared to the size of the image.

Table 1. List of classes in the training and testing sets related to the EasyInspect dataset.

List of Classes		
Classes	**Annotations-Training**	**Annotations-Testing**
Leading Edge (LE) erosion	135	67
Vortex Generator panel (VG)	126	65
Vortex Generator with Missing Teeth (VGMT)	27	14
Lightning receptor	17	7

2.2. Image Augmentations

Some types of damages on wind turbine blades are rare, and it is hard to collect representative samples during inspections. For deep learning, it is necessary to have thousands of training samples that are representative of the depicted patterns in order to obtain a good detection model [21]. The general approach is to use example images from the training set and then augment them in ways that represent probable variations in appearances, maintaining the core properties of object classes.

2.2.1. Regular Augmentations

Regular augmentations are defined as conventional ways of augmenting images for increasing the number of training samples. There are many different types of commonly-used augmentation methods that could be selected based on the knowledge about object classes and the possible occurrences of variances during acquisition. Taking drone inspection and wind turbine properties into consideration, four types of regular augmentation were chosen to be used in this work, which are listed below. These four types of regular augmentations were selected to represent real-life possibilities that could occur during drone inspection of wind turbine blades and have been proven to provide positive impacts on deep learning-based image classification tasks [22–24].

- Perspective transformation for the camera angle variation simulation.
- Left-to-right flip or top-to-bottom flip to simulate blade orientation: e.g., pointing up or down.
- Contrast normalization for variations in lighting conditions.
- Gaussian blur simulating out of focus images.

Figure 3 illustrates the examples of regular augmentations of the wind turbine inspection images containing damage examples. Figure 3e is the perspective transform of Figure 3a, where both images illustrate the same VG panels; Figure 3f is left-to-right flipping of the image patch of Figure 3b containing a lightning receptor. Figure 3g is the contrast normalized version of Figure 3c. Figure 3h is the augmented image of the lightning receptor in Figure 3d simulating de-focusing of the camera using approximate Gaussian blur.

2.2.2. Pyramid and Patching Augmentation

Drones deployed to acquire the dataset typically were equipped with very high-resolution cameras. High-resolution cameras allowed the drones to capture essential details, being at a further and safer distance from the turbines. These high-resolution images allowed the flexibility of training from rare types of damages and a wide variety of backgrounds at a different resolution using the same image.

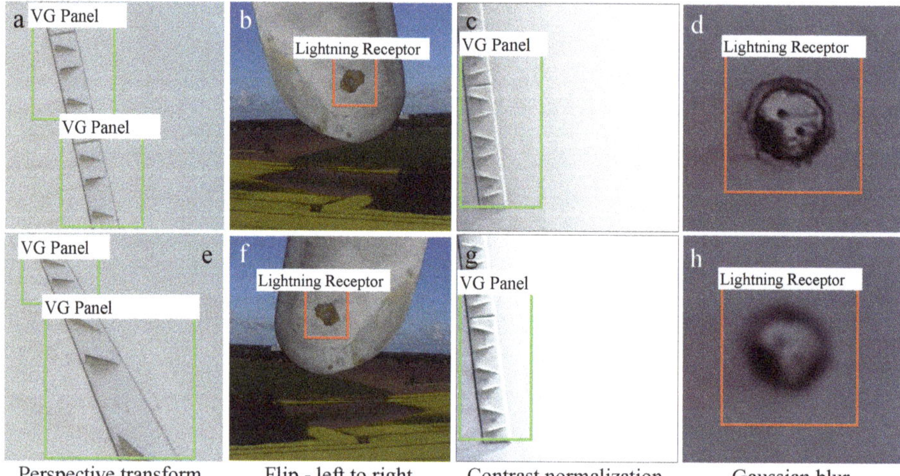

Figure 3. Illustrations of the regular augmentation methods applied to drone inspection images.

The deep learning object detection frameworks such as the faster R-CNN framework have predefined input image dimensions that typically allow for a maximum input image dimension (either height or width) of 1000 pixels [20,25]. This is maintained through re-sizing higher resolution images while keeping the ratio between width and height in situ. For high-resolution images, this limitation creates new challenges due to damages being minimal in pixel sizes compared to full image size.

In Figure 4, on the right is the pyramid scheme, and on the left is the patching scheme. The bottom level of the pyramid scheme is defined as the image size where either the height or width is 1000 pixels. In the pyramid, from top to bottom, images are scaled from $1.00\times$ to $0.33\times$, simulating from the highest to the lowest resolutions. Sliding windows with 10% overlap were scanned over the images at each resolution to extract patches containing at least one object. Resolution conversions were performed through the linear interpolation method [26]. For example, in Figure 4, top right, the acquired full resolution image is 4000×3000 pixels, where the lightning receptor only occupies around 100×100 pixels. When fed to CNN during training in full resolution, it would be resized to the pre-defined network input size, where the lightning receptor would be occupying a tiny portion of 33×33 pixels. Hence, it is rather complicated to acquire enough recognizable visual traits of the lightning receptor.

Using the multi-scale pyramid and patching scheme on the acquired high-resolution training images, scale-varied views of the same object were generated and fed to the neural network. In this scheme, the main full-resolution image was scaled to multiple resolution images ($1.00\times$, $0.67\times$, $0.33\times$), and on each of these images, patches containing objects were selected with the help of a sliding window with 10% overlap. The selected patches were always 1000×1000 pixels.

The flowchart of this multi-scale pyramid and patching scheme is shown in Figure 4. This scheme helps to represent object capture at different camera distances, allowing the detection model to be efficiently trained on both low- and high-resolution images.

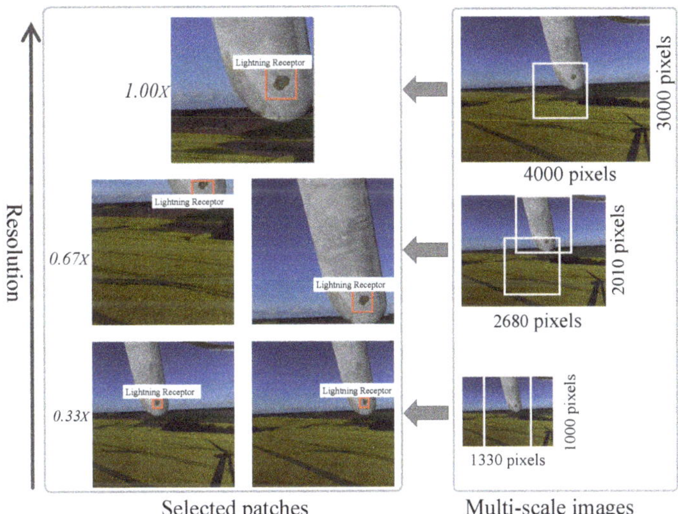

Figure 4. Proposed multi-scale pyramid and patching scheme for image augmentation.

2.3. Damage Detection Framework

With recent advances in deep learning for object detection, new architectures are frequently proposed, establishing groundbreaking performances. Currently, different stable meta architectures are publicly-available and have already been successfully deployed for many challenging applications. Among deep learning object detection frameworks, one of the best-performing methods is the faster R-CNN [20,27]. We also experimented with other object detection frameworks such as R-CNN [25], fast R-CNN [28], SSD[29], and R-FCN [30]. It was found that indeed, the faster R-CNN outperformed others in terms of accuracy when real-time processing was not needed, and deep CNN architectures like ResNet [31] were used for feature extraction. In our work, the surface damage detection and classification using drone inspection images were performed using faster R-CNN [20].

Faster R-CNN [20] uses a Region Proposal Network (RPN) [32] trained on feature descriptors extracted by CNN to predict bounding boxes for objects of interest. The CNN architecture automatically learns features such as texture, spatial arrangement, class size, shape, and so forth, from training examples. These automatically-learned features are more appropriate than hand-crafted features.

Convolutional layers in CNN summarize information based on the previous layer's content. The first layer usually learns edges; the second finds patterns in edges encoding shapes of higher complexity, and so forth. The last layer contains a feature map of much smaller spatial dimensions than the original image. The last layer feature map summarizes information about the original image. We experimented with both lighter CNN architectures such as InceptionV2 [33] and ResNet50 [31] and heavier CNN architectures such as ResNet101 [31] and Inception-ResNet-V2 [34].

The Inception [33] architecture by Google contains an inception module as one of the building blocks, and the computational cost is about 2.5-times higher than that of GoogLeNet [35]. The ResNet architecture [31] is known for its residual blocks, which help to reduce the impact of vanishing gradients from the top layers to the bottom layers during training. Inception-ResNet-V2 [34] is one of the extensions of the inception architectures that incorporates both the inception blocks and residual ones. This particular CNN network was used as the backbone architecture in our final model within a faster R-CNN framework for extracting highly discriminating feature descriptors. This network is computationally heavy and was found to provide state-of-the-art performances for object detection tasks [27].

Together with the patching and the regular augmentations, the MAP increased slightly to 38.3%. However, the pyramid scheme dramatically improved the performance of the trained model up to 70.5%. The best performing configuration was the last one with the combination of the pyramid, patching, and regular augmentation schemes, generating an MAP of 72.9%.

As shown in Figure 5a–d, the proposed combination of all the proposed augmentation methods significantly and consistently improved the performance of the model and lifts it to above 70% for all four CNN architecture. Figure 5a–d represents sequentially lighter to deeper CNN backbone architectures used for deep learning feature extraction. The CNN networks explored in this work were: Inception-V2, ResNet-50, ResNet-101, and Inception-ResNet-V2. In each individual figure (a–d): the y-axis represents the MAP of the suggestion on the test set, which are reported in percentage.

For any specific type of augmentation, MAPs, in general, were higher for the deeper networks (which were ResNet-101 and Inception-ResNet-V2) than for the lighter ones. For these two deeper networks, note that regular augmentation on top of the multi-scale pyramid and patching scheme added on average 2% gain in MAP. For lighter networks (Inception-V2 and ResNet-50), due to the limited search space, the network tended to learn and map better without the addition of regular augmentation. The results also demonstrated that for the small dataset (where some types of damages were extremely rare), it was beneficial to generate augmented images following class variation probabilities in terms of scale, light conditions, focuses, and acquisition angles.

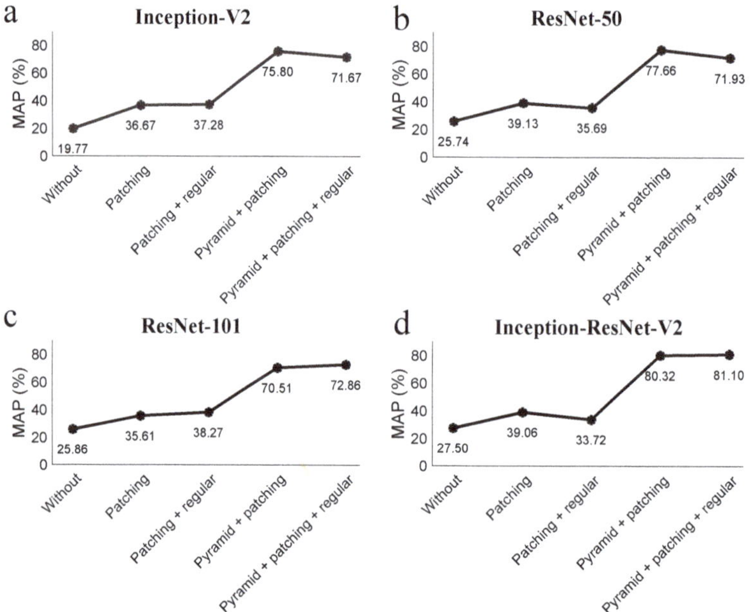

Figure 5. Comparison of the precision of trained models using various network architectures and augmentation types.

3.2. The CNN Architecture Performs Better as It Goes Deeper

When comparing the four selected CNN backbone architectures for the faster R-CNN framework, the Inception-ResNet-V2, which was the deepest, performed the best among all. If we fixed the augmentation to the combination of pyramid, patching, and regular, the MAP of the Inception-V2, ResNet-50, ResNet-101, and Inception-ResNet-V2 would be 71.67%, 71.93%, 72.86%, and 81.10%, respectively (as shown in Figure 5 and in Table 2). The number of layers in each of these networks representing the depth of the network could be arranged in the same order, as well (where Inception-V2 was the lightest and Inception-ResNet-V2 the deepest). This demonstrates that the performance regarding MAP increased as the network went deeper. The gain in performance of deeper networks comes with the cost of a longer training time and higher requirements on the hardware.

3.3. Faster Damage Detection with the Suggestive System in Inspection Analysis

The time required for automatically suggesting damage locations for new images using the trained model depends on the size of the input image and the depth of the CNN architecture used. In the presented case, the average time required for inferring a high-resolution image using Inception-V2, ResNet-50, ResNet-101, and Inception-ResNet-V2 networks (after leading the model) was respectively 1.36 s, 1.69 s, 1.87 s, and 2.11 s; whereas, for human-based analysis without suggestions, it can take around 20 s–3 min per image depending on the difficulty level for identification. With the deep learning-aided suggestion system for humans, the analysis time went significantly down (almost to two-thirds) compared to human speed without suggestions and also produced better accuracy (see Table 3).

Damages on unseen inspection images were annotated by experts having deep learning-aided suggestions as bounding boxes and without. In the case of suggestive bounding boxes, experts only needed to correct, whereas in the case of "without suggestion", they needed to draw the bounding box from scratch. While annotating 100 randomly-selected images, with suggestion, it took on average 131 s per image, whereas without suggestion, it was around 200 s per image. Human results (in terms of precision) with and without suggestions are called "Human" and "Suggestions aiding human". The precision of the deep learning trained model's suggestion is called "Suggestions". To access the precision of "Suggestions", the best-performing model Inception-ResNet-V2 within the faster R-CNN framework equipped with pyramid, patching, and regular augmentation was used.

Table 3. Summary of the experimental results.

Classes	Mean Average Precision (MAP) in Percentage		
	Suggestions	Human	Suggestions Aiding Human
LF erosion	90.62	68.00	86.60
VG panel (VG)	89.08	85.91	81.57
VG with Missing Teeth (VGMT)	73.11	80.17	91.37
Lightning receptor	71.58	98.74	81.71
Overall precision	81.10	83.20	85.31
Average processing time per image	2.11 s	200 s	131 s

4. Discussion

With the deep learning-based automated damage suggestion system, the best performing model in the proposed method produced 81.10% precision, which is within the 2.1% range of the average human precision of 83.20%. In the case of deep learning-aided suggestion for humans, the MAP improved significantly to 85.31%, and the required processing time became two-thirds that on average for each image. This result suggests that humans can benefit from suggestions by knowing where to look for damages in images, especially for difficult cases like VG panels with missing teeth.

The experimental results show that for a smaller image dataset of wind turbine inspection, the performance was more sensitive to the quality of image augmentation than the selection of CNN architecture. One of the main reasons is that most damage types can have a considerably large variance in appearance, which makes the deployed network dependent on a larger number of examples from which to learn.

The combination of ResNet and inception modules in Inception-ResNet-V2 learned difficult damage types such as missing teeth in a vortex generator with more reliability than by the other CNN architectures. Figure 6 illustrates some of the suggestion results on inspection images for testing. Figure 6a,d,f illustrates suggested lighting receptors; Figure 6b,h shows LE erosion suggestion; Figure 6c,e illustrates the suggestion of VG panels with intact teeth and those with missing teeth, respectively. The latter exemplifies one of the very challenging tasks for automated damage suggestion method. Figure 6g shows the example of when no damage is detected. The suggestion model developed in this study performed well for challenging images, providing almost human-level precision. When there was no valid class present in the image, the trained model found only the background class and presented "No detection" as the label for that image (an example is shown in Figure 3g).

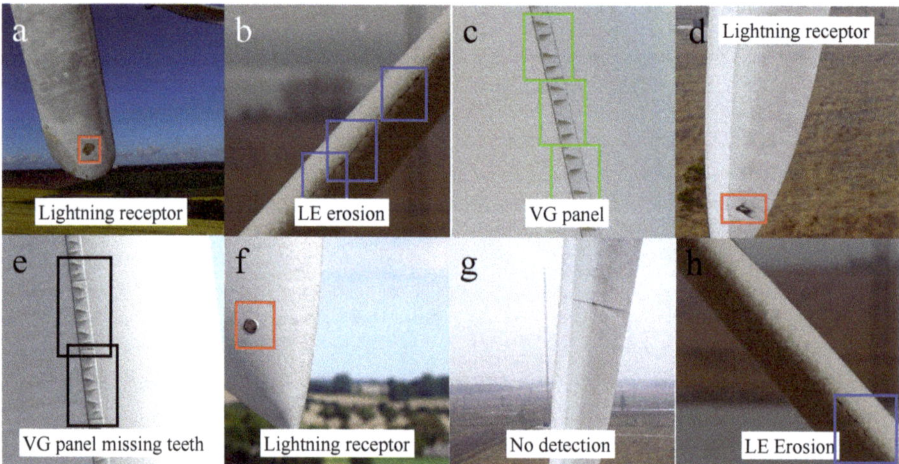

Figure 6. Suggestion results on the test images for the trained model using Inception-ResNet-V2 together with the proposed augmentation schemes.

This automated damage suggestion system has a potential cost advantage over the current manual one. Currently, drone inspections typically can cover up to 10 or 12 wind turbines per day. Damage detection, however, is much less efficient, as it involves considerable data interpretation for damage identification, annotation, classification, etc., which has to be conducted by skilled personnel. This process would incur significant labor cost considering the huge amount of images taken by the drones from the field. Using the deep learning framework for the suggestion to aid manual damage detection, the entire inspection and analysis process can be partially automated to minimize human intervention.

With suggested damages and subsequent corrections by human experts, over time, the number of annotated training examples would be increased and fed to the developed system for updating the trained suggestion model. This continuous way of learning through the help of human experts can increase the accuracy of the deep learning model (expected to provide 2–5% gain in MAP) and also reduce the required time for human corrections.

Relevant information about the damages, i.e., their size and location on the blade, can be used for further analysis in estimating wind turbine structural and aerodynamic conditions. The highest standalone MAP achieved with the proposed method was 81.10%, which is almost within human-level precision given the complexity of the problem and the conservative nature of the performance indicator. The developed automated detection system at its current state can safely work as a suggestion system for experts to finalize the damage locations from inspection data.

The required computational power and deep learning training can incur higher initial cost mainly due to the fact that acquiring training images comprising damage examples is expensive. However in the long run, a well-trained model on damage types of interest can produce a less expensive and more reliable solution to the drone inspection analysis task. With this proposed method, the majority of surface damages on the wind turbine blade can be semi-automatically recognized and reported for future actions in terms of maintenance.

5. Conclusions

This work presented a deep learning-based automated method to aid manual detection of wind turbine blade surface damages. The method can reduce human intervention by providing an accurate suggestion of damages on drone inspection images. Using the Inception-ResNet-v2 architecture inside faster R-CNN, 81.10% MAP was achieved on four different types of damages, i.e., LE erosion, vortex generator panel, vortex generator panel with missing teeth, and lightning receptor. The authors adopted a multi-scale pyramid and patching scheme that significantly improved the precision by 35% on average across the tested CNN architectures. The experimental results demonstrated that deep learning with augmentation can overcome the challenge of the scarce availability of damage samples for learning. In this work, a new image dataset of wind turbine drone inspection was published for the research community [19].

Author Contributions: R.R.P., K.B., and A.B.D. designed and supervised the research. A.S. and V.F. designed the experiments, implemented the system, and performed the research. X.C., A.N.C., and N.A.B.R. analyzed the data. A.S. wrote the manuscript with inputs from all authors.

Funding: This research was funded by the Innovation fund of Denmark through the DARWIN Project (Drone Application for pioneering Reporting in Wind turbine blade Inspections)

Acknowledgments: We would like to thank the Danish Innovation fund for the financial support through the DARWIN Project (Drone Application for pioneering Reporting in Wind turbine blade Inspections) and EasyInspect ApS for their kind contribution to this work by providing the drone inspection dataset.

Conflicts of Interest: The authors declare that they have no competing financial interests.

References

1. Aabo, C.; Scharling Holm, J.; Jensen, P.; Andersson, M.; Lulu Neilsen, E. *MEGAVIND Wind Power Annual Research and Innovation Agenda*; MEGAVIND: Copenhagen, Denmark, 2018.
2. Ziegler, L.; Gonzalez, E.; Rubert, T.; Smolka, U.; Melero, J.J. Lifetime extension of onshore wind turbines: A review covering Germany, Spain, Denmark, and the UK. *Renew. Sustain. Energy Rev.* **2018**, *82*, 1261–1271. [CrossRef]
3. Verbruggen, S. *Onshore Wind Power Operations and Maintenance to 2018*; New Energy Update: London, UK, 2016.
4. Stock-Williams, C.; Swamy, S.K. Automated daily maintenance planning for offshore wind farms. *Renew. Energy* **2018**, doi:10.1016/j.renene.2018.08.112. [CrossRef]
5. Canizo, M.; Onieva, E.; Conde, A.; Charramendieta, S.; Trujillo, S. Real-time predictive maintenance for wind turbines using Big Data frameworks. In Proceedings of the 2017 IEEE International Conference on Prognostics and Health Management (ICPHM), Dallas, TX, USA, 19–21 June 2017; pp. 70–77.
6. Zhou, K.; Fu, C.; Yang, S. Big data driven smart energy management: From big data to big insights. *Renew. Sustain. Energy Rev.* **2016**, *56*, 215–225, doi:10.1016/j.rser.2015.11.050. [CrossRef]
7. Wymore, M.L.; Dam, J.E.V.; Ceylan, H.; Qiao, D. A survey of health monitoring systems for wind turbines. *Renew. Sustain. Energy Rev.* **2015**, *52*, 976–990, doi:10.1016/j.rser.2015.07.110. [CrossRef]
8. Jha, S.K.; Bilalovic, J.; Jha, A.; Patel, N.; Zhang, H. Renewable energy: Present research and future scope of Artificial Intelligence. *Renew. Sustain. Energy Rev.* **2017**, *77*, 297–317, doi:10.1016/j.rser.2017.04.018. [CrossRef]
9. Morgenthal, G.; Hallermann, N. Quality assessment of unmanned aerial vehicle (UAV) based visual inspection of structures. *Adv. Struct. Eng.* **2014**, *17*, 289–302. [CrossRef]
10. Chen, X. Fracture of wind turbine blades in operation—Part I: A comprehensive forensic investigation. *Wind Energy* **2018**, *6*, doi:10.1002/we.2212. [CrossRef]
11. Wang, L.; Zhang, Z. Automatic detection of wind turbine blade surface cracks based on uav-taken images. *IEEE Trans. Ind. Electron.* **2017**, *64*, 7293–7303. [CrossRef]

12. Viola, P.; Jones, M. Rapid object detection using a boosted cascade of simple features. In Proceedings of the 2001 IEEE Computer Society Conference on Computer Vision and Pattern Recognition (CVPR 2001), Kauai, HI, USA, 8–14 December 2001; Volume 1.

13. Shihavuddin, A.; Arefin, M.M.N.; Ambia, M.N.; Haque, S.A.; Ahammad, T. Development of real time Face detection system using Haar like features and Adaboost algorithm. *Int. J. Comput. Sci. Netw. (IJCSNS)* **2010**, *10*, 171–178.

14. Friedman, J.; Hastie, T.; Tibshirani, R. Additive logistic regression: A statistical view of boosting (with discussion and a rejoinder by the authors). *Ann. Stat.* **2000**, *28*, 337–407. [CrossRef]

15. Liaw, A.; Wiener, M. Classification and regression by random Forest. *R News* **2002**, *2*, 18–22.

16. Cortes, C.; Vapnik, V. Support-vector networks. *Mach. Learn.* **1995**, *20*, 273–297. [CrossRef]

17. LeCun, Y.; Bengio, Y.; Hinton, G. Deep learning. *Nature* **2015**, *521*, 436. [CrossRef] [PubMed]

18. Hauberg, S.; Freifeld, O.; Larsen, A.B.L.; Fisher, J.; Hansen, L. Dreaming more data: Class-dependent distributions over diffeomorphisms for learned data augmentation. In *Artificial Intelligence and Statistics*; Cornell University (Computer Vision and Pattern Recognition): New York, NY, USA, 2016; pp. 342–350.

19. Shihavuddin, A.; Chen, X. *DTU—Drone Inspection Images of Wind Turbine*; This Dataset Set Has Temporal Inspection Images for the Years of 2017 and 2018 of the 'Nordtank' Wind Turbine at DTU Wind Energy's Test Site in in Roskilde, Denmark; Mendeley: Copenhagen, Denmark, 2018.

20. Ren, S.; He, K.; Girshick, R.; Sun, J. Faster r-cnn: Towards real-time object detection with region proposal networks. In *Advances in Neural Information Processing Systems*; Cornell University (Computer Vision and Pattern Recognition): New York, NY, USA, 2015; pp. 91–99.

21. Perez, L.; Wang, J. The effectiveness of data augmentation in image classification using deep learning. *arXiv* **2017**, arXiv:1712.04621.

22. Hussain, Z.; Gimenez, F.; Yi, D.; Rubin, D. Differential Data Augmentation Techniques for Medical Imaging Classification Tasks. In *AMIA Annual Symposium Proceedings*; American Medical Informatics Association: Bethesda, MD, USA, 2017; Volume 2017, p. 979.

23. Oliveira, A.; Pereira, S.; Silva, C.A. Augmenting data when training a CNN for retinal vessel segmentation: How to warp? In Proceedings of the 2017 IEEE 5th Portuguese Meeting on Bioengineering (ENBENG), Coimbra, Portugal, 16–18 February 2017; pp. 1–4.

24. Hu, T.; Qi, H. See Better Before Looking Closer: Weakly Supervised Data Augmentation Network for Fine-Grained Visual Classification. *arXiv* **2019**, arXiv:1901.09891.

25. Girshick, R.; Donahue, J.; Darrell, T.; Malik, J. Rich feature hierarchies for accurate object detection and semantic segmentation. In Proceedings of the IEEE Conference on Computer Vision and Pattern Recognition (CVPR), Columbus, OH, USA, 24–27 June 2014; pp. 580–587.

26. Blu, T.; Thévenaz, P.; Unser, M. Linear interpolation revitalized. *IEEE Trans. Image Process.* **2004**, *13*, 710–719. [CrossRef] [PubMed]

27. Huang, J.; Rathod, V.; Sun, C.; Zhu, M.; Korattikara, A.; Fathi, A.; Fischer, I.; Wojna, Z.; Song, Y.; Guadarrama, S.; et al. Speed/accuracy trade-offs for modern convolutional object detectors. In Proceedings of the IEEE Conference on Computer Vision and Pattern Recognition (CVPR), Honolulu, HI, USA, 21–26 July 2017.

28. Girshick, R. Fast r-cnn. In Proceedings of the IEEE International Conference on Computer Vision, Santiago, Chile, 13–16 December 2015; pp. 1440–1448.

29. Liu, W.; Anguelov, D.; Erhan, D.; Szegedy, C.; Reed, S.; Fu, C.Y.; Berg, A.C. Ssd: Single shot multibox detector. In *European Conference on Computer Vision*; Springer: Cham, Switzerland, 2016; pp. 21–37.

30. Dai, J.; Li, Y.; He, K.; Sun, J. R-fcn: Object detection via region-based fully convolutional networks. In *Advances in Neural Information Processing Systems*; NIPS: Barcelona, Spain, 2016; pp. 379–387.

31. He, K.; Zhang, X.; Ren, S.; Sun, J. Deep Residual Learning for Image Recognition. In Proceedings of the IEEE Conference on Computer Vision and Pattern Recognition (CVPR), Las Vegas, NV, USA, 26 June–1 July 2016.

32. Kong, T.; Yao, A.; Chen, Y.; Sun, F. Hypernet: Towards accurate region proposal generation and joint object detection. In Proceedings of the IEEE Conference on Computer Vision and Pattern Recognition, Las Vegas, NV, USA, 26 June–1 July 2016; pp. 845–853.

33. Ioffe, S.; Szegedy, C. Batch Normalization: Accelerating Deep Network Training by Reducing Internal Covariate Shift. In *International Conference on Machine Learning*; ICML: Lille, France, 6–11 July 2015; pp. 448–456.

Article

A Decision-making Model for Corrective Maintenance of Offshore Wind Turbines Considering Uncertainties

Sathishkumar Nachimuthu [1], Ming J. Zuo [1,*] and Yi Ding [2]

[1]   Department of Mechanical Engineering, University of Alberta, Edmonton, AB T6G 1H9, Canada; nachimut@ualberta.ca

[2]   College of Electrical Engineering, Zhejiang University, Hangzhou 310027, China; yiding@zju.edu.cn

*   Correspondence: ming.zuo@ualberta.ca; Tel.: +1-(780)-492-4466

Received: 13 February 2019; Accepted: 9 April 2019; Published: 12 April 2019

Abstract: Maintenance optimization has received special attention among the wind energy research community over the past two decades. This is mainly because of the high degree of uncertainties involved in the execution of operation and maintenance (O&M) activities throughout the lifecycle of wind farms. The increasing complexity in offshore maintenance execution demands applied research and brings forth a need to develop problem-specific maintenance decision-making models. In this paper, a mathematical model is proposed to assist wind farm stakeholders in making critical resource-related decisions for corrective maintenance at offshore wind farms (OWFs), considering uncertainties in turbine failure information.

Keywords: offshore wind farm; offshore wind turbine; maintenance; failure classification; resource decision; uncertainty

---

1. Introduction

The widespread availability and high technological maturity make wind energy a reliable renewable option to satisfy the future energy demands of the global population [1]. The limited land area and the need to reduce noise pollution is forcing the wind energy sector to shift towards offshore technologies [2,3]. Offshore wind farms (OWFs) are energy assets that have experienced a considerable growth in terms of cumulative capacity, from 4 GW to more than 18 GW over the past five years [4]. OWFs are expensive assets not only to build, but also to operate and maintain. About 23% contribution of the operation and maintenance (O&M) to the life cycle cost (LCC) makes the O&M the second major contributor for the LCC of an OWF [5]. The increased O&M cost is mainly caused by the uncertainties encountered by OWFs, which include weather, sea-state conditions, component lifetimes, etc. The high O&M cost and the unproven economic feasibility remain a hindrance for the future growth and expansion of the OWFs.

The accessibility limitations of vessels and helicopters imposed by the weather and sea-state conditions combined with the unavailability of failure data makes maintenance decision-making at OWFs, a complex and challenging task for the O&M team. A significant portion of the annual budget is wasted on many large offshore projects because of improper maintenance decisions [6]. Numerous research studies have been carried out to assist the O&M team in making maintenance decisions at OWFs. Almost 80% of the total research articles (related to offshore wind farm maintenance) have been published in the last five years, which indicates the increasing importance of O&M-related research for offshore wind farms in operation and under construction [6]. The maintenance decision problems have been analyzed from Reliability, Availability, Maintainability and Serviceability (RAMS) perspectives and many maintenance models have been developed for optimal decision-making.

Almost 98% of the models in the literature address long-term (5–20 years) and/or lifetime (which is usually 20 years) maintenance decision problems. There exist arguments in the wind research community that optimizing short-term maintenance decisions may not greatly reduce the O&M cost. It is reported in [7] that the expected total cost of one corrective maintenance trip at an OWF is $70,000–$130,000 (approximately). From the model and results of [7], it is understood that one wrong resource decision (improper vessel selection or insufficient manpower) for a corrective maintenance execution could necessitate an additional trip and account for a wastage of no less than $70,000 in the annual maintenance budget. This study shows that the maintenance decisions for offshore wind farms are critical for all time horizons (an hour, a day, a month, a year and lifetime).

Existing long-term and lifetime models are not implemented at OWFs, because the OWF stakeholders treat the models as theoretical and incomprehensive [6]. The models are touted to be complex and the stakeholders believe that it will take considerable time and require technical force to solve the models. This viewpoint of OWF stakeholders about the existing models demands a shift from theoretical research to applied research. In addition, it creates a necessity to identify maintenance decision problems (either long term or short term) that have a significant effect on the life cycle O&M costs and to provide solutions to one decision problem at a time through simple maintenance models. The corrective maintenance and its associated resource decisions (both short-term and long-term) contributes more than 60% to the life cycle O&M costs and is the highest cost driver of OWF O&M [8]. The stakeholders view of the existing maintenance models and the high cost associated with the corrective maintenance resource decisions was the motivation to identify short-term resource decision problems for corrective maintenance of the OWFs.

Few models in the literature have addressed the short-term maintenance problems at OWFs. The work reported in [9] developed an opportunistic short-term maintenance model. Whenever there is a need for corrective maintenance, the model considers the corrective maintenance trip as an opportunity to perform preventive maintenance at other turbines in the wind farm. The model is developed for two different time horizons (a day and a week) and for wind farms that follows flexible maintenance schedules. The model requires the maintenance manager to optimize the maintenance schedule in the morning of every working day and the maintenance tasks to be performed are available only after the optimization. The results of the work showed that 43% of the total preventive maintenance cost could be saved if this opportunistic maintenance with flexible everyday schedule optimization is adopted at the OWFs. The work reported in [10] developed a short-term decision-making model for scheduling resources (vessels and maintenance personnel) at the OWFs. The time horizon considered in this model is a day and it helps the OWF maintenance managers and planners to make better resource scheduling decisions each day. The model studied the impact of the number of maintenance personnel on energy loss and pointed out the importance of scheduling optimal number of maintenance personnel for daily maintenance work.

Both the short-term models [9,10] reported in the literature assumed that the information about turbine failure is always available and known for offshore turbine maintenance. With this assumption, the kind of needed repair is known, the resource decisions are certain and the maintenance team easily picks the desired resources for maintenance. The short-term models [9,10] then focused on different objectives such as opportunistic preventive maintenance [9] and resource-scheduling [10] to minimize the total maintenance costs. When the turbine failure information becomes unavailable, the resource decision-making turns out to be uncertain and the short-term models [9,10] are inapplicable to address this maintenance problem situation.

In this paper, a short-term resource decision-making model is proposed for the corrective maintenance of offshore wind turbines, considering the uncertainty in turbine failure information. The proposed model will assist multiple OWF stakeholders in making critical resource decisions for a corrective maintenance trip. The proposed model addresses the maintenance problem situation for which the information on turbine failure is not available and so it cannot be compared with the short-term models [9,10] in the literature. The paper is organized as follows: the problem description

Thus, the base case model is interpreted as OWF in which the turbines are relative new and their age is less than 5 years, that is, the turbines are operating in its first 5-year service period.

As the base case model is interpreted as OWF with turbines that are less than 5 years old, three other models are established for OWFs with increasing age of turbines with appropriate assumptions to demonstrate the powerfulness of the proposed model for different OWFs. The model 1 represents the OWF in which the turbines in operation are 5 to 10 years old. For the wind farm model 1, it is assumed that the majority of corrective maintenance corresponds to minimal replacement and it has the highest probability of occurrence. The probability of other failure classifications are then descended in the order of imperfect maintenance, perfect replacement I and perfect replacement II.

The model 2 represents the OWF in which the turbines in operation are 10 to 20 years old. For the wind farm model 2, it is assumed that the majority of corrective maintenance corresponds to perfect replacement I and it has the highest probability of occurrence. The probability of other failure classifications are then descended in the order of perfect replacement II, minimal replacement and imperfect maintenance.

Table 2. Decision Choices [11,15].

Resource Combination	Resource Elements
S_{11}	No Spare part + Access Vessel (Crew Transfer Vessel -small) + 2 maintenance personnel
S_{21}	Required Spare part + Access Vessel (Crew Transfer Vessel -small) + 3 maintenance personnel. (Use of permanent internal crane for replacement).
S_{31}	Required Spare part + Crane Vessel + Access Vessel (Crew Transfer Vessel small) + 6 maintenance personnel.
S_{41}	Required Spare part + Access Vessel (Crew Transfer Vessel -small) + Access Vessel (Jack-Up Vessel) + 6 maintenance personnel.
S_{12}	No Spare part + Access Vessel (Crew Transfer Vessel - small) + 2 maintenance personnel
S_{22}	All Class B Spare parts + Access Vessel (Crew Transfer Vessel-Large) + 3 maintenance personnel (Use of permanent internal crane for replacement).
S_{32}	All Class B and C Spare parts + build-up crane with a vessel + Access Vessel (SUVs) + 6 maintenance personnel.
S_{42}	All Class B, C and D spare parts + Access Vessel (SUVs) + Access Vessel (Jack- Up barge) + 6 maintenance personnel

The model 3 represents the OWF in which the turbines are either more than 20 years old or affected by storms or other natural disasters. For the wind farm model 3, it is assumed that the majority of corrective maintenance corresponds to perfect replacement II and it has the highest probability of occurrence. The probability of other failure classifications are then descended in the order of perfect replacement I, minimal replacement and imperfect maintenance. The reported probabilities for the base case is changed for different failure classifications to represent the wind farm models 1, 2 and 3. The probabilities of failure classifications of the base case model and the three different wind farm models are given in Table 3. The probability numbers in Table 3 are absolute values and are not in percentages.

Table 3. Probabilities of failure classifications for different OWF models [11].

Failure Classification	Probability			
	Base Case Model	Wind Farm Model 1	Wind Farm Model 2	Wind Farm Model 3
Imperfect maintenance	0.995165258	0.002353569	0.000995862	0.000995862
Minimal replacement	0.002353569	0.995165258	0.001485311	0.001485311
Perfect replacement I	0.000995862	0.001485311	0.995165258	0.002353569
Perfect replacement II	0.001485311	0.000995862	0.002353569	0.995165258

4.2. Time and Cost Inputs

The values of time elements are essential inputs to find the expected total maintenance cost. Travel time is calculated using a 14 km average distance of the wind turbines from the shore and average speed of different access vessels. The repair time for rank 1 failure classification is assumed to be 4 hours in our study. It is reported in [14] that it will take 48 hours to switch out the component in question and replace a working unit for major maintenance. This time reported in [14] is the repair time for rank 2, 3 and 4 failure classifications in our study. The reported work in [11], which defined the failure classifications, did not provide any weight data for individual spare parts. Based on the turbine components listed under each failure classification reported in [11], the maximum cargo weight of spare parts for a failure classification is considered as the cargo weight of a resource combination. The fixed cost for corrective maintenance trip from [15] is the additional trip cost in this case study. All the time and cost inputs required to find the expected total maintenance cost are given in Tables 4 and 5.

Table 4. Inputs to calculate expected total maintenance cost [11,14–16].

Resource Combination	Travel Speed (km/h)	Travel Time (h)	Repair Time (h)	Access Vessel Cost/hour	Crane/Jack up Vessel Cost	Weight of Spare Parts (tonnes)
S_{11}	37.04	0.76	4	$62.5	N/A	0
S_{21}	37.04	0.76	48	$62.5	N/A	1
S_{31}	37.04	0.76	48	$62.5	$105,259.5	50
S_{41}	37.04	0.76	48	$62.5	$119,294.1	100
S_{12}	37.04	0.76	4	$62.5	N/A	0
S_{22}	46.3	0.6	48	$93.75	N/A	11
S_{32}	18.52	1.5	48	$93.75	$105,259.5	600
S_{42}	18.52	1.5	48	$93.75	$119,294.1	600

Table 5. Inputs to calculate expected total maintenance cost [17–19].

Parameter	Values
Maintenance Personnel cost/hour	70
Cost/tonnage of spares	29.72
Revenue Loss/hour	$18,684
Fixed cost for corrective maintenance trip for offshore wind turbine	$500,000

4.3. Results

The expected total maintenance cost of each decision choice for a specific wind farm model, is represented as a 4 × 2 matrix (there are eight decision choices in this study):

$$
Z_n = \begin{bmatrix}
z_{11} & z_{12} \\
z_{21} & z_{22} \\
z_{31} & z_{32} \\
z_{41} & z_{42}
\end{bmatrix}
$$

where Z_n is the cost matrix of the wind farm model n. The elements $z'_{ij}s$ of the matrix Z_n represent the expected total maintenance cost values (in $'s) of sending respective $S'_{ij}s$ for a specific wind farm model n. That is, the element z_{11} represent the expected total maintenance cost of sending S_{11}, the element z_{21} represent the expected total maintenance cost of sending S_{21}, and so on. It is earlier stated in Section 4.1 that both A-type and B-type RC's have identical resource elements for imperfect maintenance, which indicates, the elements z_{11} and z_{12} of the matrix Z_n will have identical values. The minimum of the $z'_{ij}s$ in the matrix Z_n is selected as the optimal solution and the corresponding resource combination is identified to be the cost-effective resource combination.

Article

Wind Turbine Surface Damage Detection by Deep Learning Aided Drone Inspection Analysis

ASM Shihavuddin [1,*], **Xiao Chen** [2,*], **Vladimir Fedorov** [3], **Anders Nymark Christensen** [1], **Nicolai Andre Brogaard Riis** [1], **Kim Branner** [2], **Anders Bjorholm Dahl** [1] **and Rasmus Reinhold Paulsen** [1]

[1]   Department of Applied Mathematics and Computer Science, Technical University of Denmark (DTU), 2800 Lyngby, Denmark; anym@dtu.dk (A.N.C.); nabr@dtu.dk (N.A.B.R.); abda@dtu.dk (A.B.D.); rapa@dtu.dk (R.R.P.)

[2]   Department of Wind Energy, Technical University of Denmark (DTU), 4000 Roskilde, Denmark; kibr@dtu.dk

[3]   EasyInspect ApS, 2605 Brøndby, Denmark; vlf@easyinspect.io

*   Correspondence: shihav@dtu.dk (A.S.); xiac@dtu.dk (X.C.)

Received: 24 January 2019; Accepted: 15 February 2019; Published: 20 February 2019

Abstract: Timely detection of surface damages on wind turbine blades is imperative for minimizing downtime and avoiding possible catastrophic structural failures. With recent advances in drone technology, a large number of high-resolution images of wind turbines are routinely acquired and subsequently analyzed by experts to identify imminent damages. Automated analysis of these inspection images with the help of machine learning algorithms can reduce the inspection cost. In this work, we develop a deep learning-based automated damage suggestion system for subsequent analysis of drone inspection images. Experimental results demonstrate that the proposed approach can achieve almost human-level precision in terms of suggested damage location and types on wind turbine blades. We further demonstrate that for relatively small training sets, advanced data augmentation during deep learning training can better generalize the trained model, providing a significant gain in precision.

Dataset: doi:10.17632/hd96prn3nc.1

Keywords: wind energy; rotor blade; wind turbine; drone inspection; damage detection; deep learning; Convolutional Neural Network (CNN)

---

1. Introduction

Reducing the Levelized Cost of Energy (LCoE) remains the overall driver for the development of the wind energy sector [1,2]. Typically, the Operation and Maintenance (O&M) costs account for 20–25% of the total LCoE for both onshore and offshore wind in comparison to 15% for coal, 10% for gas, and 5% for nuclear [3]. Over the years, great efforts have been made to reduce the O&M cost of wind energy using emerging technologies, such as automation [4], data analytics [5,6], smart sensors [7], and Artificial Intelligence (AI) [8]. The aim of these technologies is to achieve more efficient operation, inspection, and maintenance with minimal human interference. However, having a technology that can be deployed for on-site blade inspection for both onshore and offshore wind turbines under unpredictable weather conditions and that can acquire high-quality data efficiently is still a challenging task. One possible solution is to use the drone-based inspection of the wind turbine blades.

The drone-based approach enables low-cost and frequent inspections, high-resolution optical image acquisition, and minimal human intervention, thereby allowing predictive maintenance at lower costs [9]. Wind turbine surface damages exhibit recognizable visual traits that can be imaged by drones with optical cameras. These damages include for example leading edge erosion, surface cracks, damaged lightning receptors, damaged vortex generators, and so forth. They are externally visible even in their early stages of development. Moreover, some of these damages, such as surface cracks, even indicate severe internal structural damages [10]. Nevertheless, internal damages such as delamination, debonding, or internal cracks are not detectable using the drone-based inspection with optical cameras. This study is limited to surface damages of the wind turbine blades.

Extracting damage information from a large number of high-resolution inspection images requires significant manual effort, which is one of the reasons for the overall inspection cost still remaining at a high level. In addition, manual image inspection is tedious and therefore error-prone. By automatically providing suggestions to experts on highly probable damage locations, we can significantly reduce the required man-hours and simultaneously enhance manual detection performance, as a result minimizing the labor cost involved with the analysis of inspection data. With regular cost-efficient and accurate drone inspection, the scheduled maintenance of wind turbines can be performed less frequently, potentially bringing down the overall maintenance cost, contributing to the reduction of LCoE.

Only very few research works have addressed the machine learning-based approaches for surface damage detection of wind turbine blades from drone images. One example, however, is Wang et al. [11], who used drone inspection images for crack detection. To automatically extract damage information, they used Haar-like features [12,13] and ensemble classifiers selected from a set of base models including logitBoost [14], decision trees [15], and support vector machines [16]. Their work was limited to detecting the crack and relied on classical machine learning methods.

Recently, deep learning technology has become efficient and popular, providing groundbreaking performances in detection systems for the last four years [17]. In this work, we addressed the problem of damage detection by deploying a deep learning object detection framework to aid human annotation. The main advantages of deep learning over other classical object detection methods are: it automatically finds the most discriminate features for the identification of objects, and it is achieved through an optimization process by minimizing the identification and localization errors.

Large size variations of different surface damages of wind turbine blades, in general, are a challenge for machine learning algorithms. In this study, we overcame this challenge with the help of advanced image augmentation methods. Image augmentation is the process of creating extra training images by altering images in the training sets. With the help of augmentation, different versions of the same image are created encapsulating different possible variations during drone acquisition [18].

The main contributions of this work are three-fold:

1. **Automated suggestion system for damage detection in drone inspection images**: implementation of a deep learning framework for automated suggestions of surface damages on wind turbine blades captured by drone inspection images. We show that deep learning technology is capable of giving reliable suggestions with almost human-level accuracy to aid manual annotation of damages.
2. **Higher precision in the suggestion model achieved through advanced data augmentation**: The advanced augmentation step called the "Multi-scale pyramid and patching scheme" enables the network to achieve better learning. This scheme significantly improves the precision of suggestions, especially for high-resolution images and for object classes that are very rare and difficult to learn.
3. **Publication of the wind turbine inspection dataset**: This work produced a publicly-available drone inspection image of the "Nordtank" turbine over the years of 2017 and 2018. The dataset is hosted within the Mendeley public dataset repository [19].

The surface damage suggestion system is trained using faster R-CNN [20], which is a state-of-the-art deep learning object detection framework. Faster R-CNN works efficiently and with high accuracy compared to other frameworks, while identifying objects in terms of the bounding box from large images. The Convolutional Neural Network (CNN) is used as the backbone architecture

2.4. Performance Measure: Mean Average Precision

All the reported performances in terms of Mean Average Precision (MAP) were measured during inference on the test images, where the inference is the process of applying the trained model to an input image to receive the detected and classified object in return. In this work, we called it suggestions if the trained model from deep learning was being used.

MAP is commonly used in computer vision to evaluate object detection performance during inference. An object proposal is considered accurate only if it overlaps with the ground truth with more than a certain threshold. Intersection over Union (IoU) is used to measure the overlap of a prediction and the ground truth where ground truth refers to the original damages identified and annotated by experts in the field.

$$\text{IoU} = \frac{P \cap GT}{P \cup GT} \tag{1}$$

The IoU value corresponds to the ratio of the common area over the sum of the proposed detection and ground truth areas (as shown in Equation (1), where P and GT are the predicted and ground truth bounding boxes, respectively). If the value is more than 0.5, the prediction or, in this case, the suggestion is considered as a true positive. This 0.5 value is relatively conservative, as it makes sure that the ground truth and the detected object have a very similar bounding box location, size, and shape. For addressing human perception diversities, we used a 0.3 IoU threshold for considering a detection as a true positive.

$$P_C = \frac{N(\text{True Positives})_C}{N(\text{Total Objects})_C} \tag{2}$$

$$AP_C = \frac{\sum P_C}{N(\text{Total Images})_C} \tag{3}$$

$$MAP = \frac{\sum AP_C}{N(\text{Classes})} \tag{4}$$

Per class precision for each image, P_C, was calculated using Equation (2). For each class, average precision, AP_C was measured over all the images in the dataset using Equation (3). Finally, MAP was measured as the mean of average precision for each class over all the classes in the dataset (see Equation (4)). Throughout this work, the MAP is reported in percentage.

2.5. Software and Codes

We used the Tensorflow [36] deep learning API for experimenting with the faster R-CNN object detection method with different CNN architectures. These architectures were compared with each other under the proposed augmentation schemes. For implementing the regular augmentations, the imgaug (https://github.com/aleju/imgaug) package was used, and for the pyramid and patching augmentation, an in-house python library was developed. Inception-ResNet-V2 and other CNN weights were initialized from the pre-trained weight on the Common Objects in COntext (COCO) dataset [37]. The COCO dataset consists of 80 categories of regular objects and is commonly used for bench-marking deep learning object detection performance.

2.6. Hardware

All the experiments reported in this work were performed on a GPU cluster machine with 11-GB GeForce GTX 1080 Graphics Cards within a Linux operating system. The initial time required for training 275 epochs (where one epoch is defined as when all the images in the training set had been used at least once for optimization of the detection model) using Inception-V2, ResNet-50, ResNet-101, and Inception-ResNet-V2 networks was on average 6.3 h, 10.5 h, 16.1 h, and 27.9 h, respectively.

2.7. Dataset

2.7.1. EasyInspect Dataset

The EasyInspect dataset is a non-public inspection dataset provided by EasyInspect ApS, which contains images (4000 × 3000 pixels in size) of different types of damages on wind turbines from different manufacturers. The four classes are LE erosion, VG panel, VG panel with missing teeth, and lightning receptor.

2.7.2. DTU Drone Inspection Dataset

In this work, we produced a new public dataset entitled DTU—Drone inspection images of the wind turbine. It is the only public wind turbine drone inspection image dataset containing a total of 701 high-resolution images. This dataset contains temporal inspection images of 2017 and 2018 covering the "Nordtank" wind turbine located at DTU Wind Energy's test site at Roskilde, Denmark. The dataset comes with the examples of damages or mounted objects such as VG panel, VG panel with missing teeth, LE erosion, cracks, lightning receptor, damaged lightning receptor, missing surface material, and others. It is hosted at [19].

3. Results

3.1. Augmentation of Training Images Provides a Significant Gain in Performance

Comparing different augmentation types showed that a combination of the regular, pyramid, and patching augmentations produced a more accurate suggestion model, especially for the deeper CNN architectures as the backbone for the faster R-CNN framework. Using CNN architecture ResNet-101 for example (as shown in Table 2 in column "all"), without any augmentation, the MAP (detailed in the Materials and Methods Section) of damage suggestion was very low with a value of 25.9%. With the help of the patching augmentation, the precision improved significantly (as for this case, the MAP increased to 35.6%). In Table 2, all the experimental results are reported in terms of MAP. VG and VGMT represent the VG panel and the VG with Missing Teeth, respectively. "All" is the overall MAP comprising all four classes.

Table 2. Experimental results for different CNN architectures and data augmentation methods.

CNN	Augmentation	MAP (%)				
		LE Erosion	VG	VGMT	Lightning Receptor	All
	Without	20.34	13.15	4.21	41.39	19.77
	Patching	37.98	53.86	32.74	22.12	36.67
Inception-V2	Patching + regular	36.90	52.58	36.08	23.54	37.28
	Pyramid + patching	88.11	86.61	58.28	70.18	75.80
	Pyramid + patching + regular	89.94	84.15	71.85	40.76	71.67
	Without	37.51	17.54	7.02	40.88	25.74
	Patching	34.48	54.92	32.17	34.96	39.13
ResNet-50	Patching + regular	35.06	47.87	24.88	34.96	35.69
	Pyramid + patching	90.19	85.15	61.43	73.85	77.66
	Pyramid + patching + regular	90.47	88.84	35.27	73.15	71.93
	Without	29.61	20.29	4.21	49.33	25.86
	Patching	34.79	47.35	25.35	34.96	35.61
ResNet-101	Patching + regular	36.06	51.41	32.16	33.46	38.27
	Pyramid + patching	88.86	87.46	41.53	64.19	70.51
	Pyramid + patching + regular	86.77	86.15	41.53	77.00	72.86
	Without	31.54	18.35	5.96	54.14	27.50
	Patching	36.30	44.43	33.37	42.12	39.06
Inception-ResNet-V2	Patching + regular	32.73	47.59	19.58	34.96	33.72
	Pyramid + patching	90.22	91.66	69.85	69.56	80.32
	Pyramid + patching + regular	90.62	89.08	73.11	71.58	81.10

34. Szegedy, C.; Ioffe, S.; Vanhoucke, V.; Alemi, A.A. *Inception-V4, Inception-Resnet and the Impact of Residual Connections on Learning*; AAAI: Palo Alto, CA, USA, 2017; Volume 4, p. 12.

35. Szegedy, C.; Liu, W.; Jia, Y.; Sermanet, P.; Reed, S.; Anguelov, D.; Erhan, D.; Vanhoucke, V.; Rabinovich, A. Going deeper with convolutions. In Proceedings of the Conference on Computer Vision and Pattern Recognition (CVPR), Boston, MA, USA, 8–10 June 2015; pp. 1–9.

36. Abadi, M.; Agarwal, A.; Barham, P.; Brevdo, E.; Chen, Z.; Citro, C.; Corrado, G.S.; Davis, A.; Dean, J.; Devin, M.; et al. TensorFlow: Large-Scale Machine Learning on Heterogeneous Systems, 2015. Software. Available online: tensorflow.org (accessed on 12 November 2017).

37. Lin, T.Y.; Maire, M.; Belongie, S.; Hays, J.; Perona, P.; Ramanan, D.; Dollár, P.; Zitnick, C.L. Microsoft COCO: Common Objects in Context. *European Conference on Computer Vision (ECCV)*; Springer: Zurich, Switzerland, 2014; pp. 740–755.

Article

Fault Simulation and Online Diagnosis of Blade Damage of Large-Scale Wind Turbines

Feng Gao, Xiaojiang Wu *, Qiang Liu, Juncheng Liu and Xiyun Yang

School of Control and Computer Engineering, North China Electric Power University, Beijing 102206, China; gaofeng@ncepu.edu.cn (F.G.); liu1990NCEPU@126.com (Q.L.); jcliu@ncepu.edu.cn (J.L.); yangxiyun916@sohu.com (X.Y.)
* Correspondence: 1172227133@ncepu.edu.cn

Received: 23 December 2018; Accepted: 2 February 2019; Published: 7 February 2019

Abstract: Damaged wind turbine (WT) blades have an imbalanced load and abnormal vibration, which affects their safe and stable operation or even results in blade rupture. To solve this problem, this study proposes a new method to detect damage in WT blades using wavelet packet energy spectrum analysis and operational modal analysis. First, a wavelet packet transform is used to analyze the tip displacement of the blades to obtain the energy spectrum. The damage is detected preliminarily based on the energy change in different frequency bands. Subsequently, an operational modal analysis method is used to obtain the modal parameters of the blade sections and the damage is located based on the modal strain energy change ratio (MSECR). Finally, the professional WT simulation software GH (Garrad Hassan) Bladed is used to simulate the blade damage and the results are verified by developing an online fault diagnosis platform integrated with MATLAB. The results show that the proposed method is able to diagnose and locate the damage accurately and provide a basis for further research of online damage diagnosis for WT blades.

Keywords: wind turbine; blade damage diagnosis; wavelet transform; operational modal analysis; modal strain energy (MSE)

---

1. Introduction

Blades are the most important component of a wind turbine (WT) and their operating status is an important factor to ensure the normal and stable operation of WTs. Since WTs are mostly located in harsh areas and the wind conditions are complex and changeable, blade faults occur more commonly with increasing operating hours. Blade faults manifest as blade damage and the reasons include fatigue due to alternating loads, lightning strokes, freezing and external impacts. If the damage is not detected and repaired in time, the blade will break or the WT may collapse and other serious accidents may occur. Therefore, online blade damage diagnosis is of great importance in terms of research value and applications [1].

The supervisory control and data acquisition (SCADA) information reflects the operating status of the WT and is easy to obtain. However, there are no data that directly reflect the state of the blades in the SCADA data. Therefore, it is important to investigate the health monitoring and fault diagnosis of the blades using intelligent algorithms such as artificial neural networks, expert systems, fuzzy logic systems and support vector machines and the WT SCADA data or vibration monitoring data. In Reference [2], a WT blade breakage monitoring method using SCADA data was proposed. A deep automatic coder (DA) model was proposed to determine impending blade damage from the SCADA data using a discriminant index. By analyzing fuzzy fault features, Yang et al. [3] detected blade faults by interpreting the data collected by the WT SCADA system; the authors used the conventional 10-minute average data in SCADA, which not only affected the diagnostic accuracy but also disregarded much of the fault training data [4].

On the other hand, it is more common to install sensors on the WT blades to monitor blade damage by using operational signals. The data directly reflect the structural damage to the blades and data processing is easy and the results are accurate. Therefore, much research has focused on the choice of sensors and signal processing. In Reference [5], an acoustic emission (AE) technique was used to obtain blade damage information and the location of the damage. Non-destructive AE methods were applied during a series of blade certification tests on a set of small WT blades [6]. However, the method requires the installation of an acoustic emitter and several receivers, which is technically difficult; in addition, the AE signals may suffer from interference from the signals from different blades or mechanical noise. In Reference [7,8], optical fiber sensors were embedded into WT blades and the blade damage was detected by processing the measured strain signal. However, optical fiber sensors can only be installed during blade manufacturing and this method is not suitable for WTs already in operation. A ceramic piezoelectric sensor mounted on the blades was used to perform a tensile test and a wavelet packet transform was used to diagnose and locate the damage [9]. Polyvinylidene fluoride resin (PVDF) film- based strain sensors were installed during a test of a full-scale WT [10]; the experimental results showed that the PVDF film-based sensors were effective for detecting damage at the trailing edge of the WT blade. In order to increase the accuracy of the damage diagnosis, the authors in Reference [11] used a hybrid sensor network consisting of capacitive film strain sensors and optical fiber sensors and using a sensor information fusion method based on a neural network. The above-mentioned experiments have achieved very good results and thin film sensors are easy to install; however, a sensor network consisting of a large number of sensors increases the installation difficulty and reduces the reliability of the system. Furthermore, it is still unknown whether the sensors affect the aerodynamic characteristics of the WT blades.

Vibration signal analysis is widely used in the field of mechanical fault diagnosis because it is a mature technology and vibration sensors are easy to install [12]. Therefore, vibration or acceleration sensors have been used commonly for damage diagnosis of blades. The authors in Reference [13–15] used the finite element analysis software ANSYS to simulate blade damage. The changes in the modal vibration shapes before and after blade damage were determined using vibration signal analysis. Because it is difficult to simulate WT operation processes in the ANSYS software, online blade damage diagnosis is not possible. However, the simulation results represent a good theoretical reference for damage diagnosis of blades using vibration sensors. In Reference [16], the characteristics of the vibration data before and after blade damage were determined under different environmental conditions and a principal component analysis was used for detecting the damage. Small-scale WT blades were used for damage experiments in Reference [17,18]. The change in the modal parameters was determined by analyzing the vibration signals of the blades. In Reference [19], a neural network was used to diagnose blade damage; the vibration signal data were obtained from excitation experiments. These types of methods are more common than simulation methods using ANSYS but they are still in the experimental stage. In many experiments, the vibration sensors were not actually installed on the blades. The use of a large number of sensors not only increases the installation difficulty but also affects the system reliability. In Reference [20], an exciter and several vibration sensors were installed on the blade of a Vestas V27 WT to monitor the operating conditions. This method was effective for monitoring the blade damage but requires an exciter for creating blade vibrations. In addition, harsh environmental conditions affect the reliability of the sensors.

In China, the most effective blade damage diagnosis method used in actual operations is still human observation. Therefore, a reliable and accurate method for online damage diagnosis is urgently needed. In this study, the GH Bladed software is used to simulate blade damage, which is then initially detected by analyzing the wavelet packet energy spectrum of the blade tip displacement. Unlike the SCADA system, Bladed is a reliable wind turbine simulation software that simulates blade faults accurately and obtains fault information without requiring a large amount of fault data. It has been shown that the wavelet packet method is more effective than the Fourier transform method. Moreover, the initial damage determination of the blades is also simpler and easier to implement using the wavelet packet decomposition method and the analysis does not complicate the diagnosis. When the blade is damaged, the modal parameters of the blade sections are determined using an operational modal analysis and subsequently, the modal strain energy (MSE) change ratio (MSECR) is calculated. The MSECR is used as an index to locate the damage. Computing the MSECR only requires the vibration signal of the blade, which can be measured by installing a small number of patch vibration sensors on the tip or embedding vibration sensors in the blade. Unlike acoustic emission technology and other methods of fault diagnosis that use sensors, an external signal excitation source does not have to be installed, which reduces the complexity and enhances the reliability of the system. Finally, a dynamic link library (DLL) is used to interface the Bladed and MATLAB software (2016a, MathWorks, Natick, MA, USA) to create an online diagnostic tool for WT blade damage. The simulation results show that this method can diagnose blade damage accurately. Moreover, the method can be applied to actual WTs by using a small number of patch-type vibration sensors on the blade tip.

2. Analysis and Simulation of Blade Damage Using GH Bladed

WT blades are typically constructed using fiber-reinforced polymeric composites and sandwich structures. Moreover, their geometries (e.g., the aerofoil chord length) gradually change along the pitch axis direction. It is, therefore, a challenging task to develop an accurate analytical model of the structural damage of WT blades. For simplification, in the present study, a WT blade is regarded as a multi-degree-of-freedom system consisting of 9 sections (U_1–U_9), as shown in Figure 1. The mass and stiffness can be adjusted in each section, as shown in Table 1.

Due to the large size and weight of large-scale WT blades, it is difficult to conduct damage experiments and the cost is high. GH Bladed provides a simulation platform that creates an approximation of an actual WT; the software contains a wealth of blade information, including length, thickness, stiffness, quality, airfoil and other data. It can also simulate the blade icing by setting the ice position and ice density. Therefore, it is convenient to use GH Bladed for fault analysis and diagnosis of blade icing. In some cases, blade icing and damage appear to be similar but the results are different. Blade icing mainly changes the quality of the blades. However, blade damage does not change the quality of blades but causes changes in the characteristic parameters such as blade stiffness. In addition, blade icing affects the operational parameters of the WT, such as the power and motor speed [21], whereas blade damage usually does not. Blade icing faults mostly occur in the cold season and cold regions and WT maintenance personnel detect blade icing using weather and WT SCADA data. Many studies have focused on blade icing diagnosis and deicing [22–24]. However, the connection between blade damage and weather is insignificant and there is no effective method for its detection.

In this study, we focus on changing the section stiffness and damping to simulate blade damage and we obtain the vibration information from GH Bladed. As the blade stiffness decreases, the vibration signal gradually increases along the major axis and the response is most pronounced at the tip of the blade. Therefore, the signal obtained from the tip is used for the initial damage detection of the blade damage. Figure 2 shows the vibration displacement of section 3, section 6 and the blade tip with severe damage in section 2; the simulation results demonstrate that the vibration signal increases gradually in the direction of the main axis. Figure 3 shows the blade tip displacement of a normal blade and a blade with severe damage.

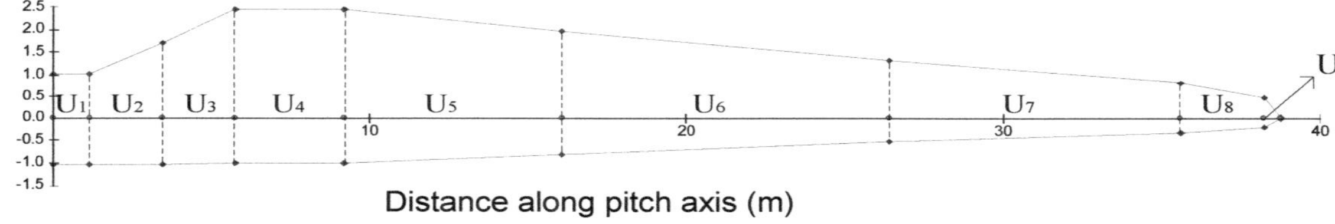

Figure 1. Blade section settings in the GH (Garrad Hassan) Bladed software

Table 1. Blade geometry and mass/stiffness information.

Section	1	2	3	4	5	6	7	8	9	10
Distance along pitch axis (m)	0	1.15	3.44	5.74	9.19	16.07	26.41	35.59	38.23	38.75
Chord (m)	2.07	2.07	2.76	3.44	3.44	2.76	1.84	1.15	0.69	0.03
Aerodynamic twist (deg)	0	0	9	13	11	7.8	3.3	0.3	2.75	4
Thickness (%)	100	100	64	40	30	22	15	13	13	13
Mass/unit length(kg/m)	1084.77	369.81	277.36	234.21	209.56	172.58	103.55	55.47	40.68	24.65
Edgewise stiffness (N·m^2)	7.47×10^9	2.61×10^9	2.09×10^9	1.43×10^9	1.29×10^9	5.65×10^8	1.22×10^8	2.43×10^7	4,518,000	8167.51
Flapwise stiffness (N·m^2)	7.47×10^9	2.43×10^9	1.41×10^9	8.34×10^8	5.56×10^8	2.09×10^8	2.95×10^7	2,259,000	113,824	3127.98

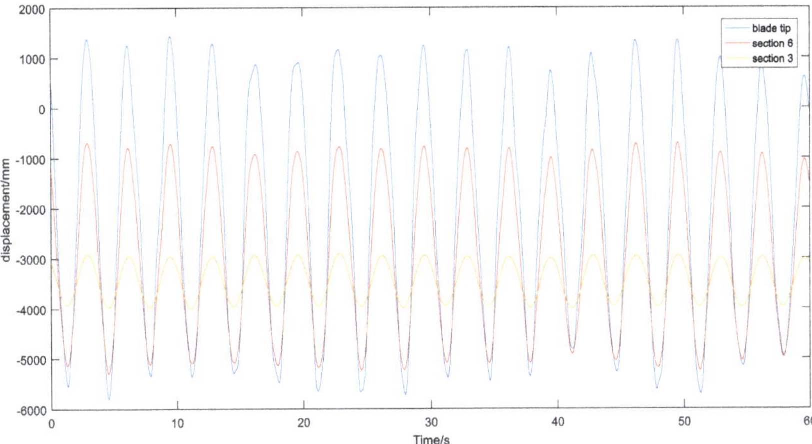

Figure 2. Displacement signal of different sections when section 2 is damaged.

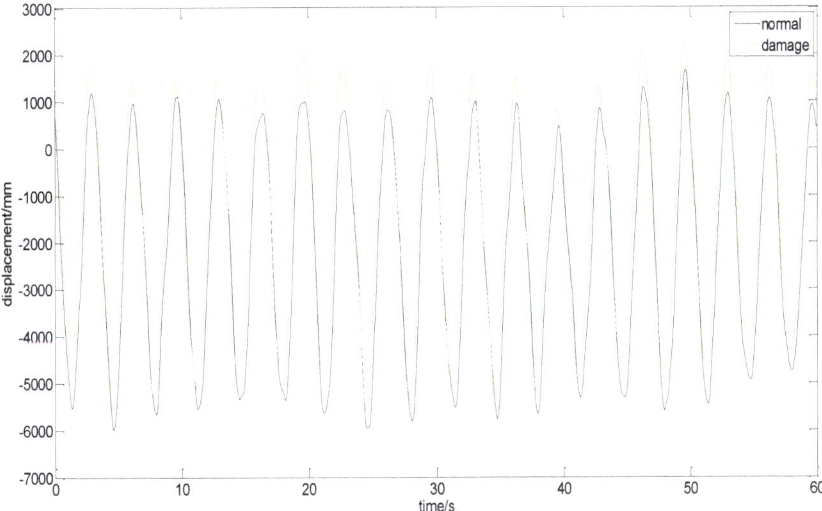

Figure 3. Blade tip displacement of normal and damaged blades.

3. Signal Analysis of the Blade Tip Displacement

3.1. Fast Fourier Transform (FFT) Analysis of Blade Tip Displacement

Fourier analysis is commonly used in traditional signal analysis; it uses a fixed window function and does not reflect the non-stationary, time-domain and frequency-domain characteristics of the signals. Figure 4 shows the results of an FFT analysis of a blade tip displacement signal with different damage degrees in section 2. In the blade damage simulation, the blade parameters are assessed before and after the blade damage and the damage is divided into four classes, namely, slight damage, moderate damage, severe damage and extreme damage. For example, slight damage is defined as 80% of the original stiffness and moderate, severe and extreme damage are defined as 70%, 60% and 50% of the original stiffness, respectively. When there is little damage, the FFT of the blade tip displacement has no apparent influence on the low-frequency part and it is difficult to determine the blade damage using this method. The information in the high-frequency part is not accurate because of the limitations of the FFT analysis method. The wavelet transform is a method that uses a fixed area but a variable window size. Wavelet packet decomposition can adaptively select the frequency bands that match the signal spectrum based on certain signal characteristics [25].

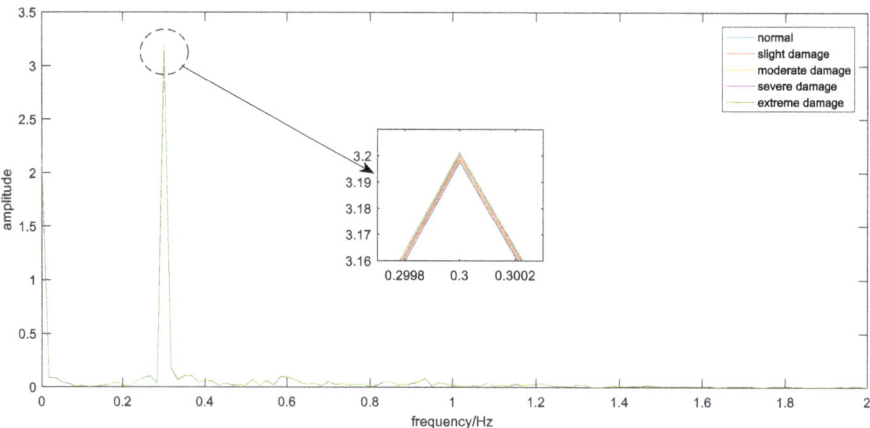

Figure 4. FFT analysis of blade tip displacement with different damage degree.

3.2. Wavelet Packet Energy Spectrum Extraction

A damaged blade exhibits abnormal vibration, which can be detected in the energy spectrum. The band energy reflects the operating status of the blade. By determining the change in the energy in different frequency bands, blade damage can be initially diagnosed. The energy-based wavelet packet decomposition results are called the wavelet packet energy spectrum of the blades.

We use a 3-layer wavelet packet decomposition as an example; the wavelet packet energy spectrum extraction method consists of the following steps:

(1) Obtaining the decomposition coefficients. The blade vibration signals are decomposed using a 3-layer wavelet packet decomposition; 8 decomposition coefficients of layer 3 from low frequency to high frequency are obtained $\left(X_3^0, X_3^1, X_3^2, \cdots, X_3^7\right)$.

(2) Reconstruction of the wavelet coefficients. We extract each sub-band range signal $S_3^j(j = 0, 1, \cdots, 7)$; the total signal is expressed as:

$$S = S_3^0 + S_3^1 + S_3^2 + \cdots + S_3^7 \tag{1}$$

(3) Calculating the signal energy of each sub-band. The reconstructed signal of each layer 3 node is expressed as: $S_3^j(j = 0, 1, \cdots, 7)$; the corresponding band energy is $E_3^j(j = 0, 1, \cdots, 7)$, which is defined as:

$$E_3^j = \int \left|S_3^j(t)\right|^2 dt = \sum_{k=1}^{N} \left|x_{jk}\right|^2 \tag{2}$$

where $x_{jk}(j = 0, 1, \cdots, 7; k = 0, 1, \cdots, n)$ represents the amplitude of the signal's discrete points.

The blade tip displacement signals were decomposed using the wavelet packet method to obtain the characteristic information of the tip displacement energy band. The details for the energy bands are shown in Table 2. The results show that the values of Band 1 increase with increasing degree of damage, although the differences are not large. The same is observed for the values of Bands 2 to 8 but the difference between the damaged blade and the normal blade is large enough to determine if the blade is damaged. Unlike the FFT analysis, this method has more significant eigenvalue changes and it is easier to determine the blade damage.

Table 2. Tip displacement energy spectrum data.

Damage Degree	Band 1	Band 2	Band 3	Band 4	Band 5	Band 6	Band 7	Band 8
Extreme damage	94.8525	2.1198	0.5638	0.1149	0.0585	0.0259	0.0139	0.0043
Severe damage	94.7138	2.1173	0.5634	0.1148	0.0585	0.0259	0.0139	0.0043
Moderate damage	94.6368	2.1126	0.5631	0.1146	0.0585	0.0258	0.0139	0.0044
Slight damage	94.5633	2.1203	0.5573	0.1141	0.0584	0.0258	0.0138	0.0043
Normal	94.2946	15.7002	1.8677	7.7063	0.4163	0.3263	0.9087	3.8432

This method only requires vibration sensors at the blade tips to measure the displacement signal, which is easier to achieve than sensor installation and signal analysis for other kinds of sensors. If patch-type wireless sensors are used, there is less impact on the aerodynamic characteristics of the blade, making this method very suitable for blade health monitoring in currently used commercial WTs. Depending on the blade monitoring requirements, a number of sensors can be installed in different sections during blade manufacturing to detect the presence of damage and also the damage location. We report on the blade damage location using multi-sensor vibration signals in the next section of this paper.

4. Blade Damage Location Based on Operational Modal Testing

In an experimental modal analysis, the modal parameters of a structure are obtained by the parameter identification of the system input and output signals collected under experimental conditions. Nevertheless, during the actual operation of a WT, no man-made excitation can be applied; therefore, an ambient excitation method is used, that is, the external wind condition is used as an excitation source. Subsequently, the operational mode test theory is applied to analyze the blade vibration signals. After a pretreatment using a random decrement technique, the output signal time series was analyzed using an autoregressive moving average (ARMA) model to solve for the modal parameters of each section. Finally, the blade damage location was determined by calculating the MSECR.

4.1. Operational Modal Test Method Applicable to WT Blades

The random decrement technique refers to removing or reducing random components from one or more stationary random response samples of a linear vibration system to obtain a free response signal under a certain initial excitation [26]. The following describes the basic principle of obtaining free vibration response signal data from a structure response signal using the random reduction method. For a linear system structure, the forced vibration response of a measuring point under any excitation can be expressed as:

$$y(t) = y(0)D(t) + \dot{y}(0)V(t) + \int_0^t h(t - \tau)f(\tau)d\tau \tag{3}$$

where $D(t)$ is the free vibration response at an initial displacement of 1 and an initial velocity of 0. $V(t)$ is the free vibration response at an initial displacement of 1 and initial velocity of 0. $y(0)$ is the initial displacement and $\dot{y}(0)$ is the initial velocity of the system vibration. $h(t)$ is the system unit impulse response function. $f(t)$ is the external excitation. We selected a suitable constant A to intercept a measured random vibration structure response signal $y(t)$; a series of different intersection moments t_i ($i = 1, 2, ..., N$) can be obtained and the response $y(t - t_i)$ from the moment t_i can be seen as a linear superposition of three parts: the free vibration response caused by the initial displacement at t_i, the free vibration response caused by the initial velocity at t_i and the forced vibration response caused by the random excitation $f(t)$. Therefore:

$$y(t - t_i) = y(t_i)D(t - t_i) + \dot{y}(t_i)V(t - t_i) + \int_{t_i}^t h(t - \tau)f(\tau)d\tau \tag{4}$$

Since the excitation $f(t)$ is stationary and the starting point does not affect its random characteristics, a series of starting points t_i of $y(t - t_i)$ can be moved to coordinate the origin to obtain the subsample function $x_i(t)$ $(i = 1, 2, ..., N)$. That is:

$$x_i(t) = AD(t) + \dot{y}(t_i)V(t) + \int_0^t h(t - \tau)f(\tau)d\tau \tag{5}$$

The statistical average of $x_i(t)$ is:

$$
\begin{aligned}
x(t) &= \frac{1}{N} \sum_{i=1}^{N} X_i(t) \approx E\left[AD(t) + \dot{y}(t_i)V(t) + \int_0^t h(t - \tau)f(\tau)d\tau\right] \\
&\approx AD(t) + E\left[\dot{y}(t_i)\right]V(t) + \int_0^t h(t - \tau)E[f(\tau)]d\tau
\end{aligned}
\tag{6}
$$

If the excitation $f(t)$ is a stationary pure random vibration with a mean value of 0 and the system vibration response $y(t)$ and $\dot{y}(t_i)$ is also a stationary random vibration with a mean of 0, then:

$$E[f(t)] = 0E\left[\dot{y}(t_i)\right] = 0 \tag{7}$$

According to the above:

$$x(t) \approx AD(t) \tag{8}$$

where $x(t)$ is called the free vibration response obtained by the random decrement method. Generally, turbulent wind is a natural random excitation in WTs and the mean vibration velocity of a blade tip should be zero, that is, $E\left[\dot{y}(t_i)\right]$ is zero. However, the mean value response of the blade tip displacement is not zero; therefore, the random decrement method cannot be directly applied to the original tip displacement signal [27]. Under the excitation of an average wind speed, the tip displacement is:

$$\int_0^t h(t - \tau)E[f(\tau)]d\tau = \int_0^t h(t - \tau)g(\overline{v})d\tau = B \tag{9}$$

where $\overline{v}$ is the average wind speed from time 0 to t, B is the mean value of the blade tip displacement, $g(\overline{v})$ is the blade excitation function for a wind speed of $\overline{v}$. In order to ensure that this method is applicable to the tip displacement signal, the signal should be pre-processed by subtracting the mean value B so that the mean value is zero. Therefore:

$$x'(t) = \frac{1}{N} \sum_{i=1}^{N}[X_i(t) - B] \approx E\left[AD(t) + \dot{y}(t_i)V(t)\right] = AD(t) \tag{10}$$

According to Equation (10), a free vibration response with an initial displacement of A and an initial velocity of 0 is obtained. The response is determined by using an ARMA model time series analysis, which is a method of using parametric models to process ordered random vibration response data for modal parameter identification [28]. The relationship between the linear system excitation and the response for N degrees of freedom can be described by a higher-order differential equation, which becomes a differential equation represented by several time series at different times, the ARMA temporal model equation, in a discrete time domain:

$$\sum_{k=0}^{2N} a_k x_{t-k} = \sum_{k=0}^{2N} b_k f_{t-k} \tag{11}$$

Equation (11) describes the relationship between the response data sequence x_t and the historical value x_{t-k}, where $2N$ is the order of the autoregressive model and the sliding mean model, a_k, b_k denote the autoregressive coefficient and the sliding mean coefficient to be identified respectively and f_t denotes white noise excitation. When $k = 0$, let $a_0 = b_0 = 1$. The ARMA equation $\{x_t\}$ has a unique smooth solution:

$$x_t = \sum_{i=0}^{\infty} h_t f_{t-i} \tag{12}$$

where h_t is the impulse response function and f_t is white noise. Therefore:

$$E[f_{t-i} f_{t+\tau-k}] = \begin{cases} \sigma^2 (k = \tau + i) \\ 0 (others) \end{cases} \tag{13}$$

where δ^2 is the white noise variance. By substituting the result of Equation (13) into Equation (12), the following is obtained:

$$R_\tau = \sigma^2 \sum_{i=0}^{\infty} h_i h_{i+\tau} \tag{14}$$

Since the linear system impulse response function h_t is the system output response when excited by a pulse signal δ_t, the expression defined by the ARMA process is:

$$\sum_{k=0}^{2N} a_k h_{t-k} = \sum_{k=0}^{2N} b_k \sigma_{t-k} = b_t \tag{15}$$

After calculating the autoregressive coefficient a_k and the sliding mean coefficient b_k, the system modal parameters can be calculated by using the expression of the ARMA model transfer function:

$$H(z) = \frac{\sum_{k=0}^{2N} b_k z^{-k}}{\sum_{k=0}^{2N} a_k z^{-k}} \tag{16}$$

The root of the denominator polynomial equation is solved using a high-order algebraic equation solving method and the obtained root is the pole of the transfer function. Their relationship with the system modal frequency ω_k and the damping ratio ξ_k is:

$$\begin{cases} z_k = e^{s_k \Delta t} = e^{(-\xi_k \omega_k + j\omega_k \sqrt{1-\xi_k^2})\Delta t} \\ z_k^* = e^{s_k^* \Delta t} = e^{(-\xi_k \omega_k - j\omega_k \sqrt{1-\xi_k^2})\Delta t} \end{cases} \tag{17}$$

The modal frequency ω_k and the damping ratio ξ_k can be obtained from Equation (18), that is:

$$\begin{cases} R_k = \ln z_k = s_k \Delta t \\ \omega_k = \frac{|R_k|}{\Delta t} \\ \xi_k = \sqrt{\dfrac{1}{1+\left(\dfrac{Im(R_k)}{Re(R_k)}\right)^2}} \end{cases} \tag{18}$$

Suppose that the k-order residue of H_{pq} (s), which is the transfer function of the p-point response excited at point q, is A_{kpq}, then the residue can be calculated as follows:

$$A_{kpq} = \lim_{z \to z_k} H_{pq}(z)(z - z_k) = \frac{\sum_{k=0}^{2N} b_k z^{-k}}{\sum_{k=0}^{2N} a_k z^{-k}}(z - z_k)|z = z_k \tag{19}$$

The modal vector can be obtained by processing the residue obtained from a series of measured response points. For a structure with n response points, the first task is determining the measurement point with the largest absolute value from the residue of n corresponding modes of the same order. Assuming that the point is the measurement point m, the normalized complex modal vector corresponding to the k-order mode can be obtained by the following formula:

$$\{\phi_k\} = \left[A_{K1q}A_{k2q}\cdots A_{kmq}\right]^T / A_{kmq} \tag{20}$$

In this way, the modal parameters such as the modal frequency, modal damping and modal vibration mode can be obtained.

4.2. Blade Damage Location Analysis Based on the MSECR

Because a blade can be simplified as a hollow cantilever beam structure, a model of the blade sections was developed based on the structural characteristics and material properties (Figure 5). Three neighboring sections are denoted as U_{n-1}, U_n and U_{n+1} with the masses of m_{n-1}, m_n and m_{n+1}, respectively. The sections U_{n-1} and U_n are connected via stiffness $k_{n-1,n}$ and damping $c_{n-1,n}$ and the sections U_n and U_{n+1} are connected via $k_{n,n+1}$ and $c_{n,n+1}$. Consequently, when an external load is applied to the blade, the dynamic response can be expressed by the following equation:

$$M\ddot{x} + C\dot{x} + Kx = F(t) \tag{21}$$

where x represents the vector of the displacement responses along the blade; M, C and K denote the equivalent structural mass, the equivalent structural damping and the equivalent structural stiffness matrices, respectively; F is the matrix of the external forces. It can be inferred that when a local defect occurs in section n, the values of $c_{n-1,n}$, $c_{n,n+1}$, $k_{n-1,n}$ and $k_{n,n+1}$ change correspondingly, whereas the damping and stiffness in the other sections may not change.

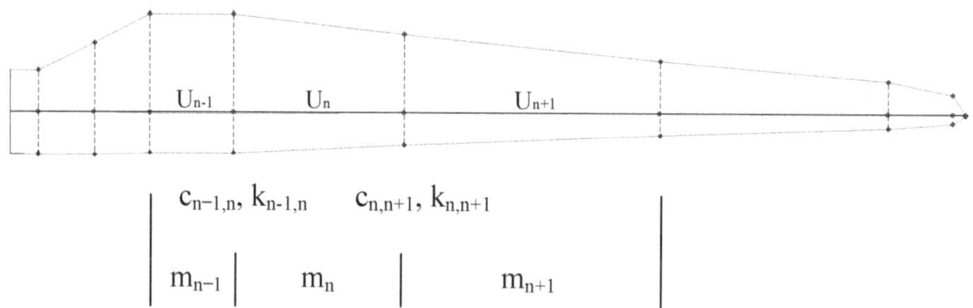

Figure 5. Model of the blade sections.

This configuration was the inspiration for developing a damage location diagnosis method for WT blades using the modal analysis of the sections. Blade icing has a significant influence on the mass characteristic parameters of the blade but has little effect on stiffness and damping. Since the values of the damping and stiffness are dependent only on the blade's structural integrity, the proposed damage location diagnosis method responds only to changes caused by structural damage. Therefore, false alarms due to ice on the blade surfaces can be avoided.

The damage to the blade structure has nothing to do with its mass, that is to say, $[\Delta M] = 0$. Therefore, structural damage is equivalent to a change in stiffness. Equation (22) shows the relationship between the structural stiffness and the modal vibration modes before and after the damage has occurred (superscript d indicates the damaged section):

$$\begin{aligned}
\left[K^d\right] &= [K] + [\Delta K] = [K] + \sum_{j=1}^{L} \partial_j [K_j] \\
\left\{\phi_i^d\right\} &= \{\phi_i\} + \{\Delta\phi_i\} = \{\phi_i\} + \sum_{j=1}^{n} c_{ij}\{\phi_j\}
\end{aligned} \tag{22}$$

In Equation (22), $-1 < \partial_j < 0$ and $0 < c_{ij} < 1$. In addition, the MSE is related to the mode shape. The *i*-order MSE of the *j-th* section before and after structural damage is defined as follows:

$$MSE_{ij} = \{\phi_i\}^T [K_j] \{\phi_i\}$$
$$MSE_{ij}^d = \{\phi_i^d\}^T [K_j] \{\phi_i^d\}$$
(23)

Usually, only a low-order modal term is calculated and the high-order modal terms are ignored; the MSE of the section before and after structural damage is defined as follows:

$$MSEC_{ij} = MSE_{ij}^d - MSE_{ij} = 2\{\phi_i\}^T [K_j] \{\Delta\phi_i\}$$
(24)

This analysis indicates that the damage location can be diagnosed by the index vectors obtained from the MSE. However, this is not sufficient to determine the damage degree. The MSECR is defined as the indicator of the damage degree:

$$MSECR_{ij} = \frac{\left| MSE_{ij}^d - MSE_{ij} \right|}{MSE_{ij}}$$
(25)

where $MSECR_{ij}$ is the MSECR for the *j-th* section with respect to the *i*-order mode.

In order to reduce the influence of the experimental modal random noise, multiple different order modalities can be used to diagnose the location of structural damage.

$$MSECR_j = \frac{1}{m} \sum_{i=1}^{m} \frac{MSECR_{ij}}{MSECR_{imax}}$$
(26)

where m is the number of used modalities and $MSECR_{imax}$ is the maximum MSECR [25].

The theoretical analysis shows that when the change in the stiffness matrix and the modal parameters before and after structural damage are used as input diagnostic information, the location of the blade damage can be obtained using the MSECR.

5. Simulation Verification

The GH Bladed software possesses high-precision WT modeling and simulation functions but the data post-processing capability is limited. In contrast, the MATLAB software has strong data analysis and post-processing capability and a rich algorithm toolbox. However, the nonlinear model of the WT created in MATLAB has low accuracy and the simulation function is also very limited. In order to verify the proposed series of blade damage diagnosis methods and perform online diagnostic simulation, the two software applications have to be combined into a fault simulation platform. Because there is no direct data communication interface between the two applications, the data interaction between GH Bladed and MATLAB was implemented by using the Bladed external DLL interface file. The schematic diagram of the integration is shown in Figure 6. The fault data were exported to MATLAB through the DLL file and the signal analysis and damage diagnosis were performed. Because the Bladed external DLL data interface allows for setting a data transmission time interval, the data transmission mode is very similar to the data acquisition process between the SCADA system and the actual sensors. Moreover, Bladed has a sensor simulation function, in which the sensor characteristics such as the signal noise, delay and fault can be specified. This greatly increases the simulation accuracy and credibility of the online monitoring and diagnosis simulation performed in this study.

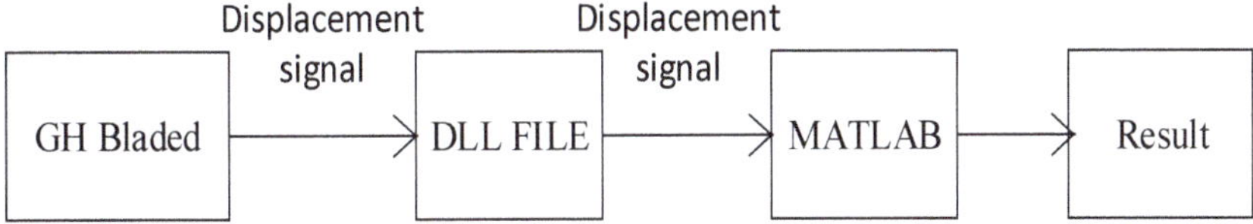

Figure 6. Flowchart of the integration of the Bladed and MATLAB data.

The experimental WT data used in GH Bladed are shown in Table 3. Figure 6 shows the flowchart of the blade damage diagnosis. The specific steps were as follows: First, the wavelet packet decomposition of the blade tip displacement signal was conducted to obtain the characteristic information of the energy bands. The blade damage was preliminarily determined based on these characteristics. For the damaged blade, the modal parameters of each section were calculated using the operational modal method and the MSECR was obtained. The MSECR represented the index to determine the location of the damage.

Table 3. Parameters of the experimental WT.

Number	Parameter Name	Value
1	Length of blade	38.75 m
2	Rated wind speed	12 m/s
3	Rated power	2 WM
4	Above-rated generator speed set-point	1500 rpm
5	Transmission ratio	83.33
6	Minimum generator speed	850 rpm
7	Pitch angle range	-2–$90°$
8	Height of tower	60 m
9	Rated generator torque	13,403 Nm
10	Maximum generator torque	14,400 Nm
11	Air density	1.225 kg/m^3
12	Cut-in wind speed	4 m/s
13	Cut-out wind speed	25 m/s
14	Number of blades	3
15	Rotor diameter	80 m

When the wavelet packet energy of the blade tip displacement signals is abnormal, the displacement signals of each section are analyzed using the operational modal analysis method to obtain the modal parameters of every section and the MSECR. Figure 7 shows the flowchart of the blade damage diagnosis. Table 4 shows the modal parameters of section 2 with severe damage. The MSECR was calculated using the data in Table 4 and the damage location was determined using MSECR as the discriminant index. Figure 8 shows the MSECR histograms of section 2 before and after the severe damage occurred at a mean wind speed of 12 m/s.

Table 4. Modal parameters of each unit.

Blade Unit	First-Order Frequency	First-Order Mode	Second-Order Frequency	Second-Order Mode
1	0.3201	-0.005207	3.8343	0.000059
2	0.2986	-0.002834	3.7893	-0.000078
3	0.2794	0.017348	3.9138	-0.000144
4	0.2677	0.073219	4.0817	0.000130
5	0.2472	0.304202	4.3765	-0.000204
6	0.2036	1.419367	4.3478	-0.001643
7	0.2349	1.096382	3.3818	-0.030174
8	0.2445	1.043385	3.0982	-0.011586
9	0.2464	1.026799	3.1026	-0.013261

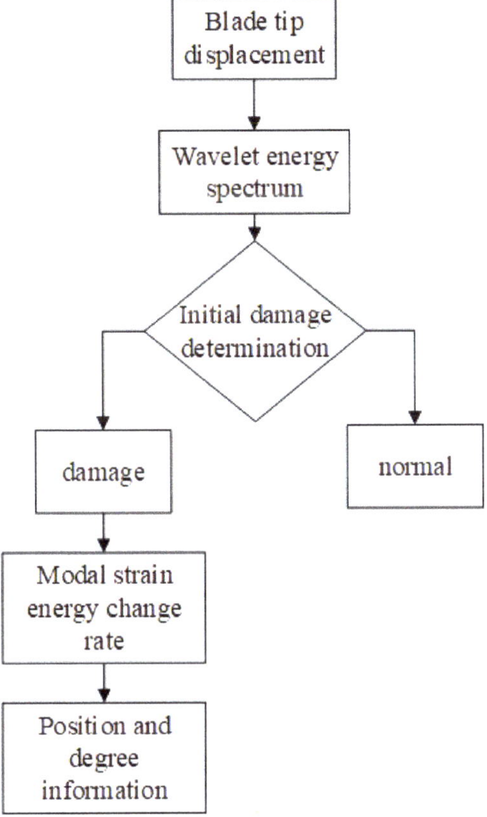

Figure 7. Flowchart of the blade damage diagnosis.

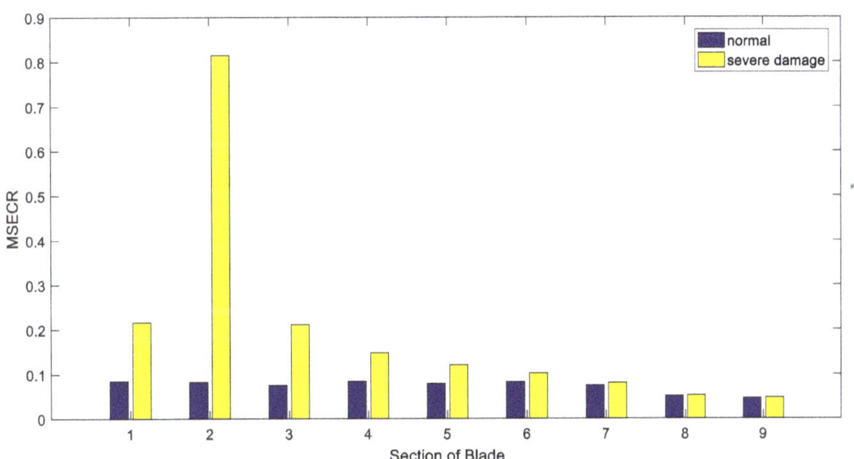

Figure 8. The MSECR of section 2 before and after damage.

In order to further illustrate the effectiveness of this method, a large number of simulation experiments were conducted using blade damage in different locations and different degrees of damage. Figure 9 shows the MSECR histograms for the different locations and degrees of damage for different wind speeds. The blade damage location can be clearly detected in Figure 9. Figure 10 shows the MSECR histograms of when two sections were damaged.

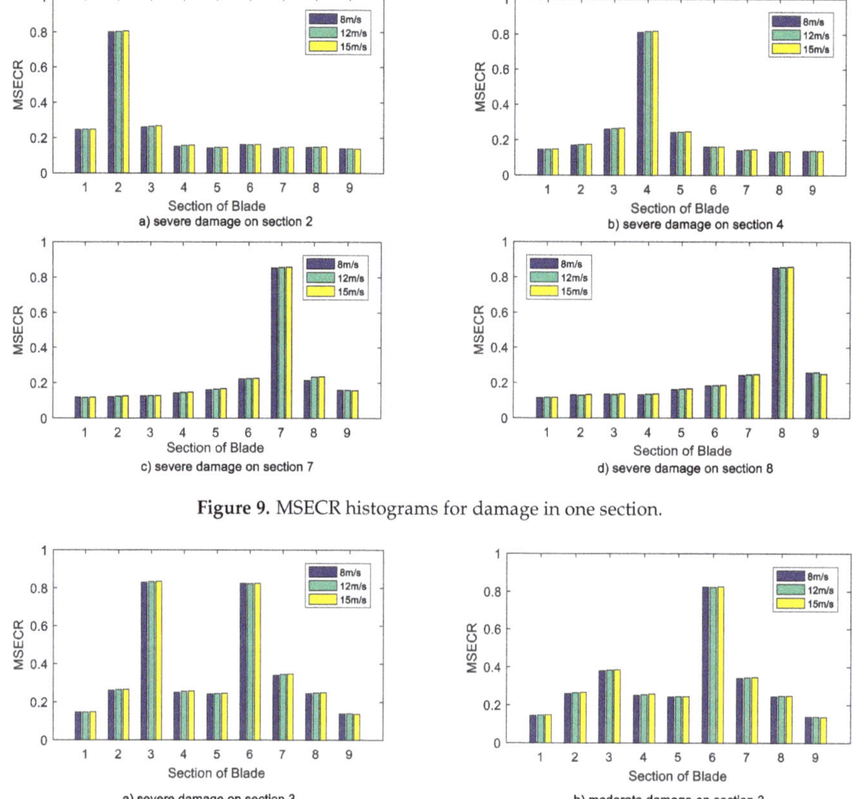

Figure 9. MSECR histograms for damage in one section.

Figure 10. MSECR histograms for damage in two sections.

Figure 9 shows the MSECR of each blade section when a single section is damaged. The simulation results show that the MSECR of the damaged section has a prominent peak, which is significantly higher than those of the undamaged sections. The MSECR of the damage section in Figure 9 is about five times as large as those of the normal sections. In addition, the MSECRs of the adjacent sections were also slightly higher than (about 50% higher than the normal sections) and there was no significant change in the MSECR of the distant sections. Therefore, the damaged section can be determined based on the MSECR. Figure 10 shows the MSECR of each blade section when two sections are damaged. It is observed that the MSECR of the damaged section is higher than that of the normal section. It is noteworthy that the MSECR provided a good indication of the degree of damage if the damage degree of the two damaged sections was large (as shown in Figure 10a). However, when the damage degree was significantly different for the two sections, the sections with the smaller damage sections may be difficult to detect (as shown in Figure 10b). As a result, it is possible to miss some minor damage when using this method for blade damage location but this does not affect the diagnosis and location of the major damage when the blade is damaged in multiple sections. Generally, blade damage consists mostly of single-section damage. And the more serious parts should be mainly considered in multiple-section damage. Environmental conditions affect the structural characteristics of the blade and the calculated value of the wavelet energy spectrum will differ under different weather conditions; however, the location of the damage is not affected. Therefore, this method can be used to locate blade damage effectively.

6. Conclusions

The proposed method for blade damage diagnosis and damage location based on the tip displacement signal, wavelet packet decomposition and operational modal analysis has the following characteristics:

(1) The wind power simulation software Bladed was used to simulate the blade damage fault of a wind turbine by determining the change in the structural characteristic parameters before and after blade damage. An integrated MATLAB and Blade simulation platform was developed for real-time online damage diagnosis.

(2) An initial on-line blade damage assessment was conducted using Fourier analysis and wavelet packet energy spectrum analysis of the tip displacement. The simulation results showed that the wavelet packet energy spectrum analysis was not only easy to implement but also provided significantly better results than the traditional Fourier analysis.

(3) A method for identifying the working modal parameters of the blades was proposed by combining the random decrement method and ARMA model parameter identification. The damage was accurately located by calculating the MSECR of each blade section without requiring additional excitation signals and the method proved effective for different degrees of damage occurring simultaneously.

Although the results of this study are based on software simulation results, the Blade software provides high accuracy and is suitable for preliminary assessment prior to empirical research. In a future study, we will build a small-blade experimental wind turbine and develop a wireless patch sensor for empirical research. The results of this study provide methodological guidance for system implementation.

Author Contributions: Conceptualization, F.G.; methodology, F.G and Q.L.; software, X.Y. and X.W.; validation, J.L. and X.W.; writing—original draft preparation, Q.L.; writing—review and editing, X W

Funding: The authors would like to acknowledge the funding support from the National Nature Science Fund Project (51677067).

References

1. Yang, B.; Sun, D. Testing, inspecting and monitoring technologies for wind turbine blades: A survey. *Renew. Sustain. Energy Rev.* **2013**, *22*, 515–526. [CrossRef]
2. Wang, L.; Zhang, Z.; Xu, J.; Liu, R. Wind turbine Blade Breakage Monitoring with Deep Auto encoders. *IEEE Trans. Smart Grid* **2018**, *9*, 2824–2833. [CrossRef]
3. Yang, W.; Court, R.; Jiang, J. Wind turbine condition monitoring by the approach of SCADA data analysis. *Renew. Energy* **2013**, *53*, 365–376. [CrossRef]
4. Helbing, G.; Ritter, M. Deep learning for fault detection in wind turbines. *Renew. Sustain. Energy Rev.* **2018**, *98*, 189–198. [CrossRef]
5. Tang, J.; Soua, S.; Mares, C.; Gan, T.H. An experimental study of acoustic emission methodology for inservice condition monitoring of wind turbine blades. *Renew. Energy* **2016**, *99*, 170–179. [CrossRef]
6. Joosse, P.A.; Blanch, M.J.; Dutton, A.G.; Kouroussis, D.A.; Philippidis, T.P.; Vionis, P.S. Acoustic emission monitoring of small wind turbine blades. *J. Sol. Energy Eng.* **2002**, *124*, 446–454. [CrossRef]
7. Sierra-Pérez, J.; Torres-Arredondo, M.A.; Güemes, A. Damage and nonlinearities detection in wind turbine blades based on strain field pattern recognition. FBGs, OBR and strain gauges comparison. *Compos. Struct.* **2016**, *135*, 156–166. [CrossRef]

8. Oh, K.Y.; Park, J.Y.; Lee, J.S.; Epureanu, B.I.; Lee, J.K. A Novel Method and Its Field Tests for Monitoring and Diagnosing Blade Health for WTs. *IEEE Trans. Instrum. Meas.* **2015**, *64*, 1726–1733.
9. Ruan, J.; Ho, S.C.; Patil, D.; Song, G. Structural Health Monitoring of wind turbine Blade using Piezo ceremic Based Active Sensing and Impedance Sensing. In Proceedings of the 11th IEEE International Conference on Networking, Sensing and Control, Miami, FL, USA, 7–9 April 2014; Volume 64, pp. 661–666.
10. Huh, Y.H.; Kim, J.I.; Lee, J.H.; Hong, S.G.; Park, J.H. Application of PVDF Film Sensor to Detect Early Damage in wind turbine Blade Components. *Procedia Eng.* **2011**, *10*, 3304–3309. [CrossRef]
11. Downey, A.; Ubertini, F.; Laflamme, S. Algorithm for damage detection in wind turbine blades using a hybrid dense sensor network with feature level data fusion. *J. Wind Eng. Ind. Aerodyn.* **2017**, *168*, 288–296. [CrossRef]
12. Sarrafi, A.; Mao, Z.; Niezrecki, C.; Poozesh, P. Vibration-based damage detection in wind turbine blades using Phase-based Motion Estimation and motion magnification. *J. Sound Vib.* **2018**, *421*, 300–318. [CrossRef]
13. Rezaei, M.M.; Behzad, M.; Moradi, H.; Haddadpour, H. Modal-based damage identification for the nonlinear model of modern wind turbine blade. *Renew. Energy* **2016**, *94*, 391–409. [CrossRef]
14. Moradi, M.; Sivoththaman, S. MEMS Multi sensor Intelligent Damage Detection for wind turbines. *IEEE Sens. J.* **2015**, *15*, 1437–1444. [CrossRef]
15. Abouhnik, A.; Albarbar, A. Wind turbine blades condition assessment based on vibration measurements and the level of an empirically decomposed feature. *Energy Convers. Manag.* **2012**, *64*, 606–613. [CrossRef]
16. Gonzalez, A.G.; Fassois, S.D. A supervised vibration-based statistical methodology for damage detection under varying environmental conditions & its laboratory assessment with a scale wind turbine blade. *J. Sound Vib.* **2016**, *366*, 484–500.
17. Wang, Y.; Liang, M.; Xiang, J. Damage detection method for wind turbine blades based on dynamics analysis and mode shape difference curvature information. *Mech. Syst. Signal Process.* **2014**, *48*, 351–367. [CrossRef]
18. Lee, K.; Aihara, A.; Puntsagdash, G.; Kawaguchi, T.; Sakamoto, H.; Okuma, M. Feasibility study on a strain based deflection monitoring system for wind turbine blades. *Mech. Syst. Signal Process.* **2017**, *82*, 117–129. [CrossRef]
19. Dervilis, N.; Choi, M.; Taylor, S.G.; Barthorpe, R.J.; Park, G.; Farrar, C.R.; Worden, K. On damage diagnosis for a wind turbine blade using pattern recognition. *J. Sound Vib.* **2014**, *333*, 1833–1850. [CrossRef]
20. Tcherniak, D.; Molgaard, L.L. Active vibration-based structural health monitoring system for wind turbine blade: Demonstration on an operating Vestas V27 WT. *Struct. Health Monit.* **2017**, *16*, 536–550. [CrossRef]
21. Shu, L.; Li, H.; Hu, Q.; Jiang, X.; Qiu, G.; McClure, G.; Yang, H. Study of ice accretion feature and power characteristics of wind turbines at natural icing environment. *Cold Reg. Sci. Technol.* **2018**, *147*, 45–54. [CrossRef]
22. Saleh, S.A.; Moloney, C.R. Development and testing of wavelet packet transform-based detector for ice accretion on wind turbines. In Proceedings of the 2011 Digital Signal Processing and Signal Processing Education Meeting (DSP/SPE), Sedona, AZ, USA, 4–7 January 2011; pp. 72–77.
23. Li, Y.; Wang, S.; Liu, Q.; Feng, F.; Tagawa, K. Characteristics of ice accretions on blade of the straight-bladed vertical axis wind turbine rotating at low tip speed ratio. *Cold Reg. Sci. Technol.* **2017**, *145*, 1–13. [CrossRef]
24. Corradini, M.L.; Ippoliti, G.; Orlando, G. A Sliding Mode Observer-Based Icing Detection and Estimation Scheme for Wind Turbines. *J. Dyn. Syst. Meas. Control* **2018**, *140*, 014502. [CrossRef]
25. Liu, T.; Li, A.Q.; Ding, Y.L.; Li, Z.J.; Fei, Q.G. Experimental study on structural damage alarming method based on wavelet packet energy spectrum. *J. Vib. Shock* **2009**, *28*, 4–9.
26. Luo, J.; Liu, G.; Huang, Z.-M. Modal parametric identification under non-stationary excitation based on random decrement method. *J. Vib. Shock* **2015**, *21*, 19–24.
27. Zhao, Y.; Xue, B.; Zhang, J.; North China University of Water Resources and Electric Power. Research on Modal Parameter Identification of Bridge Structure Based on ARMA Model. *J. North China Univ. Water Resour. Electr. Power* **2015**, *36*, 21–24.
28. Seyedpoor, S.M. A two stage method for structural damage detection using modal strain energy based index and particle swarm optimization. *Int. J. Non-Linear Mech.* **2012**, *47*, 1–8. [CrossRef]

Article

MIDAS: A Benchmarking Multi-Criteria Method for the Identification of Defective Anemometers in Wind Farms

Arkaitz Rabanal [1,‡], Alain Ulazia [2,*,‡], Gabriel Ibarra-Berastegi [3,†,‡], Jon Sáenz [4,†,‡] and Unai Elosegui [5,‡]

[1] University of the Basque Country (UPV/EHU), Otaola 29, 20600 Eibar, Spain; arabanal004@ikasle.ehu.eus
[2] Department of NE and Fluid Mechanics, University of the Basque Country (UPV/EHU), Otaola 29, 20600 Eibar, Spain
[3] Department of NE and Fluid Mechanics, University of the Basque Country (UPV/EHU), Alda, Urkijo, 48013 Bilbao, Spain; gabriel.ibarra@ehu.eus
[4] Department of Applied Physics II, University of the Basque Country (UPV/EHU), B. Sarriena s/n, 48940 Leioa, Spain; jon.saenz@ehu.eus
[5] Maxwind Technology, Portuetxe 83, 20018 Donostia, Spain; unai.elosegui@corp.hispavista.com
* Correspondence: alain.ulazia@ehu.eus
† BEGIK, Joint Research Unit (UPV/EHU-IEO) Plentziako Itsas Estazioa (PIE), University of the Basque Country (UPV/EHU), Areatza Hiribidea 47, 48620 Plentzia, Spain.
‡ These authors contributed equally to this work.

Received: 24 November 2018; Accepted: 18 December 2018; Published: 22 December 2018

Abstract: A novel multi-criteria methodology for the identification of defective anemometers is shown in this paper with a benchmarking approach: it is called MIDAS: multi-technique identification of defective anemometers. The identification of wrong wind data as provided by malfunctioning devices is very important, because the actual power curve of a wind turbine is conditioned by the quality of its anemometer measurements. Here, we present a novel method applied for the first time to anemometers' data based on the kernel probability density function and the recent reanalysis ERA5. This estimation improves classical unidimensional methods such as the Kolmogorov–Smirnov test, and the use of the global ERA5's wind data as the first benchmarking reference establishes a general method that can be used anywhere. Therefore, adopting ERA5 as the reference, this method is applied bi-dimensionally for the zonal and meridional components of wind, thus checking both components at the same time. This technique allows the identification of defective anemometers, as well as clear identification of the group of anemometers that works properly. After that, other verification techniques were used versus the faultless anemometers (Taylor diagrams, running correlation and *RMSE*, and principal component analysis), and coherent results were obtained for all statistical techniques with respect to the multidimensional method. The developed methodology combines the use of this set of techniques and was able to identify the defective anemometers in a wind farm with 10 anemometers located in Northern Europe in a terrain with forests and woodlands. Nevertheless, this methodology is general-purpose and not site-dependent, and in the future, its performance will be studied in other types of terrain and wind farms.

Keywords: wind turbine; anemometer; kernel-based multidimensional probability density function; ERA5 reanalysis

---

1. Introduction

Maintenance is a critical variable in the wind industry in order to reach competitiveness. Failure detection and diagnosis is essential, as is the study of the relation of these faults with energy production, profitability, costs, and safety. Toward this end, several advances in mathematics and computational techniques are employed in maintenance management: dynamic analysis, probabilistic methods, mathematical optimization techniques, etc. The combination of these techniques enables a multi-criteria diagnosis and decision-making processes for different problems in wind farms [1–4].

In the context of operation and maintenance (O&M), the identification of defective anemometers that are located in wind turbines is important, because they are used for the estimation of the energy produced by the turbine through its actual power curve.

For each wind energy application, the type of instrumentation required varies widely from a simple system containing only one wind speed anemometer/recorder to a very complex system designed to characterize turbulence across the rotor plane. This kind of instrumentation is very important for wind energy applications and has been discussed in detail by numerous authors [5–7], or by the measurement standards of organizations such as the American Wind Energy Association [8].

Although pressure and temperature are sometimes also measured in the wind farm to compute the air density (wind power is proportional to air density [9]), our study is focused on anemometers and wind vanes that measure wind speed and direction. The wind speed can therefore be decomposed into the zonal and meridional components (U, V; see the table of abbreviations before the References for the other parameters).

Wind-measuring instruments can be classified according to their principle of operation:

- Momentum transfer: cups, propellers, and pressure plates;
- Pressure on stationary sensors: Pitot tubes and drag spheres;
- Heat transfer: hot wires and hot films;
- Doppler effects: acoustics and laser;
- Ultrasonic devices, etc.

In recent years, it has even been possible to develop a resource assessment study using modern anemometers such as LiDAR (Doppler-effect-type laser-based anemometer) obtaining wind data at different heights at a given location [10]. SODAR (sonic detection and ranging) and meteorological masts can also be used [11], by means of on-site anemometry observation [12], and single anemometers can be used in complex terrain for resource assessment purposes [13].

In our case, cup-anemometers located at the turbines of a wind farm were analyzed, and these are the most common instruments. Baseer et al. recently studied the performance of cup anemometers installed at different mast heights [14], calculating the annual mean, median, and standard deviation. These indicators were almost the same during different years and were comparable with co-located sensors at each height. Thus, the similarity of the measurements of different cup anemometers at the same location (or almost the same location) seems to be demonstrated for modern cup anemometers.

The rotation of a cup anemometer varies in proportion to the wind speed to generate a signal. It presents an extremely linear calibration, but it can start from a zero rotation rate at zero wind to one corresponding to a sudden change. The bias in the measured mean wind speed due to the random variations in the three velocity components is overwhelmingly dominated by the fluctuations of the lateral wind velocity component [15].

The relevance of the lateral wind fluctuations in the measurement of the wind speed that determines the power curve means that wind direction cannot be ignored for the construction of the actual curve. Consequently, an evaluation method that takes into account both wind speed and wind direction (or U and V, zonal and meridional components) is necessary, and the verification cannot be reduced to wind speed alone. The stability of the atmosphere and the subsequent turbulent fluctuations is so important that a more stable atmosphere with the same average wind speed produces more energy in a wind farm [16].

In fact, although it is not the aim of this study, the bias due to lateral wind velocity fluctuations can be reduced to less than 1% by means of a special data processing of the simultaneous signals from a cup anemometer and a wind vane [15]. Bias or root mean squared error ($RMSE$) values between the anemometers of the wind farm of around 1% will be therefore referential for our purpose.

For instance, an extreme relative bias of 10% in the measurement of the anemometer can produce important deviations in the actual power curve of the turbine. Figure 1 shows a typical power curve of a wind turbine with the cut-in wind speed around 3 m/s at which the turbine starts its production and the cut-off at 25 m/s at which the turbine stops because of safety issues. The variations for these limits are important, but the changes in the $\mathbf{U}^3$ zone of the curve before reaching the rated wind speed with constant power (rated power) are also very important. This is because this kind of deformation of the power curve in the $\mathbf{U}^3$ zone is similar for other kinds of technical problems, such as the pitch misalignment of turbine blades, yaw misalignment, or instantaneous turbulent variations of the wind due to the atmospheric instability [16–18].

Recent works of the authors show real cases of energy production diminution due to pitch misalignment in wind farms and simulations with FAST that compute the fraction of power reduction at each wind speed in the $\mathbf{U}^3$ zone for different values of pitch errors. These errors can be corrected after an in situ measurement of the blade-hub configuration via laser scanner [17,19,20].

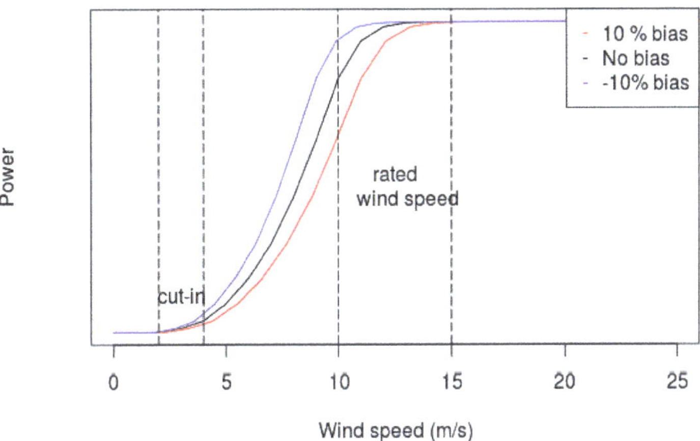

Figure 1. Power curves for different biases.

Similarly, analogous deformations of the power curve due to the errors in the yaw angle of the turbine can be described [18]. Consequently, the effects of the pitch and yaw angle deviations in the actual power curve each have their own characteristics, but they may also appear in combination with the effects of the errors in the measurement of the anemometers. That is why the identification of defective anemometers is so important for the wind energy industry.

Over time, cup anemometers' loss of performance due to aging processes can be considerable, and it can even affect the annual energy production (AEP) estimations [21]. Therefore, anemometer calibration is very important [22]. Recently, new analytical procedures based on Fourier analysis and aerodynamics have been developed to predict the degradation level of on-field anemometers [23]. This methodology might represent an alternative to the classic approaches used in the present standards of practice such as IEC 64000-12.

In this paper, we want to present the results of a joint approach for the identification of defective anemometers that has already been applied to a real-life wind farm in Northern Europe, and this paper represents the formal presentation of a fully-structured methodology to identify defective anemometers in any wind farm. The multi-criteria methodology we developed involves the combination of several techniques using a standard reference for comparison (benchmarking) that converge into the identification of the defective anemometers in a wind farm. This approach is not site-dependent and can be generalized to any wind farm in any location with any number of defective anemometers.

2. Data and Methodology

In this section, the source of data used for this study and the methodology developed are explained.

2.1. Data

2.1.1. ERA5 Reanalysis

The wind speed data at the nearest ERA5 grid point were used to assess the quality of the wind measurements at the 10 anemometers. This grid point is at a distance between 4 and 8 km from the turbines of the wind farm; without forgetting that the concept "nearest" is complex because it works in a Gaussian grid. The European Centre for Medium-Range Weather Forecasts published this advanced reanalysis in 2017, improving previous reanalyses such as ERA-Interim. The representation of the troposphere and tropical cyclones is better, as is the soil moisture, the balance of precipitation and evaporation, and the consistency between sea surface temperature and sea ice. The data assimilation system is also renewed (IFSCycle 41r2 4D-Var), and a vast amount of historical observations (satellite or in situ) are assimilated [24].

ERA5 provides hourly estimates of a large number of atmospheric, land, and oceanic climate variables. It presents a 30-km global grid and uses 137 vertical levels from the surface up to a height of 80 km. The entire dataset from 1950–present will be available soon, substantially extending the period of ERA-Interim (1979–present), with a much higher spatial and temporal resolution [24]. Furthermore, recent studies in real wind farms have shown that ERA5 can be the new "champion" of wind energy modeling [25]. ERA5's U and V wind speed components were obtained at 10 and 100 m in the nearest grid point. In order to compute the wind speed at 137 m (the anemometer's height), we first obtained the roughness of the terrain (z_0) from the log law using the average values of the wind speed module $\mathbf{U}$ at 10 and 100 m:

$$\mathbf{U}(z = 100)/\mathbf{U}(z = 10) = log(100/z_0)/log(10/z_0) \tag{1}$$

Therefore, the value of z_0 was around 500–1000 mm. According to Table 2.2 in [26], this type of terrain corresponds to forests and woodlands, which is totally consistent with the terrain of the wind farm (not mentioned for commercial reasons). There are better, but more elaborated techniques to perform the vertical extrapolation of wind data [27]. However, considering that in this study, the gridded wind data from ERA5 that is vertically extrapolated is used for relative estimations of errors with different anemometers, its use is deemed as not necessary. On the one hand, because there are other errors such as the ones derived from the representativeness error: comparison of in situ anemometer data to numerical model output at a single grid point with a spatial resolution close to 30 km; or, on the other hand, the spatial distance from the near-neighbor grid point to the anemometers in the wind farm.

Again using the log law with this value of z_0, wind speed can be raised from 10–137 m in height. Thus, the mean ratio between $\mathbf{U}(z = 137)$ and $\mathbf{U}(z = 10)$ was 1.8. That is, the wind speed increased 80% from a 10-m height to the hub height of the turbines.

2.1.2. Wind Farm

For this work, measurements obtained from 10 anemometers at the 10-turbine hub of a wind farm located in Northern Europe were analyzed. The anemometers measure both wind speed and direction. This wind farm is located in a flat area of big forests without additional effects such as land–sea interactions or mountain breezes affecting the wind field. In this paper, the 10 turbines in the wind farm are labeled as: WTG-08, WTG-09, ..., WTG-17. The period of study ran from 1 November

2016–31 August 2017 with 10-min data gathered at the 10 anemometers. Therefore, there are 43,776 cases, but not all of them are complete (there are default data or NaNs (not a number)), and the 1.9% of missing data were removed. Therefore, finally, 42,936 complete cases were considered. This is done for all the sub-techniques of MIDAS, with the exception of running plots (see Section 2.2.4), in which a temporal window of one week was used to run the time series.

The ten anemometers analyzed were located at the mentioned height (137 m, turbines' hub), and there were two clear parallel lines of turbines: WTG08, WTG09, WTG10 and WTG11, WTG12, WTG13. The other four were a group somewhat apart (Figure 2) aligned around a W–E direction, thus configuring a reasonable layout for the wind farm given that the two prevailing wind directions are NNE–NNW and SW. This layout and the position of the turbines can be seen in Figure 2, together with the ERA5's nearest grid point.

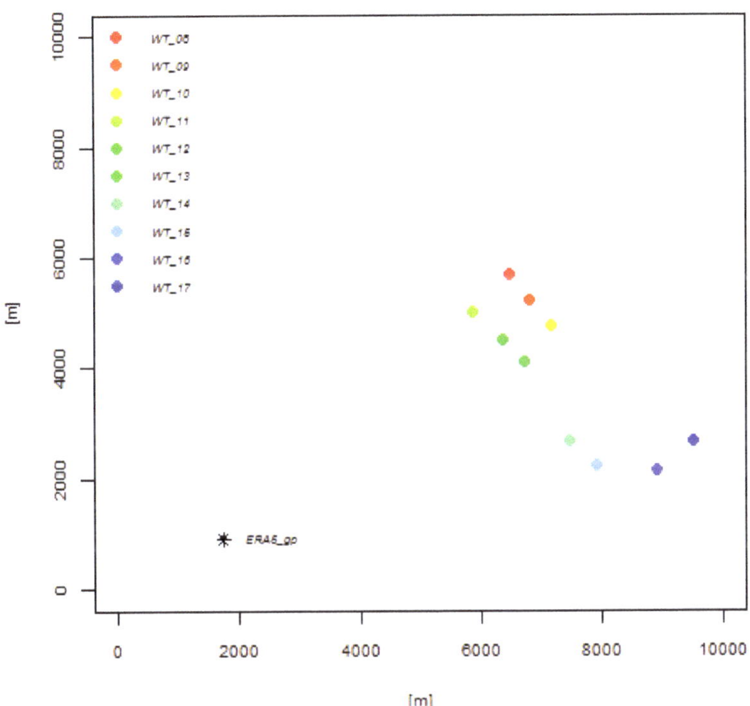

Figure 2. Layout of the wind farm and the position of the nearest ERA5 grid point.

Summarizing, the wind data measurements from the anemometers were gathered as speed + direction records every ten minutes, while ERA5 provided the hourly zonal (U) and meridional (V) wind projections. An initial lengthy phase of data arrangement for compatibility between both wind sources was needed. In it, the following tasks were carried out:

1. Wind speed + direction measurements from the 10 anemometers were transformed into U-V values
2. At the anemometers, values were taken every ten minutes, while ERA5 records were available only at hourly steps. To arrange both groups of data (ERA5, 10 anemometers) on the same timeline, only 1 in 6 measurements from the anemometers were considered.

3. Finally, ERA5 only provided wind U-V values at a height of 10 and 100 m above the terrain and not at 137 m (hub/anemometer height), so following the log law, ERA5 U-V values at 137 m were derived.

2.2. Methodology

MIDAS (multi-technique identification of defective anemometers) involves combining five different approaches: multidimensional probability density function estimators, analysis of wind roses, Taylor diagrams, the time series plots for the running correlation, running $RMSE$ and running bias, and finally, principal component analysis. These techniques are explained in detail in the following sections.

2.2.1. Multidimensional Probability Density Function Estimator

The multidimensional probability density function (MPDF) technique is a general-purpose method implemented by the authors [28] that makes it possible to compare two multidimensional probability density functions (PDFs) that were estimated by a kernel-based multivariate approach. This method provides a score between 0 (completely different PDFs) and 1 (perfect match) for multidimensional data distributions by computing the common volume under both PDFs.

In general, for the verification of models, it is customary to have a time series of measurements (approximately error-free) and model results, the data that need to be verified. However, in this case, either one (or two) of the components of wind measured by some (or none) of the wind-measuring devices might be affected by observational problems. This is a problem where the true observed state is not well known.

As such, it is similar to the problem of evaluating climate models (models that are always run beyond the first-kind predictability limit associated with weather forecasting) against observations. The climate models cannot be expected to reproduce the daily atmospheric states, but they can be expected to generate a similar probability density function [29,30]. Recently, some of the authors of this study proposed an extension of the previous evaluation strategy to multiple dimensions [28].

This extension to multiple dimensions has many advantages in the present application:

1. First, it allows us to evaluate simultaneously the probability density function of different components of the wind from different wind-measuring devices, since it supports the analysis of multiple dimensions.
2. Second, despite the fact that well-known statistical tests exist for the comparison of observed and simulated probability distribution functions (e.g., the Kolmogorov–Smirnov test or comparison of Weibull distributions [31]), it is also well known that the univariate Kolmogorov–Smirnov test cannot be easily extended to several dimensions above two or even three [32,33]. The use of a kernel-based multidimensional probability density function is a more general strategy that can be used very flexibly with a higher number of dimensions.
3. Finally, the use of a very efficient algorithm [28] makes it also computationally feasible.

To the best of our knowledge, this is the first time that the MPDF technique has been applied for wind field comparison purposes. It allows the vectorial comparison of the wind measurements at two locations and/or obtained by two different devices or from two sources. Since a wind vector is defined by its two components (U, V), in this work, the MPDF will be applied two-dimensionally.

At the beginning of this study, no information was available as to the number of faulty anemometers and which ones (if any) they were. For this reason, since we could not trust any of the measuring devices at a given moment, an external reference was adopted: the nearest grid point of ERA5. We decided to assume that wind-measuring devices should ideally be representing a similar probability density function unless any of them were affected by some kind of malfunction. To evaluate this, the MPDF was used to compare the 10 anemometers against ERA5, that is we created 11 two-dimensional PDFs corresponding to the wind vector (U, V).

The practical implementation of this MPDF technique as described in [28] was carried out in three steps:

1. First, the optimal bandwidth that must be used in the estimation of the MPDF was derived by means of smoothed bootstrap, since the Epanechnikov kernel used does not naturally lead to a simple solution by means of cross-validation.
2. Second, for the U and V components of wind-measuring devices, the corresponding two-dimensional probability density function was computed.
3. Finally, the common volume under both multidimensional PDFs was computed for the combination of wind-measuring devices.

The software was written in ANSI-C and is freely available as indicated in [28] (https://github. com/isg-ehu/unai.lopez/tree/master/density-parallel), where more specific mathematical details on every step can be found. In our case, the MPDF is bi-dimensional, and the probability can be visualized in a color-plot (U and V in the x and y axes and the probability in colors). Figure 3 shows the advantage of analyzing the problem from the point of view of a multidimensional probability density function. Both the ERA5 and in situ wind data show a probability function characterized by two local maximums. It can be seen that the different positions of these maxima would project over similar position if only the U component of wind were used. Performing the analysis of the probability density functions (and their match) in two dimensions allows better discriminating the match between model (reference, ERA5) and measured data at each anemometer.

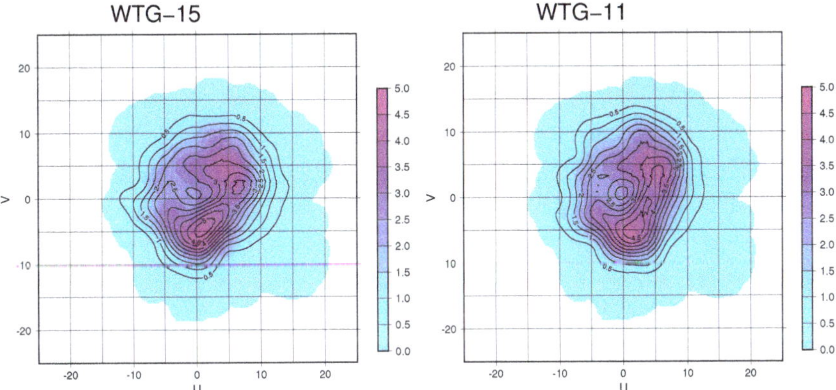

Figure 3. Two-dimensional probability density function ($\times 1000$) corresponding to wind from the nearest ERA5 grid point (color shades) and the WTG-15 anemometer and wind-vane (left, contours). The right panel represents the two-dimensional probability density function ($\times 1000$) corresponding to ERA5 data (colored shades) and the WTG-11 anemometer (contours).

As mentioned above, this score ranged from 0–1, and its values were used to establish a ranking of anemometers in the wind farm and how they behaved in comparison with the ERA5 wind field at the nearest grid point.

2.2.2. Wind Roses

A first basic visualization of wind data in each turbine can be also obtained using wind rose diagrams. This allows a fast interpretation of the predominant wind direction in the wind farm and of the farm's configuration. The diagonal disposition should be perpendicular to predominant winds in order to minimize the wake effect between turbines. In this way, we have a first visual idea about the consistency between the wind data and the wind farm configuration, and about the results obtained by the MPDF score comparing the wind roses.

2.2.3. Taylor Diagrams

The validation of the anemometers against the group of faultless anemometers was also represented using Taylor diagrams [34]. In this way, we passed from a reference at meso-scale level (ERA5) to an in-field reference in the wind farm at a micro-scale level. Hypothetically, all the suitable anemometers should present a very similar behavior, and the results of the statistical indicators of faulty anemometers evaluated against them should also be very similar.

Three statistical indicators are represented in these kinds of diagrams:

1. Root mean squared error ($RMSE$), represented by the arcs with the center in the observation point;
2. Pearson correlation coefficient, represented by the exterior arc; and
3. The ratio of standard deviations between the model and the observation ($SDratio$), represented by the interior arc in the case of $SDratio = 1$.

The trigonometric relation that exists between the three statistical indicators (i.e., the cosine law) allows this representation in a single diagram, and in our case, if all the anemometers of the wind farm were working properly, a compact cluster of points would be shown in the diagram. Therefore, any deviation of a point from this cluster would indicate a faulty anemometer.

2.2.4. Running Correlation, Running $RMSE$, and Running Bias

Taylor diagrams show three statistical indicators in a single overview. However, the temporal behavior of the indicators cannot be appreciated in this way. For this, running correlations, $RMSE$, and bias of U and V against the group of faultless anemometers were plotted along the period of study with a temporal window of one week. In these representations, we can find the moments when an anemometer starts to fail because of diverse problems that can be related to O&M issues of the wind farm. Both $RMSE$ and correlation indicate absolute values, but the bias can be both positive and negative and shows the deviation of the signal with respect to the reference value in terms of under- or over-estimation.

2.2.5. Principal Component Analysis

The study of PCs (principal components) is another common technique in time series that will be applied here. Principal components are defined as linear combinations of the original variables that explain the highest possible amount of variance with the least number of variates. The principal components are always ordered in decreasing order of explained variance so that the most common variability existing in the original dataset can be found in the first principal component (or empirical orthogonal function (EOF), as it is commonly referred to in geophysics). In this way, the computation of the first, second, and subsequent EOFs shows the contribution to the variability of the signal by each component. Under the assumption that the wind field over the farm is relatively uniform (no large spatial asymmetries are expected), the true wind field can be expected to be captured by the leading EOF, while the errors will contribute to the secondary EOFs. In order to improve the readability of results, instead of using the common orthonormal scaling of EOFs, we scaled EOFs (the loading factors affecting every anemometer in the farm) so that they represent the amount of variance at that anemometer that can be explained by each principal component. Thus, if the ith principal component allows us to describe the variability of a given j^{th} anemometer with a high fraction of variance, this means that it is particularly representing the behavior of the jth anemometer. Under the previously-mentioned assumption that the wind field is spatially uniform, all anemometers should represent the same variability, and the leading EOF should be the only one expected. If a given anemometer projects significantly onto the second EOF, we can interpret this result as being due to the fact that this anemometer is not showing the same kind of variability that is common to the rest of anemometers either because of an observational error or due to the spatial anisotropy of the wind field. In order to test the stationarity of this result, the PCs were computed not only for the whole period, but for monthly subsets, as well.

3. Results

3.1. Kernel-Based Bi-Dimensional PDF Estimator

Using this tool, the statistical distributions of the ten anemometers were compared with ERA5. A score between zero and one was obtained, indicating the similarity between the distributions of the anemometers and ERA5's wind speed distribution.

First, a common range for the U and V time series should be established for the construction of the bi-dimensional PDFs taking the minimum among the minima and the maximum among the maxima. In this case, U was between $[-22.49, 16.18]$ and V between $[-17.04, 17.29]$ m/s.

Figure 4 shows the final scores in a bar plot. All the scores were between 0.90 and 0.92, with the exception of WTG-15, which dropped to 0.88. These results preliminarily classify the anemometers between faulty ones and faultless ones; a classification that must be corroborated with the following multi-criteria steps based on different mathematical techniques. In this case, there was only one defective anemometer, which was called the worst-in-class (WIC). The others showed a similar score against ERA5, and in principle, they constituted the group of faultless anemometers that can be used as a reference in the following steps. In any case, the following diagnosis process must demonstrate this classificatory hypothesis.

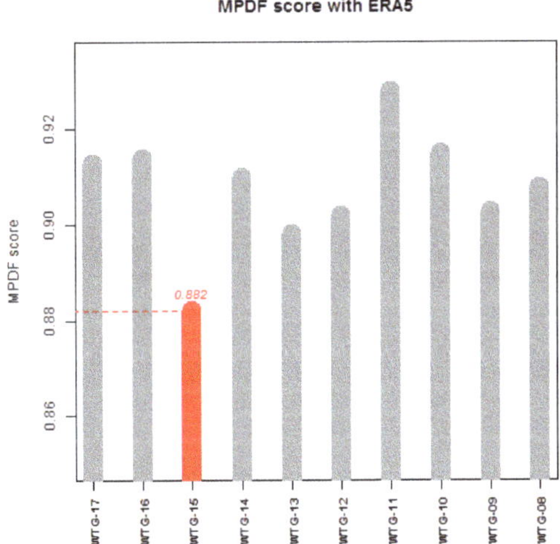

Figure 4. Results of the multidimensional probability density function (MPDF) score of the ten anemometers versus ERA5.

3.2. Wind Roses

In Figure 5, the wind roses of ERA5 (a), the WIC (b), and a faultless anemometer are shown (c). The other anemometers are not shown because the wind roses were very similar. The same color bar scale is used for the three wind roses in order to have an equal comparison.

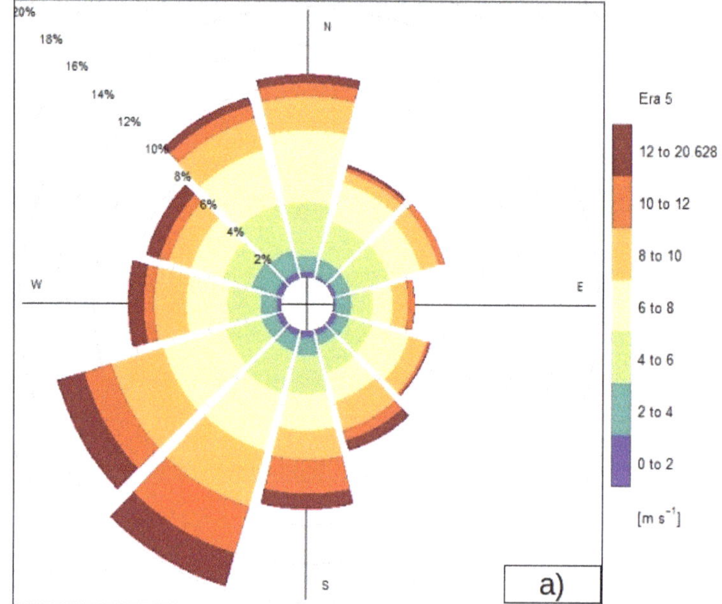

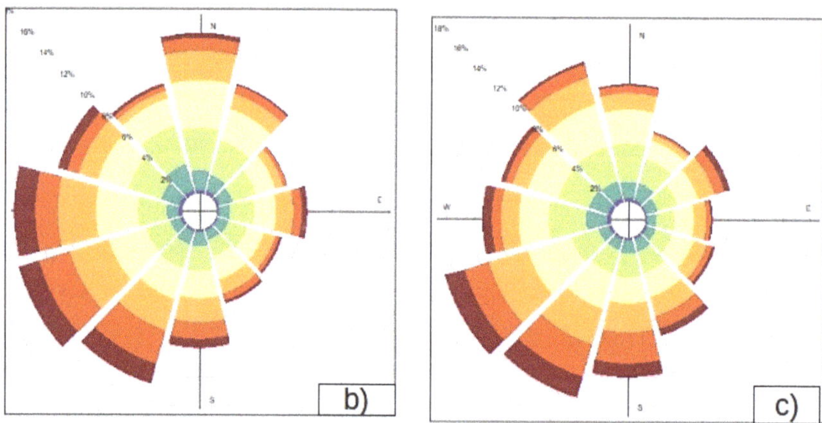

Figure 5. (**a**) Wind rose of ERA5's data; (**b**) wind rose of the worst-in-class (WIC) anemometer; (**c**) wind rose of one of the good anemometers (WTG-14).

The predominant wind directions from the SW direction were consistent with the NW–SE diagonal disposition of the farm, since they created perpendicular lines, reducing the space occupation to avoid the wake effect.

There is another very important aspect: it seems that the WIC's wind rose matched with the faultless one and the other anemometers' or ERA5's if it was rotated anticlockwise. Therefore, apart from the deviation of the WIC WTG-15 anemometer, the high similarity between ERA5's wind rose and the other anemometers' wind roses must also be emphasized. The rotation of the WIC's wind rose may mean the presence of an offset in the wind vane, which can affect the turbine in the yaw orientation.

Thus, we could have a case of vane misalignment that should be studied with the following techniques, mainly PCA analysis.

3.3. Taylor Diagrams

In Figure 6, the Taylor diagrams are shown for the zonal and meridional components, taking WTG-14 as the reference. The diagrams versus the other faultless anemometers were very similar and are not shown here.

As the diagrams show, the WIC anemometer WTG-15 (represented by the number 7) again had a lower correlation (0.90) and a higher $RMSE$ (around 3 m/s) than the other turbines (the group of the other anemometers had a correlation of 0.99 and an $RMSE$ of 1 m/s). These diagrams very clearly represent the wrong behavior that was previously identified by our kernel-based bi-dimensional density estimator and show that the data measured by this anemometer were not reliable for both U and V components.

Furthermore, the group of anemometers considered faultless in the analysis by the MPDF score again showed a very similar behavior, reinforcing the first hypothesis about their suitable behavior.

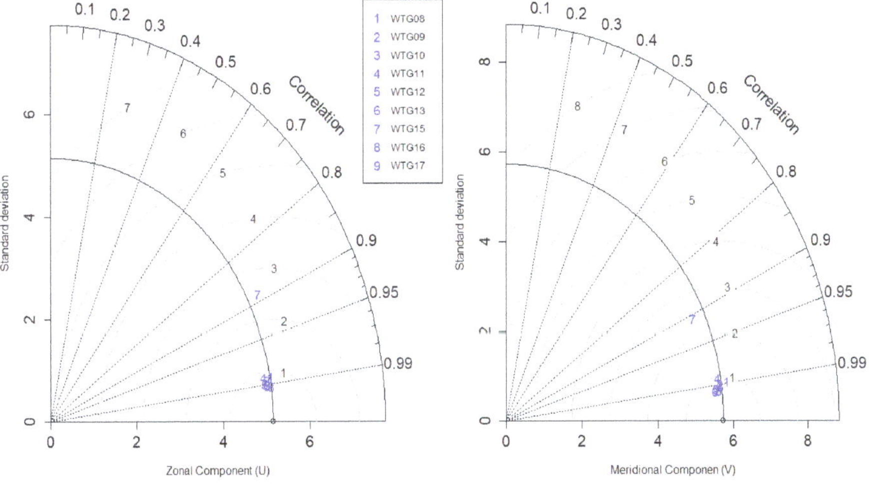

Figure 6. Taylor diagrams taking one of the faultless anemometers as the reference: (**left**) zonal component; (**right**) meridional component.

3.4. Running Statistical Indicators

In this step, we wanted to reconfirm the obtained results both in the Taylor diagram and in the MPDF score. This step is interesting mainly in its ability to identify the time frame in which the anemometer began to fail and started measuring differently from the rest.

By means of these graphs, the moments when the anemometer of WTG-15 turbine behaved erroneously were identified. To that purpose, time series plots are shown representing the values of the correlation, $RMSE$, and bias in period windows of seven days for U and V. The same colors were used for the anemometers in the six time series graphs.

Like in the Taylor diagrams, a faultless anemometer (WTG14)was chosen as the reference. We cannot show all the plots against all the faultless ones, but they were very similar. As can be seen in Figure 7, the problems of the WIC anemometer took place all along the year, reaching correlations below 0.7 and *RMSE* between 2 and 4 m/s, as well as similar positive bias for *V* (upper-estimation) and negative for *U* (underestimation). Given that the mean values of the anemometers were around 7 m/s, this means that in some moments the relative error could be above the 50% for our WIC anemometer (brown color). The others' *RMSE*s were below 1 m/s, and the biases also moved between −1 and 1 m/s, which is an important difference compared with the WIC. In general, the worst results for the three statistical indicators were shown between 1 November 2016 and 13 January 2017, at the beginning of the time series.

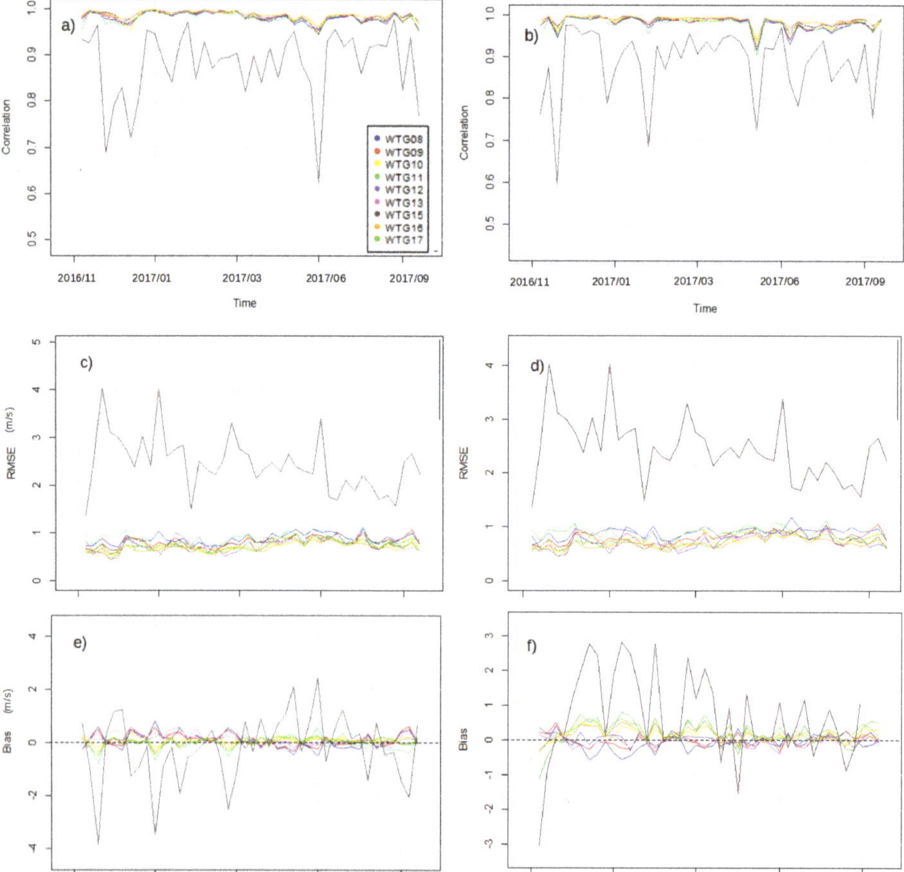

Figure 7. Running correlation of: (**a**) zonal component and (**b**) meridional component. Running *RMSE* of: (**c**) zonal component and (**d**) meridional component. Running bias of: (**e**) zonal component and (**f**) meridional component.

3.5. Principal Component Analysis

Finally, Figure 8 shows the results from the principal component analysis of zonal and meridional wind speeds. It clearly shows that the results by the MPDF scores and the Taylor diagrams were robust. This can be said because the leading principal component explained the majority of the variance at every anemometer/vane, with the clear exception of WTG-15 (the one already identified as the WIC in our dataset by the previous techniques). The results from WTG-15 were the ones appearing as the most important in the second PC. This again shows that WTG-15 was the anemometer-vane showing the most different behavior. The leading EOF for both the zonal and meridional components explained 97% of the whole variabilities of both zonal and meridional wind components, with only 2–2.5% for the second principal component, most of it concentrated in the series measured by WTG-15.

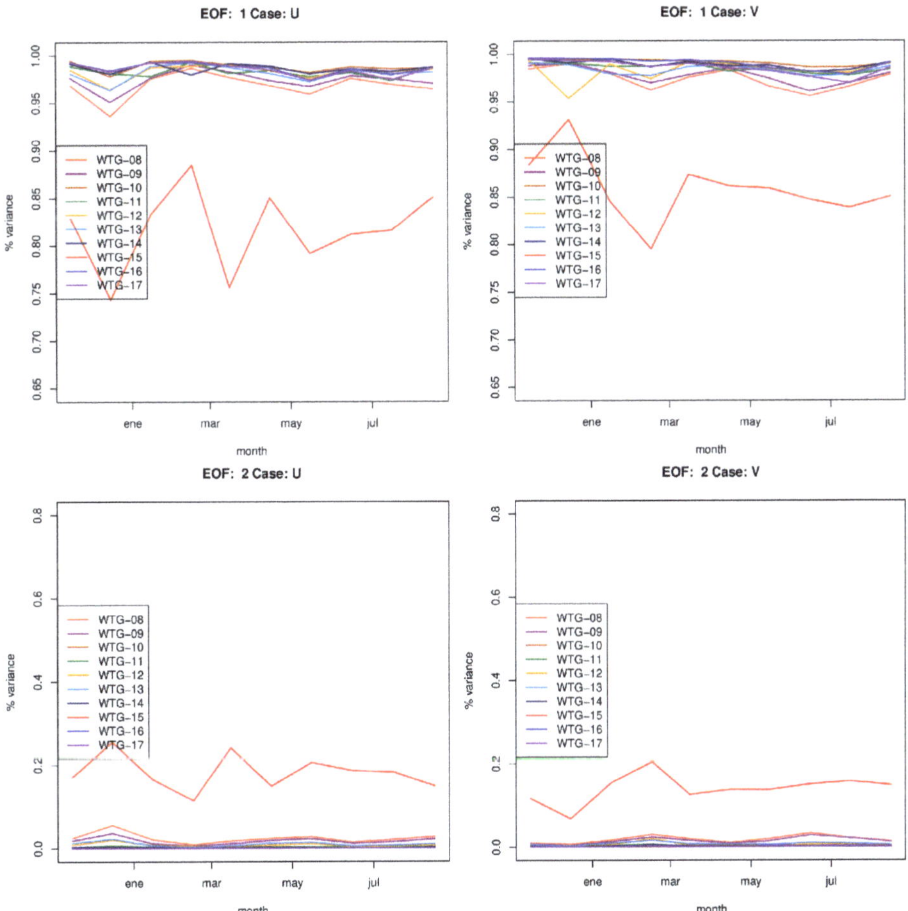

Figure 8. Principal components (leading, top row, and second, bottom row) of the zonal (left column) and meridional (right column) wind components for the anemometers in the wind farm, expressed as the fraction of variance explained at every anemometer by each principal component. EOF: empirical orthogonal function.

Thus, the variability measured by WTG-15 was to some extent decoupled from the variability measured by the rest of the anemometers. Finally, if the previous computations (based on the principal component analysis) were repeated for the magnitude of the wind speed, the leading principal component explained 96% of the variance, and WTG-15 did not appear as an outlier and faulty in the distribution of anemometers. Thus, the difference of WTG-15 with respect to the rest of the anemometers points to a stationary misalignment between them.

Since many of the results indicated that the WIC anemometer was probably affected by a misalignment, Figure 9 shows the results of explained variance corresponding to the first principal component that were obtained when a varying rotation angle was applied to the WIC anemometer for both wind components. If both velocity components of the WIC anemometer were rotated counter-clockwise by an angle of 26°, the leading principal component (representative of the true wind field) explained as much as 99% of the total variance. This confirms that the misalignment detected by comparing the wind roses corresponded to a value of 26°.

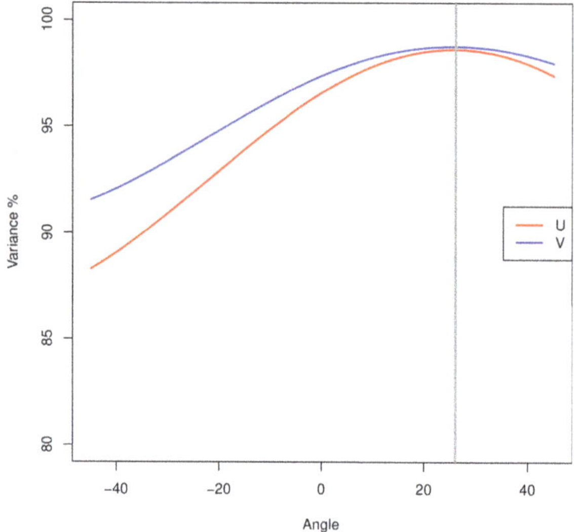

Figure 9. Variance corresponding to the leading principal component after a rotation was applied to the wind field components from the WIC anemometer before the principal component analysis (PCA) was computed. The maximum amount of variance in PC1 was obtained with an angle of 26°.

4. Discussion

In this study, we used a relatively simple formulation to take into account the vertical dependence of wind with height above the surface. Previous studies [27] have extended the surface-layer theory based on the Monin–Obukhov scaling at higher heights by using different lengths that consider in detail the stability of the atmosphere, derived from measurements in Denmark and Germany. Since the implementation of this scaling depends on the existence of atmospheric measurements at different levels, which allow on to take into account the atmospheric stability, we have preferred to keep the algorithm simple, by vertically scaling the wind from the ERA5 nearest grid point by means of a simple log law. The following considerations support our selection. First, it makes the algorithm simple to be applied to wind farms where a meteorological measurement system at multiple heights does not exist. Second, the extrapolation of wind to different vertical levels is not the only error that exists. On the one hand, wind data from ERA5 correspond to estimations of wind at a grid point representative of the characteristics of wind over a grid cell around 31 km by 31 km and hourly temporal resolution, while wind-farm observational data correspond to observations every ten minutes. Thus, there exist substantial representativity errors that are very likely higher than the error due to the vertical extrapolation using a simple log-law. Besides this representativeness problem, a second factor to consider is that there exist some kilometers between the grid point to the wind farm (from 4 km–8 km in our case).

After the initial screening provided by the visual analysis of wind roses and by the MPDF score against ERA5, it could be seen that there was one single anemometer (WIC) in the wind farm exhibiting a clearly different behavior when compared with the rest. At a second step, one of the other anemometers was chosen as a reference, and the Taylor diagrams clearly indicated a cluster of anemometers seeming the same while the position of the WIC was distant from the rest in both directional Taylor diagrams. The running diagrams corroborated the same observation, adding information about the time periods with the worst behavior of the WIC anemometer that can be valuable for O&M analysis of the wind farm.

The very defective results of the WIC anemometer (WTG-15) compared to the others could not be explained by its position in the wind farm. Furthermore, the WIC was one of the most isolated turbines in the wind farm, being out of the wake effects of the other turbines. In addition, the chosen reference WTG-14 was the nearest turbine, and consequently, there were no qualitative reasons that could explain its bad behavior in terms of micro-scale effects (e.g., terrain obstacles or turbine wakes).

The weekly relative errors in bias and *RMSE* of 50% in some cases within the temporal series emphasized the need for the adequate measurement of the wind direction if we considered the importance of lateral wind fluctuations [15]. As mentioned before, these errors can be reduced to 1% by using special data processing that combines the cup anemometer and wind vane data.

The MIDAS methodology was applied to this wind farm and is a combination of five different approaches: multidimensional probability density function estimators, analysis of wind roses, Taylor diagrams, calculation of the running correlation, running *RMSE* and running bias, and finally, principal component analysis.

This is not our case, but if the met mast is included in the SCADA data of the wind farm, it should be an important reference to develop the Taylor diagrams and the running plots. Therefore, it should be added in the cluster of wind turbine anemometers and in the initial comparison with ERA5.

While the analysis of wind roses and Taylor diagrams provided a preliminary identification of the faulty anemometer, the MPDF estimator allowed an accurate evaluation of the differences in the measured wind components. The principal component analysis made it possible to identify that the error was not in the wind vector module, but in the direction. Therefore, in this case study, a vane misalignment is the cause of the fault. Although an offset in the wind direction measurements can be detected with other numerical and experimental techniques [35], PCA offers valuable and coherent additional information in the general step-by-step perspective of the MIDAS analysis. Finally, the analysis of the running indicators made it possible to identify the exact point in time in which the faulty anemometer started to fail.

The combined approach of these methodologies allowed an in-depth identification and characterization of the faulty anemometer for this wind farm. Although developed for this specific wind farm, our methodology can hopefully be applied to any wind farm in which any number of faulty anemometers may be operating. This must in any case be tested with further studies for different wind farms with, possibly, larger spatial anisotropies in the wind field. The results in other wind farms may contribute more challenging case studies to improve some specific aspects of this general-purpose methodology, but in our view, MIDAS represents a solid methodology capable of providing an accurate diagnosis of the nature of the error (in this case, wind direction), and when those errors started to take place.

Although most wind farm locations are specifically designed to have a wind field that is as homogeneous as possible, in some cases, a few turbines might be deployed close to nearby boundary obstacles that may affect the wind vector. For this reason, this methodology involves an initial stage of an evaluation by experts to adapt this methodological approach to the specific wind farm being analyzed. In this line, although applied in this paper to a specific onshore wind farm, MIDAS could easily be applied to any offshore wind farm where a far more homogeneous wind field can be expected.

5. Conclusions and Future Outlook

MIDAS, a multi-criteria diagnosis method with a benchmarking approach for the detection of defective anemometers with different logical steps, was presented in this paper:

1. The first and main step is based on a new multi-dimensional probability density estimator computing a similarity score between the analyzed anemometers and the data offered by also the new ERA5 reanalysis. This allows a first division between defective anemometers and the group of anemometers with suitable behavior.

2. Having identified the group of suitable anemometers, these can be used as a reference for the application of other statistical and visualization techniques. In this way, we pass from a reference at a meso-scale level (ERA5) to an in situ reference at a micro-scale level in the wind farm.
3. Although the expected results against all of the faultless anemometers should be very similar, it is convenient to certify this aspect. If so, it will be enough to show the representations against one of the faultless anemometers.
4. Taylor diagrams and running plots for correlation, *RMSE,* and bias show coherent results compared to the results of the MPDF score. Furthermore, the PCA components also show very robust results.
5. The feedback of these additional validations can reinforce, as in the studied case, the validity of the first MPDF score against ERA5.

Thus, an integral method for the identification of defective anemometers was developed based on the results given by the MPDF score against ERA5 to identify the faultless and defective anemometers and establish it as the main reference for following well-known unidimensional validations. Definitively, it can be generally considered as a benchmarking method in wind farms, analogous to the method used by the authors for pitch misalignment correction [17,19,20]. Mathematically, MIDAS constitutes a robust and generalist methodology unifying several statistical techniques in a benchmarking approach, but it must be applied in other types of terrains and wind farms before having a relevant and definitive evaluation about its performance in different types of terrains and atmospheric conditions. Future research with a more extensive casuistry is needed to generalize MIDAS, because this is a paradigmatic case to show the general methodology.

In any case, it must be emphasized that, as far as we know and although they are well-known in meteorology, Taylor diagrams are used for the first time in this context of the wind energy industry's O&M, and they show a representative visualization of the deviations of the anemometers in a single diagram that is able to express three statistical indicators.

Although our objective was to present a robust and simplistic method, the results obtained can be added to those obtained using other approaches based on CFD, wake analysis by mesoscale models, or tower shadow effects [36–38].

Additionally, if pressure, temperature, and moisture data are measured in the wind farm, the third variable that defines the wind power density can be introduced in the MPDF score: the air density. This would not be relevant for an evaluation within the wind farm, but it can be used for the verification of mesoscale models or reanalysis (ERA-Interim, ERA5 in an advanced 'meso-beta scale' [39]) in the nearest grid point [24,40]. This would produce a qualitative leap for the validation of wind energy, since this method would consider all the variables in a single score.

Besides, this multi-dimensional validation can be extended to other kinds of renewable energies such as wave energy, in which the power is determined by two variables (wave height and period) and Taylor diagrams are also used for validation of meso-scale models or reanalysis against buoys [41–43].

Author Contributions: Conceptualization, A.U., G.I.-B., J.S.; methodology, A.U., G.I.-B., J.S.; software, A.R., A.U., G.I.-B., J.S.; validation, A.R.; investigation, A.R., A.U., G.I.; writing, original draft preparation, A.R., A.U.; writing, review and editing, all the authors; supervision, all the authors; project administration, U.E., A.U.; funding acquisition, U.E., A.U., J.S.

Funding: This work was financially supported by the Spanish Government through the MINECO project CGL2016-76561-R (MINECO/ERDF, UE), the University of the Basque Country through the Euskoiker PT10477 and GIU 17/002 contracts, and the project DIANEMOS of the Council of Gipuzkoa with Maxwind-Hispavista. ERA5 data were downloaded at no cost from the MARSserver of the ECMWF. Most of the calculations were carried out in the framework of R [44].

Conflicts of Interest: The authors declare no conflict of interest.

Abbreviations

The following abbreviations are used in this manuscript:

MIDAS	multi-technique identification of defective anemometers
WIC	worst-in-class anemometer
z_0	roughness of the terrain
$RMSE$	root mean squared error
MPDF	multidimensional probability density function
PDF	probability density function
U	zonal wind speed
V	meridional wind speed
U	wind speed's module
SD	standard deviation
O&M	operation and maintenance
PCA	principal component analysis
EOF	empirical orthogonal function
AEP	annual energy production
SODAR	sonic detection and ranging
CFD	computational fluid dynamics
LiDAR	light detection and ranging

References

1. Arcos Jiménez, A.; Gómez Muñoz, C.Q.; García Márquez, F.P. Machine learning for wind turbine blades maintenance management. *Energies* **2017**, *11*, 13. [CrossRef]
2. Tchakoua, P.; Wamkeue, R.; Ouhrouche, M.; Slaoui-Hasnaoui, F.; Tameghe, T.A.; Ekemb, G. Wind turbine condition monitoring: State-of-the-art review, new trends, and future challenges. *Energies* **2014**, *7*, 2595–2630. [CrossRef]
3. Gómez Muñoz, C.Q.; García Márquez, F.P. A new fault location approach for acoustic emission techniques in wind turbines. *Energies* **2016**, *9*, 40. [CrossRef]
4. Nielsen, J.S.; Sørensen, J.D. Methods for risk-based planning of O&M of wind turbines. *Energies* **2014**, *7*, 6645–6664.
5. Golding, E.W. *The Generation of Electricity by Wind Power*; Spon. Hamburg, Germany, 1956.
6. Hiester, T.; Pennell, W. *Meteorological Aspects of Siting Large Wind Turbines*; Technical Report; Battelle Pacific Northwest Labs.: Richland, WA, USA, 1981.
7. Rohatgi, J.S.; Nelson, V. *Wind Characteristics: An Analysis for the Generation of Wind Power*; Alternative Energy Institute, West Texas A&M University: Canyon, TX, USA, 1994.
8. American Wind Energy Association. *Standard Performance Testing of Wind Energy Conversion Systems*; Amerian Wind Energy Association: Washington, DC, USA, 1988.
9. Bailey, B.H. *Wind Resource Assessment Handbook*; NREL: Golden, CO, USA, 1997.
10. Rehman, S.; Mohandes, M.A.; Alhems, L.M. Wind speed and power characteristics using LiDAR anemometer based measurements. *Sustain. Energy Technol. Assess.* **2018**, *27*, 46–62. [CrossRef]
11. Khan, K.S.; Tariq, M. Wind resource assessment using SODAR and meteorological mast—A case study of Pakistan. *Renew. Sustain. Energy Rev.* **2018**, *81*, 2443–2449. [CrossRef]
12. González-Longatt, F.; González, J.S.; Payán, M.B.; Santos, J.M.R. Wind-resource atlas of Venezuela based on on-site anemometry observation. *Renew. Sustain. Energy Rev.* **2014**, *39*, 898–911. [CrossRef]
13. Song, M.; Chen, K.; He, Z.; Zhang, X. Wind resource assessment on complex terrain based on observations of a single anemometer. *J. Wind Eng. Ind. Aerodyn.* **2014**, *125*, 22–29. [CrossRef]
14. Baseer, M.A.; Meyer, J.P.; Rehman, S.; Alam, M.M.; Al-Hadhrami, L.M.; Lashin, A. Performance evaluation of cup-anemometers and wind speed characteristics analysis. *Renew. Energy* **2016**, *86*, 733–744. [CrossRef]
15. Kristensen, L. The perennial cup anemometer. *Wind Energy An Int. J. Prog. Appl. Wind Power Convers. Technol.* **1999**, *2*, 59–75. [CrossRef]
16. Han, X.; Liu, D.; Xu, C.; Shen, W.Z. Atmospheric stability and topography effects on wind turbine performance and wake properties in complex terrain. *Renew. Energy* **2018**, *126*, 640–651. [CrossRef]

17. Elosegui, U.; Egana, I.; Ulazia, A.; Ibarra, G. Pitch angle misalignment correction based on benchmarking and laser scanner measurement in wind farms. *Energies* **2018**, *11*, 3357. [CrossRef]

18. Dai, J.; Yang, X.; Hu, W.; Wen, L.; Tan, Y. Effect investigation of yaw on wind turbine performance based on SCADA data. *Energy* **2018**, *149*, 684–696. [CrossRef]

19. Elosegui, U.; Ulazia, A. Novel on-field method for pitch error correction in wind turbines. *Energy Procedia* **2017**, *142*, 9–16. [CrossRef]

20. Elosegui, U.; Elosegui, J. Method for Calculating and Correcting the Angle of Attack in a Wind Turbine Farm. 2013. PCT/ES/2013/070752. Available online: https://patentscope.wipo.int/search/en/detail.jsf?docId=WO2014068162&redirectedID=true (accessed on 12 November 2018).

21. Pindado, S.; Barrero-Gil, A.; Sanz, A. Cup anemometers' Loss of performance due to ageing processes, and its effect on Annual Energy Production (AEP) estimates. *Energies* **2012**, *5*, 1664–1685. [CrossRef]

22. Pindado, S.; Vega, E.; Martínez, A.; Meseguer, E.; Franchini, S.; Sarasola, I.P. Analysis of calibration results from cup and propeller anemometers. Influence on wind turbine Annual Energy Production (AEP) calculations. *Wind Energy* **2011**, *14*, 119–132. [CrossRef]

23. Roibas-Millan, E.; Cubas, J.; Pindado, S. Studies on cup anemometer performances carried out at idr/upm institute. past and present research. *Energies* **2017**, *10*, 1860. [CrossRef]

24. Hersbach, H. *The ERA5 Atmospheric Reanalysis*; American Geophysical Union: San Francisco, CA, USA, 2016.

25. Olauson, J. ERA5: The new champion of wind power modelling? *Renew. Energy* **2018**, *126*, 322–331. [CrossRef]

26. Manwell, J.F.; McGowan, J.G.; Rogers, A.L. *Wind Energy Explained: Theory, Design and Application*; John Wiley & Sons: Hoboken, NJ, USA, 2010.

27. Gryning, S.E.; Batchvarova, E.; Brümmer, B.; Jørgensen, H.; Larsen, S. On the extension of the wind profile over homogeneous terrain beyond the surface boundary layer. *Bound.-Layer Meteorol.* **2007**, *124*, 251–268. [CrossRef]

28. Lopez-Novoa, U.; Sáenz, J.; Mendiburu, A.; Miguel-Alonso, J.; Errasti, I.; Esnaola, G.; Ezcurra, A.; Ibarra-Berastegi, G. Multi-objective environmental model evaluation by means of multidimensional kernel density estimators: Efficient and multi-core implementations. *Environ. Model. Softw.* **2015**, *63*, 123–136. [CrossRef]

29. Maxino, C.; McAvaney, B.; Pitman, A.; Perkins, S. Ranking the AR4 climate models over the Murray-Darling Basin using simulated maximum temperature, minimum temperature and precipitation. *Int. J. Climatol. J. R. Meteorol. Soc.* **2008**, *28*, 1097–1112. [CrossRef]

30. Perkins, S.; Pitman, A.; Holbrook, N.; McAneney, J. Evaluation of the AR4 climate models' simulated daily maximum temperature, minimum temperature, and precipitation over Australia using probability density functions. *J. Clim.* **2007**, *20*, 4356–4376. [CrossRef]

31. Cassity, J.; Aven, C.; Parker, D. Applying weibull distribution and discriminant function techniques to predict damaged cup anemometers in the 2011 PHM competition. *Int. J. Progn. Heal. Manag* **2012**, *3*, 1–7.

32. Fasano, G.; Franceschini, A. A multidimensional version of the Kolmogorov–Smirnov test. *Mon. Not. R. Astron. Soc.* **1987**, *225*, 155–170. [CrossRef]

33. Lopes, R.H.; Hobson, P.R.; Reid, I.D. Computationally efficient algorithms for the two-dimensional Kolmogorov–Smirnov test. *J. Phys. Conf. Ser.* **2008**, *119*, 042019. [CrossRef]

34. Taylor, K.E. Summarizing multiple aspects of model performance in a single diagram. *J. Geophys. Res. Atmos.* **2001**, *106*, 7183–7192. [CrossRef]

35. Castellani, F.; Astolfi, D.; Piccioni, E.; Terzi, L. Numerical and experimental methods for wake flow analysis in complex terrain. *J. Phys. Conf. Ser.* **2015**, *625*, 012042. [CrossRef]

36. Bowen, A.J.; Mortensen, N.G. Exploring the limits of WAsP: The wind atlas analysis and application program. In Proceedings of the 1996 European Union Wind Energy Conference, Göteborg, Sweden, 20–24 May 1996; pp. 20–24.

37. Jiménez, P.A.; Navarro, J.; Palomares, A.M.; Dudhia, J. Mesoscale modeling of offshore wind turbine wakes at the wind farm resolving scale: A composite-based analysis with the Weather Research and Forecasting model over Horns Rev. *Wind Energy* **2015**, *18*, 559–566. [CrossRef]

38. Lubitz, W.D.; Michalak, A. Experimental and theoretical investigation of tower shadow impacts on anemometer measurements. *J. Wind Eng. Ind. Aerodyn.* **2018**, *176*, 112–119. [CrossRef]

39. Djuric, D. *Weather Analysis*; Prentice Hall: Englewood Cliffs, NJ, USA, 1994; Volume 304.

40. Dee, D.P.; Uppala, S.M.; Simmons, A.; Berrisford, P.; Poli, P.; Kobayashi, S.; Andrae, U.; Balmaseda, M.; Balsamo, G.; Bauer, D.P.; et al. The ERA-Interim reanalysis: Configuration and performance of the data assimilation system. *Q. J. R. Meteorol. Soc.* **2011**, *137*, 553–597. [CrossRef]
41. Ulazia, A.; Penalba, M.; Rabanal, A.; Ibarra-Berastegi, G.; Ringwood, J.; Sáenz, J. Historical Evolution of the Wave Resource and Energy Production off the Chilean Coast over the 20th Century. *Energies* **2018**, *11*, 2289. [CrossRef]
42. Ulazia, A.; Penalba, M.; Ibarra-Berastegui, G.; Ringwood, J.; Saénz, J. Wave energy trends over the Bay of Biscay and the consequences for wave energy converters. *Energy* **2017**, *141*, 624–634. [CrossRef]
43. Penalba, M.; Ulazia, A.; Ibarra-Berastegui, G.; Ringwood, J.; Sáenz, J. Wave energy resource variation off the west coast of Ireland and its impact on realistic wave energy converters' power absorption. *Appl. Energy* **2018**, *224*, 205–219. [CrossRef]
44. R: A Language and Environment for Statistical Computing. 2013. Available online: http://softlibre.unizar.es/manuales/aplicaciones/r/fullrefman.pdf (accessed on 15 November 2018).

Article

Pitch Angle Misalignment Correction Based on Benchmarking and Laser Scanner Measurement in Wind Farms

Unai Elosegui [1,*,†], Igor Egana [1,†], Alain Ulazia [2,†] and Gabriel Ibarra-Berastegi [3,4,†]

[1]   Maxwind, Portuetxe 83, 3, 20018 Donostia, Spain; igor.egana@maxwind.eu
[2]   Department of NE and Fluid Mechanics, University of the Basque Country (UPV/EHU), Otaola 29, 20600 Eibar, Spain; alain.ulazia@ehu.eus
[3]   Department of NE and Fluid Mechanics, University of the Basque Country (UPV/EHU), Alda, Urkijo, 48013 Bilbao, Spain; gabriel.ibarra@ehu.eus
[4]   Joint Research Unit (UPV/EHU-IEO) Plentziako Itsas Estazioa (PIE), University of Basque Country (UPV/EHU), Areatza Hiribidea 47, 48620 Plentzia, Spain
*   Correspondence: unai.elosegui@corp.hispavista.com
†   These authors contributed equally to this work.

Received: 27 October 2018; Accepted: 23 November 2018; Published: 1 December 2018

Abstract: In addition to human error, manufacturing tolerances for blades and hubs cause pitch angle misalignment in wind turbines. As a consequence, a significant number of turbines used by existing wind farms experience power production loss and a reduced turbine lifetime. Existing techniques, such as photometric technology and laser-based methods, have been used in the wind industry for on-field pitch measurements. However, in some cases, regular techniques have difficulty achieving good and accurate measurements of pitch angle settings, resulting in pitch angle errors that require cost-effective correction on wind farms. Here, the authors present a novel patented method based on laser scanner measurements. The authors applied this new method and achieved successful improvements in the Annual Energy Production of various wind farms. This technique is a benchmarking-based approach for pitch angle calibration. Two case studies are introduced to demonstrate the effectiveness of the pitch angle calibration method to yield Annual Energy Production increase.

Keywords: wind turbine; laser technology; diagnosis; pitch angle misalignment; efficiency; durability

---

1. Introduction

Ideally, there are no manufacturing or assembly errors in wind turbines, and the blades are equal in mass, center of gravity location, shape, and structure. As a consequence, the three blade root axes intersect with the rotor axis at the very same point. In addition, in the ideal scenario, the orientation of the three blades is equal, and when substantial demand is placed on them to form a fine pitch angle to maximize power production, the three blades face the wind with the very same angle of attack; this is the angle between the line of the chord at a particular blade section and the relative airflow, as in Figure 1.

However, blade manufacturing is a poorly automated process. Variances in fiber placement, bonding, and curing cause variation not only in the blade mass distribution but also in the blade profile shape. In addition, the blade pitch angle reference that is set up at the very end of the manufacturing process is subject to human error. Furthermore, hub manufacturing tolerances that move the blade axes from the intersection point on the rotor axis cause pitch angle misalignment [1–3].

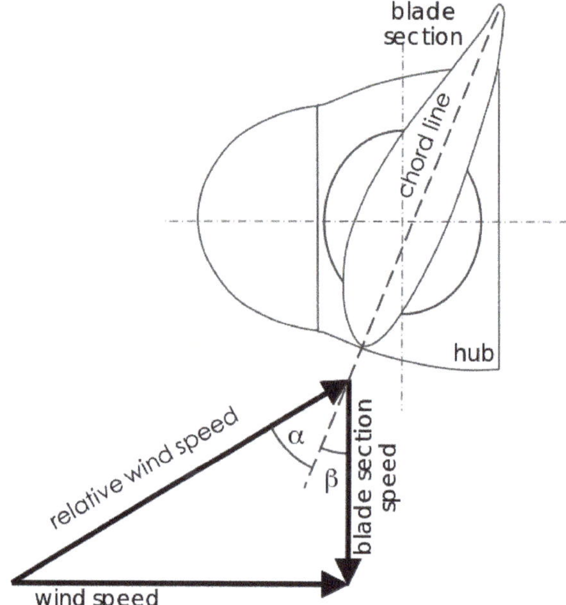

Figure 1. Angle of attack α and pitch angle β for a given blade section.

In this context, some consultancy firms offer different solutions, mainly through photometric means. The solutions currently provided by the private sector ([4] or [5]) involve such technology to calculate the pitch angle. There are some slight differences among the technologies used by these firms, but they basically carry out the measurements by placing a camera under the wind turbine or rotating with the blade itself [6], as seen in Figure 2, where the chord line is drawn. The rotation plane is set with the other two blades.

There are different techniques to measure the pitch angle. In some cases, it is calculated based on a comparison of measurements taken at the maximum chord line with the rotor plane. Another methodology is based on setting marks along the blade and establishing the blade section and chord at the given section.

However, this method faces some challenges that compromise the robustness of the results. According to our experience in the field, the authors observe the following:

1. According to this technique, pitch angles are obtained based on three different rotation planes: the plane defined by blades 2–3 for blade 1's pitch angle calculation; the plane defined by blades 3–1 for blade 2's pitch angle calculation; and the plane defined by blades 1–2 for blade 3's pitch angle calculation. Since the blade axes are not coplanar due to manufacturing and assembly tolerances, it can lead to variances larger than 0.5°.
2. To use this technique for calculations based on the same blade section, the blade axis must be placed so that it is pointing downward while being perfectly aligned with the measuring system. However, such an alignment is difficult, as it is done manually and there is no fixed reference against which the three blades of a rotor can be placed in the very same position.
3. Assuming that pitch angle differences within the rotor of a particular turbine are accurately measured, the only way to correctly determine the absolute pitch angles is by either:

(a) obtaining the pitch angle at the maximum chord line from the turbine manufacturer, although it is very rarely available due to the sensitivity of turbine design information; or

(b) placing the measuring system in the very same location with respect to the rotor in every turbine; however, it must be noted that slight errors in this position create variation in the measurement of absolute pitch angles.

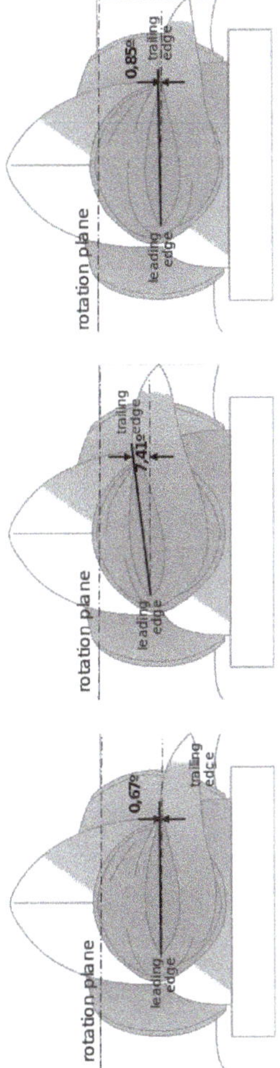

Figure 2. Photometry-based pitch angle measurement of a rotor blade set.

Another method used by other companies [7] is based on a laser device that measures the maximum chord of the profile and determines the relative misalignment between the blades. It is performed with turbines used in power production. However, this technique shows a very remarkable drawback: it does not obtain the absolute pitch angle values, which are key to ensuring power production improvement and avoiding decreases in turbine lifetime. Although setting the blades at the same relative pitch angle does remove aerodynamic imbalance, it does not guarantee that the resulting fine pitch is the one defined in the turbine design that was product certified.

Another method is based on laser scans from quite a long distance from the rotor [8]. Due to the trade-off between accuracy and scanning distance, this methodology is time-consuming. In addition, for this technique to address absolute pitch angle correction, the wind turbine's design information is required, which is seldom available.

Some recent developments in laser technology have made it suitable for long distances and outdoor conditions. However, the leading manufacturers of laser-based measuring systems ([9] and [10]) do not have experience in this particular application.

In this context [11], the parent company of [12] initiated development of the technique by exploring different laser systems. Field tests began with laser tracker technology [13], but measuring a few key points of a rotor takes an entire workday. Laser scanner technology [14,15] was finally adopted as the better option for getting far more measurements per turbine in a much shorter period of time per turbine without the hassle of accessing blade exteriors up-tower. This method was patented in 2013 with the title 'METHOD FOR CALCULATING AND CORRECTING THE ANGLE OF ATTACK IN A WIND TURBINE FARM' (Pub. No.: WO/2014/068162; International Application No.: PCT/ES2013/070752) [16]. The main claim of this patent is the replication of the pitch angle settings of the Best in Class turbine in the Worst in Class turbines, said turbines being identified through a power performance analysis.

Following the publication of this patent, Maxwind is working worldwide applying this method in several wind farms [12]. A preliminary exposition of the results in wind farms was presented last year at the ICAE2017 conference [17]. This paper is a substantial extension of that short paper: this contribution presents the quantitative consequences in power production due to pitch misalignment, describes the general correction method in which the laser scanner is an intermediate step, and, finally, shows the positive results after the correction in the last years at two wind farms.

2. Pitch Angle Misalignment

2.1. State of the Art

Pitch misalignment and its implications for energy production in wind turbines have not received much attention as a specific problem. This applies both to the developments specifically made in the framework of the private sector (usually as patents) and also the apportions from a wider number of actors as gathered in the scientific literature.

In both cases, the general approach has been to address the problem of pitch misalignment in conjunction with other issues like yaw misalignment or mechanical load balancing in the wind turbine. Load reduction method for wind turbine involves adjusting yaw alignment of wind turbine according to favorable yaw orientation, and adjusting pitch of rotor blade [18].

In addition, the analysis of pitch misalignment involves the use of general-purpose laser scanners, but the leading manufacturers of laser scanners ([9] and [10]) do not have experience in pitch misalignment measurements. Therefore, our method based on laser measurements applied to specifically measuring pitch misalignment in wind turbines is totally novel for this purpose, and difficult to compare with previous results applied in the consultancy sector.

To the best of our knowledge, there is only one study about the use of laser scanners to measure turbine blades, but it is for the determination of deformations of moving rotor blades [19]. They use multiple scanners in 1D mode to record cross sections at different positions along the rotor blades, and, after that, they compare these results with the CAD (Computer Aided Design) model of the blade. The deformations in out-of-plane and torsional direction can be derived. The nacelle is also pre-scanned to establish the coordinate system of the wind turbine as reference. Therefore, this is a dynamic scanning to measure the deviations of the blades due to bending and flapping forces when the turbine is moving under the dynamic pressure of the wind. In our particular case, the authors want to measure the inherent pitch misalignment on the hub, a problem that is not taken into account in the pre-operational tests in large certification laboratories using photometric means [20,21], or during operation via strange gauges [22] or reflective targets observed with stereo camera systems [23].

Furthermore, this inherent pitch misalignment can increase with time due to the stress and fatigue that supports the wind turbine.

As mentioned in the patent document, our correction method is post-operational. In the installation moment when the operators have to move huge blades that are implemented on the hub, there are pre-established marks to align both elements, the root of the blade and the hub. However, a final calibration is not usually performed to ensure that the position is correct. This is the original cause of the presence of an inherent pitch misalignment in wind turbines.

Pitch angle errors have no impact in terms of power production above the rated power, as blades are pitched toward the feathering position in order to limit turbine loads. However, energy production loss can be remarkably high below the rated power conditions. Thus, understanding the sensitivity of power production under pitch angle errors is paramount, as this method entails understanding how turbines are affected by pitch angle misalignment.

Coming to the scientific literature, it is clear that there is a general challenge for optimizing wind energy performance and applies to all type of turbines, including vertical rotors [24]. Again, it is important to highlight that the general approach for dealing with pitch misalignment involves a combined study in conjunction with other parameters related to wind turbine general misalignments and balance.

In fact, the sensitivity analysis of power production in relation to the pitch angle and other parameters needs complex inference strategies, such as ANFIS (Adaptive Neuro-Fuzzy Inference System), as is shown in the recent literature [25–28]. There are also data-driven approaches to detect or somehow balance pitch angle faults alone [29,30] or in combination with other parameters [31,32].

It is worth mentioning that, in the particular case of floating wind turbines, joint control of both blade pitch and platform pitch is required [33]. Computational experience has improved the power output by optimizing the set points of the blade pitch angle and generator torque [34].

Other works have applied blade design approaches to enhance the power output, both experimentally and numerically. Such methods include the use of active and passive techniques, attempts to reduce the cut-in speed, and the development of new materials [35].

For example, there are improvement techniques based on passive flow control devices, such as vortex generators (VG) or Gurney flaps (GF), that are implemented on the surface of the blade. For the best configuration cases of these devices, simulations have shown 3–10% increases in the power output, depending on the wind speed, with residual increases in the bending moment [36]. Using the case studies described in Section 4, the authors show that pitch misalignment correction can produce improvements of 16% in annual energy production (AEP) and much higher percentages for the power output below the rated wind speed.

Therefore, the influence of faults in the pitch angle, as well as active or passive control, has been widely studied in the literature, and methodologies to correctly evaluate improvements after upgrades have been developed [37]. However, these studies have not directly accounted for the possible inherent misalignment of the blade on the hub and the consequent general diminution of the power output. The technique described here offers a direct in situ way, i.e., on the wind farm, to measure this specific deviation.

In addition, pitch angle misalignment reduces the turbine lifetime. This happens when the rotor is aerodynamically imbalanced, but it also occurs with a balanced rotor showing a negative offset with respect to the correct position. The latter is further discussed in Section 2.2. However, it is important to emphasize that it is very hard to quantify this effect for extrapolating results from one turbine to another, given the complexity of the load and transient load calculations. Any assessment of this kind [27] needs to include extreme load calculations and fatigue load analysis under the wind conditions defined per IEC (International Electrotechnical Commission) wind class. Thus, this section does not aim to address every potential effect of pitch angle misalignment on the load envelope but to introduce insights into the effect of pitch angle misalignment on turbine lifetime.

2.2. Impact of Pitch Misalignment on Power Production

In order to discuss the effect of pitch angle misalignment on the power curve, this section establishes the use of the Power Curve Ratio (*PCR*). The *PCR* is defined as the ratio calculated for each wind speed used for power production by a particular turbine over the potential output power. Whereas the ultimate concern for wind farm owners is the effect on the *AEP*, *PCR* facilitates the understanding of the sensitivity of the *AEP* under pitch deviations; as a consequence, its use is suggested for diagnostic purposes.

For this analysis, static and dynamic simulations are discussed. These simulations were carried out in FAST using the publicly available turbine model *WindPACT 1.5 MW Baseline* [38]. This turbine is controlled with the standard controller described in [39], which is, to some extent, the most extended concept within the industry. This is a tool customarily used in pitch control studies [40].

Figure 3 depicts a turbine operation in three different ways: on the left-hand side is the trajectory over the power coefficient surface with the pitch angle β and tip-speed ratio (*TSR*); on the upper right-hand side is the torque demand T with respect to the rotor speed Ω, which is a feature that only depends on the control system; the lower right-hand side shows the power curve, which describes the production capability of the turbine with wind speed. These three figures reflect steady-state conditions, disregarding transients. The black thick line depicts rotor speed regulation at low rotor speeds, Ω_{low}; the blue thick line depicts operation at a maximum power coefficient with a constant *TSR*; the red thick line shows rotor speed regulation at the nominal rotor speed, Ω_N; and the thick gray line depicts operation at the nominal power with rotor speed regulation by blade pitching.

The black thick lines in Figure 3 show operation at winds slightly over cut-in conditions. *TSR* rapidly decreases as wind increases. As a consequence, the power coefficient increases with rotor speed regulation around Ω_{low} by torque modulation. The blue thick lines in Figure 3 illustrate operating at a constant *TSR* and reaching the maximum power coefficient. In these conditions, torque demand is controlled proportionally to the second power of the rotor speed Ω. Shown by the thick red lines in Figure 3, as the rotor speed reaches the nominal value Ω_N, regulation by torque demand occurs again, diminishing the power coefficient. This occurs slightly below the nominal power. The thick gray lines in Figure 3 depict the rated power operation, where the rotor speed is regulated around Ω_N by blade pitching.

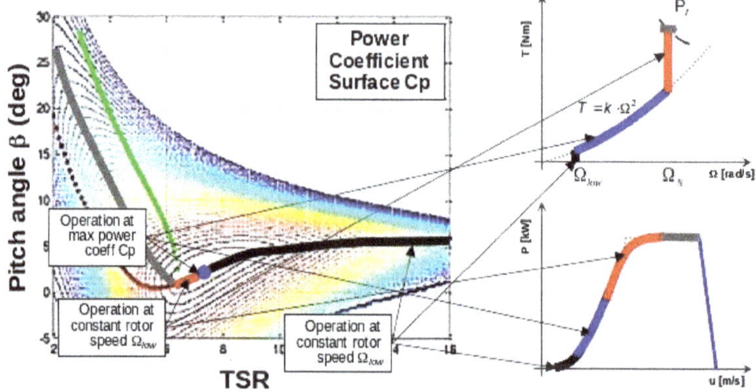

Figure 3. Power coefficient surface over the pitch angle β and Tip-Speed Ratio (*TSR*); torque demand T with rotor speed Ω; and power curve.

In this context, the power curve was simulated for different pitch misalignment conditions. It was normalized by computing the *PCR*, as explained above in this section, and the *AEP* loss. Figure 4 shows the *PCR* and *AEP* loss for different conditions of pitch misalignment resulting from steady-state simulations.

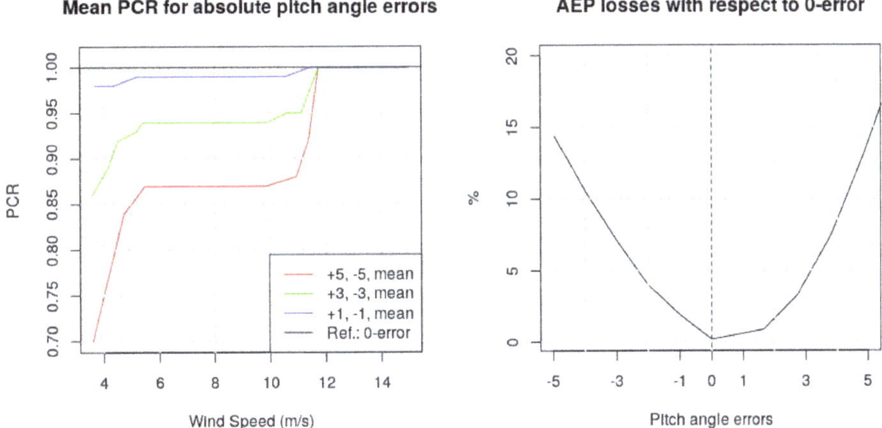

Figure 4. Steady-state Power Curve Ratio (*PCR*) and annual energy production (*AEP*) losses.

Dynamic simulations were run under turbulent wind conditions for three different pitch misalignment cases, as shown in Figure 5 and described below:

1. Case 1: One blade is in the correct position, and two blades have similar errors with opposite signs.
2. Case 2: The three blades are affected by the same pitch angle misalignment.
3. Case 3: Two blades are in the correct position, one blade has a pitch angle error.

In both static and dynamic simulations, the pitch misalignment has been introduced in the FAST input file *primary.fst* to obtain the power output time series in simulations of 600 s with a step of 0.0125 s. The initial pitch angles of the blades (parameters *BlPitch*) have been deviated from 0 according to the Case 1, 2 and 3, and pitch control has been deactivated to keep these values below the rated wind speed (*PCMode*: 0).

Before the dynamic simulation, a turbulent input of the wind speed has been created using TurbSim for the same period of simulation. This has been done at each wind speed with a step of 0.5 m/s between the cut-in wind speed and the rated wind speed. In this way, the power curve due to a given misalignment is determined. After that, for the computation of *AEP*, a typical Weibull distribution of wind speed with the form parameter $k = 2$ has been implemented on the deviated power curve due to misalignment. An average wind speed $\overline{U}$ of 4 m/s has been used, and the subsequent Weibull's scale parameter [41]:

$$c = \frac{\overline{U}}{\Gamma(1+1/k)}.$$

(1)

On behalf of clarity and due to the approximate results for negative and positive pitch errors with the same absolute value, the average behavior of the two values of errors for each wind speed are shown in the *PCR* figures on the left in the Case 1, 2, and 3 (Figure 5a–c). On the right, the corresponding *AEP* losses are shown for each case with respect to a reference without pitch misalignment (0-error), and considering misaligned cases with integer negative and positive values of pitch angle deviation. Although the mean value is used for the *PCR* curves, the *AEP* figures do provide evidence of the asymmetry between negative and positive values in favor of higher values of the last ones.

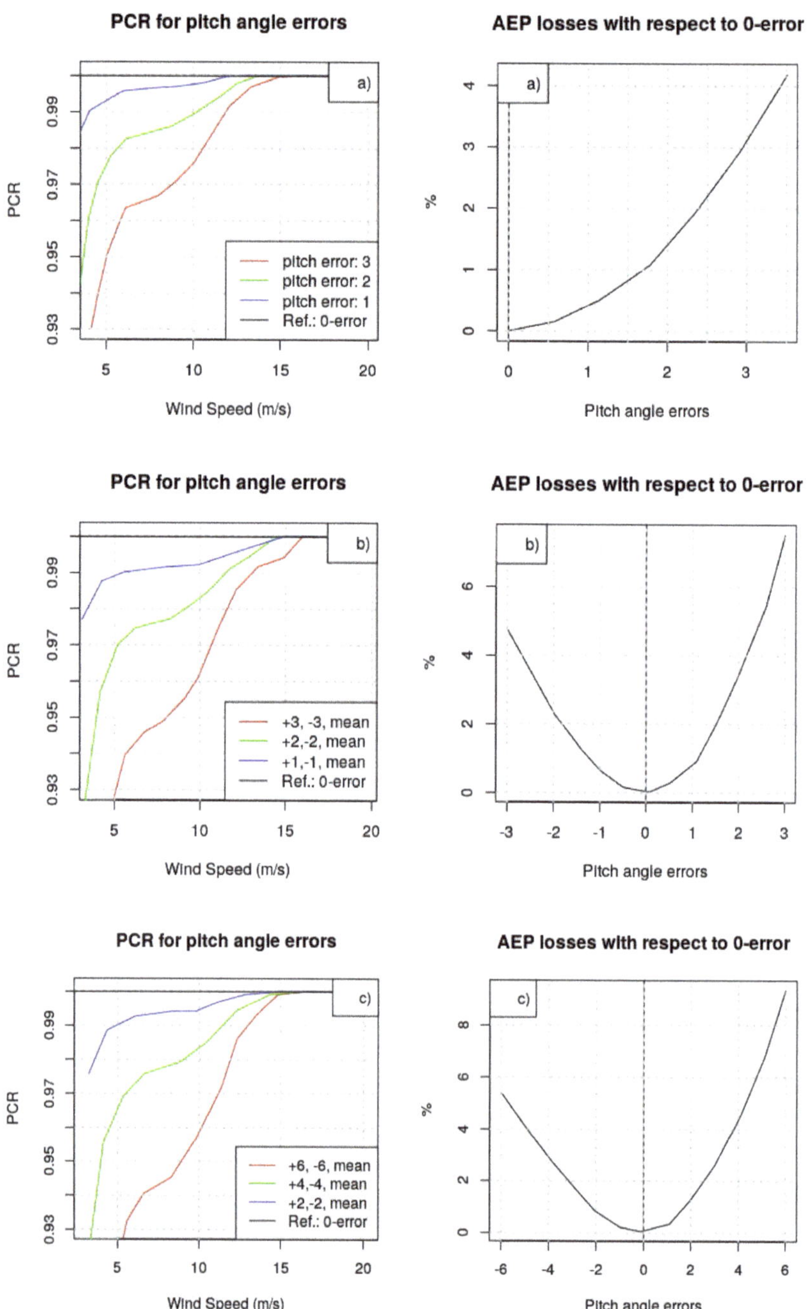

Figure 5. Dynamic FAST simulation: (**a**) Case 1; (**b**) Case 2; (**c**) Case 3.

Similar patterns to those of the static results are observed, and they are consistent with [42], although this reference did not suggest using this method for the detection of pitch angle misalignment. For the *PCR* functions, the curves exhibit a small slope in the quadratic zone caused by wind speed variability around its mean value. In the quadratic zone, the turbine is operating around the maximum power coefficient (C_p) point, but not just at its maximum. In general terms, from the static and dynamic simulations, it can be stated that:

1. The static results match the dynamic results.
2. The *PCR* functions exhibit a well-defined shape that is dependent on the operation zone. This will be useful to identify pitch errors in field measurements.
3. The total *AEP* losses can be approximated by adding the *AEP* losses on each blade. The total wind turbine *AEP* is notably sensitive to blade assembly errors. A small pitch discrepancy of 2° in only one blade is able to reduce approximately 1% of the *AEP* value.
4. An absolute misalignment of 2° in the three blades also causes an *AEP* loss; in this case, the loss is 3.5%.
5. Because of the nonlinear dependence of the *AEP* loss on the pitch error, it is better to have small pitch errors in all three blades than it is to have only one blade with a high error.
6. Understanding that power production is proportional to the third power of the wind speed, yaw misalignment yields a completely different pattern that is virtually flat on the low end and middle of the wind speed range.

2.3. Impact of Pitch Misalignment on Turbine Lifetime

Since the consequences of pitch angle misalignment on turbine lifetime depend on the turbine design [43], it is not possible to extrapolate results from one turbine to another. Such an assessment [44] should include not only thorough extreme load calculations, but also thorough fatigue load analysis under the wind conditions defined per IEC wind class.

Thus, this section does not aim to address every potential effect of pitch angle misalignment on the load envelope but at introducing some insights into the effect of pitch angle misalignment on turbine lifetime.

2.3.1. Effect of Even Positive Pitch Angle Offset

This subsection analyzes the effect of the three blades suffering from an offset toward higher pitch angles, as shown in Figure 6.

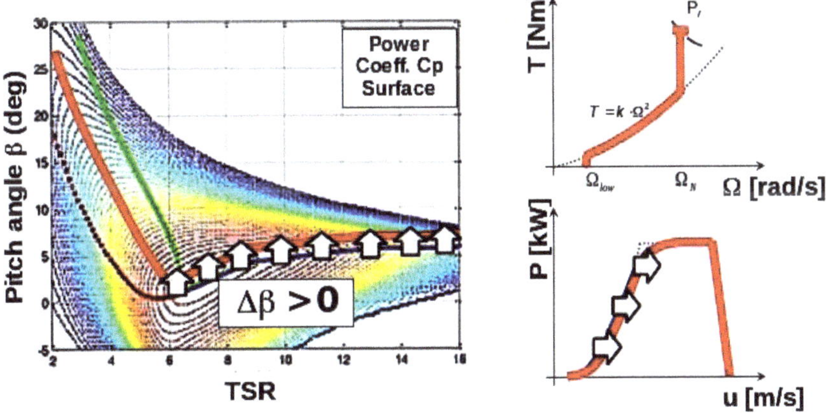

Figure 6. Effect of pitch angle offset toward higher angles in the turbine operation.

Should the three blades be equally shifted toward higher pitch angles, then the maximum power coefficient is not reached, as shown by the red trajectory over the power coefficient surface in Figure 6. The immediate consequence is that performance in terms of power production is decreased, as the power curve is shifted toward the right-hand side. Rated power is reached at higher wind speeds, and power curtailment occurs by pitching the blades toward the feathering position, beginning with higher pitch angles. Rotor speed controller robustness should be more than capable of coping with this uncertainty, although it is true that the performance at crossover frequencies would be affected, yielding a higher peak in the sensitivity function.

2.3.2. Effect of Even Negative Pitch Angle Offset

This subsection analyzes the effect of the three blades suffering from an offset toward lower pitch angles, as shown in Figure 7.

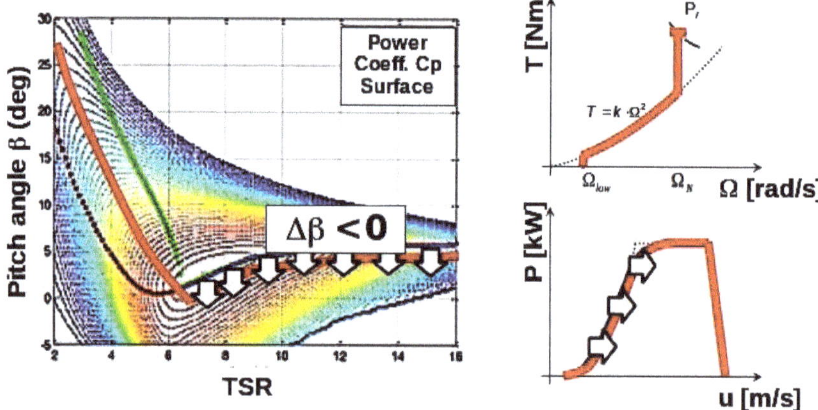

Figure 7. Effect of pitch angle offset toward lower angles in turbine operation.

Should the three blades be equally shifted toward lower pitch angles, then the maximum power coefficient is not reached, as shown by the red trajectory over the power coefficient surface in Figure 7. Thus, power production performance is decreased, as the power curve is shifted toward the right-hand side. Rated power is reached at higher wind speeds, and power curtailment occurs by pitching the blades toward the feathering position, beginning with lower pitch angles. This has a twofold effect:

- The peak of the steady-state curve of the thrust force over the rotor increases.
- Performance of the rotor speed controller by blade pitching is diminished due to lower plant gain. As a consequence, the controller bandwidth is reduced, creating much higher dynamic loads.

It is worth mentioning that loads at this particular operational point are extremely important for turbine integrity. As a consequence, those two effects can compromise turbine lifetime for key components, including the blades, hub, mainframe, yaw bearing, tower, and foundation.

Any pitch angle correction that does not ensure absolute pitch angle correction does not guarantee a power production increase and is a potential cause of turbine lifetime decrease.

3. Methodology

3.1. Novel on-Field Method for Pitch Calculation and Compensation

This section describes the methodology used for the detection and correction of pitch angle misalignment. The specific details about the use of the laser scanner and other aspects can be found in the original patent document [16].

Laser scanning is widely used for different technological purposes because it is a contactless 3D measurement method. It enables the measurement of distances and corresponding angles with a frequency of up to 1 MHz for moving objects, but most common applications capture static objects. The acquisition of points usually results from the rotation of the laser around the horizontal and vertical axis (3D). For referencing into a global coordinate system, other sensors such as GPS, INS or cameras are necessary [19,45,46]. Figure 8 shows the workflow of our method in which laser measurement is an intermediate step. First, a production performance assessment is carried out. The primary purpose of this analysis is to determine the turbines affected by pitch angle misalignment, and also to select the best turbine in terms of pitch angle settings: the outperforming turbine is the Best in Class turbine, whereas the Worst in Class turbines are those in which the pattern described in Section 2.1 is found. A secondary goal of such an analysis is to assess the AEP gaps between the Worst in Class turbines and the Best in Class turbine.

Laser measurements are taken on the Best in Class turbine and on the Worst in Class turbines, whereas the rest of the turbines are excluded from further site study activities. The benefit of this approach is that all efforts are focused on turbines where a significant AEP increase is guaranteed. Those measurements are then executed to determine the pitch angles of each blade of said turbines. Pitch correction angles are proposed based on measurements so that the pitch angle settings of the Best in Class turbine are used in the Worst in Class turbines.

Finally, a production performance assessment shows the effect of the pitch angle correction. Success is measured as the reduction in the AEP gaps between the Worst in Class turbines and the Best in Class turbine.

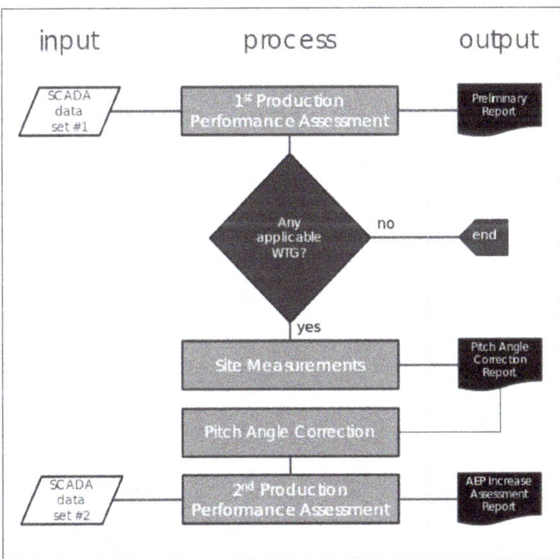

Figure 8. Pitch angle detection and correction workflow.

3.2. Production Performance Assessment

As briefly explained earlier, an initial power performance assessment aiming to detect pitch angle misalignment patterns is carried out. For this purpose, the power curve of each turbine on the wind farm is calculated from the 10-minute average data as per industry standards [44], with some remarks:

1. Wind speeds measured above the turbine nacelles are used. Although anemometers are not calibrated or certified, multi-megawatt wind turbines are nowadays equipped with good anemometers that offer reliable measurements. This particularly holds for sonic anemometers, which are only affected by nacelle aerodynamics and the blade passing wake. In any case, this type of disturbance is consistent throughout the entire wind farm, so the comparison between the results obtained from the Best in Class turbine and the Worst in Class turbines will be consistent. Nevertheless, the correlation between wind speed measurements for different wind turbines is also checked.

2. Output power: data in which the power might be compromised by effects not purely related to aerodynamics are ruled out, including:

    (a) Turbine and complex terrain disturbances, as described in [47],

    (b) Ten-minute periods in which the average power does not show the capability of the turbine to produce energy, e.g., due to starts and stops, maintenance work, power curtailment operation, etc.

After computing the power curves for each turbine, a Best in Class turbine is selected. Such a turbine is the best-performing turbine in terms of pitch angle misalignment. For the rest of the turbines, the Power Curve Ratio (*PCR*) is computed as the ratio of each of their power curves over the power curve of the Best in Class turbine.

3.3. On-Field Pitch Measurement and Calculation

Every detail of this novel method cannot be presented in this paper due to commercial issues, but, as mentioned, the careful description of the use of laser scanner can be found in the first author's original PCT/ES2013/070752 international patent [16].

Therefore, a qualitative step-by-step description is presented here to explain the main procedures of the misalignment measurement and correction, after the identification of the Best in Class (BIC) turbine. Furthermore, the academic value of the present contribution is enhanced by the final results, since real energy production improvements on specific wind farms are quantitatively shown. The general method is described below.

Blades are scanned for the Best in Class and Worst in Class turbines. A laser scanner is used in order to capture rotor geometry and measure the pitch angle of each blade. The pitch angle is measured at a particular blade section. Consequently, it is possible not only to measure pitch angle differences within a rotor, but also to make accurate comparisons with other turbines of the blade, a remarkable advantage of this technique.

For example, a characteristic schema of the 3D laser scanning reference system in the patent document is the Figure 9 in the page ([16], p. 25). GXYZ establishes the reference system with the origin (G) and the coordinate axes (XYZ, 1, 2, 3) for the laser measurement. Thirteen is the nacelle, 14 the hub, and 5 is the junction plane between both, which must be captured by the laser. Four is the circular section defined by three points on the junction of the blade and the hub, which are also located by means of the scanner.

This 3D configuration established the reference to measure the position of the airfoil's chord and consequently the pitch angle. For that, different laser shots are executed from different positions in order to get a complete 3D image of the blade. Figure 10 shows the measurement moment in a wind turbine and several targets used in the procedure. A more detailed description can be found in the 'Description' and 'Claims' section of the original patent document [16].

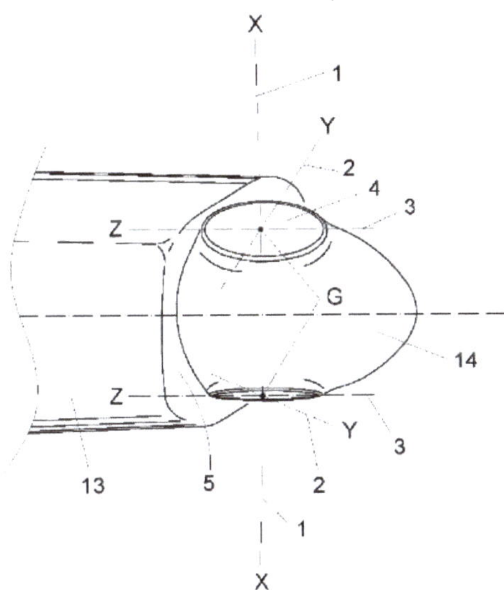

Figure 9. Schema of the reference system for the measuring procedure.

Figure 10. Photo of the measurement moment in a wind farm.

Since pitch angle calculation uses a hub-based coordinate system, pitch angles measured in different rotors can be fairly compared. The method concludes by proposing pitch angle corrections for the Worst in Class turbines so that their blades mimic the Best in Class turbine blade settings. It is the authors' experience that Best in Class turbines always have balanced rotors (relative differences lower than $\pm 0.14°$), so this method corrects both absolute and relative pitch angle misalignments.

It is also worth mentioning that this technique obtains intermediate results in the form of other measurements of rotor quality. Although the three blade axes and the hub rotating axis should be ideally merging to the same point, manufacturing tolerances of the hub, pitch bearings, and blades prevent the actual intersection of the four axes. Intersection points of the three blade axes with the hub axis yield a triangle whose surface can be used to quantify this error. It is also the authors' experience that all Best in Class turbines detected in power performance assessments always present the best alignment of these four axes, whereas the underperforming turbines often—but not always—show this type of misalignment. Note that, depending on the direction in which the blade axes lean, the effect of misalignment can be described as a decrease in the effective swept area. Should this be the case, little can be done by pitching the blades.

4. Results

This section introduces two case studies: one for a very small wind farm, and another for a wind farm that is significantly larger. The authors show the improvements by means of our correction recommendations below the rated power conditions, for which the power output improvement after implementing these recommendations was remarkable. This improvement was quantified by PCR gap percentage (PCR_{gap}) for the power output of a given turbine T at a wind speed U ($P_T(U)$), with respect to the power output at that speed of the Best in Class (BIC) turbine on the farm ($P_{BIC}(U)$):

$$PCR_{gap} = \left(1 - \frac{P_T(U)}{P_{BIC}(U)}\right) \times 100. \tag{2}$$

Apart from the PCR_{gap} correction interval at each wind speed of the power curve, the total AEP improvement is also shown in each case study. For that, the AEP gap (percent) is defined analogously.

4.1. Case Study 1

This is a small wind farm located in Catalonia, with two variable-speed 1.5 MW turbines mounting a 77 m diameter rotor. The AEP difference between them was 6.68%. At low wind speeds near the cut-in speed, the PCR_{gap} defined in Equation (2) was around 5%, and it maintained this value for medium wind speeds (5–7 m/s). There was no improvement at the rated wind speed because, as mentioned above, the effect of pitch misalignment disappears when the rated power of the generator is reached.

Pitch angle correction recommendations for the Worst in Class turbine A2 are shown in Table 1, considering that the Best in Class turbine setting for the chosen blade section was 83.15°.

Table 1. Blade angle measurements and consequent recommendations.

Blades	Measurement	Recommendation
Blade 1	83.06°	+0.1°
Blade 2	81.64°	+1.5°
Blade 3	84.34°	−1.2°

After the application of the recommendations, important energy production improvements were achieved, reaching a 2.46% increase in AEP with respect to the Best in Class turbine A1. It is worth mentioning that the remaining patter is related to yaw misalignment, not to pitch angle errors. Figure 11 shows the PCR_{gap} before and after pitch correction.

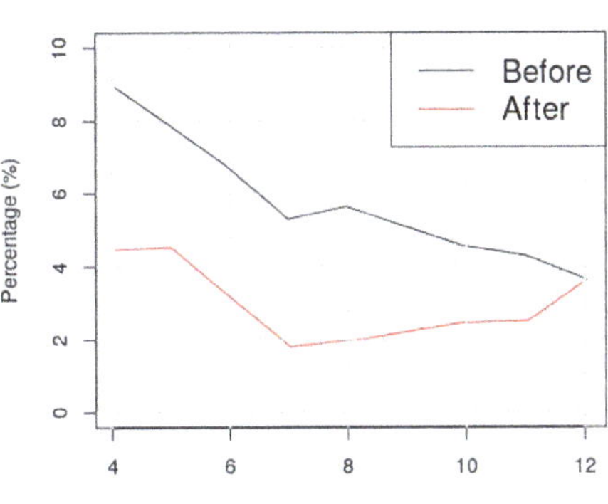

Figure 11. PCR_{gap} reduction after calibration for turbine A2.

4.2. Case Study 2

This is a wind farm comprising a larger number of turbines located in Soria, Spain, with 20 variable-speed 800 kW turbines mounting a 56 m diameter rotor. The *AEP* gap of turbine ES10 was 16.3%, whereas the *AEP* gap of turbine ES11 was 15%. It should be emphasized that, for low wind speeds near the cut-in speed, ES10 and ES11's PCR_{gap} correction was around 25%. At medium wind speeds (around 5–7 m/s), the PCR_{gap} was 10% for ES10 and around 5% for ES11. Thus, these are significant increases compared to the results obtained for other kinds of aerodynamic improvements that use passive control devices or other techniques [36].

The results and pitch angle correction recommendations are shown in Table 2, and they indicate that the Best in Class turbine setting for the chosen blade section was 100.3°.

Table 2. Blade angle measurements and consequent recommendations.

Turbine-Blades	Measurement	Recommendation
ES10-Blade 1	99.68°	+0.61°
ES10-Blade 2	100.30°	Leave as is
ES10-Blade 3	94.4°	5.89°
ES11-Blade 1	98.12°	+2.17°
ES11-Blade 2	98.08°	+2.21°
ES11-Blade 3	97.10°	+3.16°

After the application of the recommendations, the production gaps for ES10 and ES11 with respect to ES01 were removed, as shown in Figure 12. It is worth mentioning that the reason for turbine ES10 outperforming the Best in Class ES01 is that the latter has some minor yaw misalignment. On the other hand, turbine ES11 falls slightly short of achieving the performance of ES01 due to yaw misalignment.

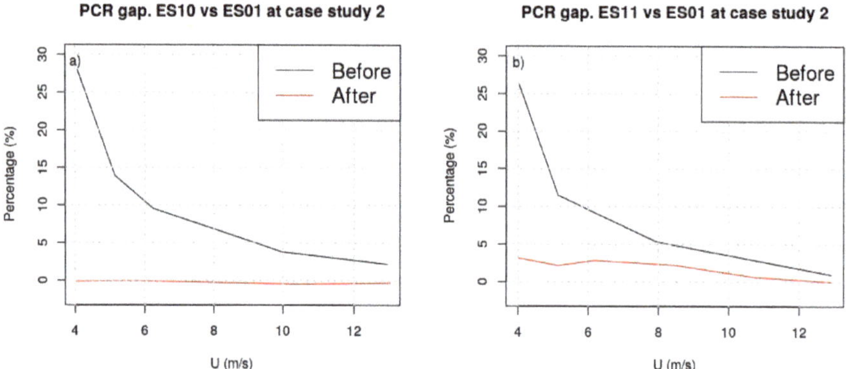

Figure 12. PCR_{gap} reduction after calibration for turbines ES10 (**a**) and ES11 (**b**).

5. Discussion

In this paper, a new methodology is introduced to detect, measure, and correct pitch angle misalignment in wind turbine rotors. Two case studies are discussed demonstrating the reduction in power production gaps between the Worst in Class turbines and the Best in Class turbines. Since this issue has an impact on a wind project's AEP and on turbine lifetime, its correction has a significant impact on wind project profitability. To summarize, the features of this technique are:

1. No product design information is needed from the turbine manufacturer, as the Best in Class turbine is used to define the pitch angle settings for the Worst in Class turbines.
2. Mimicking the pitch angle settings of the Best in Class turbine in the Worst in Class turbines ensures that product certification is not compromised.
3. Power performance benchmarking guarantees an increase in power production. This AEP increase can be used to compute the increase in revenue, and thus to calculate Return On Investment of this service.
4. There is no lost revenue for wind farm owners, as laser measurements in the affected turbines are carried out in idling conditions, under very low wind speeds.
5. Although pitch angle corrections cannot address hub manufacturing tolerances, the measurement results of this technique offer valuable information for turbine manufacturers.
6. At low wind speeds around 5 m/s, a PCR_{gap} of 25% can be reached in some cases, that is, the power output can be improved by 25% at those wind speeds.
7. These corrections in the power curve can produce general improvement in energy production that reaches even 15% for AEP.
8. For a referential turbine of 1 MW (our turbines are 0.8 and 1.5 MW), this improvement implies 450 MWh more production annually, 3000 being the typical number of full load hours.
9. If the 2018 spot-market price of 1 MWh in Spain is considered ($\pm$62 € /MWh according to [48], this increment of energy production supposes $\pm$28,000 € more annually per installed MW of turbines like the studied worst ones.
10. In addition, these empirical percentages of AEP loss are quite coherent with the results obtained in the simulations of the Section 2.2 using FAST for different pitch misalignment combinations of the three blades, which show losses that can reach the 10% in the worst cases.

6. Conclusions and Future Outlook

Although the use of laser scanner and the 3D exploration of the blade on the hub is an important step, the method presented here is generated by other steps, and the most interesting aspect is the comparative procedure that must be established in each wind farm analyzing the SCADA data and identifying the BIC turbine to set a reference for pitch error correction of the other turbines. This benchmarking-based approach is the fundamental perspective. Out of this benchmarking method, the main drawback of other techniques is that they do not obtain absolute pitch angle values which are able to improve the energy production and avoid fatigue. Although aerodynamic imbalance can be removed by establishing same relative pitch angles, the correct implementation of the certified pitch of the manufacturer is not assured. The benchmark approach offers an independent way to establish a new reliable reference in the wind farm.

This new technique was used onshore, but it is also suitable for offshore locations, where the only fixed platform available nearby is located at the tower base. Other techniques require completing laser measurements from distant places orthogonal to the rotor, and the lack of a fixed point in front of the rotor makes other techniques simply not feasible. In general, the method presented here requires less space under the turbine and, therefore, the necessary time and workload can be reduced in complex terrains or in offshore platforms. As mentioned above [8], laser scans from long distances is time-consuming, due to the trade-off between accuracy and distance.

An additional reduction of complexity is also given by the lack of necessity of information from the turbine manufacturer because the BIC turbine of the wind farm is mimicked in the benchmark procedure. Other techniques consider that the blade axes are coplanar, an erroneous supposition due to manufacturing and assembly tolerances, and it adds complexity to these techniques. In our case, the merging of the blade axes and the rotor axis can be studied, and, according to our experience, the best merging has been found in the BIC turbines. This fact simplifies the theoretical background of the method and gives coherence to the benchmarking perspective. In addition, the misalignment is measured in idling conditions gaining also simplicity without energy losses.

Thus, the obtained information is another advantage of this technique. The reference is not the manufacturer's original configuration, but the real BIC turbine and the rotor plane configuration of the three blades. In this way, the original design and the results of this benchmarking technique can be compared by the manufacturers to improve their future wind turbines and the on-field construction of them, mainly in the final step of the implementation of the blades on the hub.

Future work aims at improving the diagnostic capability of this preliminary performance assessment. For the time being, only turbines affected by pitch angle misalignment are identified, whereas, by understanding turbine dynamics, it is also possible to gain further insight into the conditions of the analyzed turbines before taking the actual laser measurements.

In addition, the identification of defective anemometers in wind turbines is very important for the implementation of this pitch correction technique, since the cause of the low production of the turbine can be the pitch misalignment or a defective measurement by the anemometer, with a consequent deviation in the real power curve. Therefore, pre-evaluation of the SCADA data of the wind farm and a comparative analysis of the anemometers are essential to ensuring that all anemometers are working properly, thus accepting the plausible hypothesis that pitch misalignment explains the bad behavior of a given turbine.

In this way, the objective of future research is to better understand the improper performance of anemometers through the use of statistical methods, including Pearson correlation, *RMSE*, and standard deviation in Taylor Diagrams to visualize all these parameters in a single plot and to identify the deviating anemometers from the group [49]. The authors are developing a novel method for the identification of defective anemometers based on a Kernel Multidimensional Probability Density Function that improves on classical unidimensional methods, such as the Kolmogorov–Smirnov test [50]. This method will be applied in a bidimensional manner for the zonal and meridional components (U, V) of wind and will identify the Best in Class anemometer of the wind farm, resulting in an initial identification of the defective anemometers.

Author Contributions: Conceptualization, U.E. and I.E.; Methodology, U.E.; Software, U.E. and I.E.; Validation, U.E. and I.E.; Investigation, A.U. and G.I.-B.; Writing—Original Draft Preparation, Unai Elosegui and Igor Egana; Writing—Review and Editing, all authors; Supervision, all authors; Project Administration, U.E., A.U., and G.I.-B.; Funding Acquisition, U.E. and A.U.

Funding: This work is funded by the Council of Gipuzkoa (Gipuzkoako Foru Aldundia, Basque Country, Spain) within the R&D subsidy for the project DIANEMOS on the identification of defective anemometers in wind turbines, and the University of the Basque Country (UPV/EHU, GIU 17/002).

Acknowledgments: The authors would like to show gratitude to Jose Luis Azpeitia from the Centre IK4-Tekniker for his valuable contribution with FAST simulations.

Conflicts of Interest: The authors declare no conflict of interest.

Abbreviations

The following abbreviations are used in this manuscript:

AEP	Annual energy production
PCR	Power curve ratio of a given turbine with respect to the BIC turbine
BIC	Best in class turbine
PCR_{gap}	Power curve ratio gap in percent
C_p	Power coefficient
U	Wind speed
T	Axis torque
Ω	Angular velocity of the rotor
Ω_{low}	Low rotor speed
Ω_N	Nominal rotor speed
β	Pitch angle
TSR	Tip speed ratio
FAST	An aeroelastic computer-aided engineering tool for horizontal axis wind turbines
$RMSE$	Root mean square error

References

1. Schubel, P.; Crossley, R.; Boateng, E.; Hutchinson, J. Review of structural health and cure monitoring techniques for large wind turbine blades. *Renew. Energy* **2013**, *51*, 113–123. [CrossRef]
2. Veers, P.S.; Ashwill, T.D.; Sutherland, H.J.; Laird, D.L.; Lobitz, D.W.; Griffin, D.A.; Mandell, J.F.; Musial, W.D.; Jackson, K.; Zuteck, M.; et al. Trends in the design, manufacture and evaluation of wind turbine blades. *Wind Energy* **2003**, *6*, 245–259. [CrossRef]
3. McGugan, M.; Pereira, G.; Sørensen, B.F.; Toftegaard, H.; Branner, K. Damage tolerance and structural monitoring for wind turbine blades. *Philos. Trans. R. Soc. A* **2015**, *373*, 20140077. [CrossRef] [PubMed]
4. Berlin Wind. Available online: http://www.berlinwind.com/ (accessed on 15 March 2017).
5. cp.max Rotortechnik. Available online: http://cpmax.com/en/cpmax-rotortechnik.html (accessed on 15 March 2017).
6. Heilmann, C.; Grunwald, A.; Melsheimer, M. Blade-root based multi-camera system for induced blade twist measurements at wind turbine rotors during operation. In Proceedings of the 13th German Wind Energy Conference, Bremen, Germany, 17–18 October 2017; p. 38.
7. Wohlert, T. *Measuring Rotor Blades With Lasers*; Technical Report; WindTech International: Groningen, The Netherlands, 2016.
8. Wind-Consult. Available online: https://www.wind-consult.de/cms/ (accessed on 15 October 2018).
9. FARO. Available online: http://www.faro.com/products/3d-surveying/laser-scanner-faro-focus-3d/overview (accessed on 15 March 2017).
10. Leica Geosystems. Available online: http://www.leica-geosystems.us/en/HDS-Laser-Scanners-SW_5570.htm (accessed on 15 March 2017).
11. Hispavista Labs. Available online: https://hispavistalabs.com/ (accessed on 15 March 2017).

12. MAXWIND. Available online: https://www.maxwindtech.com/ (accessed on 15 March 2017).

13. Cheng, W.W.L.X.F.; Cichang, C. Wind Rotor Blade Measurement with Laser Tracker. *Aerosp. Manuf. Technol.* **2009**, *6*, 008.

14. Yang, S.; Allen, M.S. Output-only modal analysis using continuous-scan laser Doppler vibrometry and application to a 20 kW wind turbine. *Mech. Syst. Signal Process.* **2012**, *31*, 228–245. [CrossRef]

15. Ghoshal, A.; Sundaresan, M.J.; Schulz, M.J.; Pai, P.F. Structural health monitoring techniques for wind turbine blades. *J. Wind Eng. Ind. Aerodyn.* **2000**, *85*, 309–324. [CrossRef]

16. Elosegui, U.; Elosegui, J. METHOD FOR CALCULATING AND CORRECTING THE ANGLE OF ATTACK IN A WIND TURBINE FARM. PCT/ES/2013/070752. Available online: https://patentscope.wipo.int/search/en/detail.jsf?docId=WO2014068162&redirectedID=true (accessed on 15 November 2018).

17. Elosegui, U.; Ulazia, A. Novel on-field method for pitch error correction in wind turbines. *Energy Procedia* **2017**, *142*, 9–16. [CrossRef]

18. Moroz, E.; Pierce, K. Methods and Apparatus for Reduction of Asymmetric Rotor Loads in Wind Turbines. EP1612413-A2, 30 June 2004.

19. Grosse-Schwiep, M.; Piechel, J.; Luhmann, T. Measurement of rotor blade deformations of wind energy converters with laser scanners. *J. Phys. Conf. Ser.* **2014**, *524*, 012067. [CrossRef]

20. IWES. Rotor Blade Testing. Available online: http://www.iwes.fraunhofer.de/en/labore/Rotor_Blade.html (accessed on 15 November 2018).

21. Márquez, F.P.G.; Tobias, A.M.; Pérez, J.M.P.; Papaelias, M. Condition monitoring of wind turbines: Techniques and methods. *Renew. Energy* **2012**, *46*, 169–178. [CrossRef]

22. Papadopoulos, K.; Morfiadakis, E.; Philippidis, T.; Lekou, D. Assessment of the strain gauge technique for measurement of wind turbine blade loads. *Wind Energy* **2000**, *3*, 35–65. [CrossRef]

23. Paulsen, U.S.; Erne, O.; Möller, T.; Sanow, G.; Schmidt, T. Wind turbine operational and emergency stop measurements using point tracking videogrammetry. In Proceedings of the SEM Annual Conference and Exposition on Experimental and Applied Mechanics, Albuquerque, NM, USA, 1–4 June 2009.

24. Yang, Y.; Guo, Z.; Song, Q.; Zhang, Y.; Li, Q. Effect of Blade Pitch Angle on the Aerodynamic Characteristics of a Straight-bladed Vertical Axis Wind Turbine Based on Experiments and Simulations. *Energies* **2018**, *11*, 1514. [CrossRef]

25. Petković, D.; Ćojbašič, Ž.; Nikolić, V. Adaptive neuro-fuzzy approach for wind turbine power coefficient estimation. *Renew. Sustain. Energy Rev.* **2013**, *28*, 191–195. [CrossRef]

26. Petković, D.; Ćojbašič, Ž.; Nikolić, V.; Shamshirband, S.; Kiah, M.L.M.; Anuar, N.B.; Wahab, A.W.A. Adaptive neuro-fuzzy maximal power extraction of wind turbine with continuously variable transmission. *Energy* **2014**, *64*, 868–874. [CrossRef]

27. Petković, D.; Pavlović, N.T.; Ćojbašič, Ž. Wind farm efficiency by adaptive neuro-fuzzy strategy. *Int. J. Electr. Power Energy Syst.* **2016**, *81*, 215–221. [CrossRef]

28. Nikolić, V.; Mitić, V.V.; Kocić, L.; Petković, D. Wind speed parameters sensitivity analysis based on fractals and neuro-fuzzy selection technique. *Knowl. Inf. Syst.* **2017**, *52*, 255–265. [CrossRef]

29. Kusiak, A.; Verma, A. A data-driven approach for monitoring blade pitch faults in wind turbines. *IEEE Trans. Sustain. Energy* **2011**, *2*, 87–96. [CrossRef]

30. Yin, S.; Wang, G.; Karimi, H.R. Data-driven design of robust fault detection system for wind turbines. *Mechatronics* **2014**, *24*, 298–306. [CrossRef]

31. Yang, X.; Li, J.; Liu, W.; Guo, P. Petri net model and reliability evaluation for wind turbine hydraulic variable pitch systems. *Energies* **2011**, *4*, 978–997. [CrossRef]

32. Naik, K.A.; Gupta, C.P. Output Power Smoothing and Voltage Regulation of a Fixed Speed Wind Generator in the Partial Load Region Using STATCOM and a Pitch Angle Controller. *Energies* **2017**, *11*, 58. [CrossRef]

33. Olondriz, J.; Elorza, I.; Jugo, J.; Alonso-Quesada, S.; Pujana-Arrese, A. An Advanced Control Technique for Floating Offshore Wind Turbines Based on More Compact Barge Platforms. *Energies* **2018**, *11*, 1187. [CrossRef]

34. Kusiak, A.; Zheng, H. Optimization of wind turbine energy and power factor with an evolutionary computation algorithm. *Energy* **2010**, *35*, 1324–1332. [CrossRef]

35. Rehman, S.; Alam, M.M.; Alhems, L.M.; Rafique, M.M. Horizontal Axis Wind Turbine Blade Design Methodologies for Efficiency Enhancement—A Review. *Energies* **2018**, *11*, 506. [CrossRef]

36. Fernandez-Gamiz, U.; Zulueta, E.; Boyano, A.; Ansoategui, I.; Uriarte, I. Five megawatt wind turbine power output improvements by passive flow control devices. *Energies* **2017**, *10*, 742. [CrossRef]
37. Astolfi, D.; Castellani, F.; Terzi, L. Wind Turbine Power Curve Upgrades. *Energies* **2018**, *11*, 1300. [CrossRef]
38. Poore, R.; Lettenmaier, T. *Alternative Design Study Report: WindPACT Advanced Wind Turbine Drive Train Designs Study; November 1, 2000–February 28, 2002*; Technical Report; National Renewable Energy Lab. (NREL): Golden, CO, USA, 2003.
39. Burton, T.; Jenkins, N.; Sharpe, D.; Bossanyi, E. *Wind Energy Handbook*; John Wiley & Sons: Hoboken, NJ, USA, 2011.
40. Imran, R.M.; Hussain, D.; Chowdhry, B.S. Parameterized Disturbance Observer Based Controller to Reduce Cyclic Loads of Wind Turbine. *Energies* **2018**, *11*, 1296. [CrossRef]
41. Manwell, J.F.; McGowan, J.G.; Rogers, A.L. *Wind Energy Explained: Theory, Design and Application*; John Wiley & Sons: Hoboken, NJ, USA, 2010.
42. Myrent, N.J.; Bilal, N.; Adams, D.; Griffith, D.T. Aerodynamic sensitivity analysis of rotor imbalance and shear web disbond detection strategies for offshore structural health prognostics management of wind turbine blades. In Proceedings of the 32nd ASME Wind Energy Symposium, National Harbor, MD, USA, 13–17 January 2014; p. 0714.
43. Schubel, P.J.; Crossley, R.J. Wind turbine blade design. *Energies* **2012**, *5*, 3425–3449. [CrossRef]
44. *Wind Energy Generation Systems. Structural Design*; Technical Report, IEC 61400-1-7; International Electrotechnical Commission: Geneva, Switzerland, 2017.
45. El-Sheimy, N. An overview of mobile mapping systems. In Proceedings of the FIG Working Week 2005, Cairo, Egypt, 16–21 April 2005; pp. 16–21.
46. Boeder, V. HCU-HMSS: A Multi Sensor System in Hydrographic Applications. In Proceedings of the 2nd International Conference on Machine Control & Guidance, Bonn, Germany, 9–11 March 2010; pp. 65–74.
47. *Wind Energy Generation Systems. Power Performance Measurements of Electricity Producing Wind Turbines*; Technical report, IEC 61400-12-1; International Electrotechnical Commission: Geneva, Switzerland, 2017.
48. Red Eléctrica de España. Available online: https://www.esios.ree.es/es?locale=es (accessed on 1 October 2018).
49. Taylor, K.E. Summarizing multiple aspects of model performance in a single diagram. *J. Geophys. Res. Atmos.* **2001**, *106*, 7183–7192. [CrossRef]
50. Lopez-Novoa, U.; Sáenz, J.; Mendiburu, A.; Miguel-Alonso, J. An efficient implementation of kernel density estimation for multi-core and many-core architectures. *Int. J. High Perform. Comput. Appl.* **2015**, *29*, 331–347. [CrossRef]

Article

Ice Detection Model of Wind Turbine Blades Based on Random Forest Classifier

Lijun Zhang [1,*], Kai Liu [1], Yufeng Wang [1] and Zachary Bosire Omariba [1,2]

[1]   National Center for Materials Service Safety, University of Science and Technology Beijing, Beijing 100083, China; s20161185@xs.ustb.edu.cn (K.L.); s20171176@xs.ustb.edu.cn (Y.W.); zomariba@egerton.ac.ke (Z.B.O.)

[2]   Computer Science Department, Egerton University, Egerton 20115, Kenya

*   Correspondence: ljzhang@ustb.edu.cn; Tel.: +86-10-6232-1017

Received: 14 September 2018; Accepted: 21 September 2018; Published: 25 September 2018

Abstract: When wind turbine blades are icing, the output power of a wind turbine tends to reduce, thus informing the selection of two basic variables of wind speed and power. Then other features, such as the degree of power deviation from the power curve fitted by normal sample data, are extracted to build the model based on the random forest classifier with the confusion matrix for result assessment. The model indicates that it has high accuracy and good generalization ability verified with the data from the China Industrial Big Data Innovation Competition. This study looks at ice detection on wind turbine blades using supervisory control and data acquisition (SCADA) data and thereafter a model based on the random forest classifier is proposed. Compared with other classification models, the model based on the random forest classifier is more accurate and more efficient in terms of computing capabilities, making it more suitable for the practical application on ice detection.

Keywords: ice detection; wind turbine blades; SCADA data; random forest classifier; power curve; confusion matrix

1. Introduction

With the gradual depletion of traditional fossil fuels such as coal, oil and natural gas, the development and use of new energy such as wind power has received increasing attention making wind power one of the fastest growing energy sources in the world [1]. In 2017, the newly installed capacity of wind power worldwide reached 52,492 MW, and the cumulative installed capacity reached 539,123 MW. Among them, the newly installed capacity of wind power in China accounted for 37%, and the cumulative installed capacity accounted for 35%. The newly installed capacity of wind power accounts for more than 15% of the total installed capacity in recent years, and the cumulative installed capacity accounts for a steady increase. The Global Wind Energy Council (GWEC) predicts that as costs drop and the market begins to recover at the end of this decade; global wind power installed capacity will increase by more than 50% over the next five years. According to GWEC, as countries around the world develop renewable energy sources to achieve emission reduction targets, wind energy costs continue to decline, and by the end of 2022 installed capacity global wind power is expected to increase to 840 GW [2].

Wind power however faces some challenges that restrict its development with cost making of the list of the important issue. According to the study of Department of Energy (DOE), United States, 20% revenue growth of wind farms by 2030 will come from improvement of wind turbine working status and reduction of maintenance costs. Using the appropriate maintenance and maintenance strategy to reduce the cost of operation and maintenance is an important way to increase wind farm income [3].

Land-based wind farms are established mostly based on high altitude mountainous areas. These regions experience low temperature and high humidity, which makes it possible for wind turbine blades to form varying degrees of icing easily, especially in winter. However, there are shutdown events caused by wind turbine blade icing, which seriously threatens the normal operation of wind power plants. Any wind turbine blade icing will cause power loss, mechanical failure, equipment failure, and safety issues [4]. The freezing of wind turbine blades changes the aerodynamic performance of the blades, which yield into power generation loss. Equally, the uneven distribution of ice from the blade changes the original mass distribution, making the wind turbine to run unstably, and causing severe damage to the blade in varying degrees, which not only lead to huge economic losses, but also have serious security risks [5]. Therefore, a reliable detection method for icing wind turbine blade is very important, especially in the early stage icing detection.

Now there are many standards and guidelines that have been developed by the IEC (International Electro technical Commission), and it has helped us a lot in analyzing turbine faults [6]. For the problem of ice detection in the blades, the existing methods mainly use the mechanism of icing to conduct theoretical analysis and research and establish the physical model of icing, then according to the monitoring data, make a judgment whether the wind turbine blades are frozen at the current moment. Davies et al. studied three methods of creating a power threshold curve to distinguish the ice growth period from the non-icing period to identify the power loss caused by icing [7]. Wang et al. proposed a numerical simulation method for three-dimensional wind turbine blade icing and compared it with experimental results to verify the effectiveness of the method [8]. Shu et al. studied the characteristics of leaf icing and the severity of icing on the power characteristics of wind turbines under natural icing conditions [9]. Blasco et al. performed a quantitative analysis of the power loss of a representative 1.5 MW wind turbine under various icing conditions, attempting to reduce the loss of wind farms in cold regions by formulating some control strategies [10]. Based on the analysis of supervisory control and data acquisition (SCADA) data, Li et al. proposed a method for detection of blade icing based on logistic regression [11]. Aral et al. proposes and demonstrates Phase-based Motion Estimation (PME) and a motion magnification algorithm to perform non-contact structural damage detection of a wind turbine blade [12]. Yu et al. developed a simple method to detect damage based on a discrete mathematical model for fan blades using changes in natural frequencies combined with a fluid-structure analysis [13]. The above research generally requires additional sensor placement for wind turbine blades. The disadvantages such as inconvenient practical application and increased wind farm operation and maintenance cost make them unable to be widely used in practice.

Vibration signal analysis [14] and SCADA system data analysis are two different aspects. The former pays more attention to the analysis of the equipment mechanism, while the latter tends to analyze the data. Both have their own advantages. Wind turbine blades work at high altitude, and they are inconvenient to measure the vibration acceleration signal offline. The amount of acceleration signal on the line is larger, which is inconvenient to transmit and store. Therefore, more and more people are committed to using SCADA data to predict and diagnose wind turbine faults. The SCADA system is the most widely used and technologically advanced data acquisition and monitoring system, in fault diagnosis of a large amount of wind power equipment [14–19]. This system collects environmental parameters and operating parameters of wind power equipment, which can fully characterize the operational status of the wind turbine. More and more people use it for data modeling and analysis to mine information of equipment fault and blade icing detection, etc. [20–24].

When the wind turbine blade is early frozen and detected by the model in this paper, this warning information will be fed back to the wind farm owners (or managers, or controllers). At this time, the wind farm has not experienced a serious accident. This early warning information of early icing gave them time to deal with the icing of the wind turbine blades. During this time, they could use other methods to reasonably arrange when to take measures such as deicing the blades, to prevent loss and damage due to severe icing of wind turbine blades.

This paper analyzes the SCADA data of a wind farm, combines the mechanism analysis and data analysis of the wind turbine icing to extract features that are sensitive to wind turbine icing, and then uses a random forest-based classification algorithm to achieve the detection of wind turbine blade icing. The first section introduces the related theories of icing of wind turbine blades and the research ideas of this paper; the second section introduces the related theories of the model based on the random forest classifier and the model assessment method selected for this paper; the third section is the data preprocessing which extracts the sensitive characteristics of early icing of wind turbine blades by analyzing the SCADA data; the fourth section is optimization and comparison of the model based on the random forest classifier with the results of other classifiers; and the last section is the conclusion.

2. Materials and Methods

2.1. Ice Detection on Wind Turbine Blades

2.1.1. Theory and Process of Icing

Icing is a physical phenomenon with a complete and specialized theory and research system. Blade icing of wind turbines is a type of atmospheric icing. The international standard ISO12494: 2017 [25] describes in detail the definition, scope, classification, principle, characteristics, and effects of such an icing. For wind turbines, atmospheric icing refers to the process of icing in the air frozen or adhered to objects exposed in the atmosphere under certain atmospheric conditions, including water droplets, rain, drizzle, snow, and other forms.

There are three forms of blade icing: cloud ice, sedimentation ice and accumulation of frost. Cloud ice refers to icing condensed from sub-cooled water droplets floating in clouds; sedimentation ice refers to icing caused by freezing rain or wet snow under low temperature conditions; frost accumulation refers to the direct phase change of water vapor. The icing process usually occurs at low temperatures. Among them, cloud ice and sedimentation ice are more common in wind turbine icing, and once it occurs, it will have a serious impact on the wind turbine and cause more damage.

2.1.2. Ice Detection Method

Ice detection analysis on blades is generally composed of several parts such as physical principle analysis, icing process analysis, feature extraction, detection model establishment and result presentation. This paper adopts blade detection model construction process based on a random forest classification, as shown in Figure 1.

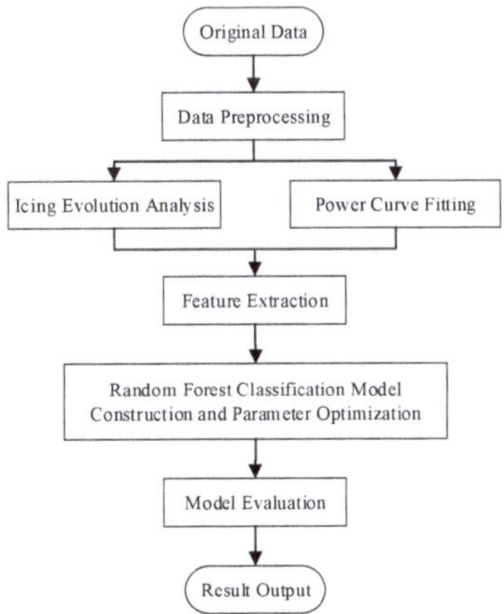

Figure 1. Flow chart of model construction.

Severe icing detection during the actual operation of the wind turbine is easily established, but automatically deicing by the wind turbine deicing system is also a challenge. However, the icing of the wind turbine blade is a slow process. In the early days of icing, the impact on the wind turbine is generally small and difficult to find. Besides, early icing will cause certain changes to the shape of the blades, which will cause water droplets in the atmosphere to stick to and freeze at the surface of the blades. Eventually, the probability of serious icing is greatly increased. The treatment of early icing is easier and has less impact on the wind turbine. It has a certain early warning effect of the occurrence of severe icing. Therefore, the detection of early icing is very important.

2.2. Model Based on Random Forest Classifier

2.2.1. Random Forest Classifier

The random forest [26] is a machine learning algorithm first published in 2001 by Breiman, L. which combines bagging ensemble learning theory [27] proposed in 1996, with the stochastic subspace method proposed by Ho, T. in 1998 [28]. This model adopts bootstrapping re-sampling technology to randomly select n samples from the original training sample set N and put it back randomly to generate a new training sample set to train a decision tree. Then, the above steps generate m decision trees to form a random forest. The classification results of new data-based upon the score formed by how many classification trees vote. Its essence is an improvement in the decision tree algorithm, with multiple decision trees merged together. The establishment of each tree depends on the independently extracted samples. Figure 2 shows the basic structure of the random forest classifier.

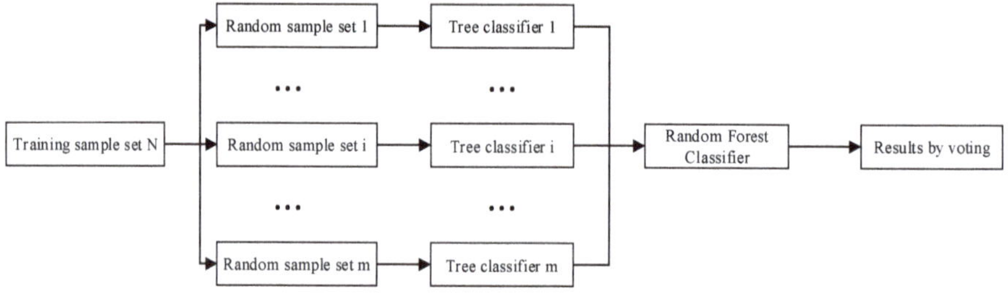

Figure 2. Basic structure of random forest classifier.

The classification ability of a single tree may be small, but after randomly generating many decision trees, a test sample through the statistics of the classification of each tree is selected to obtain the most likely classification.

1. The steps for the basic construction process of a random forest are: Use the bootstrapping method from the original training set to select n samples randomly in m times to generate m training sets.
2. For the newly generated m training sets, train m decision tree classification models.
3. For a single tree, every time a new node based upon the information gain or information gain ratio or the Gini is split to select the best split method.
4. Split each tree according to step 3 until the training sample is correctly classified at a certain node or reaches the maximum depth of the tree.
5. Organize the resulting multiple decision trees into the random forest classifier and the final classification results are determined by voting.

The randomness of each tree corresponding to the sampling of the training set and the way in which part of the features are selected when splitting to form a new node. The random forest does not need to be pruned and almost no over fitting occurs, and have good tolerance for noise and outliers, high stability, and strong generalization ability. In addition, the random forest is suitable for parallel computing, and even for large samples and high latitude data, they have the higher training speed and the achieve efficient calculation.

This paper used a model based on the random forest classifier to identify early icing data from normal data to achieve the goal of predicting early icing failure, and then to determine if there would be icing failure in the next period.

2.2.2. Model Assessment Method

The confusion matrix [29] is a classical method for evaluating the results of classification models. Table 1 shows the confusion matrix representation.

Table 1. Confusion matrix representation.

Confusion Matrix	Prediction of Icing	Prediction of Non-Icing
Actual icing	*TP*	*FN*
Actual non-icing	*FP*	*TN*

where: *TP* indicates the proportion of all actual icing samples predicted to be icing samples; *TN* indicates the proportion of all actual non-icing samples predicted to be non-icing samples; *FP* indicates the proportion of all actual non-icing samples predicted to be icing samples; *FN* indicates the proportion of all actual icing samples predicted to be non-icing samples.

In addition, based upon the confusion matrix the precision of the test results and the recall rate assessed further to evaluate the model classification results [26].

$$Precision = \frac{TP}{TP + FP} \tag{1}$$

$$Recall = \frac{TP}{TP + FN} \tag{2}$$

3. Data Preprocessing

3.1. Data Sources and Introduction

In this paper, the test data are driven from the first China Industrial Big Data Innovation Competition [30], which contains two wind turbines SCADA data in a wind farm provided by Goldwind for predicting icing failures on blades. The SCADA data of each wind turbine contains 28 variables such as the time stamp, operating condition parameters, environmental parameters, and status parameters. The acquisition time was two months and with the sample size of about 580,000. Table 2 shows the statistical information of SCADA data. In addition, more detailed information of SCADA data can be seen in the Appendix A, Table A1.

Table 2. Statistical information of supervisory control and data acquisition (SCADA) data.

Data Set	Sample Size	Time Range
Wind turbine 15#	393,886	November & December 2015
Wind turbine 21#	190,494	November 2015

In addition, the organizers of the event conducted preliminary processing on the data, which removed severely frozen data and made the data not continuous; the data was also standardized, thus lost the physical meaning of the original data. Standardization means that making the mean of every variable in data is 0 and the variance is 1. The contest organizer has already set the labels for the data-icing and non-icing (due to the authority of the data owner and the supervisor, the accuracy of the data label is guaranteed, so the basis for judging whether the data is frozen or not is also credible); we only need to process the data that has been tagged.

3.2. Features Extraction

Some indicators in the raw data given by the contest organizers are sensitive to icing, and some indicators are almost not related to icing. So, the first step in this paper on the data is to pick out the icing-sensitive indicators from the raw data indicators. However, relying solely on these indicators does not well identify early icing data from non-icing data, this paper further processed the data and obtained some better indicators of icing and non-icing. In general, it is through the screening and supplementation of indicators to achieve better characterization of early icing with fewer features, which not only reduces the running time of the model but also gives better results.

This section will introduce the process of data preprocessing, including the screening of basic features and the construction of other features, with giving some figures to make features more intuitively judgment—whether it is easier to distinguish between icing and non-icing.

Extraction features, especially quantitative features [31] are very essential for the fault diagnosis of equipment. On the one hand, because the inertia of the wind turbine blade will reduce the correlation between the instantaneous power and the instantaneous wind speed, taking the average value from the data over a certain time span can reduce the inertial effect to some extent. On the other hand, in the original data, about 8 samples are collected every minute, but because the data provider has deleted some data, the sample interval time in the data is not fixed. So, the data are resampled in one-minute intervals, the specific process is as follows. According to the timestamp, the SCADA data grouped every minute for the time span, and then the mean of each group sample is taken as the new sample characteristics.

$$V = \frac{1}{n}\sum_{i=1}^{n} wind\_speed_i \tag{3}$$

where V is the average wind speed—the new sample characteristics; n is the number of wind_speed in one minute. The solutions of average power P and other new variables are the same as Equation (3).

Then the data is filtered.

1. Filter unspecified data. In the given raw data set, the data covers normal sample data, icing sample data, and other unspecified data, according to the status tag. Unspecified data will affect the classification of normal sample data/icing sample data, due to the uncertainty of its information; it will be classified as invalid data.

2. Filter samples below 80% of full power. Taking wind turbine 21# as an example, it can be found that when the wind turbine is at more than 80% full power icing status data does not exist by comparison of the original instantaneous power-wind speed scatter plot (shown in Figure 3), and the processed average power-wind speed scatter plot (shown in Figure 4). In other words, the wind turbine power cannot reach 80% of full power after the blade's freeze. Filter the samples below 80% of full power can make it easier to identify early icing data.

In following figures, the green points are in the wind turbine normal state and the red points are in the icing state. The blue lines in Figures 3 and 4 represent the dividing lines representing 80% of full power.

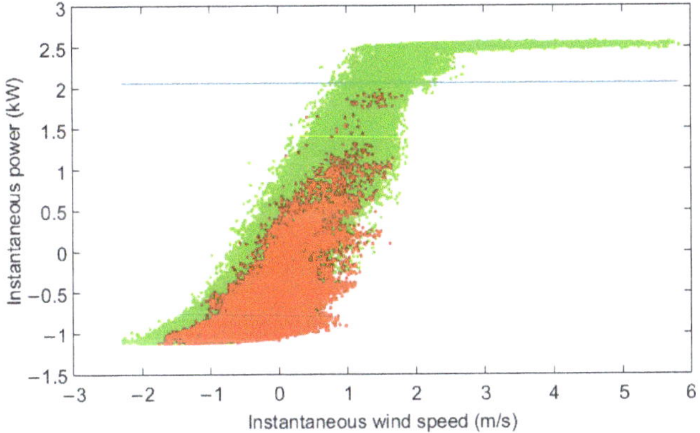

Figure 3. Original instantaneous power-wind speed scatter plot.

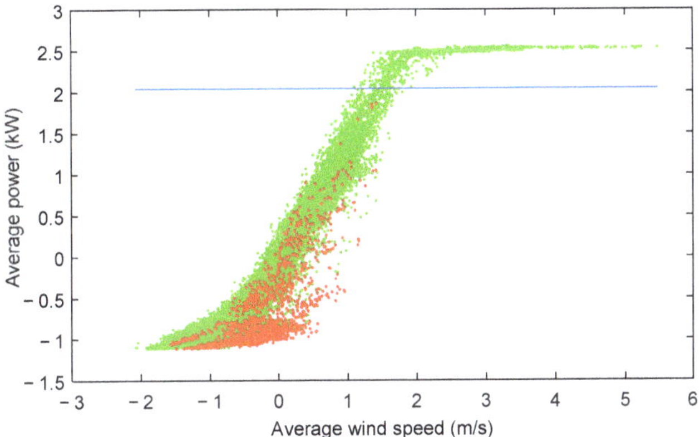

Figure 4. Average power-wind speed scatter plot.

Filter unspecified data and samples below 80% of full power and normalize the remaining data (making the scale from 0 to 1), then plot the average power and average wind speed as a scatter plot (shown in Figure 5).

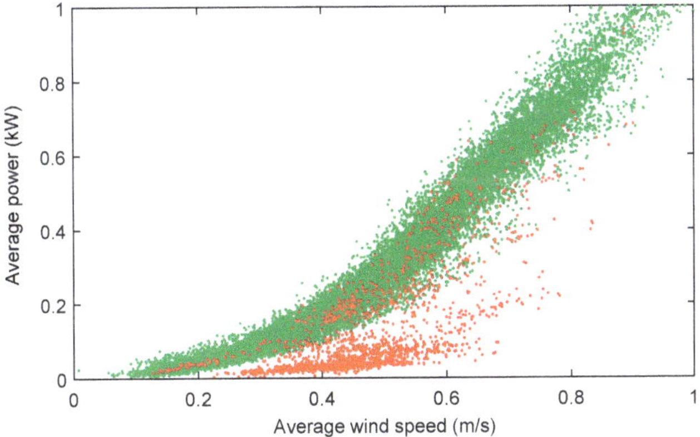

Figure 5. Average power-wind speed with removing more than 80% full power data scatter plot.

Wind turbines are devices that convert wind energy into mechanical energy and then into electrical energy, where wind speed and power are regarded as the two basic features of icing prediction. When the blades freeze, the shape and aerodynamic characteristics of the blades will change, reducing the power output. Therefore, when the wind turbine blades freeze, the relationship of the output power and the wind speed will be changed.

In the non-icing condition, the wind machine operates according to the wind turbine power characteristic curve in the normal mode (the green part of Figure 5). After the icing formation, the actual operation state of the wind turbine will deviate, and the power cannot reach the rated power. When the normal state sample data is used, the abnormal point eliminated, the power characteristic curve of the wind turbine is fitted to obtain a baseline model of the power characteristic curve [32], and then this model is used to predict the output power at the corresponding wind speed. The baseline model obtained by curve fitting is shown in Figure 6.

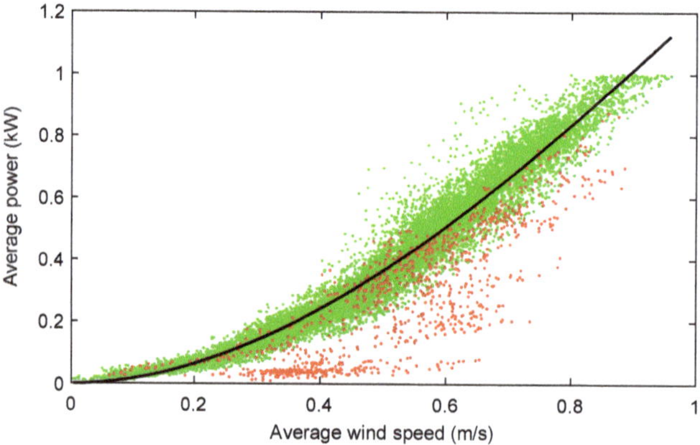

Figure 6. Fitted power curve.

From Figure 6 the icing sample is more deviating from the baseline model than the normal sample, thus constructing another feature of icing prediction, which can distinguish then better: the degree of deviation from the output power.

$$C = \frac{P_{pre} - P_{real}}{P_{pre}} = 1 - \frac{P_{real}}{P_{pre}} \tag{4}$$

where P_{real} is the actual measured output power and P_{pre} is the output power estimated by the actual wind speed and power curve.

After calculating the power degree by Equation (4), to facilitate visual observation of whether the variable is helpful for model classification, we draw a figure about relationship between the power degree and the average wind speed, as shown in Figure 7. As can be seen from Figure 7, there are more red dots (icing samples) that are distinguished from green dots (non-icing samples).

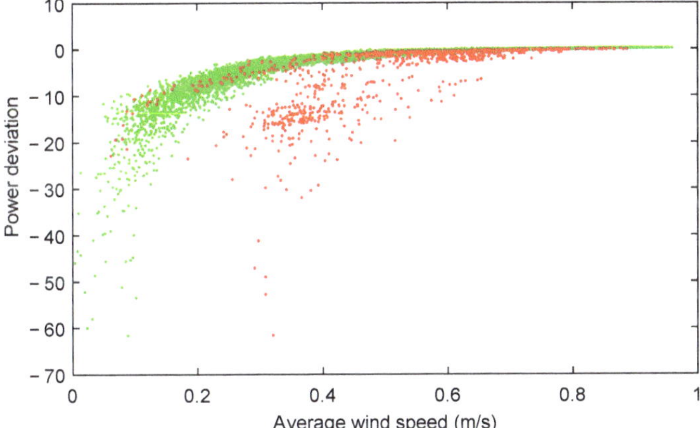

Figure 7. Relationship between degree of deviation and average wind speed.

In the early stage of icing, the operation state of the wind turbine is similar to the normal state, and it is difficult to separate the icing state from the normal state. However, the detection of early icing conditions is a very important process, and the healthy operation of the wind turbine unit is of utmost importance. It can minimize the loss to the unit due to icing on the blades. Because icing is a cumulative process, instantaneous characteristics such as the wind speed, the power, and the degree of deviation make it difficult to characterize fully icing conditions, especially in the early icing part. Therefore, it is necessary to analyze the evolution of the icing process and extract features that can characterize icing changes to better distinguish early icing conditions and achieve early icing prediction.

This paper mainly extracts features of early icing based upon the characteristics of degree of deviation. The icing process of the wind turbine contains certain periodicity, thus calling for serialization of the original data. Calculate the average rate of change (ΔC) of the degree of deviation at the corresponding time in each time segment.

$$\Delta C = \frac{C_t - C_{t_1}}{t - t_1} \tag{5}$$

where ΔC represents the average rate of change of the current degree of deviation; C_t represents the current degree of deviation; C_{t_1} represents the degree of deviation from the previous moment; $t - t_1$ represents the time span (t is the current time after digitization and t_1 is the previous time after digitization).

Then, according to the sliding window method, take ten minutes as the window length and one minute as the moving step length to obtain the maximum value maxC to the degree of deviation and the cumulative value sumΔC of the ΔC within 10 min before the current time.

First, this paper selects two basic features from the 28-dimensional features in the given data, and then adds four additional features based upon the mechanism of the wind turbine operation and icing. Finally, Table 3 represents the six groups of icing prediction features obtained.

Table 3. Features of ice detection on wind turbine blades.

Features	Description
V	Average wind speed
P	Average power
C	Degree of deviation of P
ΔC	Average rate of change of C
maxC	Maximum value of C
sumΔC	Cumulative value of ΔC

4. Results

4.1. Classification Model Optimization

This paper mainly adjusts two important parameters of the model based on the random forest classifier to optimize the model: the number of trees and the maximum depth of the tree. Divide 70% of sample data of wind turbine 15# into the training set, 30% of sample data into the test set, adjust the number of decision tree and maximum depth of a tree in the random forest classifier, and then calculate the output from the model. The confusion matrix is chosen as the evaluation index, and the calculation results are shown in Table 4.

Table 4. Results of the model based on random forest classifier optimization.

Confusion Matrix/%		Number of Trees			
		10	20	30	40
Maximum depth of the tree	5	$\begin{pmatrix} 64.8 & 35.2 \\ 17.4 & 82.8 \end{pmatrix}$	$\begin{pmatrix} 66.7 & 33.3 \\ 15.6 & 84.4 \end{pmatrix}$	$\begin{pmatrix} 67.8 & 32.2 \\ 14.3 & 85.7 \end{pmatrix}$	$\begin{pmatrix} 68.5 & 31.5 \\ 13.1 & 86.9 \end{pmatrix}$
	10	$\begin{pmatrix} 77.6 & 22.4 \\ 12.3 & 87.7 \end{pmatrix}$	$\begin{pmatrix} 80.1 & 19.9 \\ 8.8 & 91.2 \end{pmatrix}$	$\begin{pmatrix} 84.2 & 15.8 \\ 7.9 & 92.1 \end{pmatrix}$	$\begin{pmatrix} 88.5 & 11.5 \\ 5.6 & 94.4 \end{pmatrix}$
	15	$\begin{pmatrix} 86.7 & 13.3 \\ 9.8 & 90.2 \end{pmatrix}$	$\begin{pmatrix} 89.8 & 10.2 \\ 4.4 & 95.6 \end{pmatrix}$	$\begin{pmatrix} 90.6 & 9.4 \\ 3.2 & 96.8 \end{pmatrix}$	$\begin{pmatrix} 91.0 & 9.0 \\ 3.5 & 96.5 \end{pmatrix}$
	20	$\begin{pmatrix} 90.6 & 9.4 \\ 5.6 & 94.4 \end{pmatrix}$	$\begin{pmatrix} 91.3 & 8.7 \\ 2.4 & 97.6 \end{pmatrix}$	$\begin{pmatrix} 91.8 & 8.2 \\ 2.5 & 97.5 \end{pmatrix}$	$\begin{pmatrix} 91.9 & 8.1 \\ 2.6 & 97.4 \end{pmatrix}$
	25	$\begin{pmatrix} 91.2 & 8.8 \\ 2.8 & 97.2 \end{pmatrix}$	$\begin{pmatrix} 92.2 & 7.8 \\ 2.5 & 97.5 \end{pmatrix}$	$\begin{pmatrix} 92.1 & 7.9 \\ 2.6 & 97.4 \end{pmatrix}$	$\begin{pmatrix} 92.0 & 8.0 \\ 2.5 & 97.5 \end{pmatrix}$
	30	$\begin{pmatrix} 91.1 & 8.9 \\ 2.8 & 97.2 \end{pmatrix}$	$\begin{pmatrix} 92.1 & 7.9 \\ 2.6 & 97.4 \end{pmatrix}$	$\begin{pmatrix} 92.1 & 7.9 \\ 2.6 & 97.4 \end{pmatrix}$	$\begin{pmatrix} 92.1 & 7.9 \\ 2.7 & 97.3 \end{pmatrix}$

From Table 4, random forest model parameters selections are: the number of trees 20, and the maximum depth 25.

4.2. Test Results

After the optimization of the model based on the random forest classifier in the previous section, the parameters of the model based on the random forest classifier are determined.

The next four groups of tests are about the different classification results of the model based on the random forest classifier between data of wind turbines 15# and 21#. The train and test set details and the classification results from each test are as follows. In addition, in the following tables, the indicator, the running time, refers to the total time the model takes to training and predicts the data on the experimental computer.

In the Test No. 1, 70% of the sample data of the wind turbine 15# was divided into training sets, and 30% of the sample data was divided into a test set. The results are shown in Table 5.

Table 5. Result of Test No. 1 (Running time: 27.0 s).

Confusion Matrix	Prediction of Icing	Prediction of Non-Icing
Actual icing	92.2%	7.8%
Actual non-icing	2.5%	97.5%

In the Test No. 2, 70% of the sample data of the wind turbine 21# divided into training sets, and 30% of the sample data is the test set, as shown in Table 6.

Table 6. Result of Test No. 2 (Running time: 14.2 s).

Confusion Matrix	Prediction of Icing	Prediction of Non-Icing
Actual icing	85.6%	14.4%
Actual non-icing	5.0%	95.0%

Test No. 3 takes all the sample data of wind turbine 21# into the training set and the sample data of 15# wind turbine into test set, as shown in Table 7.

Table 7. Result of Test No. 3 (Running time: 38.1 s).

Confusion Matrix	Prediction of Icing	Prediction of Non-Icing
Actual icing	82.2%	17.8%
Actual non-icing	18.2%	81.8%

Test No. 4 takes all the sample data of wind turbine 15# into the training set and the sample data of wind turbine 21# into the test set, as shown in Table 8.

Table 8. Result of Test No. 4 (Running time: 26.8 s).

Confusion Matrix	Prediction of Icing	Prediction of Non-Icing
Actual icing	89.8%	10.2%
Actual non-icing	8.8%	91.2%

Consequently, the classification results between different classification models are also compared. This paper selects the logistic regression classifier, the GBDT (Gradient Boosting Decision Tree) classifier and the random forest classifier for comparison. In these tests, 70% of the sample data of the wind turbine 15# is the training set, and 30% of the sample data is set as the test set.

Test No.5 used a logistic regression classification model. After the optimization, the classification threshold was set to 0.86, as results shown in Table 9 demonstrate.

Table 9. Result of Test No. 5 (Running time: 86.7 s).

Confusion Matrix	Prediction of Icing	Prediction of Non-Icing
Actual icing	88.8%	11.2%
Actual non-icing	8.8%	91.2%

The GBDT classification model was used in the Test No. 6. In addition, after the model optimization, the classification result is presented in Table 10.

Table 10. Result of Test No. 6 (Running time: 56.8 s).

Confusion Matrix	Prediction of Icing	Prediction of Non-Icing
Actual icing	76.6%	23.4%
Actual non-icing	5.7%	94.3%

The precision and recall of the six tests computed separately, and the obtained results are shown in Table 11.

Table 11. Precision and recall tests.

Test Number	Train Data	Test Data	Model	Precision	Recall
No. 1	70% of 15#	30% of 15#	RF	97.4%	92.2%
No. 2	70% of 21#	30% of 21#	RF	94.5%	85.6%
No. 3	all of 21#	all of 15#	RF	81.9%	82.2%
No. 4	all of 15#	all of 21#	RF	91.1%	89.8%
No. 5	70% of 15#	30% of 15#	LR	97.4%	88.8%
No. 6	70% of 15#	30% of 15#	GBDT	93.1%	76.6%

Where RF means random forest classifier, and LR means logistic regression classifier.

As we all known, for a classification model, when both precision and recall have higher values at the same time without considering other factors, the model is thought to have a better performance. Through the comparison of test results we draw, the following summary is concluded:

1. Considering precision and recall at the same time, the results of the model based on the random forest classifier are better than other models (as the result of Table 11 shown), so that the model based on the random forest classifier has high accuracy in ice detection of wind turbine blades and can identify the early failure.
2. The trained model based on the random forest classifier still performs well on a new test set (as the results of Tests No. 3 and No. 4 shown), so that the model has good generalization ability, thus it has a strong adaptability to the new sample data.
3. The data of wind turbine 15# is more than the data of 21#. From the results of these tests (as the results of Tests No. 1, No. 2, No. 3, and No. 4), when only focusing on precision and recall without considering the running time, increasing the sample size of the training set can improve the classification performance of the model.
4. Considering the running time of these models, the model based on the random forest classifier shorter running time than the logistic regression and the GBDT classification models (as the results of Tests No. 5 and No. 6), so that the model based on the random forest classifier has more efficient calculation ability.

5. Conclusions

To detect wind turbine blade icing, a model based on the random forest classifier was proposed. The model with high accuracy and good generalization ability was verified by the data of the China Industrial Big Data Innovation Competition.

1. The features extracted in this paper well reflect the characteristics of icing failure. Two basic variables of wind speed and output power from SCADA data and more instantaneous and statistical features extracted from a power curve can be used to characterize the early icing on wind turbine blades. The various models selected in this paper have good results on this data set. So, it has high practical application value for ice detection in wind farms, which, as reflected in the effective prevention of severe ice formation in the blades, increase of wind farm's profit, and reduction of safety accident.
2. The model based on the random forest classifier has very high accuracy and good generalization ability for ice detection on a wind turbine's blades, which is used to identify the early icing. Compared with other classification models, the model based on the random forest classifier has higher accuracy and more efficiency in terms of computing capabilities, making it more suitable for the practical application of ice detection.

3. In the identification of new data, the accuracy of the model has reached more than 80%, which shows that there is room for further improvement. In the future, it can expand better features or use other models to ensure the best results in both accuracy and generalization.

Author Contributions: Conceptualization, L.Z.; Data curation, K.L.; Methodology, K.L. and Y.W.; Supervision, L.Z.; Writing–original draft, K.L.; Writing–review and editing, L.Z. and Z.B.O.

Funding: This research was funded by the National Key Research and Development Program of China (No. 2016YFF0203800), the Fundamental Research Funds for Central Universities of China (No. FRF-BD-18-001A) and the National Natural Science Foundation of China (No. 51775037).

Acknowledgments: The authors would like to thank the anonymous reviewers for their valuable comments and suggestions that helped improve the quality of this manuscript.

Conflicts of Interest: The authors declare no conflict of interest.

Appendix A

Table A1. Variable name and description.

Number	Name	Description
1	time	timestamp
2	wind_speed	wind speed
3	generator_speed	generator speed
4	power	grid side active power
5	wind_direction	wind direction
6	wind_direction_mean	mean of wind direction
7	yaw_position	yaw position
8	yaw_speed	yaw speed
9	pitch1_angle	pitch 1 angle
10	pitch2_angle	pitch 2 angle
11	pitch3_angle	pitch 3 angle
12	pitch1_speed	pitch 1 speed
13	pitch2_speed	pitch 2 speed
14	pitch3_speed	pitch 3 speed
15	pitch1_moto_tmp	pitch motor 1 temperature
16	Pitch2_moto_tmp	pitch motor 2 temperature
17	Pitch3_moto_tmp	pitch motor 3 temperature
18	acc_x	X-direction acceleration
19	acc_y	Y-direction acceleration
20	environment_tmp	environment temperature
21	int_tmp	cabin temperature
22	pitch1_ng5_tmp	ng5 1 temperature
23	pitch2_ng5_tmp	ng5 2 temperature
24	pitch3_ng5_tmp	ng5 3 temperature
25	pitch1_ng5_DC	ng5 1 charger DC current
26	pitch2_ng5_DC	ng5 2 charger DC current
27	pitch3_ng5_DC	ng5 3 charger DC current
28	group	data group identification

References

1. Leite, G.D.N.P.; Araújo, A.M.; Rosas, P.A.C. Prognostic techniques applied to maintenance of wind turbines: a concise and specific review. *Renew. Sustain. Energy Rev.* **2018**, *81*, 1917–1925. [CrossRef]

2. Global Wind Energy Council (GWEC). Available online: http://gwec.net/global-figures/graphs/ (accessed on 20 June 2018).

3. Oh, K.Y.; Nam, W.; Ryu, M.S.; Kim, J.Y.; Epureanu, B.I. A review of foundations of offshore wind energy convertors: Current status and future perspectives. *Renew. Sustain. Energy Rev.* **2018**, *88*, 16–36. [CrossRef]

4. Gantasala, S.; Luneno, J.C.; Aidanpää, J.O. Influence of icing on the modal behavior of wind turbine blades. *Energies* **2016**, *9*, 862. [CrossRef]

5. Shu, L.; Liang, J.; Hu, Q.; Jiang, X.; Ren, X.; Qiu, G. Study on small wind turbine icing and its performance. *Cold Reg. Sci Technol.* **2017**, *134*, 11–19. [CrossRef]

6. British Standards Institution. Overhead Transmission Lines-Design Criteria. IEC 60826: 2017. Available online: https://webstore.ansi.org/RecordDetail.aspx?sku=BS+IEC+60826%3A2017 (accessed on 18 September 2018).

7. Davis, N.N.; Byrkjedal, Ø.; Hahmann, A.N.; Clausen, N.; Mark, Ž. Ice detection on wind turbines using the observed power curve. *Wind Energy* **2016**, *19*, 999–1010. [CrossRef]

8. Wang, Z.; Zhu, C. Numerical simulation for in-cloud icing of three-dimensional wind turbine blades. *Simulation* **2017**, *94*, 31–41. [CrossRef]

9. Shu, L.; Li, H.; Hu, Q.; Jiang, X.; Qiu, G.; McClure, G.; Yang, H. Study of ice accretion feature and power characteristics of wind turbines at natural icing environment. *Cold Reg. Sci. Technol.* **2018**, *147*, 45–54. [CrossRef]

10. Blasco, P.; Palacios, J.; Schmitz, S. Effect of icing roughness on wind turbine power production. *Wind Energy* **2017**, *20*, 601–617. [CrossRef]

11. Li, N.; Yan, T.; Li, N.; Kong, D.; Liu, Q.; Lei, Y. Ice detection method by using SCADA data on wind turbine blades. *Power Gener. Technol.* **2018**, *39*, 58–62.

12. Aral, S.; Zhu, M.; Christopher, N.; Peyman, P. Vibration-based damage detection in wind turbine blades using Phase-based Motion Estimation and motion magnification. *J. Sound Vib.* **2018**, *421*, 300–318.

13. Yu, M.Y.; Fu, S.; Gao, Y.B.; Zheng, H.; Xu, Y.G. Crack detection of fan blade based on natural frequencies. *Int. J. Rotating Mach.* **2018**, *2018*, 1–13. [CrossRef]

14. Qin, Y.; Zou, J.Q.; Cao, F.L. Adaptively detecting the transient feature of faulty wind turbine planetary gearboxes by the improved kurtosis and iterative thresholding algorithm. *IEEE Access.* **2018**, *6*, 14602–14612. [CrossRef]

15. Tautz-Weinert, J.; Watson, S.J. Using SCADA data for wind turbine condition monitoring—A review. *IET Renew. Power Gener.* **2017**, *11*, 382–394. [CrossRef]

16. Dai, J.; Yang, W.; Cao, J.; Liu, D.; Long, X. Ageing assessment of a wind turbine over time by interpreting wind farm SCADA data. *Renew. Energy* **2017**, *116*, 199–208. [CrossRef]

17. Chen, N.; Yu, R.; Chen, Y.; Xie, H. Hierarchical method for wind turbine prognosis using SCADA data. *IET Renew. Power Gener.* **2017**, *11*, 403–410. [CrossRef]

18. Dao, P.B.; Staszewski, W.J.; Barszcz, T.; Uhl, T. Condition monitoring and fault detection in wind turbines based on cointegration analysis of SCADA data. *Renew. Energy* **2017**, *116*, 107–122. [CrossRef]

19. Dai, J.; Yang, X.; Hu, W.; Wen, L.; Tan, Y. Effect investigation of yaw on wind turbine performance based on SCADA data. *Energy* **2018**, *149*, 684–696. [CrossRef]

20. Bangalore, P.; Patriksson, M. Analysis of SCADA data for early fault detection, with application to the maintenance management of wind turbines. *Renew. Energy* **2017**, *115*, 521–532. [CrossRef]

21. Alvarez, E.J.; Ribaric, A.P. An improved-accuracy method for fatigue load analysis of wind turbine gearbox based on SCADA. *Renew. Energy* **2018**, *115*, 391–399. [CrossRef]

22. El-Asha, S.; Zhan, L.; Lungo, G.V. Quantification of power losses due to wind turbine wake interactions through SCADA, meteorological and wind LiDAR data. *Wind Energy* **2017**, *20*, 1823–1839. [CrossRef]

23. Cao, M.; Qiu, Y.; Feng, Y.; Wang, H.; Li, D. Study of wind turbine fault diagnosis based on unscented Kalman filter and SCADA data. *Energies* **2016**, *9*, 847. [CrossRef]

24. Sun, P.; Li, J.; Wang, C.; Lei, X. A generalized model for wind turbine anomaly identification based on SCADA data. *Appl. Energy* **2016**, *168*, 550–567. [CrossRef]

25. Atmospheric Icing of Structures. Available online: https://www.iso.org/standard/72443.html (accessed on 6 June 2018).

26. Breiman, L. Random forests. *Mach. Learn.* **2001**, *45*, 5–32. [CrossRef]

27. Breiman, L. Bagging predictors. *Mach. Learn.* **1996**, *24*, 123–140. [CrossRef]

28. Ho, T. The random subspace method for constructing decision forests. *IEEE Trans. Pattern Anal. Mach. Intell.* **1998**, *20*, 832–844.

29. Deng, X.; Liu, Q.; Deng, Y.; Mahadevan, S. An improved method to construct basic probability assignment based on the confusion matrix for classification problem. *Inf. Sci.* **2016**, *340*, 250–261. [CrossRef]

30. Industrial Big Data Innovation Competition. Available online: http://www.industrial-bigdata.com (accessed on 6 June 2018).

31. Cui, L.; Wu, N.; Ma, C.; Wang, H. Quantitative fault analysis of roller bearings based on a novel matching pursuit method with a new step-impulse dictionary. *Mech. Syst. Signal Process.* **2016**, *68–69*, 34–43. [CrossRef]
32. Yang, X.; Xue, W.; Bao, L. Application of IEC standards in the measurement of practical power curve of wind turbine generator. *Power Syst. Clean Energy* **2012**, *28*, 87–91.

MDPI

St. Alban-Anlage 66

4052 Basel

Switzerland

Tel. +41 61 683 77 34

Fax +41 61 302 89 18

www.mdpi.com

Energies Editorial Office

E-mail: energies@mdpi.com

www.mdpi.com/journal/energies